Polymer
Microscopy

Polymer Microscopy

Second edition

LINDA C. SAWYER
Hoechst Celanese Corporation
Summit, NJ
USA

and

DAVID T. GRUBB
Cornell University
Ithaca, NY
USA

CHAPMAN & HALL

London · Glasgow · Weinheim · New York · Tokyo · Melbourne · Madras

Published by
Chapman & Hall, 2–6 Boundary Row, London SE1 8HN, UK

Chapman & Hall, 2–6 Boundary Row, London SE1 8HN, UK

Blackie Academic & Professional, Wester Cleddens Road, Bishopbriggs, Glasgow G64 2NZ, UK

Chapman & Hall GmbH, Pappelallee 3, 69469 Weinheim, Germany

Chapman & Hall USA, 115 Fifth Avenue, New York, NY 10003, USA

Chapman & Hall Japan, ITP-Japan, Kyowa Building, 3F, 2-2-1 Hirakawacho, Chiyoda-ku, Tokyo 102, Japan

Chapman & Hall Australia, 102 Dodds Street, South Melbourne, Victoria 3205, Australia

Chapman & Hall India, R. Seshadri, 32 Second Main Road, CIT East, Madras 600 035, India

First edition 1987

Second edition 1996

© 1987, 1996 L. C. Sawyer and D. T. Grubb

Typeset in 10/12 pt Palatino by Techset Composition Ltd., Salisbury, Wiltshire.
Printed in Great Britain by the Alden Press, Oxford

ISBN 0 412 60490 6

A catalogue record for this book is available from the British Library

Library of Congress Catalog Card Number: 94-74686

∞ Printed on acid-free text paper, manufactured in accordance with ANSI/NISO Z39.48-1992 (Permanence of Paper).

Contents

Color plates appear between pages 274 and 275

Appendices   379

Preface to the second edition

The major objective of this text is to provide information on the basic microscopy techniques and specimen preparation methods applicable to polymers. This book will attempt to provide enough detail so that the methods described can be applied, and will also reference appropriate publications for the investigator interested in more detail. Some discussion will consider polymer structure and properties, but only as this is needed to put the microscopy into context.

We recognize that scientists from a wide range of backgrounds may be interested in polymer microscopy. Some may be experienced in the field, and this text should provide a reference work and resource whenever a new material or a new problem comes to their attention. The scientist, engineer or graduate student new to the field needs more explanation and help. Some may need to know more about the intrinsic capabilities of different microscopes, others may know all about microscopes and little about polymer fibers. This text includes sections designed for all of these groups, so that some portions of the text are not for every reader. The organization of chapters and section headings should lead the reader to the information needed, and an extensive index is provided for the same purpose.

The first edition of this book was published in 1987. In the past eight years there have been major changes in microscopy that have had an effect on research into many materials, including polymers. It was no surprise when Chapman & Hall invited us to provide a second edition which would include the new techniques, as we were already applying them to our own research. This second edition follows the same basic principles as the first, with the addition of new material throughout the text as well as in a new chapter.

Chapter 1 provides a brief introduction to polymer materials, processes, morphology and characterization. Chapter 2 is a concise review of the fundamentals of microscopy, where many important terms are defined. Chapter 3 reviews imaging theory for the reader who wants to understand the nature of image formation in the various types of microscopes, with particular reference to imaging polymers. All of these chapters are mere summaries of large fields of science, to make this text complete. They contain many references to more specialized texts and reviews. Chapters 4 and 5 contain the major thrust of the book. Chapter 4 covers specimen preparation, organized by method, with enough detail to conduct such preparations*. The references are chosen to provide the best detail and support. Chapter 5 describes the use of these methods in the study of specific types of polymers. The organization is by the form of the material, as fiber, film, composite or engineering resin. The emphasis is on applications, and particularly on applications where more than one specimen preparation method or microscopy technique is used.

Chapter 6 is a new chapter for this edition. It is titled 'New techniques in polymer microscopy'. It contains a detailed discussion of recently developed optical, electron and scanning probe microscopies now being used to study polymers. Some, for example low voltage scanning electron microscopy, are modifications of older existing techniques. Others, such as atomic force microscopy are entirely novel. In this case, it is important to realize that the fields are still developing very rapidly. Image formation is not yet fully understood, and improved instruments and new techniques appear frequently. Polymers may still require special efforts in specimen preparation or imaging. Chapter 7 (which was Chapter 6 in the first edition) describes how other techniques for investigating polymer structure should be considered as part of a problem solving approach to microscopy.

The selection of the authors for this text came from a desire for a comprehensive review of polymer microscopy with emphasis on methods and techniques rather than on the results obtained. The synergism provided by two authors with very different backgrounds was important. One author (LCS) has an industrial focus and a background in chemistry, while the other (DTG) is in an academic environment and has a background in polymer physics. As in the first edition, David Grubb's contribution has been in Chapters 2 and 3; additionally he has contributed the new Chapter 6. Linda Sawyer has been responsible for Chapters 1, 4, 5 and 7.

Linda C. Sawyer
Chatham, New Jersey
and *David T. Grubb*
Ithaca, New York
August, 1994

---

* The specimen preparation methods used for microscopy of polymers involve the use of many toxic chemicals as well as the use of instruments which can be radiation hazards. It is well beyond the purpose of this text to provide the information required for the proper and safe handling of such chemicals and instruments and the researcher is encouraged to obtain the required safety information prior to their use.

Acknowledgements

Special thanks go to the Microscopy group staff whose efforts over the years provided many of the micrographs used in both the first and second editions of this book, especially Madge Jamieson, Roman Brozynski and Rong T. Chen. The support of the Hoechst Celanese Corporation as well as the technical information center and computer services group are also gratefully acknowledged. Many colleagues, too numerous to mention by name, are also acknowledged for their comments and overall support. Micrographs for the second edition are gratefully acknowledged from Olga Shaffer (Lehigh University), Barbara Wood (DuPont), David Martin (University of Michigan) and Deborah Vezie (MIT). Acknowledgement is given to David Grubb for collaborating in this project and for taking the lead role in the production of the new chapter in the second edition, and who, by adding his biases to my own, has continued to help make this book useful to both the industrial and academic communities. I would also like to acknowledge the support of a very understanding colleague and husband, David Sawyer, who provided technical advice, micrographs, and who prepared all the micrographs for publication.

A text on microscopy of necessity relies on many figures; these have been kindly supplied by many people, especially my coworkers at Hoechst Celanese during the last two decades. Acknowledgement and special thanks are given specifically below to coworkers who provided most of the figures. Thanks are due to the many authors supplying micrographs, as noted in the figure captions. Figures not specifically noted here or in the captions were supplied by the authors.

Madge Jamieson for Figures 4.5, 4.25B, 4.26, 4.29, 4.31–32, 4.34–35, 4.38, 4.42, 5.5–7, 5.12A–B, 5.18B–D, 5.23, 5.25A, 5.30, 5.32, 5.35–36, 5.43A, 5.49, 5.59, 5.71, 5.99, 5.113–114. **Roman Brozynski** for Figures 4.8–9, 4.33, 4.36–37, 5.10, 5.14, 5.26–27, 5.37–41, 5.45, 5.48, 5.50–51, 5.63–64, 5.67, 5.72, 5.92–93, 5.103. **Rong T. Chen** for Figures 5.25B–C, 5.43B–D, 5.115–5.118.

Linda C. Sawyer

CHAPTER ONE

Introduction to polymer morphology

1.1 POLYMER MATERIALS

1.1.1 Introduction

Organic polymers are materials that are widely used in many important emerging technologies of the twentieth century. Feedstocks for synthetic polymers are petroleum, coal and natural gas, which are sources of ethylene, methane, alkenes and aromatics. Polymers are used in a wide range of everyday applications, in clothing, housing materials, appliance housings, automotive and aerospace parts and in communication. Materials science, the study of the structure and properties of materials, is applied to polymers in much the same way as it is to metals and ceramics: to understand the relationships between the manufacturing process, the structures produced and the resulting physical and mechanical properties. This chapter is an introduction to *polymer morphology*, which must be understood in order to develop relations between the structure and properties of these materials. An introduction by Young [1] is but one reference from the vast literature on the topic of polymer morphology. Subsequent sections and chapters have many hundreds of references cited as an aid to the interested reader. The emphasis in this text is on the elucidation of polymer morphology by microscopy techniques.

Polymers have advantages over other types of materials, such as metals and ceramics, because their low processing costs, low weight and properties such as transparency and toughness form unique combinations. Many polymers have useful characteristics, such as tensile strength, modulus, elongation and impact strength which make them more cost effective than metals and ceramics. Plastics and engineering resins are processed into a wide range of fabricated forms, such as fibers, films, membranes and filters, moldings and extrudates. Recently new technologies have emerged resulting in novel polymers with highly oriented structures. These include polymers which exhibit liquid crystallinity in the melt or in solution, some of which can be processed into materials with ultrahigh performance characteristics. Applications of polymers are wide ranging and varied and include the examples shown in Table 1.1. A listing of the names and abbreviations of some common polymers is shown in Appendix I for reference. Appendix II is a listing of acronyms commonly used for analytical techniques. Appendix III provides a listing of common fibers and Appendix IV is a listing of common plastics and a few

Table 1.1 Polymer applications

Fibers	Polyethylene, polyester, nylon, acetate, polyacrylonitrile, polybenzobisthiazole, polypropylene, acrylic, aramid
Films/ Packaging	Polyethylene, polyester, polypropylene, polycarbonate, polyimide, fluoropolymers, polyurethanes, poly(vinyl chloride)
Membranes	Cellulose acetate, polysulfone, polyamide, polypropylene, polycarbonate, polyimide, polyacrylonitrile, fluoropolymers
Engineering resins	Polyoxymethylene, polyester, nylon, polyethersulfone, poly(phenylene sulfide), acrylonitrile–butadiene–styrene, polystyrene
Adhesives	Poly(vinyl acetate), epoxies, polyimides
Emulsions	Styrene–butadiene–styrene, poly(vinyl acetate)
Coatings	Epoxies, polyimides, poly(vinyl alcohol)
Elastomers	Styrene–butadiene rubber, urethanes, polyisobutylene, ethylene–propylene rubber

applications. Finally, general suppliers of accessories, microscopes and x-ray microanalysis equipment are found in Appendices V–VIII, respectively.

1.1.2 Definitions

Polymers are macromolecules formed by joining a large number of small molecules, or monomers, in a chain. These *monomers*, small repeating units, react chemically to form long molecules. The repetition of monomer units can be linear, branched or interconnected to form three dimensional networks. *Homopolymers*, composed of a single repeating monomer, and *heteropolymers*, composed of several repeating monomers, are two broad forms of polymers. Copolymers are the most common form of heteropolymers. They are often formed from a sequence of two types of monomer unit. Alternating copolymers can be simple alternating repeats of two monomers, e.g.

–A–B–A–B–A–B–A–B–, or random repeats of two monomers, e.g. –A–A–A–B–B–A–B–B–A–A–B–, whereas block copolymers include long sequences of one repeat unit, e.g. –A–B–B–B–B–B–B–B–A–A–B–B–B–B–A–. There are several forms of block copolymers, including AB and ABA, where A and B each stand for a sequence of several hundred monomers. If monomer B is added while A chains are still growing the result is a graded block copolymer. Additionally each component of block copolymers can be amorphous or crystalline. Amorphous block copolymers generally form characteristic domain structures, such as those in styrene–butadiene–styrene block copolymers. Crystalline block copolymers, such as polystyrene–poly(ethylene oxide), typically form structures that are characteristic of the crystallizable component. Block copolymers which have the second component grafted onto the backbone chain are termed graft copolymers. Graft copolymers of industrial significance, high impact polystyrene (HIPS) and acrylonitrile–butadiene–styrene (ABS), have rubber inclusions in a glassy matrix. Many common homopolymers can also be found as repeat units in heteropolymers, such as polyethylene, as in polyethylene–polypropylene copolymer. Modified polymers, copolymers and polymer blends can be tailored for specific end uses which will be discussed.

There are three major polymer classes: thermoplastics, thermosets and rubbers or elastomers. Polymers that are typical of each of these classes are listed in Table 1.2. *Thermoplastics* are among the most common polymers and these materials are commonly termed 'plastics'. Linear or branched thermoplastics can be reversibly melted or can be dissolved in a suitable solvent. In some cases thermoplastics are crosslinked in processing so as to provide heat stability and limit flow and melting during use. In *thermosets* there is a three dimensional network structure, a single highly connected molecule, which imparts rigidity and intractability. Thermosets are heated to form rigid structures but once set they do not melt upon prolonged heating nor do they dissolve in solvents. Thermosets generally have

Table 1.2 Major classes of polymers

Crystallizable thermoplastics	Glassy thermoplastics
Polyacetal	Polystyrene
Polyacetal	Poly(methyl methacrylate)
Polyamide	Poly(vinyl chloride)
Polycarbonate	Poly(vinyl acetate)
Polycarbonate	
Poly(ethylene terephthalate)	
Polyethylene	
Polypropylene	
Thermosets	**Elastomers**
Epoxy	Polybutadiene
Phenolic	Ethylene–propylene
Polyester (unsaturated)	copolymers
	Styrene–butadiene rubber
	Ethylene–vinyl acetate
	Styrene–butadiene copolymers

only short chains between crosslinks and exhibit glassy brittle behavior. Thermosets are used as high performance adhesives (e.g. epoxies).

Polymers with long flexible chains between crosslinks are *rubbers* and *elastomers* which, like the thermosets, cannot be melted. Elastomers are characterized by a three dimensional crosslinked network which has the well known property of being stretchable and springing back to the original form. Crosslinks are chemical bonds between molecules. An example of a crosslinking reaction is the vulcanization of rubber, where the sulfur molecules react with the double bonded carbon atoms creating the structure. Multiphase polymers, combinations of thermoplastics and elastomers, take advantage of the ease of fabrication of thermoplastics and the increased toughness of elastomers, providing engineering resins with enhanced impact strength.

The chemical composition of macromolecules is important in determination of properties.

Variations in the stereochemistry, the spatial arrangement, also result in very different materials. Three forms of spatial arrangement are isotactic, syndiotactic and atactic. The *isotactic* (i) forms have pendant group placement on the same side of the chain, while in *syndiotactic* polymers there is a regular, alternating placement of pendant groups with respect to the chain. *Atactic* (a) polymers have disordered sequences or a random arrangement of side groups. Isotactic polypropylene (iPP) crystallizes and has major uses, whereas atactic PP cannot crystallize, is sticky (has a low thermal transition) and finds little application.

1.2 POLYMER MORPHOLOGY

In polymer science the term morphology generally refers to form and organization on a size scale above the atomic arrangement but smaller than the size and shape of the whole sample. The term 'structure' refers more to the local atomic and molecular details. The characterization techniques used to determine structure differ somewhat from those used to determine morphology. Examples of polymer morphology include the size and shape of fillers and additives, and the size, distribution and association of the structural units within the macrostructure. However, as is probably clear from the overlapping definitions, the terms 'structure' and 'morphology' are commonly used interchangeably. The characterization techniques are complementary to one another and both are needed to determine fully the morphology and microstructure and to develop structure–property relationships.

X-ray, electron and optical scattering techniques and a range of other analytical tools are commonly applied to determine the structure of polymers. X-ray diffraction, for example, permits the determination of interatomic ordering and chain packing. The morphology of polymers is determined by a wide range of optical and electron microscopy techniques, which are the major subject of this text. Finally, there are many other analytical techniques that provide important information regarding polymer structure, such as neutron scattering, infrared spectroscopy, thermal analysis, mass spectroscopy, nuclear magnetic resonance, etc., which are beyond the scope of this text.

Polymers are considered to be either *amorphous* or *crystalline* although they may not be completely one or the other. Crystalline polymers are more correctly termed *semicrystalline* as their measured densities differ from those obtained for perfect materials. The degree of crystallinity, measured by x-ray scattering, also shows these polymers are less than completely crystalline. The general morphology of crystalline polymers is now well known and understood and has been described by Geil [2], Keller [3], Wunderlich [4], Grubb [5, 6], Uhlmann and Kolbeck [7], Bassett [8, 9] and Seymour [10]. The text by Bassett [8] is a good overview of the topic for the beginner in the field. There is no measurable order by x-ray scattering techniques, an absence of crystallographic reflections, in noncrystalline or amorphous polymers. Characterization of semicrystalline polymer morphology can require an understanding of the entire texture. This extends from the interatomic structures and individual crystallites, to the macroscopic details and the relative arrangement of the crystallites in the macrostructure. The units of organization in polymers are lamellae or crystals and spherulites [11]. Bulk polymers are composed of lamellar crystals which are typically arranged as spherulites when cooled from the melt.

1.2.1 Amorphous polymers

Amorphous polymers of commercial importance include polymers which are glassy or rubbery at room temperature. Many amorphous thermoplastics, such as atactic polystyrene and poly (methyl methacrylate), form brittle glasses when cooled from the melt. The glass transition temperature, T_g or glass–rubber transition, is the temperature above which the polymer is rubbery and can be elongated and below which the polymer behaves as a glass. Thermal analysis of amorphous polymers shows only a glass transition temperature whereas crystalline polymers also exhibit a crystalline melting temperature.

Commercially important glassy polymers include polymers which are crystallizable but which may form as amorphous materials. These noncrystalline polymers are formed by rapid cooling of a polymer from above the melting transition temperature. They yield by forming a necked zone where the molecules are highly oriented and aligned in the draw direction. Other important polymers amorphous at room temperature include natural rubber (polyisoprene) and other elastomers. These exhibit a high degree of elasticity, stretching considerably in the elastic region and then fracturing with no plastic deformation.

Plastic deformation in glassy polymers and in rubber toughened polymers is due to crazing and shear banding. *Crazing* is the formation of thin sheets perpendicular to the tensile stress direction which contain fibrils and voids. The fibrils and the molecular chains in them are aligned parallel to the tensile stress direction. Crazes scatter light and can be seen by eye if there are many of them as whitened areas. Crazing is often enhanced by rubber inclusions which impart increased toughness to the polymer and reduce brittle fracture by initiating or terminating crazes at the rubber particle surface. *Shear banding* is a local deformation, at about 45° to the stress direction, which results in a high degree of chain orientation. The material in the shear band is more highly oriented than in the adjacent regions. The topic of yielding and fracture will be further explored (Section 4.8). Overall, the mechanical behavior of amorphous polymers depends upon the chemical composition, the distribution of chain lengths, molecular orientation, branching and crosslinking.

1.2.2 Semicrystalline polymers

Semicrystalline polymers exhibit a melting transition temperature (T_m), a glass transition temperature (T_g) and crystalline order, as shown by x-ray and electron scattering. The fraction of the crystalline material is determined by x-ray diffraction, heat of fusion and density measurements. Major structural units of semicrystalline polymers are the platelet-like crystallites, or *lamellae*. The dominant feature of melt crystallized specimens is the *spherulite*. The formation of polymer crystals and the spherulitic morphology in bulk polymers has been fully described by Keith and Padden [12], Ward [13], Bassett [8, 9] and many others. Single crystals can be formed by precipitation from dilute solution as shown later (see Fig. 4.1). These crystals can be found as faceted platelets of regular shape for regular polymers. They have a less perfect shape when formed from polymers with a less perfect structure. The molecular chains are approximately normal to the basal plane of the lamellae, parallel to the short direction. Some chains fold and re-enter the same crystal. In polyethylene the lamellae are on the order of several micrometers across and about 10–50 μm thick, independent of the length of the molecule. Bulk crystallized spherulites can range from about 1 to 100 μm or larger. Small angle x-ray scattering (SAXS) and electron diffraction data have confirmed the lamellar nature of single crystals in bulk material.

1.2.2.1 *Crystallization under quiescent conditions*

When a polymer is melted and then cooled it can recrystallize, with process variables such as temperature, rate of cooling, pressure and additives affecting the nature of the structures formed. Two types of microstructure observed for semicrystalline bulk polymers are spherulites and row nucleated textures. Bulk crystallized material is composed of microscopic units called *spherulites*, which are formed during crystallization under quiescent conditions. The structures exhibit radially symmetric growth of the lamellae from a central nucleus with the molecular chain direction perpendicular to the growth direction. The plates branch as they grow. The molecular chains therefore run perpendicular to the spherulite radius. The crystallite or lamellar thickness in the bulk polymer depends upon the molecular weight of the polymer, crystallization conditions and thermal treatment. The size and number of spherulites is controlled by nucleation. Spherulites are smaller and more numerous if there are more growth nuclei and larger if slow cooled or isothermally crystallized. In commercial

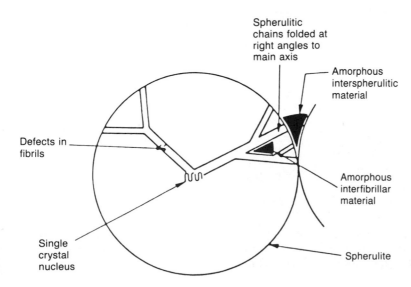

Fig. 1.1 Schematic of spherulite structure. (From Ward [13]; used with permission.)

processes additives are commonly used to control nucleation density. When crystallizing during cooling, the radial growth rate of the spherulites is an important factor in determining their size. The morphology of isothermally crystallized polyethylene (PE) melts has revealed the nature of the lamellae [14] by a sectioning and staining method for transmission electron microscopy (TEM) (see Fig. 4.15), which will be described later.

A schematic of the spherulite structure is shown in Fig. 1.1 [13]. The structure [12, 13, 15] consists of radiating fibrils with amorphous material, additives and impurities between the fibrils and between individual spherulites. Although the shape of the growing spherulite is round, as shown in a polarized light micrograph (Fig. 1.2), spherulites generally impinge upon one another, resulting in polyhedral shapes (Fig. 1.3, see color plate section). When thin melt quenched films, or sections of a bulk polymer are viewed in crossed polarizers the spherulites appear bright because they are anisotropic and crystalline in nature. Isotropic materials exhibit the same properties in all directions, whereas anisotropic materials exhibit a variation in properties with

direction. Polarized light micrographs of a sectioned, bulk crystallized nylon (Fig. 1.3, color section) show the size range of spherulites obtained by bulk crystallization. A more complete discussion of polarization optics will be found later (Section 3.4), but for this discussion it is clear that the size of individual spherulites can be determined by analysis of polarized light micrographs. Average spherulite sizes are determined by small angle light scattering techniques.

1.2.2.2 *Crystallization under flow*

When a bulk polymer is crystallized under conditions of flow a *row nucleated*, or 'shish kebab', structure can be formed. Typically, the melt [16] or solution [17] is subjected to a highly elongational flow field at a temperature close to the melting or dissolution temperature. A non-spherulitic, crystalline microstructure forms from elongated crystals aligned in the flow direction and containing partially extended chains. At high flow rates these microfibers, some 20 nm across, dominate the structure. At lower flow rates the backbones are overgrown by folded chain platelets. This epitaxial growth on the surface of the

Fig. 1.2 Polarized light micrograph of this polyoxymethylene film cooled from melt shows recrystallization and formation of spherulites. The shape of the growing, birefringent spherulites is round.

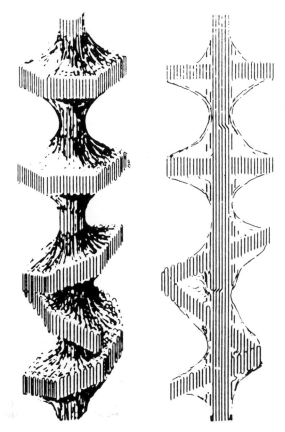

Fig. 1.4 Schematic of a shish kebab structure. (From Pennings [17]; used with permission.)

fibers, fully describing their preparation, structure and properties. High modulus fibers are found in applications such as fiber reinforced composites for aerospace, military and sporting applications. Industrial uses are for belts and tire cords. Extended chain crystals can also form when polymers are crystallized very slowly near the melting temperature but they are weak and brittle.

1.2.3 Liquid crystalline polymers

Rigid and semirigid polymer chains form anisotropic structures in the melt or in solution which result in high orientation in the solid state without drawing [22]. Liquid crystalline melts (thermotropic) or solutions (lyotropic) are composed of sequences of monomers with long rigid molecules. Aromatic polyamides and polyamide-hydrazides are examples of two polymers which form liquid crystal solutions. Aromatic copolyesters and polyazomethines form nematic liquid crystalline melts at elevated temperature. Melt or solution spinning processing of anisotropic liquid crystalline polymers results in an extended chain structure in the fiber or film. Heat treatment improves the orientation and the high modulus and tensile strength properties of these materials.

extended chain produces folded chain lamellae oriented perpendicular to the strain or flow direction [15]. At still lower flow rates the large lamellar overgrowths do not retain this orientation. Dilute solutions of polymers stirred during crystallization are known to form this shish kebab structure where the shish is the elongated crystals in the row structure and the kebabs are the overgrown epitaxial plates, as shown in the schematic in Fig. 1.4 [17].

High modulus fibers and films are produced from extended chain crystals in both conventional polymers, notably PE, and liquid crystalline polymers. Carter and Schenk [18], Jaffe and Jones [19], and Zachariades and Porter [20] have reviewed the topic of high modulus organic

1.3 POLYMER PROCESSES

Commercial processes used to manufacture polymer materials include: fiber spinning, drawing and annealing; film extrusion and stretching; rod extrusion, molding and compounding. The process and chemical composition give specific properties to the polymer. Development of relations between the structure and properties of the polymers requires an understanding of the process as well as characterization of the resulting structure and morphology. The objective of this section is to outline the nature of some relevant processes, the important process variables and the relation of those variables to the structure of the final product. For this discussion, two process categories will be discussed: (1)

extrusion of fibers and films and (2) extrusion of molding of fabricated plastics.

1.3.1 Extrusion of fibers and films

Polymer fibers are found in textile applications (Appendix III) for clothing and household items, such as sheeting and upholstery, and also for industrial uses, such as cords, ropes and belts. Fibers are produced by a melt or solution spinning process which results in oriented materials. Spun fibers are taken up on a bobbin and may be further oriented by drawing on-line or by a post-treatment process resulting in high tensile strength and modulus. Requirements for textile fibers are that the crystalline melting temperature must be above 200°C, so that the textile fabric can be ironed, and yet below 300°C, to permit conventional, melt spinning processing. Alternatively, the polymer must be dissolved in a solvent from which it can be spun by such processes as wet and dry spinning. Films are formed by extrusion followed by stretching to orient the structure. Polymer films are used in many applications including packaging and electronic recording and in membranes for many separations applications. Deformation processes that impart orientation to polymers can result in anisotropic mechanical properties. The increase in molecular alignment can result in increased stiffness and strength. The effects of orientation are dependent on the nature of the starting materials and on whether they are isotropic or anisotropic.

1.3.1.1 *Orientation methods*

There are a variety of well known methods used to orient materials, as reviewed by Holliday and Ward [21], Ciferri and Ward [22] and Hay [23]. Ductile thermoplastics can be *cold drawn* near room temperature whereas thermoplastics which are brittle at room temperature can only be drawn at elevated temperatures. Thermosets are oriented by drawing the precursor polymer prior to crosslinking, resulting in an irreversible orientation. Rubbers can be reversibly elongated

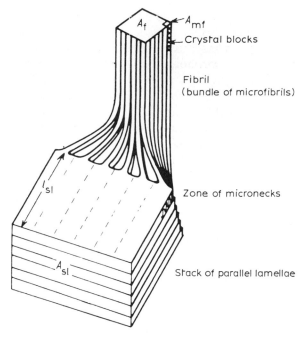

Fig. 1.5 Schematic of cold drawing process with transformation of the lamellar texture into a microfibrillar structure. (From Peterlin [24]; used with permission.)

at room temperature. The orientation in oriented rubbers is locked in place by cooling, while heating drawn thermoplastics causes recovery.

1.3.1.2 *Cold drawing*

Cold drawing is a solid transformation process, conducted near room temperature and below the melting transition temperature of the polymer, if it is crystalline. The process yields a high degree of chain axis alignment by stretching or drawing the polymer with major deformation in the neck region. Deformation of the randomly oriented spherulitic structure in thermoplastics, such as in PE and nylon, results in a change from the stacked lamellae (c. 20 nm thick and 1 μm long) to a highly oriented microfibrillar structure (microfibrils 10 nm wide and very long) with the molecular chains oriented along the draw direction, as shown in the schematic in Fig. 1.5 [24, 25]. Tie molecules, links between the adjacent crystal plates in the spherulites, also appear to orient and yet still connect the stacked plates in the final fibrillar structure. Additionally, the drawing process causes orientation in the non-crystalline amorphous component of the polymer. The degree of orientation of the crystalline component is characterized by wide angle x-ray scattering (WAXS) and birefringence. Clearly, the deformation process causes a major change in the microstructure of the polymer resulting in improved strength and tensile modulus.

1.3.1.3 *Fiber and film formation*

Textile fibers, formed by melt extrusion or solution spinning, as shown in the schematic in Fig. 1.6 [26], are crystallized and then drawn, creating a highly oriented structure similar to that described for the drawing of bulk polymers. Films are also formed by extrusion and drawing processes with uniaxial or biaxial structures. In tensile drawing stress is applied, resulting in thinning and elongation of the crystal and rotation of the molecules or bundles in the draw direction; increasing the draw ratio is known to increase the Young's modulus and the breaking strength by improving the degree of molecular alignment or extension. The diameter of the microfibrillar texture is also affected by the draw ratio with thinner microfibrils at higher draw ratios [27]. The high speed spin–draw fiber process also yields fibers with high modulus and tensile strength when temperature, draw ratio and speed of the process are well controlled. Heat treatments are used to impart desired structures and properties in the fiber and annealing of fiber forming thermoplastics yields a highly crystalline morphology. Uniaxially oriented fibers have a high degree of molecular symmetry and high cohesive energy associated with the high degree of crystallinity. Films are

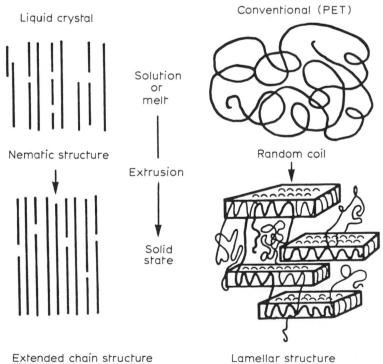

Liquid crystal

Nematic structure

Extended chain structure
- High chain continuity
- High mechanical properties

Solution or melt

Extrusion

Solid state

Conventional (PET)

Random coil

Lamellar structure
- Low chain continuity
- Low mechanical properties

Fig. 1.6 Schematic of fiber formation process for conventional and liquid crystalline polymers. (From Calundann and Jaffe [26], Hoechst Celanese Research Company.)

formed by similar uniaxial or biaxial processes which impart high strength properties in either one or two directions, respectively. Processes include: film extrusion, drawing, stretching, extrusion at high pressure through a die and crystallization under flow.

1.3.2 Extrusion and molding

Polymer morphology in extrudates and moldings is affected by process variables, such as melt and mold temperature, pressure and shear and elongational flow. Process variables can affect the physical properties of the material during processing, which in turn affects the resulting morphology. The morphology of the fabricated product in turn influences performance and mechanical properties. Pressure increases, for instance, can increase both the melting temperature and the glass transition temperature of a polymer, with the result that the polymer solidifies more quickly. In a crystalline polymer the nucleation density can increase, resulting in a decrease in spherulite size with increased pressure in injection molding.

Structures that are typically observed in molded parts and extrudates include anisotropic textures. The higher orientation in extrusion can result in highly oriented rods or strands, at high draw ratios and/or small diameters, or in structures with an oriented skin and a less oriented central core in thicker strands. This skin–core texture is due to a combination of temperature variations between the surface and the bulk and the flow field in both extrusion and molding processes. For instance, the flow fields in a molded part are shown schematically in Fig. 1.7 [26]. Extensional flow along the melt front causes orientation. Solidification of the polymer on the cold mold surface freezes in this orientation. Flow between the solid layers is affected by the temperature gradient in the mold and the resulting flow effects [28, 29] result in a rapidly cooled and well oriented skin structure and a slowly cooled, randomly oriented core. Extensional flow along the melt front results in molecular orientation parallel to the knit lines when two melt fronts meet. The resulting knit, or weld, line is a region of weakness in the molded part and weld line fractures are commonly encountered. The local orientation is quite important as tensile strength and impact strength properties are known to be higher in the orientation direction.

Typically, in a semicrystalline polymer there are three zones within the molded part: an

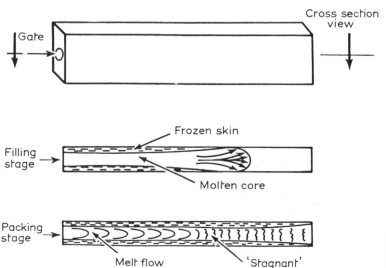

Fig. 1.7 The pattern of the flow directions in the injection molding process is shown in the schematic diagram. (From Calundann and Jaffe [26], Hoechst Celanese Research Company.)

oriented, nonspherulitic skin; a subsurface region with high shear orientation, or a transcrystalline region; and a randomly oriented spherulitic core. The thickness of the skin and shear zone is known to be an inverse function of the melt and mold temperature with decreased temperatures resulting in increased layer thickness. In the skin, the lamellae are oriented parallel to the injection direction and perpendicular to the surface of the mold. Amorphous polymers also show a thin surface oriented skin on injection molding. When amorphous polymers are heated to the glass transition temperature and then relaxed they exhibit shrinkage in the orientation direction and swelling in the other directions.

1.3.2.1 *Muliphase polymers*

Many amorphous thermoplastics are brittle, limiting their range of applications. Toughening with rubber is well known to enhance fracture resistance and toughness. Many major chemical industries are based on toughened plastics, such as ABS, HIPS and ionomers [30–34]. Important issues in the design of fracture resistant polymers are compatibility, deformation, toughening mechanisms and characterization. Particle size distribution and adhesion to the matrix must be determined by microscopy to develop structure–property relationships.

Rubber toughened polymers are usually either copolymers or polymer blends. In random copolymers a single rubber or matrix phase can be modified by the addition of the second component. Graft and block copolymers have modified properties due to the nature of the rubber–matrix interface. In graft copolymers grafting provides a strong bond between the rubber and matrix in the branched structure where one monomer forms the backbone and the other monomer forms the branch. Graft copolymers are usually produced by dissolving the rubber in the plastic monomer and polymerizing it to form the graft. Typical block copolymers are polystyrene–polybutadiene and polyethylene–polypropylene. They have the monomers joined

end-to-end along the main chain which results in a bonding of the two phases.

1.3.2.2 *Composites*

Polymer composites are engineering resins or plastics which contain particle and fibrous fillers. Specialty composites, such as those reinforced with carbon fibers, are used in aerospace applications while glass fiber reinforced resins are used in automobiles. Composite properties depend upon the size, shape, agglomeration and distribution of the filler and its adhesion to the resin matrix. It is well known that long, well bonded fibers result in increased stiffness and strength whereas poor adhesion of the fibers can result in poor reinforcement and poorer properties. Reinforcements for polymers include carbon and glass fibers, and fillers such as glass beads and mineral particles such as talc, clay and silica. Inorganic fillers are also used in elastomers which are not strong enough themselves. Small particles (carbon blacks or silicas) are added during manufacture. The toughness and abrasion resistance imparted is very important in rubber applications such as tires. The theory, processes and characterization of short fiber reinforced thermoplastics have been reviewed by Folkes [35].

Compounding is the process of introducing fibers or particles into a resin prior to molding, to enhance tensile strength or modulus, wear resistance, viscosity and color etc. [36, 37]. Fillers are blended with polymers to modify physical properties or to reduce the polymer level in expensive materials. Reinforcements modify strength and modulus as the particles or fibers bear a fraction of the applied load. Process parameters, such as melt temperature and mold pressures, relate to the structure in the final material. High speed and pressure are known to result in a glossy surface finish and high melt temperature is used to reduce viscosity and minimize fiber breakage. Fillers such as titanium dioxide and clay are used in paints and adhesives as pigments and to toughen them.

1.4 POLYMER CHARACTERIZATION

1.4.1 General techniques

A very wide range of analytical techniques is used to characterize polymeric materials (for example, see refs 4, 38, 39). The primary characterization of an organic material must be chemical. Elemental analysis by wet chemistry or spectroscopy may be useful in a few cases, for example to determine the degree of chlorination in chlorinated PE, but most chemical analysis is at the level of the functional group. Ultraviolet/ visible spectroscopy and mass spectroscopy (MS) of fragments broken from the polymer chain are often used. Even more common spectroscopies are infrared (IR) absorption, Raman, and nuclear magnetic resonance (NMR), which is very important. All of these can distinguish specific chemical groups in a complex system. Raman may be used on small particles and inclusions much more easily than IR, but IR and in particular Fourier transform IR (FTIR) has advantages of sensitivity and precision. NMR also gives local information, on a very fine scale, about the environment of the atoms investigated.

Once the chemistry of the molecule is known, the next important characteristic is the molecular weight distribution (unless the material is a thermoset or elastomer with infinite molecular weight). The molecular weight distribution is determined by a range of solution methods of physical chemistry, viscometry, osmometry, light scattering and size exclusion chromatography. Chemical and physical characterization methods overlap in the polymer field, for NMR of solid samples can determine the mobility of atoms in various regions and the orientation of molecules. IR and Raman are also sensitive to orientation and crystallinity of the sample.

There are two further types of physical characterization. They either involve scattering of light, neutrons or x-rays, or the formation of images of the polymer by microscopy – the subject of this text. Electron diffraction logically belongs in the first group, but is always performed in an electron microscope, so it is associated with microscopy. This technique shares with microscopy the ability to determine the structure of a local region, while other scattering methods determine the average structure in a large sample volume.

1.4.2 Microscopy techniques

There is a wide range of microscopy instruments available which can resolve details ranging from the millimeter to the subnanometer size scale (Table 1.3). The size and distribution of spherulites can be observed by optical techniques, but more detailed study requires electron microscopy. Single lamellar crystals can be seen with phase contrast optical microscopy but need TEM for detailed imaging and measurements. Polarized light microscopy is an optical technique that enhances contrast in crystalline materials. Phase contrast optical techniques enhance contrast between polymers that are transparent but which have different optical properties, such as refractive index and thickness. Reflected light techniques reveal surface structures whereas observation of internal textures of thin polymer slices is possible by transmitted light. Combinations of these microscopy techniques provide images of the morphology of polymer materials. An introductory text [40] and a more recent monograph [41] further describe microscopy techniques and provide descriptions and definitions of microscopes and relevant principles.

Microscopy is the study of the fine structure and morphology of objects with the use of a

Table 1.3 Characterization techniques: size ranges

Wide angle x-ray scattering (WAXS)	0.01–1.5 nm
Small angle x-ray scattering (SAXS)	1.5–100 nm
Transmission electron microscopy (TEM)	0.2 nm–0.2 mm
Scanning probe microscopy (STM, AFM . . .)	0.2 nm–0.2 mm
Scanning electron microscopy (SEM)	4 nm–4 mm
Optical microscopy (OM)	200 nm–200 μm
Light scattering (LS)	200 nm–200 μm

microscope. Resolution and contrast are key parameters in microscopy studies which will be discussed further. The specimen and the preparation method also affect the actual information obtained as the contrast must permit structures to be distinguished. Optical bright field imaging of multiphase polymers has the potential of resolving details less than one micrometer across; however, if the polymers are both transparent they cannot be distinguished due to a lack of contrast. There are variations among microscopes in available resolution, magnification, contrast mechanisms and the depth of focus and depth of field. Optical microscopes produce images with a small depth of focus whereas scanning electron microscopes have both a large depth of focus and large depth of field.

Review articles that describe the application of microscopy techniques to the study of polymer materials include Hobbs [42, 43] who discusses applications, and Grubb [6] and Thomas [44], who review the characterization of polymers, with emphasis on TEM techniques. White and Thomas [45] review SEM of polymers.

1.4.3 Specimen preparation methods

The range of specimen preparation methods is nearly as broad as the materials which must be prepared for observation. Metals and ceramics are prepared by well known, standard methods and biological materials have been prepared by methods specifically developed for their observation. Polymer materials are a bit newer than either of these materials, with a wide range of material forms and types and potential problems that are similar to metals and ceramics and to biological materials. Polymers, in common with biological materials, have low atomic number, display little scattering and thus have little contrast in the TEM. In addition, they are highly beam sensitive which must be taken into account. Like metals and ceramics, polymers can also be filled with hard inorganic materials. Overall, the types of methods developed for polymers are a composite of those methods known for metals

and biological materials and are specially adapted for macromolecules. Goodhew [46, 47] described many of the metallurgical methods for preparation of specimens for optical, scanning and transmission electron microscopy.

There are many characterization problems for microscopy where quite simple preparation methods are applied at least in the initial stages of morphological study. However, most polymers must be prepared with three things in mind: (1) to isolate the surface or bulk, (2) to enhance contrast and (3) to minimize radiation damage. For surface study, simple cutting out of the specimen or more tedious replication procedures supply the specimen of interest. Bulk specimens are obtained by cutting, fracturing, polishing or sectioning. A major problem with polymers sectioned for TEM is their inherent lack of contrast. Polymers have a low scattering power which results in low contrast. Methods which are employed to enhance contrast include staining, etching, replication, shadowing and metal decoration. Conductive coatings are applied to enhance contrast and to minimize radiation damage. Methods used include metal shadowing, formation of carbon support films, conductive coatings for SEM, metal coatings for optical microscopy and gold decoration. Specimens for SEM need to be electrically conductive in order to produce secondary electrons and to minimize charge buildup as polymers are generally nonconductive. Replication depends on the application of conductive coatings for their formation. Other methods adapted for microscopy preparation of soft, deformable materials are freeze drying, critical point drying and freeze fracture-etching which have been primarily developed by biologists. Methods developed specifically for polymers have been reviewed [6, 42–44, 48].

The specimen preparation methods used for microscopy of polymers involves the use of many toxic chemicals as well as the use of instruments which can be radiation hazards. It is well beyond the scope of this text to provide the information required for the proper and safe handling of such chemicals and instruments and the researcher is encouraged to obtain the required safety information prior to their use.

1.4.4 Applications of microscopy to polymers

The increased use of optical and electron microscopy applied to polymer research has been the result of widespread acceptance of the techniques and extended property requirements of the polymer materials. It is known that the structures present in a polymer reflect the process variables, and further that they greatly influence the physical and mechanical properties. Thus, the properties of polymer materials are influenced by their chemical composition, process history and the resulting morphology. Morphological study involves two aspects prior to the study itself: selection of instrumental techniques and development of specimen preparation methods. Structural observations must be correlated with the properties of the material in order to develop an understanding of the material.

What then are the key specimen preparation methods for studying polymer materials by microscopy techniques? This topic could be organized in one of two ways, that is by each specific microscopy technique or by each preparation method. The approach that has been chosen is to describe each specimen preparation type for all microscopies in order to minimize overlap and also to make it simpler to use for reference. Those preparation methods chosen for discussion are the typical ones found to be of major utility in the industrial laboratory. They cover the full range of study of the industrial scientist, that is everything from rapid failure analysis to process optimization studies and fundamental research. The fundamental studies must often be fitted into a limited time framework that requires good choices of methods and techniques on a wide range of materials.

Key issues in any microscopy study are that the polymer process must be understood and the structure characterized, in order to develop structure–property relationships. Yet, there are many questions for even the experienced materials scientist. Where do you start characterization

of a new material? Are there any protocols that work most of the time in order to solve each problem by the best technique in the shortest time? How do you minimize artifacts in conducting microscopy experiments? There are no easy answers, but the approach that is described here is to understand the image formation process, to study and understand the advantages and the drawbacks of the preparation methods, to know the instruments, both theoretically and practically, and finally, to conduct the microscopy study, and collect the observations and relate the characterization to the process and the physical and mechanical properties.

1.4.5 New microscopy techniques

New microscopy techniques continue to be developed, and recent years have been a particularly active time for this. Some new techniques are extensions and modifications of existing technology, others are completely new. A 1989 issue of the *Journal of Microscopy* provides a good review of a wide range of types of microscopy [49–53]. The reviews cover the history of the techniques and their future prospects as seen at that time. Now many of these future prospects have become real commercialized systems, out of the hands of instrument developers and into the hands of polymer microscopists. These newly available techniques include laser confocal scanning microscopy (LCSM) [49, 54–56], low voltage high resolution scanning electron microscopy (HRSEM) [57–60], high pressure SEM (HPSEM) [50] and high resolution transmission electron microscopy (HREM) [51, 52, 61, 62]. HREM is not really new, but it is very difficult to apply the method to polymers. This has slowed its transfer to polymer microscopy. Similarly, electron energy loss spectroscopy (EELS) has been known for some time, but only recently has it been routinely applied to polymers [63, 64].

More recent is the invention of the scanning probe microscopes, which include the scanning tunneling microscope (STM) [65] and atomic force microscope (AFM) [66]. These instruments can resolve individual atoms, and can operate at atmospheric pressure or even under water. The AFM requires no specimen preparation beyond the creation or exposure of a surface of interest. With such capabilities it is not surprising that development and commercialization has been very rapid. The new instruments have made dramatic changes in the imaging of all materials, and that includes biological and synthetic polymers [67]. The theory and practical aspects of these new technologies will be addressed in several chapters in this text.

REFERENCES

1. R. J. Young, *Introduction to Polymers* (Chapman and Hall, London, 1981).
2. P. H. Geil, *Polymer Single Crystals* (Interscience, New York, 1963).
3. A. Keller, *Rep. Prog. Phys.* **31** (1968) 623.
4. B. Wunderlich, *Macromolecular Physics*, Vol. 1, *Crystal Structure, Morphology, Defects*, and Vol. 2, *Crystal Nucleation, Growth, Annealing* (Academic Press, New York, 1973).
5. D. T. Grubb, *J. Microsc. Spectrosc. Electron.* **1** (1976) 671.
6. D. T. Grubb, in *Developments in Crystalline Polymers – 1* (Applied Science, London, 1982) p. 1.
7. D. R. Uhlmann and A. G. Kolbeck, *Sci. Am.* **233** (1975) 96.
8. D. C. Bassett, *Principles of Polymer Morphology* (Cambridge University Press, Cambridge, 1981).
9. D. C. Bassett, *CRC Crit. Rev. Solid State Mater. Sci.* **12**(2) (1984) 97.
10. R. B. Seymour, Ed., *The History of Polymer Science and Technology* (Marcel Dekker, New York, 1982).
11. H. D. Keith, *Kolloid Z. Z. Polym.* **231** (1969) 421.
12. H. D. Keith and F. J. Padden, *J. Appl. Phys.* **34** (1963) 2409.
13. I. M. Ward, Ed., *Structure and Properties of Oriented Polymers* (Applied Science, London, 1975).
14. D. T. Grubb and A. Keller, *J. Polym. Sci., Polym. Phys. Edn.* **18** (1980) 207.
15. A. Peterlin, in *Structure and Properties of Oriented Polymers*, edited by I. M. Ward (Halsted, Wiley, New York, 1975) p. 36.
16. M. R. Mackley and A. Keller, *Polymer* **14** (1973) 16.
17. A. J. Pennings, *J. Polym. Sci., Polym. Symp.* **59** (1977) 55.
18. G. B. Carter and V. T. J. Schenk, in *Structure and Properties of Oriented Polymers*, edited by I. M. Ward (Halsted, Wiley, New York, 1975) p. 454.
19. M. Jaffe and R. S. Jones, in *High Technology Fibers, Part A, Handbook of Fiber Science and Technology*, Vol. III, edited by M. Lewin and J. Preston (Marcel Dekker, New York, 1985).
20. A. Zachariades and R. S. Porter, Eds, *The Strength and Stiffness of Polymers* (Marcel Dekker, New York, 1983).
21. L. Holliday and I. M. Ward, in *Structure and Properties of Oriented Polymers*, edited by I. M. Ward (Halsted, Wiley, New York, 1975) p. 1.
22. A. Ciferri and I. M. Ward, Eds, *Ultra-high Modulus Polymers* (Applied Science, London, 1979).
23. I. L. Hay, in *Methods of Experimental Physics: Polymers, Part C, Physical Properties*, edited by R. A. Fava (Academic Press, New York, 1980) p. 137.
24. A. Peterlin, *Adv. Polym. Sci. Eng.* (1972) 1.
25. A Peterlin, in *Ultra-high Modulus Polymers*, edited by A. Ciferri and I. M. Ward (Applied Science, London, 1979) p. 279.
26. G. Calundann and M. Jaffe, in *Proc. Robert A. Welch Conf. on Chemical Research, XXVI, Synthetic Polymers* (Houston, Texas, Nov. 15–17, 1982) p. 247.
27. G. Capaccio, A. G. Gibson and I. M. Ward, in *Ultra-high Modulus Polymers*, edited by A. Ciferri and I. M. Ward (Applied Science, London, 1979) p. 1.
28. Z. Tadmor, *J. Appl. Polym. Sci.* **18** (1974) 1753.
29. Z. Tadmor and C. G. Gogos, *Principles of Polymer Processing* (Wiley–Interscience, New York, 1979).
30. C. B. Bucknall, *Toughened Plastics* (Applied Science, London, 1977).
31. N. A. J. Platzner, *Copolymers, Polyblends, and Composites*, ACS Adv. Chem. 142 (American Chemical Society, Washington DC, 1975).
32. J. A. Manson and L. H. Sperling, *Polymer Blends and Composites* (Plenum, New York, 1976).
33. R. W. Hertzberg and J. A. Manson, *Fatigue of Engineering Plastics* (Academic Press, New York, 1980).
34. G. Pritchard, Ed., *Developments in Reinforced Plastics – 3* (Elsevier-Applied Science, London, 1984).
35. M. J. Folkes, *Short Fibre Reinforced Thermoplastics* (John Wiley, Chichester, 1982).
36. D. Hull, *An Introduction to Composite Materials* (Cambridge University Press, New York, 1981).
37. S. M. Lee, in *Developments in Reinforced Plastics – 3*, edited by G. Pritchard (Elsevier-Applied

Science, London, 1984).

38. E. W. Fischer, Structure of amorphous organic polymers in bulk, in *Proc. Conf. Non-Crystalline Solids* (Clausthal-Zellerfeld, Sept. 1976).

39. R. G. Vadimsky, in *Methods of Experimental Physics* Vol. 16B, edited by R. A. Fava (Academic Press, New York, 1980) p. 185.

40. T. G. Rochow and E. G. Rochow, *An Introduction to Microscopy by Means of Light, Electrons, X-rays or Ultrasound* (Plenum, New York, 1978).

41. D. A. Hemsley, *The Light Microscopy of Synthetic Polymers* (Oxford University Press, Oxford, 1984).

42. S. Y. Hobbs, *J. Macromol. Sci., Rev. Macromol. Chem. C* **19**(2) (1980) 221.

43. S. Y. Hobbs, in *Plastics Polymer Science and Technology* edited by M. D. Bayal (Wiley–Interscience, New York, 1982) p. 239.

44. E. L. Thomas, in *Structure of Crystalline Polymers*, edited by I. H. Hall (Elsevier-Applied Science, London, 1984) pp. 79.

45. J. R. White and E. L. Thomas, *Rubber Chem. Technol.* **57**(3) (1984) 457.

46. P. J. Goodhew, Specimen preparation in materials science, in *Practical Methods in Electron Microscopy*, edited by A. M. Glauert (North Holland-American Elsevier, Amsterdam, 1973).

47. P. J. Goodhew, *Specimen Preparation for Transmission Electron Microscopy* (Oxford University Press, Oxford, 1984).

48. D. Hemsley, in *Developments in Polymer Characterization*, edited by J. V. Dawkins (Applied Science, London, 1978) p. 245.

49. H. J. Tanke, *J. Microscopy* **155** (1989) 405.

50. D. McMullan, *J. Microscopy* **155** (1989) 373.

51. B. Ralph, *J. Microscopy* **155** (1989) 339.

52. P. B. Hirsch, *J. Microscopy* **155** (1989) 361.

53. A. Howie, *J. Microscopy* **155** (1989) 419.

54. T. Wilson, *Theory and Practice of Scanning Optical Microscopy* (Academic Press, London, 1984).

55. J. B. Pawley, Ed., *Handbook of Biological Confocal Microscopy* (Plenum Press, New York, 1990).

56. T. Wilson, Ed., *Confocal Microscopy* (Academic Press, London, 1990).

57. L. Reimer, *Image Formation in Low Voltage Scanning Electron Microscopy* (SPIE, Bellingham, 1993).

58. J. H. Butler, D. C. Joy, G. F. Bradley, S. J. Krause and G. M. Brown, in *Microscopy: The Key Research Tool* (Electron Microscopy Society of America, Woods Hole, 1992) p. 103.

59. L. Reimer, *Scanning Electron Microscopy, Physics of Image Formation and Microanalysis* (Springer, Berlin, 1985).

60. J. I. Goldstein, D. E. Newbury, P. Echlin, D. C. Joy, A. D. Romig, Jr., C. E. Lyman, C. Fiori and E. Lifshin, *Scanning Electron Microscopy and X-ray Microanalysis*, 2nd Edn (Plenum Press, New York, 1992).

61. J. C. H. Spence, *Experimental High Resolution Transmission Electron Microscopy and Associated Techniques*, 2nd Edn (Oxford University Press, Oxford, 1988).

62. L. Reimer, *Transmission Electron Microscopy, Physics of Image Formation and Microanalysis*, 2nd Edn (Springer, Berlin, 1989).

63. M. M. Disko, in *Transmission Electron Energy Loss Spectrometry in Materials Science*, edited by M. M. Disko, C. C. Ahn and B. Fultz (TMS Books, Warrendale, PA, 1992) p. 1.

64. L. L. Ban, M. J. Doyle, M. M. Disko and G. R. Smith, *Polymer Comm.* **29** (1988) 163.

65. G. Binnig, H. Rohrer, C. Gerber and E. Weibel, *Phys. Rev. Lett.* **49** (1982) 57.

66. G. Binnig, C. F. Quate and C. Gerber, *Phys. Rev. Lett.* **56** (1986) 930.

67. G. J. Leggett, M. C. Davies, D. E. Jackson, C. J. Roberts and S. J. B. Tendler, *TRIP* **1**(4) (1993) 115.

CHAPTER TWO

Fundamentals of microscopy

2.1 INTRODUCTION

Microscopy is the study of the fine structure and morphology of objects with the use of a microscope. Microscopes range from optical microscopes, which resolve details on the micrometer level, to microscopes that can resolve individual atoms in suitable samples, but all form magnified images of the specimen. Some instruments give information about a surface and not the specimen interior, but preparation methods may create an internal surface that can be imaged. Apart from this, the size and visibility of the polymer structure to be characterized generally determines which instrument is to be used. For example, the size and distribution of spherulites can be observed by light-optical techniques, but a study of their internal structure requires a higher resolution method, such as transmission electron microscopy. Combinations of the various microscopy techniques generally provide the best insight into the morphology of polymer materials.

Microscopes are commonly identified by the radiation used, as optical or electron microscopes. A more fundamental distinction is whether the image is formed all at once, by lenses, or sequentially, point by point, by scanning. In computer terms, these would be described as parallel and serial transmission of information respectively. A TV image is an example of an image formed by scanning. The conventional optical microscope and the trans-

mission electron microscope are examples of the other type, where an image is formed all at once by lenses. The recently developed scanning probe microscopes (such as the scanning tunneling microscope) move a solid sharp probe over the specimen surface. These microscopes have no lenses and build up their images point by point. The scanning electron microscope (and the scanning transmission electron microscope) forms a sequential image, but it uses a beam of electrons, focused by lenses, as the scanning probe. Some concepts of lens imaging are relevant to these microscopes. The confocal optical microscope described in Section 6.2 is another intermediate type; it forms an image with lenses, but an aperture limits the area viewed at any instant, and the image is built up by scanning. Table 2.1 shows some basic properties of different types of microscope, for comparison. They are divided into lens-imaging and scanning-imaging classes.

Key parameters of microscope images are resolution and visibility. *Resolution* is the minimum distance between two object features at which they can still be seen as two features. The visibility of a feature is described by its *contrast*, which is the fractional change in image brightness that it causes. If two features have high contrast but are separated by less than the resolution, they will appear as one feature in the image, visible but not resolved. If their contrast is very low, below about 0.05, they will not be visible at all. Small features in the

Table 2.1 Properties of various types of microscope

Instrument	Optical microscope		Transmission electron microscope	Scanning electron microscope		Atomic force microscope
	OM	lens-imaging	TEM	SEM	scanning-imaging	AFM
Lateral Resolution	300 nm		0.5 (0.2)* nm	4 (1) nm		4 (0.3) nm
Magnification	2–2000		200–2×10^6	20–1×10^5		1000–2×10^6
Can observe	surface, or bulk if transparent		'bulk', but thin films, $< 0.2\ \mu m$	surfaces		surfaces
Specimen environment	ambient, or transparent fluid		high vacuum	high vacuum (30 torr in HPSEM)		ambient, high vacuum or fluid
Radiation damage	none		severe	rarely serious		none
Specimen preparation	easy		very difficult	easy		easy
Chemical analysis	no, unless μ Raman		yes, x-ray and electron energy loss	yes, x-ray		no
Can detect molecular orientation	yes		yes	no		no

* Values in parentheses are limits obtained with special equipment or techniques.

specimen that have low contrast may not be observable, even if they are larger than the resolution limit of the instrument. Table 2.1 lists an indication of resolution but not contrast, as this depends on the sample. The *field* is the area of the specimen included in the image, so the number of independent data points in the image is (field/resolution)^2.

It is beyond the scope of this text to describe the design features and operation of specific microscopes and their attachments. Any attempt to discuss microscope operation or construction in specific detail could rapidly become outdated. Manufacturers or their representatives are the best source for operating instructions for their own microscopes. Lists of manufacturers are given in Appendices VI and VII.

2.1.1 Lens-imaging microscopes

Some definitions important for these lens-imaging microscopes should be given now. The *depth* of field is the depth or thickness of the specimen that is simultaneously in focus. As explained in Chapter 3, transmission electron microscopes have a depth of field that is normally greater than the specimen thickness, so all of the specimen is in focus at once. Optical microscopes have a depth of field comparable to their resolution, and this is often much less than the sample thickness. (The *depth of focus* is the depth of the image that is in focus and is not important in microscopy.) *Bright field* (BF) in transmission is when the direct unscattered beam is allowed to reach the image plane. An image field which contains no specimen is then bright. In reflection, BF is when specular reflection from the specimen surface is allowed to reach the image plane, so that a flat specimen is bright in BF. *Dark field* (DF) is the opposite imaging mode, where only scattered radiation is allowed to form the image. In transmission an image field with no specimen is dark in DF.

The examples of lens-imaging microscopes are

optical microscopes and transmission electron microscopes. Electromagnetic coils are the lenses for electron beams, and glass lenses are used for light. Magnetic lenses are focused by changing their power, glass lenses by changing their position. These differences and the high vacuum systems needed for electron beams tend to hide the close similarity of function and arrangement of the various components. The parallels of construction are particularly close if a transmitted light optical microscope is compared to a TEM as in Fig. 2.1. The source of illumination is normally a small hot filament in both cases and the radiation emitted from the filament is efficiently collected by one *condenser lens*. A second condenser lens controls the transfer of the radiation to the specimen plane. In TEM construction these are simply called condenser lens 1 (C_1) and condenser lens 2 (C_2), but the operator's controls for these lenses may be labelled 'spot size' and 'brightness'. In transmission optical microscopy the second lens is called the substage condenser, or simply 'the condenser'. The first lens, which is built into the base

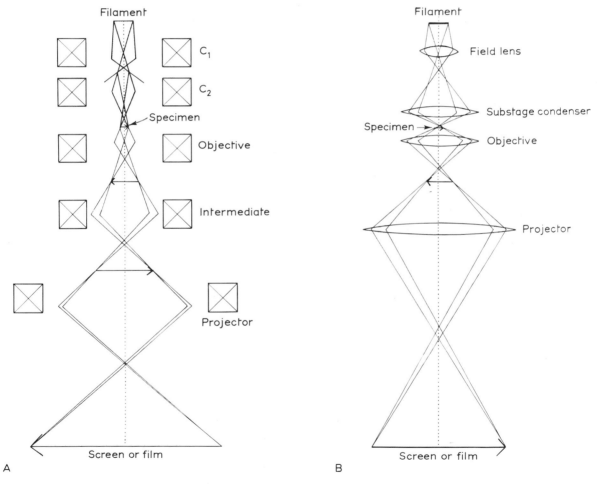

Fig. 2.1 The optics of (A) a transmission electron microscope and (B) an optical microscope using transmitted light. In both cases illumination is focused onto the specimen by two condenser lenses. The objective lens forms a real magnified image of the specimen, and this is further magnified by the following lenses to form an image on a screen or film. The electron microscope has more magnifying lenses.

unit or is part of a separate free standing illuminator, may be called the auxiliary or lamp condenser, or the field lens.

There is an aperture associated with each condenser lens, and the apertures and lenses control the area illuminated and the angular divergence of the illumination. The details of the illumination system are described in Section 3.1.5. In both optical and electron microscopes the resolution and contrast of the image may be degraded if the illumination is not properly adjusted.

After the radiation has passed through the specimen, the scattered radiation is collected by an *objective lens*. This lens is the most critical, and imperfections in it will affect the image quality directly. An aperture associated with the objective lens, the objective aperture, is often used in transmission electron microscopy. In optical microscopy the size of the objective lens usually acts as the limiting aperture.

The eyepiece lens in an optical microscope forms a virtual image for the eye to focus on. When an optical microscope is set up for photo-microscopy, a *projector lens* is used in place of an eyepiece. This forms a real image on the photographic film. The TEM is like this, but there are at least two lenses after the objective, the *intermediate lens* and the projector lens. Each produces a real and magnified image, the projector lens producing its image on the fluorescent imaging screen or on the film. Two lenses are used to give the greater total magnification required for electron microscopy. Most electron microscopes have a fourth imaging lens (I_2 or diffraction). This allows the magnification to be one million times, and it makes it simple to control the size of the diffraction pattern. Modern TEMs may have yet more lenses, but they also have computer controls that hide these complications from the operator.

2.1.2 Scanning-imaging microscopes

In scanning-imaging microscopes a small probe is passed over the specimen surface, and a signal is collected that relates to some local property of the specimen. The probe may be a sharp point, or a narrow beam of electrons or photons, and the signals and detectors are very diverse. All the instruments have a common feature, a detector output signal that varies with time and is displayed like a TV image. The electron beam in the display tube moves in synchrony with the probe on the specimen and is modulated in position or brightness by the signal from the microscope. The magnification is then simply the linear size of the image divided by the size of the region scanned on the specimen, as shown schematically in Fig. 2.2. Magnification can be altered without having any effect on other imaging conditions. The spatial resolution of the microscope is determined by the size of the specimen region from which the signal is derived. There is a limit to the number of lines that can be usefully displayed. This was typically 500 for analog systems, and may be 1024 or 2048 for digital systems today. A similar number of distinguishable spots on each line gave $500^2 = 250,000$ *pixels* – picture elements – per image. In that case the maximum useful magnification is (display screen size, L)/($500 \times$ microscope resolution).

The signal from scanning microscopes may be readily digitized, processed and stored elec-

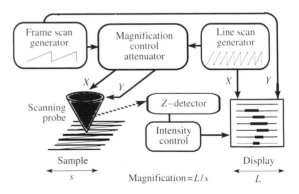

Fig. 2.2 Schematic of imaging by scanning. The scanning probe may be an electron beam, in the SEM, or a solid probe in the STM. The detector may detect any of a wide variety of signals, including in STM the height of the probe, Z. The magnification is simply the length scanned on the display output divided by the length scanned on the object.

tronically. Modern systems are built around the computer, and may have mouse control of most operating parameters and electronic output to video printers. The advantages of integration with computers are so great, particularly for image analysis and image processing, that many lens-imaging microscopes are fitted with TV cameras to turn their output into such a signal. If the signal, or image brightness, is digitized to 256 levels, then each 1024 square image will take 1 Mbyte of storage space, if it is not compressed. Storage requirements quickly build up, so photographs of the screen output are still often used for archival purposes.

2.2 OPTICAL MICROSCOPY

2.2.1 Introduction

In the conventional optical microscope (OM) an object is illuminated and light that it scatters or transmits is collected by a system of lenses to form an image. The image can reveal fine detail in the specimen at a range of magnifications from $2\times$ to $2000\times$. Resolution of about 0.5 μm is possible, limited by the nature of the specimen, the objective lens and the wavelength of light. Many texts deal with the subject at a range of levels. Spencer [1] describes the fundamental science in a non-mathematical manner. Others are more comprehensive or have more practical information [2–8]. Some concentrate on a particular topic, such as polarized light microscopy [9], the identification of materials [10], or the study of polymers [11].

The information obtained in the OM normally concerns the size, shape and relative arrangement of visible features. Local measurement of optical constants such as the refractive index (Section 2.2.2) and the birefringence (Section 2.2.3) is also possible. Many techniques are used to enhance contrast, and thus make more of the structure visible. Images are typically recorded photographically, on video tape for the study of dynamic processes, or in a computer system for digital image analysis.

Simple microscopes have only one imaging lens (though this may have several elements) and

operate at low magnification, like a magnifying glass. The optical microscopes in the laboratory are generally *compound microscopes*, with more than one imaging lens. They operate at higher magnification and higher resolution giving more detail on smaller specimens. They fall into two categories, stereomicroscopes and research microscopes. Binocular stereomicroscopes [12] provide two different images of the specimen through the two eyepieces; these are views from slightly different directions. The observer sees this as a three-dimensional image, very useful for the examination of bulk specimens. The stereomicroscope is a good starting point for investigations of the nature of the material. It also helps to identify regions of the specimen for further study. It should be noted that many people are unable to form proper three-dimensional images from stereoscopic views, using instead visual cues of size and shape. These can be misleading in microscopic images. (This is more serious in stereoscopic TEM where normal visual cues are often absent from the image.)

Typical compound research microscopes are also binocular, but the images in the eyepieces are identical. The two images are provided to reduce eyestrain. The microscopes can be equipped for both transmitted and incident light. In *transmitted light* techniques a light beam passes from a condenser lens system through the specimen and into the imaging system, the objective and eyepiece lenses. Opaque materials or samples too thick to provide information by transmitted light techniques can be imaged using *incident light*. Here the light passes through the objective lens and is reflected from the specimen surface back into the objective. At low magnification or for dark field imaging a separate illuminating lens may be used, but in all cases the light is reflected from the specimen to reveal surface topography.

2.2.2 Objective lenses

The *objective lens* or *objective* is the most important part of the optical system. All optical objectives have information engraved on them, and it is helpful to be able to interpret this. An old lens may only show its focal length, but a modern lens may have something similar to 'Pol 25/0.55 0.17/160'. The letters indicate the type of lens, here strain-free, suitable for use with polarized light. Different manufacturers may use different abbreviations, but most are readily understood. The most obscure is Fl or Fluor. This means the lens is a high quality general purpose lens, once (but no longer) made using the mineral Fluorite. Such lenses are also called semi-apochromats.

The first two numbers are always the magnification and the *numerical aperture* (NA) of the lens. Numerical aperture is defined as ($n \sin \alpha$), where n is the refractive index of the medium in front of the objective and α is half the angular range of light that the lens can accept. The resolution of the lens, if it is perfect, is proportional to $1/(NA)$, and the depth of field is proportional to $1/(NA)^2$. If the lens is 'dry', it has air in front of it, $n = 1$, and the maximum NA is 0.95. An immersion oil of $n = 1.5$ between lens and sample can significantly increase the resolution limit by increasing the possible NA to 1.3. A rule of thumb is that the highest useful magnification is about 700–1000 times the numerical aperture of the objective lens. Higher magnifications are 'empty' as the images do not contain any extra information, but they may be necessary for observers with poor eyesight.

The second two numbers are the correct thickness of cover glass to use between lens and specimen in mm and the *tube length*, also in mm. Cover glasses are standardized at 0.17 mm thick, so the number is normally 0.17 or 0, indicating whether or not the objective was designed for use with a cover glass. High power dry objectives are very sensitive to cover glass thickness variations, and may have an adjustment ring that is set to suit the actual value. The tube length is not important for the user unless objectives are bought for one microscope and used with another. Poor results will be obtained unless both instruments have the same tube length. The tube length was originally the physical length of the tube separating the objective and eyepiece lenses, but no more.

Modern microscopes commonly have a tube length of infinity, which has many advantages for modular design.

2.2.3 Imaging modes

2.2.3.1 *Bright and dark field*

Bright field is the normal mode of operation of an optical microscope. The contrast in transmitted light is based on variations of optical density and color within the material. Carbon black agglomerates, pigment particles and other fillers are clearly observed in polymers in bright field as the matrix polymers are typically transparent in thin sections. Dark field transmitted light is less common, but has higher contrast. In reflected light the visibility of details in polymer samples is often poor due to low surface reflectivity, scatter from within the specimen and glare from other surfaces. A metal coating on the surface (vapor deposited or sputtered) will increase the surface reflectivity. This enhances brightness and contrast in reflected light. Dark field in reflected light may be used to increase the contrast of surface roughness. If the specimen is transparent and not coated, dark field allows observation of subsurface features and details in reflected light.

2.2.3.2 *Phase contrast*

Thin sections of polymer blends can give bright field images with little or no contrast between the components. Transmitted light phase contrast converts the refractive index differences in such a specimen to light and dark image regions. Small differences in thickness are also made visible [1–7, 9]. Accessories for normal (Zernike) phase contrast include a special condenser with an opaque central plate, and a matching phase ring in the back focal plane of the objective. Unscattered light passes through the phase ring, and its phase is altered compared to the scattered light. Interference between the scattered and direct rays causes changes in the image intensity. In a two-component transparent sample light is scattered from the interfaces.

The phase contrast image has characteristic bright halos around fine structure, due to some scattered light passing through the ring in the phase plate. A method that produces images without a halo effect is Hoffman modulation contrast [4, 5, 13]. In this technique, scattered light is changed in amplitude, rather than phase, by a modulator disk in the back focal plane of the objective. The illumination is limited by a slit, not a ring. Earlier versions of phase contrast used similar methods [14], but Hoffman modulation contrast uses a rotatable polarizer under the condenser to control the illumination, and an asymmetric modulator disk. This makes for a very flexible system. The images are sharp and three-dimensional with shadows and textures that give an appearance of oblique illumination.

2.2.3.3 *Interference microscopy*

In interference microscopy the illumination is split into two beams with different paths, and the two beams are recombined so that they interfere. The interference pattern can be used to measure the specimen thickness in transmission, or the surface profile of the specimen in reflection. In reflected light microscopy the beam splitter is usually a half-silvered mirror; one beam is reflected off the specimen and the other off a flat reference mirror [3–6]. Transmitted light is more complex as the beam splitter may be a doubly refracting crystal (Wollaston prism), and the two beams can be displaced vertically or horizontally [2–7]. In some systems the displacement is large and one beam does not pass through the specimen. For example, the Mach–Zehnder interference system (Leitz) [4, 15] consists of a double microscope system, where the specimen and the reference beams are separated by several millimeters. The Jamin–Lebedeff system (Zeiss) [1, 3, 5, 15] uses a single microscope system. The reference beam is displaced by more than its diameter but both reference and measuring beam can be seen in the same field of view. The reference beam is set to pass through a featureless area of the specimen.

Differential interference contrast (Nomarski contrast or DIC) also has the illumination split into two beams [1, 4, 5]. One beam is displaced at the specimen plane, and the beams are recombined to give contrast by interference. The important difference between this and other interference methods is that in DIC the beams are displaced a very small distance, much smaller than the beam diameter. Thus a region of constant properties shows no contrast because both beams see the same material, but a sudden change of thickness or refractive index gives strong contrast. In transmission the technique is similar to Hoffman modulation contrast, containing the same information as phase contrast without the halo, and with apparent relief and shadows as in oblique illumination. A single birefringent plate can produce the polarization and the separation of the beams, and in reflected light the same plate recombines the reflected light. In transmission a matched pair of condenser and objective lenses is needed for each magnification. DIC is therefore most often used in reflected light, where surface topography is highlighted and a pseudo three-dimensional image produced. As for bright field, this reflected image can be enhanced by deposition of metal on the surface. Using white light in DIC gives attractive micrographs in both reflection and transmission mode, as the contrast then appears as changes in interference color.

2.2.4 Measurement of refractive index

Refractive index n is easily measured in the optical microscope and helps to identify unknown materials [7, 9, 10, 15]. Small particles are mounted in a liquid of known refractive index and observed in transmission. The liquid can be changed until $n(\text{liquid}) = n(\text{particle})$, when a transparent or translucent particle will have very low contrast. To find out whether the change should increase or decrease n, the sample can be illuminated obliquely. The particles then act as rough lenses, converging lenses if $n(\text{liquid}) < n(\text{particle})$. Light striking one side of the particle will be diverted toward the axis, and that side will appear bright. This shading will be reversed if the *n*(liquid) > *n*(particle). Alternatively, axial illumination of particles with sharp boundaries gives a narrow band of light near the edge, the *Becke line*. Light is scattered towards the side of greater *n*, so the line appears on this side when the focus is above the particle (overfocus), and on the other side at underfocus.

Phase contrast increases the contrast due to differences in *n*, and so allows a more accurate determination. Interference contrast in transmission gives the optical path length, and so the average refractive index through the specimen thickness. The Becke line method gives the surface refractive index. Fibers often give different results by the two methods because of the variation in refractive index across a fiber cross section.

2.2.5 Polarized light

Polarizing microscopy is the study of the microstructure of objects using their interactions with polarized light [4–9, 15–18]. The method is widely applied to polymers and to liquid crystals [20, 21]. Wood [16] gives a very basic introduction to the subject and Hartshorne and Stuart [9] is comprehensive (see also Section 3.3). A *polarizing microscope* is a transmitted light microscope that has, besides all the basic features, a rotatable stage, a *polarizer* in the illumination system and an *analyzer* between objective and eyepiece. One or both of the polarizer and analyzer must be rotatable. The polarizer and the analyzer are both *polars*, that is, devices that selectively transmit light polarized in one specific plane. Most polars are made from a Polaroid filter, which is *dichroic* – it selectively absorbs light of one polarization state. By far the most common arrangement in the polarizing microscope is *crossed polars*. The transmitted polarization planes of the two polars are set to be perpendicular or 'crossed' so that the analyzer does not transmit light transmitted by the polarizer. With no specimen, or with an isotropic specimen, the field of view will be dark in crossed polars. The polars are usually set to transmit light polarized in the directions given by

3–9 o'clock and 6–12 o'clock, imagining a clock face on the specimen. These directions are referred to as $0°$ and $90°$.

Optically anisotropic, birefringent materials may appear bright between crossed polars. When well oriented specimens such as fibers are rotated on the rotatable stage, they go through four *extinction positions* of minimum intensity, and four positions of maximum intensity. In the extinction positions the orientation direction is aligned parallel to one polarization direction, at $0°$ or $90°$. Maximum intensity is at the $45°$ positions. Circularly polarized light, obtained and analyzed by the addition of two crossed quarter-wave plates into the light path, eliminates these extinction positions (Section 3.3). All anisotropic specimens are bright between crossed circular polars regardless of their orientation.

Birefringent materials can be considered to split light which passes through them into two plane polarized waves that vibrate in planes at right angles to one another. These transmitted waves have different velocities and refractive indices n_1 and n_2. The direction of vibration with the larger refractive index is called the *slow direction*. The sample *birefringence* is the difference between the two refractive indices, (n_1-n_2) or Δn. If the sample has a clear reference direction it is used to define the sign of the birefringence. If the reference direction is the slow direction the birefringence is said to be positive. Examples of reference directions are the length of a fiber and the radius of a spherulite.

Birefringence is often measured to obtain quantitative data on the degree of molecular orientation in the sample. Most birefringence is due to the orientation of optically anisotropic elements. These can be amorphous or crystalline chains in a polymer, aligned by deformation such as drawing. However, there is also form birefringence, which arises when the material contains at least two phases, having different refractive indices, and with some dimension about the wavelength of light. Form effects can contribute to birefringence in copolymers and semi-crystalline polymers; it must be allowed for in the calculation of molecular orientation. Elastic strain can also cause birefringence, and this should not be confused with permanent orientation effects.

Birefringence can be measured directly, by measuring the two refractive indices of the sample and taking the difference, but this is usually inaccurate. Normally the specimen thickness and *retardation* are measured. Retardation is defined as $[(n_1 - n_2) \times (\text{specimen thickness})]$ in nm, and is measured using a *compensator*, which is a crystal plate of known retardation. The specimen to be measured is set to the $-45°$ position between crossed polars, and a compensator is inserted in its slot. This is above the specimen but below the analyzer at $+45°$. The compensator is adjusted until the specimen is dark, when its retardation is exactly cancelled by the compensator. If this adjustment is impossible, the sample must be rotated $90°$ to $+45°$.

In white light, anisotropic structures may appear brightly colored when viewed in crossed (or parallel) polars. These *polarization* or *interference colors* depend on the retardation (Section 3.3). An estimate of sample retardation can be made from the standard sequence of colors, published as the Michel–Levy chart in many texts [4, 7, 9, 17, 18]. Color can also be used to find the sign of birefringence when a *first order red plate* is inserted as a compensator in white light.

2.3 SCANNING ELECTRON MICROSCOPY

2.3.1 Introduction

The scanning electron microscope (SEM) forms an image by scanning a probe across the specimen, and in the SEM the probe is a focused electron beam. The probe interacts with a thin surface layer of the specimen, a few micrometers thick at most. The detected signal commonly used to form the TV-type image is the number of low energy *secondary electrons* emitted from the sample surface. Scanning electron microscopy is fully described in several texts [22–26], and its use with polymers has been reviewed by White and Thomas [27].

A simple analog of a scanning microscope, which may make its operation easy to understand, is a flashlight and a light meter in a dark room. The intensity of reflected light and thus the light meter response is large when the flashlight beam falls on a pale wall. When it falls on dark drapes, or out of a window, there is a small signal on the light meter. Scanning the spot of light systematically over the wall and recording the signal maps out the dark regions in the scanned area. In the room analog, dark regions are due to reduced reflectivity or to gaps in the reflecting surface. In the SEM these correspond to *compositional* and *topographic* contrast respectively because the first depends on the composition of the sample and the second on its shape.

Figure 2.3 is a block diagram of a normal SEM showing the electron optical column of three condenser lenses used to form the probe, and the common source for display scanning and probe scanning, as in Fig. 2.2. Compared to reflected light optical microscopy, the SEM has higher resolution and a much larger depth of field. The specimen chamber in the SEM is large and samples up to several inches in diameter can be accommodated. Specimen preparation is generally quite simple, if the materials can withstand drying and high vacuum. Nonconductive materials, such as most polymers, require conductive coatings or the use of low accelerating voltages to prevent them from charging up in the electron beam. SEM images are easy to interpret qualitatively. They appear as though the specimen is viewed from the source of the scanning beam, and illuminated by a light at the detector position.

Imaging by scanning allows any radiation from the specimen, or any result of its interaction with the beam, to be used to form the image [23–26, 28]. The appearance of the image will depend on the interaction involved and the detector and signal processing used. The spatial resolution, limited by the size of the specimen region from which the signal is derived [29, 30], varies considerably as shown in Fig. 2.4. It is related to the *interaction volume*, the region where the beam interacts with the specimen. A direct

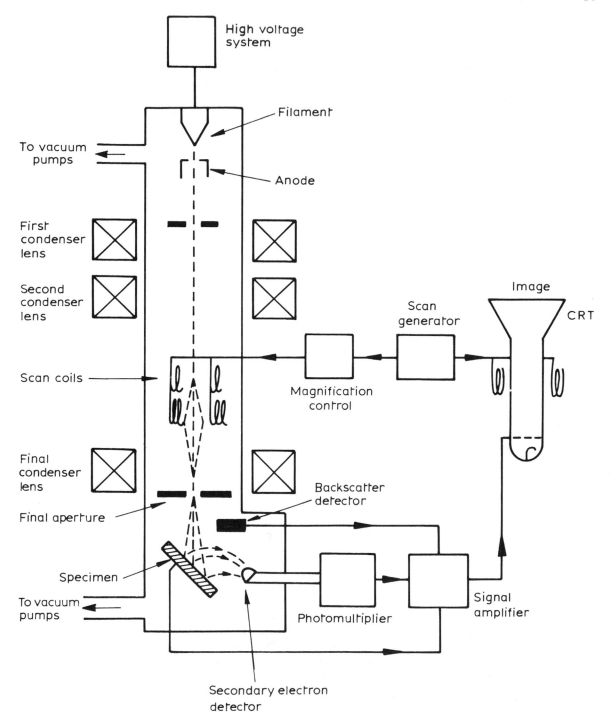

Fig. 2.3 Schematic diagram of a scanning electron microscope. Two pairs of deflection coils are shown in the SEM column. This double deflection allows the scanning beam to pass through the final aperture. Four pairs are actually used, for double deflection in the x- and y-directions.

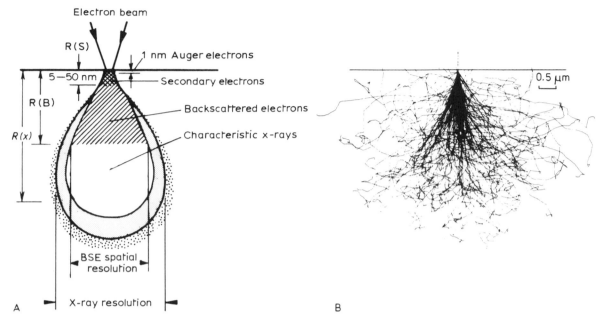

Fig. 2.4 (A) Schematic of the interaction of an incident electron beam with a solid specimen. Backscattered electrons can escape from much greater depths than secondary electrons $R(B) \gg R(S)$. X-rays are produced in a larger volume and have less resolution. (B) This calculation of electron paths in carbon shows that the neat boundaries and contours in (A) are only an indication of the statistics involved [23].

demonstration has been made of the interaction volume of 20 kV electrons in PMMA, using its radiation sensitivity [31]. After exposure to a beam of electrons, the material was cross sectioned, polished and etched. The result is pear-shaped holes up to 10 μm deep, Fig. 2.5, showing where the beam interacted with the PMMA (and reduced its molecular weight). Calculation of the interaction of the electron beam with a solid shows that the interaction

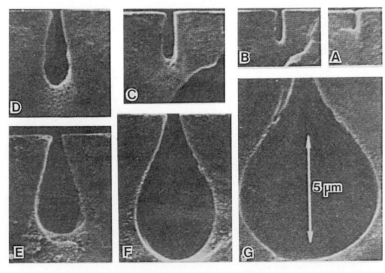

Fig. 2.5 The electron interaction volume in poly(methyl methacrylate) for 20 kV electrons is shown directly by etching away irradiated material. The incident electron dose is the same in (A–G), but the etching time is increased, so the material irradiated less is removed [31].

volume increases at high accelerating voltage and for low atomic number of the specimen [23–26].

2.3.2 Imaging signals

2.3.2.1 *Backscattered electrons*

Three important signals from the specimen are backscattered electrons, secondary electrons and x-rays; x-rays are dealt with in Section 2.7. *Backscattered electrons* are primary beam electrons that have been elastically scattered by nuclei in the sample and escape from the surface. The fraction escaping varies from 0.06 for carbon to 0.5 for gold at 20 keV [23] so a backscattered electron image (BEI) has compositional contrast. Backscattered electrons have a high energy and they can come from depths of 1 μm or more within the specimen at high beam voltages. They then leave the surface from a wide area, and this means that the resolution in BEI is low, generally about 1 μm, but dependent on beam voltage and sample composition. BEI combined with x-ray microanalysis is a powerful method for determining the local chemical composition of a material.

Backscattered electrons travel in straight lines after leaving the specimen, so a detector placed to one side will give topographic contrast. The topographic contrast in BEI with a single detector of limited area is extreme, with complete darkness in the shadows, like pictures of the lunar landscape. Mixing of signals from more than one detector may produce a good image. The detectors for BEI are placed close to the specimen to increase collection efficiency, and the image is affected strongly by the position, which is usually above the specimen, near to the final aperture. The detectors may be either silicon diodes of large area, or scintillators (Robinson detectors) [32, 33]. In either case the sensitive area may be semicircular, annular or 90° quadrants. Atomic number contrast is obtained by adding the signals from all sectors, and topographic contrast by subtraction of signals [34].

2.3.2.2 *Secondary electrons*

Direct *secondary electrons* are produced by the interaction of the primary beam with the specimen. They are emitted from the specimen with low energy, less than 50 eV, so they can only come from the top few nanometers of the material [23–26]. If the beam falls on a tilted surface or onto an edge, more secondaries will escape from the specimen (Fig. 2.6). If the beam falls into a valley or pit, fewer secondaries escape because less of the interaction volume is near the

 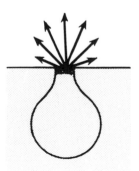 

Fig. 2.6 A tilted surface produces more secondary emission because more of the interaction volume is near the surface, and because the tendency for electrons to scatter forwards allows more to escape. In a hollow, some secondaries will re-enter the specimen and not be detected.

surface, or because electrons are reabsorbed by the specimen. These secondary electrons produced by the primary beam are responsible for high resolution topographic images, as they come from an area defined by the beam size. However, other sources also produce secondary electrons. One source is the backscattered electrons as they leave the specimen, another is the backscattered electrons striking the chamber wall [23, 24]. These secondary electron signals will have the resolution of the BEI signal, and may degrade the image.

The detector in the SEM that is normally used for imaging is the Thornley–Everhart scintillator/photomultiplier. This is a low-noise, high speed efficient detector, but it detects a small fraction of the backscattered electrons as well as the secondary electrons. The image it produces is often described as a secondary electron image (SEI), but this is not strictly accurate. In many cases, at moderate and low resolution, the backscattered electrons improve the appearance of the image by adding a directionality to the topographic contrast. To reduce noise at the best resolution, special techniques may be used to limit the signal to secondary electrons produced in the specimen [35].

Some polymers, such as PMMA, are quite sensitive to the beam. Extended exposure to the electron beam will cause mass loss and pitting or depression of the surface in the rastered area. Semi-crystalline polymers lose crystallinity on exposure to the electron beam and this can change specimen morphology in thin films. Low accelerating voltages reduce charging effects, beam damage and beam penetration, giving greater sensitivity to surface detail. In the past signal collection was difficult, the electron sources were less bright and stray fields disturbed the image at accelerating voltages less than 5 kV. More recently SEM instruments have been modified to give good SEI performance at low voltages, down to less than 1 kV [26–38]. At low voltage, the resolution of BEI is much the same as that of SEI and the number of secondaries emitted by the specimen increases (Section 6.3.1).

2.3.3 SEM optimization

The major issues to consider in the optimization of SEM operation for stable specimens are noise, depth of field and resolution. Parameters the operator can vary are the beam voltage, beam current, final aperture size and working distance. Standard conditions for maximum resolution are:

(1) high accelerating voltage;
(2) small probe size, obtained with short working distance and small final aperture;
(3) slow scans (long exposure times are required for low noise, as the small probe size means that the beam current is low).

The tungsten filament source has a limit to the beam current that can be obtained in a small spot. Replacing it with a lanthanum hexaboride (LaB_6) gun provides a beam about thirty times brighter. This results in an improved signal to noise ratio, and allows better resolution at high scan speeds. The increased current is also useful for analytical microscopy. A disadvantage is that this gun requires a better vacuum that must be provided by the addition of an ion pump to the gun chamber. The various types of field emission guns require even better vacuum and produce a much smaller probe. The small bright probes give higher resolution, and allow high scan speeds with a good signal to noise ratio. Their disadvantages are cost and a lower maximum total beam current.

The conditions for high resolution imaging with secondary electrons are also conditions that result in maximum beam damage to sensitive specimens. Instrument optimization for these materials involves consideration of the specimen and its interaction with the electron beam (Sections 2.5, 3.4). Reduced damage in the SEM is obtained with low beam currents and low accelerating voltages. Low voltage increases surface detail and reduces bright edge effects as well as decreasing charging and specimen damage. However, this is at the expense of a larger minimum probe diameter. This may lower spatial resolution, depending on the sample. Many SEM studies involve rough surfaces at magnifications below about 10,000, and then depth of field and minimum degradation are more important than resolution. Conditions for maximum depth of field are:

(1) long working distance;
(2) small final aperture size (*c.* 100 μm);
(3) large probe size (low resolution).

Methods for minimizing radiation damage in the SEM are discussed in Section 2.6.1.

2.3.4 Special SEM types

The optimization discussed above considers only the parameters available to the operator of a conventional instrument. Two specialized instruments which extend the range of SEM, and will be discussed fully later (Section 6.2), are the high resolution or field emission SEM (HRSEM or FESEM) and the high pressure SEM (HPSEM). This latter instrument is commonly called the environmental SEM or ESEM, but ESEM is a trademark of a particular commercial product. The size of the electron beam probe can be substantially reduced if a much brighter (but more expensive) field emission source is used instead of the regular tungsten of LaB_6 filament. The probe size is the limiting factor for the highest resolution signal, the direct secondary electron emission. In the HRSEM a short focal length final condenser lens is used to give the highest possible resolution, less than 1 nm, but this requires that the specimen is inside the lens. The specimen size and specimen stages may be restricted in this instrument. A more regular SEM arrangement with the specimen below the lens is more commonly used, and with field emission sources this makes a FESEM. Comparatively high resolution and brightness are maintained at low accelerating voltage in the FESEM, 5 nm and at 3 kV in high contrast samples [38, 39]. The low voltage reduces specimen damage and more important, removes or minimizes the need to coat non-conducting samples. A very thin metal coating works very well. The high quality SEM images can replace the transmission electron microscopy of replicas [40] and compete with atomic force microscopy in some cases [41, 42].

The 'high pressure' in the HPSEM is a specimen chamber pressure of up to 30 torr, high only in comparison to its normal value of a few microtorrs. A good vacuum must be maintained at the electron source, so there is a series of pressure limiting apertures on the optic axis of the instrument to limit gas flow. These apertures form a series of chambers of intermediate pressure, each separately pumped. The specimen must be close to the final aperture (a small *working distance*) and to the detector so that the electrons penetrate the gas. At 10 torr pressure a resolution of 5 nm can be obtained in ideal high contrast samples, and there is no charging even at 'normal' accelerating voltages of 10–30 kV because the positive ions from the gas neutralize surface charges on the specimen. This is particularly important for dynamic microscopy (Section 2.8) where the exposed surface is changing in the microscope.

2.4 TRANSMISSION ELECTRON MICROSCOPY

2.4.1 Imaging in the transmission electron microscope (TEM)

2.4.1.1 *Conventional TEM*

Conventional transmission electron microscopes (CTEM, or TEM) are electron optical instruments analogous to light microscopes, where the specimen is illuminated by an electron beam. This requires operation in a vacuum since air scatters electrons. High resolution is possible because of the short wavelength of the electrons. Typical instruments use accelerating voltages from 40–120 kV and microscopes are becoming more common in the range of 200–400 kV. National facilities have instruments in the 1000 kV range, which have potentially the highest instrumental resolution (see Fig. 6.10). Reimer [43] is the most thorough modern text describing TEM, but it is highly technical. Other texts [44–48] may be more accessible, particularly for practical techniques, and for the TEM of polymers there are several reviews [48–50].

TEM image contrast is due to electron scattering. Electrons scattered to large angles by the sample do not contribute to the image in bright field. In ordered or crystalline material this gives *diffraction contrast*, strongly dependent on crystal orientation. In amorphous materials *mass thickness contrast* results, where the image brightness depends on the local mass thickness (thickness × density). Darker regions in the bright field image are regions of higher scattering. Contrast is greater at low accelerating voltages and at small objective aperture diameters. If the scattered electrons do contribute to the image, they may still produce an effect due to *phase contrast*. This is important in high resolution work, and when the sample is deliberately imaged out of focus to improve the contrast.

Dark field (DF) images normally have much higher contrast than bright field (BF) images, but are much weaker in intensity. DF images from amorphous samples are particularly low in intensity and are rarely used in CTEM. This is because the electrons are scattered in all directions, and the objective aperture can collect only a few of them. Scattered electrons may be collected for the DF image by displacing the objective aperture, but it is much better to tilt the incident beam. The electron diffraction pattern of a crystalline specimen is displayed on the fluorescent screen, and when the required scattered beam is seen, the incident beam is tilted. The tilt is chosen to bring the incident beam to the original position of the scattered beam, and the opposite scattered beam to the center, to pass through the aperture. (So if the (110) diffracted beam is seen when the beam is not tilted, the DF image is formed from the ($\bar{1}10$) beam.) DF of crystalline materials gives information unavailable in BF. When the DF image is formed from one spot in the diffraction pattern, bright regions in the image show only ordered areas with the correct orientation, so crystallite dimensions can be measured and their orientation determined. DF imaging of polymers can be difficult because

the images may be unstable and yet require long exposures to record [48–52].

Polymers have low atomic number and scatter electrons weakly, giving poor contrast in the TEM. They are highly beam sensitive; radiation damage causes destruction of crystalline order, chain scission or cross linking, mass loss and dimensional changes (Sections 2.5, 3.4). Increasing the accelerating voltage and cooling the specimen can help to reduce the damage, and specimen preparation methods to increase the contrast (Chapter 4) make damage less important.

2.4.1.2 *STEM*

In the scanning transmission electron microscope (STEM), as in the SEM, a fine electron beam or probe is formed and scanned across the specimen. But in the STEM the specimen is thin and the intensity of a transmitted signal is detected, amplified and synchronized with a display of the image. The image can be processed to give a wide range of structural and chemical information. Collecting the direct transmitted beam gives a BF image and an annular detector produces a DF image.

The resolution in STEM is limited by the probe size and fine probes are formed using high brightness sources. A 'dedicated STEM' has no imaging lenses, so the instrument is 'dedicated' in that it can only operate as a STEM. It uses a field emission gun, requiring ultra-high vacuum [53, 54]. A TEM with a probe forming condenser lens and a scanning attachment can be used as either TEM or STEM. This instrument may be called the C/STEM or AEM (Analytical Electron Microscope), as the probe can add high resolution x-ray analysis (described in Section 2.7) to TEM [55, 56]. It usually has a standard tungsten or lanthanum hexaboride electron source. A third possible instrumental arrangement for STEM imaging is an SEM with a transmission attachment, or SEM-based STEM, although these are less common.

The capabilities of the three instruments are quite different although the principles involved are similar. The lower beam voltage of the SEM makes very thin specimens necessary. This and the comparatively large probe diameter make the SEM-based STEM impractical, though easy to set up [57]. The AEM is much more common than the dedicated STEM. The latter may have excellent image resolution of 0.2 nm, and analysis capability at that resolution, but they are much more expensive than the AEM, and are difficult to operate at such high resolution. (STEM has been described by an expert as 'Scientist's Time Eating Monster' [58]; he was only half joking.)

2.4.1.3 *Comparison of TEM and STEM*

The fine probe of the STEM can obtain chemical and structural analysis from very small regions, smaller than obtainable in the TEM, and this is the reason for most purchases of the instruments. However, few if any polymers can withstand the radiation environment of a stationary finely focused beam of high energy electrons, which rapidly affect even stable inorganic compounds [54]. Originally the C/STEM had an advantage over TEM for imaging polymers, and this was their video output signal, amplified and processed on-line. The processed signal allows a bright, high contrast image to be formed with a low beam current, which would produce no visible image on the fluorescent screen of the TEM. Now that it is simple to outfit a TEM with an image intensifier and video camera, or other low light level imaging system, the difference between the instruments is less striking.

One remaining advantage of the STEM for radiation sensitive polymers is that only the scanned, imaged, area of the specimen is irradiated. In TEM it is difficult to limit the irradiation of adjacent areas. This is especially important for diffraction experiments. Microdiffraction can be readily conducted in a STEM with a probe that leaves adjacent regions undamaged. In polymers the microdiffraction area must be comparatively large, attainable by both instruments, but the STEM allows better control. Another possible advantage is that

STEM imaging has higher resolution in very thick films of polymers (several micrometers thick). The advantage over the TEM is greatest for disordered, low atomic number materials [59] and least for single crystals of heavier atoms [60]. Advantages of STEM for polymer imaging are:

(1) simultaneous image display is possible (e.g. BF and DF) and one may mix multiple signals;
(2) there is no change in focus with magnification;
(3) irradiation is limited to the area examined;
(4) thicker specimens may be imaged than in the TEM;
(5) higher efficiency in dark field of non-crystalline materials.

Disadvantages of STEM for polymers are:

(1) C/STEM images have much worse resolution than the TEM (unless a field emission gun is used);
(2) dedicated STEMs require very good vacuum in the specimen chamber (this may prevent use of many polymers, and certainly slows specimen exchange);
(3) dedicated STEMs are not suitable for diffraction studies on polymers;
(4) high resolution analysis and microdiffraction are not possible in radiation sensitive materials;
(5) the STEM has higher cost and is more difficult to use than the TEM.

The C/STEM or AEM instrument has been used for several polymer studies [61–63].

2.4.2 Diffraction techniques

Electron diffraction is an important technique for the study of crystalline materials [64, 65]. It is regularly used to identify crystal structures and local orientation. The directions in which electrons are diffracted from a specimen relate to the atomic spacings and orientation of the material (Section 3.2). A crystal has a regular arrangement of atoms and so in the TEM it will produce a diffraction pattern consisting of sharp spots.

Polycrystalline materials have many spots, which together form continuous rings. Small or imperfect crystals give fuzzy spots or rings.

In *selected area electron diffraction* (SAED) in the TEM the intermediate or selected area aperture is used to select a region of the specimen for diffraction. A near parallel beam of electrons illuminates the specimen. Generally the region contributing to the pattern is several micrometers in diameter. This is a large area compared to that in STEM microdiffraction, but very much smaller than that needed for x-ray diffraction. The intermediate aperture is below the specimen, so a large area of the specimen is irradiated during SAED. This is extremely undesirable for polymer specimens that are damaged by the beam and other techniques must be used.

The beam can be focused using the condenser lenses to limit the area irradiated, but the intensity produced by focusing the beam is too high. A strongly excited first condenser lens and a very small second condenser lens aperture reduce the intensity. With these conditions a near focused beam illuminates a small region of the specimen with a near parallel electron beam. Here the diffraction area is 'selected' by the incident beam diameter as the aperture is above the specimen.

Convergent beam microdiffraction uses a convergent rather than a near parallel beam, and this makes it possible to limit the beam to extremely small regions. The diffracting area is limited spatially by the beam diameter, but few polymers can withstand the focused beam. For polymers a larger area must be used to produce sufficient signal at a dose consistent with limiting beam damage.

2.4.3 Phase contrast and lattice imaging

In phase contrast scattered beams are allowed to pass through a large objective aperture and recombine with the unscattered beam to form the image. This would give no contrast if the objective lens was perfect, and perfectly in focus. The lens is not perfect, and often defocused, causing the scattered beams to be phase shifted.

When the beams recombine, the phase shift causes a change of intensity. Phase contrast is always present, but can often be ignored except at high resolution or in cases of deliberate defocus at low resolution [43, 50, 66a] (Section 3.1.4). At low magnifications and good focus, a large objective aperture will give very low image contrast, and a small objective aperture, excluding scattered beams, will produce mass thickness contrast (or diffraction contrast if the specimen is crystalline). In the light microscope, lenses are much more nearly perfect and the analogous method uses a phase plate to produce a fixed phase shift in scattered beams. In the TEM the situation is more complex; the phase shift depends on the scattering angle and the defocus. The nature of the relation between defocus and phase must be well known in order to interpret the images accurately.

If the scattered beam is a sharp spot diffracted from a single crystal, the phase contrast image when it is recombined with the unscattered beam is an image of the crystal lattice planes that produce the scattering by Bragg diffraction. If several beams are recombined, the result is an image of the crystal lattice. This specialized phase contrast technique is applied to the study of atomic scale structure in crystalline specimens of metals and ceramics. It is called high resolution electron microscopy (HREM) and allows the direct imaging of defects and interfaces on the atomic scale [66b–68]. The technique is very difficult to apply to the study of polymeric materials because of their instability in the electron beam. Useful lattice images have been recently obtained from a range of polymers [69, 70] and the topic is discussed further in Section 6.4.

2.5 SCANNING PROBE MICROSCOPY

Scanning probe microscopy (SPM) is the general name for the use of any of a variety of microscopes that have one basic feature in common. The common feature is that the image is produced by scanning a solid probe on or just above the surface of a specimen, and detecting some signal from the interaction of the probe with the surface. As in the SEM, where the probe is a focused beam of electrons, the resolution of the image is controlled by the region of interaction, and therefore the probe is made to have a very fine tip. The interaction region is so small in several forms of the SPM that they have atomic resolution. SPM has grown very quickly since the invention that began this field in 1982 [71]. New types of microscopes and new commercial designs appear frequently and there is a rapidly expanding range of literature. The field is much less mature than the others dealt with in this book, for the SEM is the most recent of the other types of microscopes, and it became commercially available over thirty years ago. Information about scanning probe microscopy given here, in Section 6.3 and in Appendix VIII may become out of date quite quickly. Instrumental design has already progressed from the elaborate original research instruments to tabletop models costing little more than a top-quality optical microscope. Current books on SPM [72–74] are very detailed and more useful for instrument designers than for the user, who may be better served by review and encyclopedia articles [75–77].

The first SPM to be developed was the *scanning tunneling microscope* (STM) [71] where the probe is a conductor set at a small bias voltage difference from a conducting sample, and the signal is a current that passes between them. This quantum mechanical tunneling current has a measurable value when the two conductors are a very small distance apart, typically less than 1 nm. In the STM the region of interaction in the specimen can be limited to a single atom, so the microscope can show the atomic arrangement on surfaces. Changing the bias voltage gives equally detailed information about surface electronic states. The fine scanning and the distance between probe and specimen are controlled by piezoelectric drivers that allow rapid and extremely precise control. The two basic modes of operation are to scan at constant height and observe the variations in current, or to employ a feedback loop to maintain a constant current by changing the height. These two modes are not

specific to the STM, but generally applicable to SPM, if they are described as *constant height* and *constant signal* modes (Fig. 2.7). The constant height mode allows higher scanning speeds, but requires that the surface be very smooth in the region scanned. If a protrusion is greater than the set constant height, the probe tip crashes into the sample.

The next SPM to be developed [78] and currently the most important for the study of polymers is the atomic force microscope (AFM) [76, 77]. Here the probe is mounted on a thin cantilever and deflection of the cantilever is the detected signal. The principle is similar to that of a profilometer, where a stylus is dragged over a surface. The AFM is more sensitive, so the force between probe and sample is much smaller, < 10 nN. This gives a very small contact area and thus high resolution. The high sensitivity also allows the detection of *noncontact* longer range forces, such as van der Waal or electrostatic forces, with the tip 10 nm or more away from the surface. The spatial resolution is then not as good

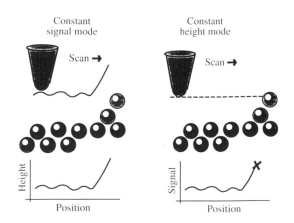

Fig. 2.7 Schematic of the imaging modes in SPM. In constant signal mode, feedback control is used to keep the interaction of probe and sample at constant value. This results in the probe moving at constant distance above the surface, and the output is the controlled vertical position. In constant height mode the probe is scanned without feedback, and the signal is the varying interaction between probe and sample. As shown, if the sample is not sufficiently smooth, this interaction may be destructive.

as in the contact mode, but there is less danger of surface forces damaging the specimen.

Although much is made of the atomic resolution of these microscopes, a very important part of their use is in accurate metrology of surfaces on length scales up to several micrometers. Both can detect roughness in deposited films on the sub-nanometer vertical scale, or measure the exact size of fabricated structures in microelectronics. The limitation of STM is that the sample must be conducting, while the AFM is more complex and intrinsically more susceptible to outside influences. Non-conducting polymers can be metal coated for the STM, but this may affect the fine surface detail on the nanometer scale. High resolution has been obtained in the STM with non-conducting molecules adsorbed onto conducting substrates, individually or as ordered arrays [79, 80], but interpretation of the images is not always simple.

Keeping the AFM probe stationary and measuring the force as a function of height measures the surface forces and the mechanical properties of the sample surface on a very fine scale. Conceptually simple modification of the AFM allows other sample properties to be detected. If the probe tip is ferromagnetic then we have a *magnetic force microscope* (MFM) sensitive to stray magnetic fields and thus showing the magnetic domain structure of the sample [81]. If it is charged, then electrostatic forces due to local static charge on the specimen are detected. If lateral deflection of the cantilever beam can be measured, we have a *lateral* or *frictional force microscope* (LFM or FFM) which is capable of tribological studies at scales down to the atomic. Frictional forces can be displayed as an image and correlated with surface features or surface chemistry [82]. These microscopes may be treated as a group and called *scanning force microscopes* (SFM).

The active part of an SPM, that is the probe, its piezoelectric drivers and mounting are rather small, a few inches across in all. Thus one solution to the difficulty of locating a particular area to be viewed at high resolution is to place the whole microscope in the specimen chamber of an SEM or on top of an inverted optical microscope. Other arrangements allow observation at high vacuum for clean surfaces, or in water for biological samples in their natural state. A vacuum, dry gas or a liquid environment is also used to reduce the effects of adsorbed water at the capillary formed by the specimen and the probe tip. Design for observation of materials at low or high temperature is particularly tricky because any differential thermal expansion would easily ruin the alignment.

There are now many other types of SPM at various stages of development. Few have so far been applied to polymer science. Some are listed here only to give an idea of the capabilities that may become more widely available in the future. A full list of the types of SPM can become an overwhelming set of acronyms. There is the *near field optical microscope* (NFOM or SNOM) [83] where the scanning probe is an optical fiber or aperture that acts as a fine source of photons or a high resolution detector. This allows optical microscopy at a resolution of 5–10 nm, not limited by the wavelength of light because of the narrow gap between probe and specimen. So far the uses of this instrument have been directed at *in vivo* observation of biological systems. If the probe is fabricated to have a thermocouple at its tip, the microscope becomes the *scanning thermal profiler* (STP) and can determine the temperature or thermal conductivity of a sample at better than 100 nm resolution [84]. A probe that is a micropipette can measure the local ionic conductance of a membrane and make a *scanning ion conductance microscope* (SICM) [85].

Some of these SPM systems can be combined with others – a conducting tip on a cantilever can be operated as a combined STM and AFM. Measurement of both vertical and lateral forces gives a FFM and AFM. Tunneling from a thermocouple tip gives STM and STP, and the combined effect measures local chemical potential [86].

Both the STM and the AFM have been used for nano-fabrication, deliberately moving atoms around on the surface of the sample. This shows that large local forces can be generated in the microscopes. Polymers are sensitive to damage by these local forces, particularly in the AFM, because they are mechanically soft and weak. In an extreme case where the local forces are too high, the scanning probe will plough a groove through the specimen, destroying its original surface. Measurements on soft polymeric or biological systems must be set up to keep the local forces small.

2.6 MICROSCOPY OF RADIATION SENSITIVE MATERIALS

This section relates to electron microscopy and particularly to TEM; the critical feature for the TEM of polymers is radiation sensitivity, and beam-sensitive materials require very special treatment [50, 51, 87, 88]. Discussion of the optimal strategy for dealing with radiation sensitive specimens generally deals with the choice of accelerating voltage and other tactics. First, however, stand back and view the overall strategy. The best way to avoid the bad effects of irradiating a radiation sensitive polymer is not to irradiate it. The following three questions should therefore be answered with a firm 'NO' before considering procedures to minimize radiation damage.

(1) Can information be obtained from a sample after heavy radiation doses in the microscope?
(2) Can techniques other than electron microscopy be applied to the problem?
(3) Can the material be made less sensitive, e.g. by staining or etching?

Question (1). This defines the radiation sensitivity of the material in an operational way that depends very strongly on the sample and the information required from it. Thus a fluoropolymer may degrade rapidly and have no known staining procedure. When the microstructure of the homopolymer is studied, this will be a very sensitive material. No information will be available after a large radiation dose. The same polymer could be a minority second component in a blend, with the TEM used to determine the size and arrangement of the component. Radiation damage may now help rather than hinder the investigation by increasing contrast between the two components. This blend sample would not require procedures to minimize radiation damage.

Question (2). It is possible that optical or scanning probe microscopy or one of the many other characterization methods (Chapter 7) could give the required information. The problems of radiation sensitivity are more severe in the TEM than in the SEM, so the high resolution SEM may be an alternative involving less difficulty.

Question (3). A shadowed carbon replica of an etched sample contains little or no polymer. If it contains the information required, it is an excellent way of producing a 'less sensitive' sample. Stains by definition increase the contrast of the specimen, but do not necessarily increase the stability of the image. Some stains, such as iodine, are rapidly driven off by irradiation in vacuum while others are completely stable.

An example where such methods of avoiding radiation damage have been successfully applied is melt crystallized polyethylene. The lamellar microstructure of a normal sample is extremely sensitive to radiation. Low dose techniques in the TEM produced low contrast images difficult to interpret. By staining [89, 90], and by etching [91–93], completely stable convincing high contrast images of the lamellae have been produced. High resolution SEM [39] also shows the lamellar structure and may require much less specimen preparation.

2.6.1 SEM operation

Electrons of 30 keV energy do not penetrate far into the surface of the specimen. Unless the irradiated material evaporates, so that the beam drills a hole in the sample, the unaffected bulk holds the damaged thin surface layer in place and the effect on the image is limited to fine scale details. To reduce these effects, the thickness of the surface layer that is irradiated should be reduced. Either the metal coating can be made thicker or the accelerating voltage reduced. Both reduce the total energy deposited in the polymer but both also reduce the resolution of the image in the SEM. Neither should be considered without evidence of damage (or charging) under normal operating conditions. When they are tried, the best conditions will have to be found by trial and error, since the response depends strongly on the materials in the sample, and on the information being sought. A low voltage of 2–5 kV with a thin (<10 nm) layer of metal is usually satisfactory. If the resolution is sufficient, operation at 1–3 kV may be even better [38, 40, 94]. Conditions that can minimize damage in the SEM are as follows:

(1) low accelerating voltage;
(2) thicker surface coatings;
(3) minimum beam current and exposure time that has low enough noise;
(4) low magnification;
(5) beam blanking between exposures.

Items (1) and (2) on this list were discussed above. Items (3) and (4) are to limit the radiation dose as much as possible, by reducing the number of incident electrons and spreading them out over a large area. The last item is to make the radiation more effective, by stopping it unless data is being collected.

2.6.2 Low dose TEM operation

Assume now that it is really necessary to form an image from some transitory feature that is destroyed by radiation. Noise is the primary problem, so it is most important to use all the available electrons to form an image – 'make every shot count'. The way to do this is to use *low dose methods* [95–98] where focusing and other adjustments are done while the beam passes through an adjacent area of the specimen. The beam is moved between its focusing position and its exposure position with the beam deflection coils. A shutter above the specimen prevents radiation reaching it during film transport. A TV viewing system with image intensifier is necessary, not for recording the image, but to make the operations easier, by making it unnecessary for

the operator to have complete dark adaptation. Some commercial microscopes have 'low dose' accessories to automate the procedures, but manual operation is always possible. The main points to remember are:

(1) keep the beam from falling on the area to be imaged unless the image is being recorded;
(2) match the specimen dose, the magnification and the sensitivity of the recording medium.

When the low dose technique works well, the resolution in the image of a beam sensitive material approaches its theoretical limit (Section 3.4). Obtaining optimum imaging conditions beyond this point means improving this theoretical limit. The limit depends on the following four factors, which will be discussed further:

(1) required signal to noise ratio;
(2) maximum electron dose that the specimen can withstand;
(3) image contrast;
(4) fraction of incident electrons that form the image.

Image processing can reduce the signal to noise ratio required in the original image. It may be used to average over many repeating units of the structure or to remove noise in frequency bands that cannot correspond to information. In either case, processing makes the image detail more visible [99–102], and this improves the resolution in a noisy image.

2.6.2.1 *Maximum radiation dose*

Three methods have been found to increase the number of electrons required to destroy the image feature. The first is the use of *higher accelerating voltages*, as the probability that a high energy electron deposits energy in a sample is proportional to the time it spends there. Higher energy electrons are faster, deposit less energy and do less damage. The improvement is proportional to beam voltage up to about 200 kV, then levels off as the electron velocity approaches that of light. There is a 40% improvement in ideal resolution on going from 100 to

200 kV, but to get another 40% improvement, 1 MV is required [103, 104].

The quoted improvement in resolution cannot be achieved because interactions producing the signal of interest are also reduced at high voltage. For a given sample thickness the contrast in bright field will be reduced. A further problem is that film sensitivity is lower at high voltage. Special thick emulsions or other techniques have to be used to regain sensitivity at 1 MV [105, 106]. High voltage electron microscopes (HVEM) are generally much less convenient to operate than the modern conventional TEM. This is important in practice for beam sensitive materials.

These problems mean that for most polymers it is not worthwhile using beam voltages above 200–300 kV to reduce radiation damage. Some types of specimens might be better at higher voltage [107], but in practice little polymer work has been done above 200 kV. The penetration of thick specimens is the main advantage of the HVEM [108].

Cryomicroscopy is a second method for increasing the dose required to destroy image features. A low irradiation temperature increases the dose required to destroy structure in many organic materials. Polymers that undergo scission are relatively unaffected. Many cross-linking polymers last 2–3 times longer at liquid nitrogen temperatures and 4–5 times longer at liquid helium temperatures [103, 109–111]. Modern side entry liquid nitrogen cooled stages are easy to use and do not contaminate. Cold stage operation should therefore be routine for the many beam sensitive specimens that show a temperature effect. The improvement on using liquid helium as the coolant has not been found enough to outweigh the practical difficulties involved.

Low specimen temperatures reduce the damage rate, and crystalline material undergoes chemical change more slowly than amorphous material. Both effects can be explained by the reduced mobility of the molecules. Therefore it is to be expected that a physical restraint on the escape of material from the specimen should have a beneficial effect. *Specimen coating* is thus the third method that might be used to increase the allowed dose. The coating normally used is a thin film of carbon on both sides of the specimen, which combines good physical properties with a comparatively small loss of contrast in the image.

Reimer [112] found a moderate reduction in mass loss rate for coated PMMA. Salih and Cosslett [113] coated an aromatic organic crystal and found that the dose needed to destroy the diffraction pattern increased by five times. Fryer and Holland [114] found factors of three to ten times improvement for a wide range of organic crystals. No reports have been published for diffraction lifetimes from coated polymer films, but with a very simple technique promising significant gains, it is worth a try.

2.6.2.2 *Increased contrast or signal*

The choice of imaging conditions affects both the contrast and the fraction of incident electrons that form the image. Thus dark field images have higher contrast but fewer electrons in the image, and these factors tend to cancel. When there is a choice of BF or DF, CTEM or STEM there is no clear winner in terms of the resolution of a beam sensitive specimen. A detailed knowledge of the scattering properties of the image feature is needed for exact calculations of the optimum imaging mode [115]. In most polymer structures, this detailed knowledge is not available.

Without exact calculations it is difficult to give any general rule on what imaging mode is best as many factors may come into play. A weakly diffracting object, for example, may give a very low contrast in bright field. Even if it exceeds the signal to noise criterion it may not be visible to the eye, which needs about 5% contrast. If it is amorphous, going to STEM where the contrast can easily be amplified then makes sense, whatever the optimum for resolution. A crystalline material may be better in TEM DF. Commonly the sample and the information required from it require specific imaging conditions. Staining, even with labile and impermanent heavy atoms, can clearly improve the noise limited resolution by increasing the contrast.

2.7 ANALYTICAL MICROSCOPY

Chemical analysis in the optical microscope has normally been performed by measuring the physical properties of the sample and comparing them to tabulated values [7, 10]. Melting point and refractive index are most commonly used. Coupling a laser Raman spectrometer to an optical microscope to make an instrument called the 'Raman microprobe' has made quantitative microanalysis possible for a very wide range of materials that includes polymers [116, 117]. In TEM and STEM, the energy spectrum of electrons passing through the specimen may be used to get information about the elemental composition. The technique is called electron energy loss spectroscopy (EELS), and it is most useful for light elements [43, 56, 118]. EELS has been little used in polymer systems; the electron dose required to obtain a spectrum can be high, and the polymer composition is changed by radiation damage.

Very much more common is elemental analysis using the x-rays emitted from the specimen in the SEM [23–26, 119–123] and the TEM [23, 43, 55, 56]. X-ray analysis is most useful for heavier elements, so in polymers it is often applied to find out the nature of fillers and contaminants. Microanalysis of heavier atoms within the polymer itself (such as chlorine in polyvinyl chloride) is more difficult. It is limited by the sensitivity of polymers to radiation and heat damage from the intense electron beam needed for microanalysis.

2.7.1 X-ray microanalysis

When a high energy electron beam impinges upon a specimen, x-ray photons are produced. They fall into two classes. *Characteristic* x-rays have well defined energies that are characteristic of the atoms in the specimen. These x-rays form sharp peaks in the x-ray energy spectrum and contain analytical information. They are emitted by atoms in the specimen as they return to their ground state, after an inner shell electron has been removed by an interaction with a high energy beam electron. *Continuum* x-rays have a wide range of energies and form the background in the x-ray energy spectrum. They are produced when incident high energy electrons are slowed by scattering near the atomic nucleus and carry no useful information.

The SEM, TEM and STEM can be fitted with x-ray detectors for elemental analysis of the specimen. (The electron probe microanalyzer (EPMA) or x-ray microanalyzer (XRM) is basically an SEM designed for x-ray microanalysis.) The x-ray signals are produced from almost the entire interaction volume in the specimen (Fig. 2.4) when the beam energy is well above the energy of the x-rays. In thick specimens the interaction volume is generally much wider than the beam diameter, due to beam broadening in the specimen. The interaction volume then determines the spatial resolution of the technique, and the volume increases for low atomic number specimens and with the beam voltage. The peak to background ratio and the signal sensitivity increase with the electron beam voltage, so optimum voltage is a compromise between peak to background ratio and spatial resolution. The backscattered electron image is produced from signals coming from almost the same excitation volume as the x-ray signal, Fig. 2.4. This is helpful, as BEI can be used to scan the specimen quickly for atomic number differences to guide the slower x-ray mapping.

Two different types of detectors are used to measure the x-ray intensity as a function of wavelength or energy. In an *energy dispersive x-ray spectrometer* (EDS) x-rays generated from the sample enter a solid state, lithium drifted silicon detector (a reversed bias p–n junction). They create electron hole pairs that cause a pulse of current to flow through the detector circuit. The number of pairs produced by each x-ray photon is proportional to its energy. The pulses produced are amplified, sorted by size with a multichannel analyzer and displayed as an energy spectrum. Typically all elements of atomic number above 11 (sodium) are detected simultaneously. There may be problems of elemental overlap in some materials. Ultrathin window (UTW) and windowless detectors have extended the range down to include carbon, atomic number 6, but contamination and lack of sensitivity are problems. Polymer samples tend to outgas and seriously contaminate the unprotected detectors.

In a *wavelength dispersive x-ray spectrometer* (WDS) the x-rays fall on a bent crystal and are reflected only if they satisfy Bragg's law. The crystal is set to focus x-rays of one specific wavelength onto a detector and rotates to scan the wavelength detected. Only one element can be detected at a time with one crystal. The resulting WDS spectra are quite sharp and elemental overlap is minimal due to the good signal to noise ratio. In WDS, typically in an EPMA, accurate quantitative analysis is possible if the specimen is flat and standards are used for calibration. Computer analysis allows for such complicating features as the x-ray absorption and fluorescence, which depend on the elemental composition that is being determined.

Comparison of the two x-ray techniques is shown in Table 2.2. Microanalysis in the SEM is best conducted by a combination of these two techniques to take advantage of the strengths of both. Several spectrometers can be mounted on the microscope. Computer hardware and software are available which control both types of spectrometer and combine the data on a single system. Microanalysis in the TEM and STEM is conducted by EDS analysis and EELS.

2.7.2 X-ray analysis in the SEM versus AEM

Energy dispersive x-ray spectroscopy can be conducted in the SEM, STEM and in the AEM whereas wavelength dispersive spectroscopy is conducted only in the SEM or the EPMA. For light element analysis, from boron to sodium, the WDS technique is preferred to ultrathin window EDS for polymers, so the AEM should not be used unless its spatial resolution is required. If thin specimens are used in the AEM, high magnification images and diffraction information can be accompanied by EDS with spatial resolution about 10–100 nm. EDS of solid specimens in the SEM has resolution of a few μm. Just as for STEM imaging, this difference is due to the small interaction volume in thin films, where the beam does not spread out. Thin specimens also limit the need for absorption or fluorescence corrections, permitting the application of quantitative analysis techniques.

There are three major problems with microanalysis of thin films in the AEM:

(1) spurious x-rays are produced which can be detected and confused with x-rays from the specimen;
(2) beam currents in the AEM are small (compared to the EPMA). This and the small excitation volume give very low x-ray count

Table 2.2 Comparison of EDS and WDS microanalysis

Energy dispersive	*Wavelength dispersive*
Interfaced with SEM, TEM, STEM	Interfaced with SEM, EPMA
Simultaneous detection of elements	Quantitative detection of one element at a time
Rapid analysis – about 100 s	Slow analysis – from 5 mins to hours
Spatial resolution good in TEM, AEM, STEM, poor in SEM	Spatial resolution poor (1–5 μm)
Background counts from backscattered electrons reduce sensitivity	Peak/background ratio 10 to 50 times better than EDS, good sensitivity
Serious peak overlap problems – results may be ambiguous	Good energy resolution, little peak overlap
Single detector	Need several crystals to cover range of elements
Detection limit $Z > 11$ (regular window)	Detection limit $Z > 3$
$\qquad Z > 5$ (ultra-thin window)	

rates. Low levels of an element present in the specimen are difficult to detect;

(3) preparation of uniform thin specimens for microanalysis is important but is difficult to accomplish.

Polymer specimens are particularly difficult to analyze in the AEM. Generally, there are small amounts of heavy elements in a polymer. These low levels are difficult to detect in a material that changes readily in the electron beam. These difficulties preclude routine quantitative analysis of polymers in either the SEM or AEM although microanalysis techniques can be applied. The major consideration for the polymer microscopist is that changes occur in the polymer during study.

2.7.3 Elemental mapping

So far it has been assumed that the result of microanalysis is the elemental composition of a small region of the specimen. This is obtained from the x-ray spectrum produced when the electron beam is stationary. Often it is more useful to show the concentration of a specific element as a function of position on the specimen. This is *elemental mapping*. The map is formed by using the intensity of x-ray emission in a specific energy range to modulate the intensity on a display as the beam scans the specimen. The energy region, or window, is set to include the characteristic x-ray energy of the element of interest.

Briefly, the major issue in elemental mapping is to have a high enough count rate for good counting statistics. The higher the concentration of the element in the region scanned the less time is required for good counting statistics. Elements present at low concentrations require long counting times during which sensitive specimens can be damaged. In principle the spatial resolution in elemental mapping is much better in the AEM than in the SEM because of the small interaction volume. In practice, the signal is small in the AEM and only larger regions that have better statistics may be visible. Elemental mapping using EDS in the SEM is easy to accomplish.

Mapping of individual particles that are not in a matrix is straightforward, and particles below 1 μm in size can often be identified by EDS or WDS analysis. However, thick specimens with particles in a matrix, or regions of differing atomic number, are difficult to analyze. The two dimensional elemental maps are produced from x-ray signals from a large interaction volume that extends well below the specimen surface. The deep features contributing to the elemental map will not be visible in the image.

Systems now available allow simultaneous acquisition of maps for several different elements. Digital maps with colors assigned for each element permit a more rapid and detailed analysis. Superposition of the color maps is useful in determining associations between elements. This technique is more than simple elemental mapping and approaches more definitive compositional studies. It is important for beam sensitive specimens because of its speed. Multiple maps may be obtained in the time previously used for mapping a single element.

2.8 QUANTITATIVE MICROSCOPY

2.8.1 Stereology and image analysis

A common problem in microscopy is that three-dimensional information on microstructure is required, but the objects (sections or surfaces) and their images are two-dimensional. *Stereology* is the field that provides the mathematical methods that allow one to go from two to three dimensions [124, 125]. The mathematical methods are quite complex. A simple example of the problem is a material containing spheres dispersed in a matrix. If the spheres are of uniform diameter, an image of a thin section will contain circular structures of varying diameter. If the spheres are of a range of sizes, a thin section will appear much the same. Analysis of the size distribution of the circles is needed to distinguish the two cases.

Image analysis is the discipline that involves making the kinds of measurements used in stereology [126–128]. In a simplistic sense, image analysis is the measurement of geometric features in images, and it can be done with a ruler, time and patience. The state of the art today has automated, quantitative instruments available to conduct such analyses. Advances closely parallel those in small computer systems, and so have been extremely rapid. Literature describes systems that have been used [129, 130] but power, speed and ease of use continues to improve rapidly in commercial systems. Analysis capability may be built into the computer that controls the microscope, but it is also convenient to have a 'stand alone' system that analyzes digital micrographs from many sources.

The parameters that are commonly measured include particle numbers, diameters, areas, perimeters and ferets. A flexible system can be programmed to measure anything, such as arc lengths in diffraction patterns. In the example described above, an image of circular structures of varying diameters, image analysis methods are used to measure a statistical number of these structures. Stereological formulae provide the size distribution of the spheres.

It is important that image analysis is not misapplied. First, any problems with specimen preparation and microscopy must be solved before image analysis, as the analysis is no better than the image. Second, the image must be carefully calibrated if it is to provide accurate data. Finally, the images analyzed must be representative. As in any statistical technique, a sufficiently large and representative population of features must be measured, so it is important to consider the number of times samples should be collected for the microscope, and the number of different images taken from each to obtain a random sample. For example, in a typical particle size analysis, the number of particles measured must be larger for broader size distributions.

2.8.2 Calibration

It should be obvious that the size of object features derived from image analysis and stereology depends on knowing the actual magnification. Image analysis, whether done by ruler or by computer, depends totally on knowing the relation between linear measurements on the micrograph and in the object.

In optical microscopy any change in the objective lens, tube ocular, or distance from the object to the image plane changes the magnification. Stage micrometers with accurate markings are required to provide calibration. It is good practice to include an image of the micrometer in every set of micrographs. In the SEM or TEM it is easy to read the magnification indicated on the instrument, and the same number may be automatically printed on the micrographs. The problem is that these numbers may not be accurate. Magnifications should be checked following every routine maintenance of the microscope using standard specimens. Critical studies involving quantitative analysis should have calibration standards run during the study.

One example of the need for calibration will be mentioned briefly; it arises when latex particle diameters are measured from TEM micrographs, a common procedure. Standard or control particles are added to the sample suspension before preparation for microscopy. They are used as internal standards to correct for changes during specimen preparation, during observation in the microscope, and during reproduction of the images. In addition, a grating replica is used to calibrate the TEM and further standards are used to calibrate the image analyzer.

2.8.3 Image processing

Image processing and image analysis are similar fields that are often confused. In image analysis features in an image are measured, whereas in image processing the image itself is modified in some way [98, 99]. Confusion arises because it is common to modify an image by computer processing, enhancing contrast and/or suppressing unwanted detail, before analyzing it. The same computer system is generally used for both tasks. Important processing functions such as erosion and dilation have been set up to modify particle images so the particles are not touching and can be counted in image analysis.

Image processing is now common even when the image is not to be analyzed. Simple image processing has always been built into SEMs. More complex processing has become more common in all microscopes as the convenience of computer systems has increased and the cost of digital storage for digitized images has fallen dramatically. As these changes continue, image processing is becoming a more standard procedure. For example any image stored and displayed digitally can be displayed in 'false color' where a palette of colors replaces monochrome intensity as an indicator signal level. This example is often frivolously used to make a pretty colored picture, but it has a serious use. The human eye is more sensitive to changes of tint than to changes of intensity, so a false color image can convey more information.

2.9 DYNAMIC MICROSCOPY

Most imaging of materials with microscopes is static, that is, the specimen is not intended to change during observation. Dynamic experiments, where the specimen microstructure can be monitored as it changes, can also be conducted with optical or electron microscopes. The specimen is observed during deformation or environmental change [131–133]. Photographic recording is not suitable for dynamic experiments unless the changes are slow, so video cameras and recorders are used that permit acquisition of dynamic images in real time. A sequence of images can also be digitally stored. The quality of most analog video recordings is such that digital storage is best if there is to be any later quantitative analysis. Until digital storage capacity and transfer rates improve by a few more orders of magnitude, analog tape may be required, particularly for those cases when an event occurs rapidly at some arbitrary time into the experiment. This requires storing images for a long period, then selecting the few frames where something is happening. A good quality 3/4" videotape recorder should then be used.

When an analog tape is the eventual recording medium, a digital image storage buffer may be used to aid the transfer from the scanning rate of the SEM or STEM to standard video rates. Normally the video camera used in the TEM or OM produces standard TV signals, but again digital processing can be used, in this case to integrate over a longer period and reduce noise while collecting data and before recording the images. As video standards vary across the world, an analog tape is unlikely to play in another country.

2.9.1 Stages for dynamic microscopy

2.9.1.1 *Tensile stages*

Tensile stages are used to observe the small scale deformation of materials in tension. Most such experiments are conducted in the SEM or OM [134, 135], though tensile stages exist for the TEM. More detailed changes in crystallography and composition can be monitored in the TEM, but the small specimen thickness can affect the processes of deformation. Other modes of deformation such as bending and shearing can be studied with suitably modified stages in the OM or SEM. Problems of adhesion and interfacial strength can also be studied in this way. Most of the SEM manufacturers and electron microscope suppliers (see Appendices VI–VII) provide tensile stages for their microscopes, thus the specific design of such accessories will not be considered here.

The major disadvantages of conducting dynamic tensile experiments in the microscope are that the experiments are time-consuming. It is not usually possible to observe a statistically significant number of specimens, and neither stress–strain measurements nor microscopy is optimal. A common use of tensile stages in the study of polymers is to study the tensile failure of fibers and yarns (Section 5.1) [134]. Charging of polymer specimens has been a major problem in the SEM; metal coating will not work if the sample is deformed significantly or broken to expose new surfaces. Anti-static coating has been sprayed onto the yarn, but it hides the specimen surface details. Here low voltage operation of the SEM is really advantageous.

2.9.1.2 *Hot and cold stages*

Hot and cold stages can be attached to optical and electron microscopes [136, 137]. Cold stages in TEM reduce specimen contamination and often increase the lifetime of beam sensitive specimens. In biology they keep hydrated samples frozen and stable in the vacuum of the TEM and SEM (for example [138]), and this has some applications in polymers [139]. Hot stages are more commonly used for the dynamic microscopy of polymers, where changes in microstructure and composition are observed as a function of temperature and thermal history. Heating a polymer sample in an electron microscope will increase the rate of radiation damage. At high temperatures the polymers may out-gas or degrade and evaporate, and this may contaminate and damage the microscope. The high pressure SEM is the natural instrument to use when this may be a problem. Since the applied temperature changes are usually not very great, heating by the electron beam in the TEM, although it may only be a few degrees, makes the specimen temperature uncertain.

A major application of dynamic microscopy is the study of polymer structure, and its development as a function of temperature, in a hot stage in the optical microscope [10, 132, 140]. One possible result from such a thermo-optical experiment is the microstructure of the specimen as a function of temperature and time. Direct observation can determine the nature of phase changes and at the same time measure the transformation temperatures. The kinetics of crystal growth can also be directly observed and numerically analyzed with the aid of image processing [141]. A liquid crystalline polymer may have several phase changes as it is heated, and changes in appearance or birefringence can be correlated with calorimetric data [142]. The calorimetric data is acquired by differential scanning calorimetry (DSC) or differential thermal analysis (Section 7.4). With a combined hot-

stage DSC it is possible to observe the sample in the OM and simultaneously obtain the DSC trace.

REFERENCES

1. M. Spencer, *Fundamentals in Light Microscopy* (Cambridge University Press, Cambridge, 1982).
2. W. G. Hartley, *Hartley's Microscopy* (Senecio Publishing, Charlbury, 1981).
3. S. Bradbury, *An Introduction to the Optical Microscope* (Oxford University Press, Oxford, 1984).
4. T. G. Rochow and E. G. Rochow, *An Introduction to Microscopy by Means of Light, Electrons, X-rays, or Ultrasound* (Plenum, New York, 1978).
5. R. Telle and G. Petzow, in *Materials Science and Technology*, Vol. 2A, edited by E. Lifshin (VCH Publishers, Weinheim, 1993).
6. J. H. Richardson, *Optical Microscopy for Materials Sciences* (Marcel Dekker, New York, 1971).
7. C. W. Mason, *Handbook of Chemical Microscopy*, 4th Edn (Wiley, New York, 1983).
8. G. L. Clark, Ed., *The Encyclopedia of Microscopy* (Reinhold, New York, 1961).
9. N. H. Hartshorne and A. Stuart, *Crystals and the Polarizing Microscope* (Arnold, London, 1970).
10. W. C. McCrone and J. G. Delly, *Particle Atlas*, Vol. 1 (Ann Arbor Science, Ann Arbor, 1973).
11. D. A. Hemsley, Ed., *Applied Polymer Light Microscopy* (Elsevier Applied Science, London, New York, 1989).
12. G. E. Schlueter and W. E. Gumpertz, *Am. Lab.* (April 1976) 61.
13. R. Hoffman, *J. Microsc.* **110** (1977) 209.
14. A. H. Bennett, H. Jupnik, H. Osterberg and O. W. Richards, *Phase Microscopy* (Wiley, New York, 1951).
15. F. A. Jenkins and H. E. White, *Fundamentals of Optics*, 4th Edn (McGraw Hill, New York, 1976).
16. E. A. Wood, *Crystals and Light*, 2nd Edn (Dover, New York, 1977).
17. P. F. Kerr, *Optical Mineralogy* (McGraw Hill, New York, 1959).
18. P. Gay, *An Introduction to Crystal Optics* (Longmans, London, 1967).
20. G. Friedel, *Ann. Phys.* **18** (1922) 273.
21. N. H. Hartshorne, *The Microscopy of Liquid Crystals* (Microscope Publications, Chicago, 1974).

22. O. C. Wells, *Scanning Electron Microscopy* (McGraw Hill, New York, 1974).

23. J. I. Goldstein, D. E. Newbury, P. Echlin, D. C. Joy, A. D. Romig Jr., C. E. Lyman, C. Fiori and E. Lifshin, *Scanning Electron Microscopy and X-ray Microanalysis*, 2nd Edn (Plenum, New York, 1992).

24. L. Reimer, *Scanning Electron Microscopy, Physics of Image Formation and Microanalysis* (Springer, Berlin, 1985).

25. D. E. Newbury, D. C. Joy, P. Echlin, C. Fiori and J. I. Goldstein, *Advanced Scanning Electron Microscopy and X-ray Microanalysis* (Plenum, New York, 1986).

26. C. E. Lyman, *Scanning Electron Microscopy, X-ray Microanalysis and Analytical Microscopy* (Plenum, New York, 1990).

27. J. R. White and E. L. Thomas, *Rubber Chem. Technol.* **57** (1984) 457.

28. D. C. Joy, *J. Microsc.* **136** (1984) 241.

29. C. J. Catto and K. C. Smith, *J. Microsc.* **98** (1973) 417.

30. A. N. Broers, *Ultramicroscopy* **8** (1982) 137.

31. T. E. Everhart, R. F. Herzog, M. S. Chang and W. J. DeVore, in *Proc. 6th Int. Conf. on X-ray Optics and Microanalysis*, edited by G. Shinoda, K. Kohra and T. Ichinokawa (University of Tokyo, Tokyo, 1972).

32. V. N. E. Robinson, *Scanning* **3** (1980) 15.

33. R. Autrata, *Scann. Microsc.* **3** (1989) 739.

34. J. Lebiedzik, *Scanning* **2** (1979) 230.

35. K. R. Peters, in *Electron Beam Interactions with Solids* (SEM Inc. AMF, O'Hare, 1984) p. 363.

36. J. Pawley, *J. Microsc.* **136** (1984) 45.

37. T. Nagatani, S. Saito and M. Yamada, *Scann. Microsc.* **1** (1987) 901.

38. L. Reimer, *Image Formation in Low-Voltage Scanning Electron Microscopy* (SPIE, Bellingham, 1993).

39. D. L. Vezie, W. W. Adams and E. L. Thomas, *Polymer* (1995). In press.

40. T. Tagawa, J. Mori, S. Aita and K. Ogura, *Micron* **9** (1978) 215.

41. D. W. Schwark, D. L. Vezie, J. R. Reffner, E. L. Thomas and B. K. Annis, *J. Mater. Sci. Lett.* **11** (1992) 352.

42. L. C. Sawyer, R. T. Chen, M. G. Jamieson, I. H. Musselman and P. E. Russell, *J. Mater. Sci.* **28** (1993) 225.

43. L. Reimer, *Transmission Electron Microscopy, Physics of Image Formation and Microanalysis*, 2nd Edn (Springer, Berlin, 1989).

44. A. M. Glauert, Ed., *Practical Methods in Electron Microscopy* (North-Holland, Amsterdam, 1972).

45. J. W. Edington, *Practical Electron Microscopy in Materials Science* (Van Nostrand Reinhold, New York, 1976) reprinted in 1991 by TechBooks.

46. J. R. Fryer, *The Chemical Applications of Transmission Electron Microscopy* (Academic Press, London, 1979).

47. S. Amelinckx, in *Materials Science and Technology*, Vol. 2A, edited by E. Lifshin (VCH Publishers, Weinheim, 1993).

48. R. G. Vadimsky, in *Methods of Experimental Physics*, Vol. 16B, edited by R. A. Fava (Academic Press, New York, 1980).

49. D. T. Grubb, in *Developments in Crystalline Polymers – 1*, edited by D. C. Bassett (Applied Science Publishers, London, 1982).

50. E. L. Thomas, in *The Structure of Crystalline Polymers*, edited by I. H. Hall (Elsevier Applied Sci., London, 1984) p. 79.

51. J. R. White, *J. Mater. Sci.* **9** (1974) 1860.

52. J. R. White, *Polymer* **16** (1975) 157.

53. A. V. Crewe, *Q. Rev. Biophys.* **3** (1970) 137.

54. L. W. Hobbs, *Ultramicroscopy* **23** (1987) 339.

55. D. C. Joy, A. D. Romig and J. I. Goldstein, Eds., *Principles of Analytical Electron Microscopy* (Plenum, New York, 1986).

56. E. L. Hall, in *Materials Science and Technology*, Vol. 2A, edited by E. Lifshin (VCH Publishers, Weinheim, 1993).

57. E. Oho, T. Sasaki, K. Adachi, Y. Muranaka and K. Kanaya, *Proc. XIth Congr. Electr. Microsc.* Vol. 1 (Kyoto 1986) p. 421.

58. J. Silcox, in a lecture on STEM, Cornell University, Feb. 1994.

59. V. E. Cosslett, *Phys. Stat. Sol.* **55** (1979) 545.

60. H. L. Fraser, I. P. Jones and M. H. Loretto, *Phil. Mag.* **35** (1977) 159.

61. A. Low, D. Vesely, P. Allan and M. Bevis, *J. Mater. Sci.* **13** (1978) 711.

62. E. J. Roche, R. S. Stein and E. L. Thomas, *J. Polym. Sci., Polym. Phys. Ed.* **18** (1980) 1145.

63. E. S. Sherman, W. W. Adams and E. L. Thomas, *J. Mater. Sci.* **16** (1981) 1.

64. P. B. Hirsch, A. Howie, R. B. Nicholson, D. W. Pashley and M. J. Whelan, *Electron Microscopy of Thin Crystals*, 2nd Edn (Butterworths, London, 1965).

65. P. Goodman, Ed., *Fifty Years of Electron Diffraction* (Riedel, Dordrecht, 1981).

66a. M. J. Miles and J. Petermann, *J. Macromol. Sci. Phys.* **B16**(2) (1979) 243.

66b. P. Buseck, J. Cowley, L. Eyring, Eds., *High-Resolution Transmission Electron Microscopy and Associated Techniques* (Oxford University Press, Oxford, 1988).

67. J. C. H. Spence, *Experimental High Resolution Transmission Electron Microscopy*, 2nd Edn (Oxford University Press, Oxford, 1988).

68. J. M. Gibson, *MRS Bulletin* **16** (3) (1991) 27.

69. J.-F. Revol, H. D. Chanzy, Y. Deslandes and R. H. Marchessault, *Polymer* **30** (1989) 1973.

70. J. R. Ojeda and D. C. Martin, *Macromolecules* **26** (1993) 6557.

71. G. Binnig, H. Rohrer, C. Gerber and H. Weibel, *Phys. Rev. Lett.* **49** (1982) 57.

72. D. Sarid, *Scanning Force Microscopy* (Oxford University Press, Oxford, 1991).

73. M. S. Guntherodt and R. Wiesendanger, *Scanning Tunneling Microscopy I, II and III*, Springer Series in Surface Science, Vols. 20, 28 and 29 (Springer, Berlin, 1992).

74. C. J. Chen, *Introduction to Scanning Tunneling Microscopy* (Oxford University Press, Oxford, 1993).

75. J. Jahanmir, B. G. Haggar and J. B. Hayes, *Scanning Microscopy* **6** (1992) 625.

76. E. Meyer, *Progr. Surf. Sci.* **41** (1992) 3.

77. N. J. DiNardo, in *Materials Science and Technology*, Vol. 2B, edited by E. Lifshin (VCH Publishers, Weinheim, 1994).

78. G. Binnig, C. F. Quate and C. Gerber, *Phys. Rev. Lett.* **56** (1986) 930.

79. T. P. Beebe Jr., T. E. Wilson, D. F. Ogletree, J. E. Katz, R. Balhorn, M. B. Salmeron and W. J. Siekhaus, *Science* **243** (1989) 370.

80. I. Fujiwara, C. Ishimoto and J. Seto, *J. Vac. Sci. Technol.* **B9** (1991) 1148.

81. P. Grutter, D. Rugar and H. J. Mamin, *Ultramicroscopy* **47** (1992) 393.

82. C. M. Mate, G. M. McClelland, R. Erlandsson and S. Chiang, *Phys. Rev. Lett.* **59** (1987) 1942.

83. E. Betzig and J. K. Trautman, *Science* **257** (1992) 189.

84. C. C. Williams and H. K. Wickramasinghe, in *IEEE 1986 Ultrasonics Symp.* (IEEE Piscataway NJ 1986) p. 393.

85. P. K. Hansma, B. Drake, O. Marti, S. A. C. Goud and C. B. Prater, *Science* **243** (1989) 641.

86. C. C. Williams and H. K. Wickramasinghe, *J. Vac Sci. Technol.* **B9** (1991) 537.

87. D. T. Grubb, *J. Mater. Sci.* **9** (1974) 1715.

88. M. S. Isaacson, in *Principles and Techniques of Electron Microscopy*, Vol. 7, edited by M. Hayat (Van Nostrand, New York, 1977).

89. G. Kanig, *Kolloid Z. u Z. Polymere* **251** (1973) 176.

90. G. Kanig, *J. Crystal Growth* **48** (1980) 303.

91. D. C. Bassett and A. M. Hodge, *Polymer* **19** (1978) 469.

92. D. C. Bassett and A. M. Hodge, *Proc. Roy. Soc. A*, **377** (1981) 23, 39, 61.

93. D. C. Bassett, in *CRC Critical Reviews in Solid State and Mat. Sci.* Vol. 12, Issue 2 (CRC Press, New York, 1984).

94. L. C. Sawyer and M. Jaffe, *J. Mater. Sci.* **21** (1986) 1897.

95. R. C. Williams and H. W. Fischer, *J. Mol. Biol.* **52** (1970) 121.

96. K. H. Herrmann, J. Menadue and H. T. Pearce-Percy, in *Electron Microscopy 1976*, edited by D. G. Brandon (Tal International, Jerusalem, 1976).

97. Y. Fujiyoshi, T. Kobayashi, K. Ishizuka, N. Uyeda, Y. Ishida and Y. Harada, *Ultramicroscopy* **5** (1980) 459.

98. M. Ohtsuki and E. Zeitler, *Ultramicroscopy* **1** (1975) 163.

99. D. L. Misell, *Image Analysis, Enhancement and Interpretation* in *Practical Methods* in *Electron Microscope Images* (Springer-Verlag, Berlin, 1980).

100. P. W. Hawkes, Ed., *Computer Processing of Electron Microscope Images* (Springer-Verlag, Berlin, 1980).

101. M. van Heel, *Ultramicroscopy* **7** (1982) 331.

102. D. Van Dyck and W. Coene, *J. Microsc. Spectrosc. Electron.* **13** (1988) 463.

103. K. Kobayashi and K. Sakaoku, *Lab. Invest.* **14** (1965) 1097.

104. L. E. Thomas, C. J. Humphreys, W. R. Duff and D. T. Grubb, *Radiat. Effects* **3** (1970) 89.

105. M. V. King and D. F. Parsons, *Ultramicroscopy* **2** (1977) 371.

106. V. E. Cosslett, Image Recording and Image Processing in High Voltage Electron Microscopy, in *Proc. 5th Int. Conf. High Voltage Electron Microscopy* (Kyoto) p. 27.

107. A. M. Glauert, *J. Cell Biol.* **63** (1974) 717.

108. M. V. King, D. F. Parsons, J. N. Turner, B. B. Chang and A. J. Ratkowski, *Cell. Biophys.* **2** (1980) 1.

109. D. T. Grubb and G. W. Groves, *Phil. Mag.* **24** (1971) 815.

110. R. M. Glaeser and K. A. Taylor, *J. Microsc.* **112** (1978) 127.

111. D. L. Dorset and F. Zemlin, *Ultramicroscopy* **17** (1985) 229.
112. L. Reimer, *Z. Naturforsch.* **B14** (1959) 566.
113. S. M. Salih and V. E. Cosslett, *Phil. Mag.* **30** (1974) 255.
114. J. R. Fryer and F. Holland, *Ultramicroscopy* **11** (1983) 67.
115. D. L. Misell, *J. Phys. D; Appl. Phys.* **10** (1977) 1085.
116. D. J. Gardiner and P. R. Graves, Eds., *Practical Raman Spectroscopy* (Springer-Verlag, Berlin, 1989).
117. J. G. Grasselli and B. J. Bulkin, Eds., *Analytical Raman Spectroscopy* (Wiley, New York, 1991).
118. R. F. Egerton, *Electron Energy Loss Spectroscopy in the Electron Microscope* (Plenum Press, New York, 1986).
119. S. J. B. Reed, *Electron Microprobe Analysis* (University Press, Cambridge, 1975).
120. J. A. Chandler, *X-ray Microanalysis in the Electron Microscope* (North-Holland, Amsterdam, 1977).
121. K. F. J. Heinrich, *Electron Probe Microanalysis* (Van Nostrand, New York, 1981).
122. K. F. J. Heinrich and D. E. Newbury, *Electron Probe Quantitation* (Plenum Press, New York, 1991).
123. E. Lifshin, in *Materials Science and Technology*, Vol. 2B, edited by E. Lifshin (VCH Publishers, Weinheim, 1994).
124. E. E. Underwood, *Quantitative Stereology* (Addison-Wesley, Reading, MA, 1970).
125. J. C. Russ, *Practical Stereology* (Plenum Press, New York, 1986).
126. R. T. DeHoff and F. N. Rhines, *Quantitative Microscopy* (McGraw Hill, New York, 1968).
127. J. C. Russ, *Computer-Assisted Microscopy – The Measurement and Analysis of Images* (Plenum Press, New York, 1990).
128. H. E. Exner, in *Materials Science and Technology*, Vol. 2B, edited by E. Lifshin (VCH Publishers, Weinheim, 1994).
129. L. R. Jarvis, *J. Microsc.* **150** (1988) 83.
130. F. J. Guild and J. Summerscales, *Composites* **24** (1993) 383.
131. A. W. Agar, R. H. Alderson and D. Chescoe, in *Practical Methods in Electron Microscopy*, Vol. 2, edited by A. M. Glauert (North-Holland, Amsterdam, 1974).
132. E. P. Butler and K. F. Hale, in *Practical Methods in Electron Microscopy*, Vol. 9, edited by A. M. Glauert (North-Holland, Amsterdam, 1981).
133. J. A. Reffner, *Amer. Lab.* (April 1984) 29.
134. J. W. S. Hearle, J. T. Sparrow and P. M. Cross, *The Use of the Scanning Electron Microscope* (Pergamon, Oxford, 1972).
135. E. J. Noonan, A. van Riessen, F. C. O'Hara and M. Verrall, *J. Computer-Assisted Microsc.* **4** (1992) 179.
136. G. R. Loppnow and R. A. Mathis, *Rev. Sci. Instrum.* **66** (1989) 2628.
137. K. A. Taylor, R. A. Milligan, C. Raeburn and P. N. T. Unwin, *Ultramicroscopy* **13** (1984) 185.
138. P. K. Vinson, J. R. Bellare, H. T. Davis, W. G. Miller, L. E. Scriven, *J. Colloid Interf. Sci.* **142** (1992) 74.
139. M. S. Silverstein, Y. Talmon and M. Narkis, *Polymer* **30** (1989) 416.
140. N. H. Hartshorne, *The Microscope* **23** (1975) 177.
141. Yu Long, Z. H. Stachurski and R. A. Shanks, *Metals Forum* **16** (1992) 259.
142. J. A. Moore and Ji-Heung Kim, *Macromolecules* **25** (1992) 1427.

CHAPTER THREE

Imaging theory

3.1 IMAGING WITH LENSES

3.1.1 Basic optics

The basic optics of the optical microscope and the conventional TEM are similar. Condenser lenses illuminate the object to be imaged with a flood of radiation, and imaging lenses form the radiation leaving the object into a magnified image. Both electrons and light may be considered as particles or as propagating waves in space. The wave has an amplitude and a phase, though only the intensity which equals (amplitude)^2 can be directly observed. Although wave optics gives the most rigorous derivation, it is simpler to consider both geometrical and wave optics to understand the formation, contrast and resolution of microscope images.

A reader unfamiliar with any optics should consult a textbook for more information. Welford [1] is at first year college level, and Jenkins and White [2] is one of several standard advanced college texts. Martin [3] concentrates on microscopy, and there many are more detailed and technical [4, 5]. Spencer [6] provides an elementary introduction to microscope optics, and optical microscopy texts already cited in Chapter 2 (refs 3–9 there) are also useful sources. All of these describe light optics and light microscopy, as TEM texts largely assume this knowledge and are more advanced.

Matter slows both light and electrons, decreasing their wavelength λ; the *refractive index*, n is defined as:

$$n = \frac{(\text{wavelength in vacuum})}{(\text{wavelength in material})}.$$

The *optical path length* of a wave in a material of thickness t is nt, as the material contains nt/λ wavelengths as would a path of length nt in vacuum. The *optical path difference* Δ due to the presence of this material is then $(n-1)t$, and the phase difference produced is $(2\pi/\lambda)\,\Delta$. Table 3.1 shows how the optical properties of polystyrene depend on the incident radiation. The refractive index of polystyrene for electrons is calculated from values for carbon, corrected for the lower density of polystyrene (see Section 3.1 of ref. 5).

If the phase of the wave can be calculated from its phase at nearby points and times, the radiation is *coherent*; monochromatic and parallel light, for example, is coherent. The phase and

Table 3.1 Optical properties of polystyrene

	Light Hg green	Electrons 100 keV	Electrons 1 MeV
Wavelength, λ (nm)	547	0.0037	0.0009
$n-1$	0.4	2.5×10^{-5}	3.8×10^{-6}
Film thickness (nm) for $\Delta = \lambda/4$	340	37	60

amplitude of completely *incoherent* light vary randomly in space and time. When two coherent waves of amplitudes a and b come together they *interfere* with each other. The result is a wave intensity $(a + b)^2$ if they are exactly in phase (constructive interference), and intensity $(a - b)^2$ if they are completely out of phase (destructive interference). In general the waves must be added by vector sum rules. Incoherent waves interfere momentarily, but over any period of observation the phase effects average out. The resulting average intensity is the sum of the intensities of the two waves, $(a^2 + b^2)$.

3.1.1.1 Diffraction

When an object scatters coherent waves, interference produces a variation in intensity as a function of their direction, the *diffraction pattern* of the object. An object with regular periodicity d in one dimension has a pattern with maximum intensity when the angle between incident and scattered radiation, ϕ takes the values given by:

$$d \sin \phi = m\lambda \quad (m = \pm 1, 2, 3, \ldots).$$

A three-dimensional periodicity, such as atoms in a crystal, gives maxima at different angles given by:

$$2d \sin(\phi/2) = m\lambda \quad (m = \pm 1, 2, 3, \ldots).$$

This is Bragg's law, usually written with the angle between incident and scattered radiation as 2θ instead of ϕ.

If the object has an exactly sinusoidal variation of absorption, thickness or refractive index in one dimension, diffracted beams appear only when $d \sin \phi = \pm\lambda$ (i.e. $m = \pm 1$). This is important because of Fourier's Theorem, which states that any (single valued) function of a variable x can be expanded as a sum of sines and cosines of multiples of x. Thus any phase or intensity variation in the sample can be considered as a sum of sinusoidal variations of different wavelength, each giving a certain intensity at a single characteristic angle ϕ. The intensity at a point in the diffraction pattern corresponds to the strength of a variation of some sample property

with a particular direction and spatial frequency. Closely spaced structures have high spatial frequencies and produce intensity at high angles in the diffraction pattern. The image is a map of the object as a function of position and the diffraction pattern is a map of the variations in the object as a function of spatial frequency.

This concept of the diffraction pattern as a map in spatial frequency space (reciprocal lattice space and Fourier space are other names), is somewhat abstract and mathematically complex [3–7]. It is nevertheless extremely useful, as it gives a physical insight into many facets of microscope optics.

3.1.1.2 Image formation

Figure 3.1 shows the geometry of image formation by the objective lens. It is schematically shown as a thin lens, though in both optical and TEM instruments the lens is actually longer than its focal length. The TEM objective has a focal length of about 2 mm and it magnifies the object, forming a real image 20–50× enlarged and 50–100 mm away. Optical objectives act in the same way and may be described by their focal length or their magnification M, which ranges from four to one hundred times. (Their focal lengths are about $160/M$ in millimeters.) It is apparent from the figure that the angular divergence of rays from a point on the object is greater than the divergence from a point on the image ($\alpha > \alpha'$). This demonstrates a general principle in geometrical optics, that divergence angles go down as the magnification goes up. As we will see later, lens aberrations increase with ray angle, so the aberrations at the objective lens are the most serious. This is why the objective lens is most important in controlling the resolution of the image.

Figure 3.1 shows how rays leaving from the same point on the object meet at the image plane, while rays leaving in the same direction meet at the *back focal plane*. The back focal plane is the plane where parallel rays are brought together, and the focal length f is the distance from the lens to this plane. A diffraction pattern thus appears

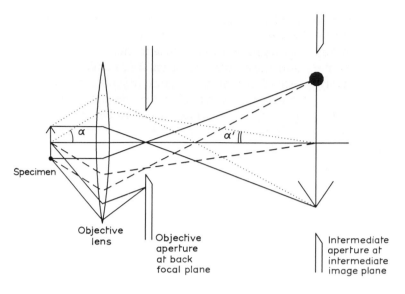

Fig. 3.1 Image formation by the objective lens. Rays leaving the specimen in a given direction meet at a point in the back focal plane of the lens. The aperture there limits the angular acceptance of the imaging system to α. The intermediate aperture selects the specimen area which contributes to the image.

at the back focal plane and in the TEM this is where the objective aperture is placed, to limit the angular acceptance of the imaging system. The distance from specimen to lens is only a little over f, so the image plane is about Mf from the lens.

3.1.2 Resolution

In all microscopy a primary concern is the spatial resolution that can be obtained. If the details are points, the resolution may be specified as 'point-to-point' resolution. An object with regularly repeating details such as a set of lines will give a different value for resolution. Convenient test objects are ruled lines in optics and lattice fringes (sheets of atoms) in TEM. Near the resolution limit, an image will be formed with reduced contrast. A precise measure of resolution requires a quantitative test to determine if the detail can be distinguished.

The contrast of a feature is defined as $|(I_0 - I)|$ divided by I_0 where I_0 is the background intensity and I is the intensity at the detail [7, 8]. Several other definitions are possible, for

example:

$$\log(I_0/I) \qquad \text{(Hall [9])}$$

$$(I_0 - I)/(I_0 + I) \quad \text{(De Palma and Long [10])}$$

$$(I_{\max} - I_{\min})/I_{\max} \qquad \text{(Goldstein } et\ al.\ [11])$$

Thus care is required in comparing results from different sources. The best method of determining resolution is to measure the (contrast in the image)/(contrast in the object) as a function of detail separation. This is the *modulation transfer function* (MTF), also called the contrast transfer function, which describes how the modulation or contrast in the object is transferred to the image. Clearly when MTF falls to some very small value no object detail is reproduced in the image.

The MTF may be measured using test objects of known size and contrast, as for testing lithographic processes. Another method is to use a random object that contains details of all sizes and find which details are reproduced in the image. A diffraction pattern of the image of the random object, obtained by computation from a digitized image or by optical diffraction shows which modulations have been transferred. The resolution is then taken as the detail size at which the MTF first falls to some arbitrary value. The Rayleigh criterion for resolution is that the intensity between two bright spots should be no more than 80% of the peak value [2, 3]. This is derived from human vision, and corresponds to a criterion of an MTF of $(1 - 0.8)/(1 - 0) = 0.2$. Contrast of 0.05 is needed in the image detail for it to be visible to the eye, so objects of contrast > 0.25 will be visibly resolved when MTF ≥ 0.2. In the TEM, MTF can be a complicated function of feature size [7, 12, 13]. For high resolution work, a plot of this function is more informative than any single number called resolution.

3.1.2.1 *Limitations to resolution*

Three factors may limit the resolution of an image: the diffraction limit, lens aberrations and noise. Noise can be a problem in scanning microscopy. A very bright source is required if rapid, TV scan rates are to be used at high resolution. In CTEM noise is a problem for radiation sensitive polymers (Section 3.4.4) because a limited number of electrons can be used to form the image.

The diffraction limit depends on the wavelength of the radiation and the angular acceptance of the objective. The diffracted beams caused by a periodicity d come off at angles given by $\sin \phi = \lambda/d$ or greater. Therefore the lens must accept a semi-angle $\alpha \geq \phi$, for the periodicity to affect the image. Thus a lens cannot resolve details smaller than $\lambda/\sin \alpha$. If oblique but still coherent illumination is considered, some rays diffracted by $\phi = 2\alpha$ will be accepted into the imaging system (Fig. 3.2) so that the smallest resolvable detail on the object, d_{min}, will be $\lambda/(2 \sin \alpha)$. (This is the Abbe or diffraction theory of imaging). Calculation for incoherent illumination or self-luminous objects gives the very similar diffraction limited resolving power of $0.61\lambda/\sin \alpha$ [2–5].

The wavelength λ used in the formulae above refers to its value in the space between specimen and objective. If, as is more common in optical microscopy, λ is taken to be the free space wavelength; $d = 0.6\lambda/NA$. NA, the numerical aperture, is $(n \sin \alpha)$, where n is the refractive index of the medium in front of the objective. The lens aberrations of concern are chromatic aberration, where the focal length of the lens depends on wavelength, and the five Seidel geometric aberrations:

(1) spherical aberration – axial;
(2) astigmatism – off axis and axial;
(3) coma – off axis;
(4) field curvature – off axis;
(5) distortion – off axis.

3.1.2.2 *Optical resolution*

All optical microscope lenses are corrected for chromatic and spherical aberration and will give

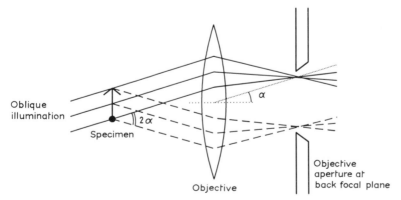

Oblique illumination

Specimen

2α

α

Objective

Objective aperture at back focal plane

Fig. 3.2 With oblique illumination, an angular acceptance of α will just permit rays diffracted through an angle of 2α to enter the imaging system. The spacing which produces these diffracted rays will be the diffraction limited resolution.

Table 3.2 Diffraction limited resolution d for objective lenses in the optical microscope, $\lambda = 0.5\ \mu$m

Objective magnification	NA	$d = 0.6\lambda/NA$ (μm)	Total system magnification	Image resolution (μm)
10×	0.1	3	100	300
60×	0.75	0.4	600	240
100×	1.25	0.22	1000	220

a resolution close to the diffraction limit described above. Table 3.2 shows the diffraction limit for typical high and low power objective lenses.

Table 3.2 also shows how the object resolution obtained matches the total magnification M of the system and the resolution of the detector – in this case, the eye. The image resolution obtained (object resolution × magnification) is 200–300 μm. For visual observation the (virtual) image will be at the standard viewing distance of 25 cm from the eye. The detail therefore subtends an angle of 10^{-4} rad at the eye, near but comfortably within the eye's angular resolution of 3×10^{-5} rad. Higher magnification would be empty, producing no further information.

3.1.2.3 *TEM resolution*

Electron lenses have serious aberrations that control the resolution of the instrument. The off-axis Seidel aberrations vanish for rays parallel to the axis. In the TEM, the angular range of the rays is severely restricted, so these aberrations will not be important. The axial astigmatism can be corrected by adjustment of stigmators that cancel the residual non-circularity of the objective. This leaves chromatic and spherical aberration as the most important.

Chromatic aberration

Chromatic aberration affects the image when electrons that do not all have the same energy contribute to it. Modern microscopes have highly stable accelerating voltage, but there is an intrinsic energy spread of about 1 eV in electrons

leaving a tungsten filament source. The chromatic limit to resolution due to this is only 0.2 nm at 100 keV, not important as a limit for polymer microscopy.

For polymers the major effect of chromatic aberration appears when a thick specimen causes the transmitted electrons to lose energy. Just how serious this is depends on the exact nature of the material as well as the specimen thickness. However, a carbonaceous material of density 10^3 kg m^{-2} (specific gravity 1) will cause a 100 keV electron to lose about 0.4 eV per nanometer of thickness on average [7, 14–16]. Thus a 100 nm thick foil causes 40 eV loss, for a chromatic limit to resolution of 6 nm. This chromatic aberration is the basis for the rule of thumb (often given for biological specimens) that one should not expect to resolve details smaller than one tenth of the specimen thickness.

Spherical aberration

Spherical aberration occurs because rays passing through the outer portions of the lens are diverted too much, and come to a focus short of the ideal focal plane (Fig. 3.3). It causes the image of a point at the ideal (Gaussian) focal plane to become a circle of radius $M\alpha^3 C_s$, corresponding to a resolution on the object of about $\alpha^3 C_s$. C_s is the spherical aberration coefficient, typically 1–2 mm for 100 keV lenses. This resolution limit is proportional to α^3, while the diffraction resolution limit is proportional to $1/\alpha$ (for small angles where $\sin \alpha \approx \alpha$), so there is an optimum value of α that gives the best resolution. It is given by:

$$\alpha_{\mathrm{opt}}^3 C_s = \lambda/\alpha_{\mathrm{opt}}; \quad \alpha_{\mathrm{opt}} = (\lambda/C_s)^{1/4}.$$

At this divergence the resolution is:

$$d_{min} = (C_s\lambda^3)^{1/4}$$

Focus considerations

It is apparent from the ray diagram of Fig. 3.3 that there is a plane, A–A′, nearer to the lens than the geometrical or Gaussian focus plane, where the resolution is improved. It is called the 'plane of least confusion' or Scherzer focus [3, 4, 7], and it is close to where the rays from the outermost parts of the lens intersect the axis. From Fig. 3.3 the distance $\delta Z'$ between this plane and the Gaussian image plane is approximately the radius of the image disc in the Gaussian plane divided by α'. The radius is $M\alpha^3 C_s$ and $\alpha' = \alpha/M$, so

$$\delta Z' \approx C_s(\alpha M)^2.$$

Reducing the lens power, *underfocusing*, to move the image plane by $\delta Z'$ or the object by δZ (where $\delta Z = \delta Z'/M^2$) improves resolution. A detailed calculation gives:

Resolution : $\qquad d_{min} = 0.43(C_s\lambda^3)^{1/4}$

Optimum divergence : $\quad \alpha_{opt} = 1.41(\lambda/C_s)^{1/4}$

At defocus : $\qquad \delta Z_{opt} = (C_s\lambda)^{1/2}$

(Defocus is conventionally given as a motion of the object, as one would focus an optical microscope). Comparing Scherzer focus to geometric focus, the resolution is improved by a factor of about two, if the included divergence angle is increased by $\sqrt{2}$. Taking values of $\lambda = 0.004$ nm (for 100 keV electrons) and $C_s = 2$ mm these become $\alpha_{opt} = 10$ mrad (0.5°) and $d_{min} = 0.3$ nm. This is not high resolution for the TEM, but is much higher than normally obtainable in polymers. Radiation damage (Section 3.4) is more important than microscope optics for these materials.

Once a resolution limit has been defined, the depth of field can be determined. The depth of field is the range of object distances where defocus produces a spreading of a point image equal to the image resolution limit, Md. Figure 3.4 takes the whole imaging lens system to be a single thin lens. It shows that the depth of field is the resolution of the object, d, divided by the divergence angle α. Depth of focus is the range of image plane positions which produces the same spreading of the image, and is $M^2\times$ (depth of field).

It can be seen in Table 3.3 that the depth of field at high magnifications in the optical microscope is very limited, less than the resolving power. The depth of field in the TEM is as small in absolute terms, but very much greater than the resolution and greater than a useful specimen thickness at that resolution. Thus all the thickness of a TEM specimen is in focus at once, and the depth of field is never a problem.

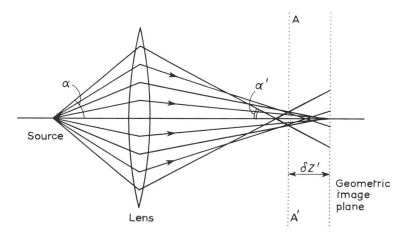

Fig. 3.3 Spherical aberration causes rays at larger angles to the axis of the lens to come to a focus short of the ideal focal plane. The smallest image of a point source appears in the plane A–A′, at the Scherzer focus position.

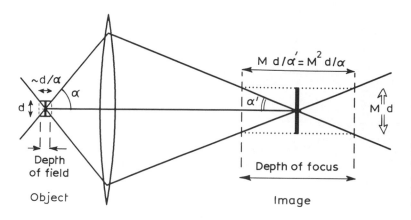

Fig. 3.4 The depth of field depends on the resolution d and the angular acceptance of the lens, α. The object may be moved by d/α and an image of size Md can still be formed.

3.1.3 Electron diffraction

The diffraction pattern has already been defined as a map of spatial frequencies in the object. Many polymers are amorphous, so the distances between atoms can have a range of values. An unoriented amorphous material will give a diffraction pattern consisting of a few broad rings (Fig. 3.5A). Unoriented polycrystalline materials give sharp rings (Fig. 3.5C). A full analysis of the intensity of the ring pattern gives the *radial distribution function* (RDF) of inter-atomic spacings in the material [17], but such analysis is rarely performed. When an amorphous material is oriented the ring pattern becomes arced. The *meridional* reflections are those intensifying on a line parallel to the draw direction, and they are associated with spacings along the molecular chain. Those stronger on a line perpendicular to the draw direction are called *equatorial* and are associated with inter-molecular spacings (Fig. 3.5B). These associations are also normally correct for crystalline materials.

It is the 'off-axis' reflections (those on neither line) that prove that there is crystalline order in a fiber (Fig 3.5D).

A large and perfect crystal scatters electrons into a diffraction pattern of sharp spots. General interpretation of this pattern requires a knowledge of crystallography. There are many texts in this field [18–20], with some specifically aimed at microscopists [21, 22]. Books on crystal optics (Section 2.3) contain basic summaries [23, 24]. There are also many texts on diffraction from materials [25–27], some concentrating on electron diffraction [28, 29]. The most common use of a crystal diffraction pattern is to find the orientation of a crystal of known structure. Wunderlich [30] contains a listing of many polymer crystal structures.

In simple terms we can regard the crystal as a set of lattice planes, reflecting radiation according to Bragg's law. The diffraction angle is then 2θ where $2d \sin \theta = 1$. For electrons $\lambda \ll d$ so θ is very small. This means that lattice planes will diffract only if they are almost parallel to the

Table 3.3 Objective lens parameters

Objective lens	α (rad)	resolution (μm)	System magnification	Depth of field (μm)	Depth of focus
Optical 10×	0.1	3	100	30	30 cm
Optical 100×	0.93	0.22	1000	0.15	15 cm
TEM $C_s = 2$ mm	0.0066	6×10^{-4}	100 000	0.1	1 km!

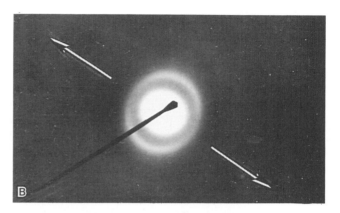

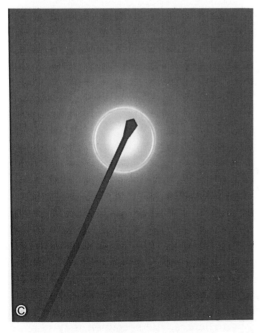

Fig. 3.5 Electron diffraction patterns from (A) amorphous carbon, (B) oriented amorphous polystyrene (tensile direction indicated by arrows) and (C) a polycrystalline PE film. The sharpness of the rings in (C) indicates crystalline order. Highly oriented polyethylene is shown in the diffraction pattern (D) (tensile direction indicated by arrows). The off-axis spots prove the presence of three dimensional order.

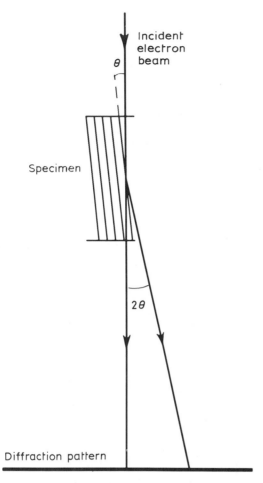

Fig. 3.6 A schematic of electron diffraction from a set of lattice planes, showing that the planes must be nearly parallel to the incident beam direction. For 100 keV electrons, θ would typically be ten times smaller than shown in the figure.

atoms in the crystal – that is, determines the crystal structure. Intensities are difficult to measure accurately because of background from inelastic scattering of electrons. Electrons interact so strongly with matter that an electron can be scattered into one diffracted beam and then from that to another, even in a very thin crystal. This multiple scattering makes full theoretical treatment of electron diffraction complex [27–29]. Although obtaining an accurate electron diffraction pattern from a polymer crystal may be difficult because of its instability, the small thickness and low atomic number of most polymer crystals can make the analysis more straightforward. The small size of available single crystals rules out the equivalent x-ray experiment. Many structural determinations have been made of polymers and other organic crystals [34–38].

Polymer crystals are frequently small and imperfect, so that the diffraction spots are fuzzy, or they are arcs from an oriented polycrystalline texture. The degree of perfection of crystals of known structure can be determined from measurement of diffraction line widths and intensities. The analysis used for X-ray diffraction [25, 39, 40] can be transferred directly to electron diffraction. One can distinguish between crystal size effects and the effects of disorder within the crystals, but often a simple estimate of the mean crystal size, $0.9\lambda/$(angular breadth), is used. Any measure of disorder can only be an upper limit, unless careful precautions are taken to account for the effects of radiation damage.

3.1.4 Contrast mechanisms

There are two parts to a contrast mechanism. The first is the interaction between the incident radiation and the specimen. The second part is the optical system that makes the interaction produce an intensity change in an image. When imaging with lenses, the specimen may change the amplitude of the wave passing through it, or the direction or phase of the wave. Contrast in optical microscope images is rarely interpreted

beam (Fig. 3.6). The geometry of an electron diffraction pattern is thus easier to analyze than the equivalent x-ray diffraction pattern. The geometry of a single pattern is enough to determine orientation, or to distinguish between different crystallographic phases [31, 32]. The unit cell and symmetry of an unknown phase may also be determined from the geometry of diffraction patterns [30, 33].

Proper interpretation of the intensities of spots in a diffraction pattern gives the positions of the quantitatively, and is dealt with in Sections 2.2.3 and 2.2.5. This section therefore concentrates on TEM.

3.1.4.1 *Amplitude or scattering contrast*

A TEM specimen thick enough to absorb significantly would degrade the resolution very seriously through chromatic aberration. However, because the imaging system in the TEM has a limited acceptance angle, any intensity scattered to an angle >1° will be excluded from the image and will appear to be absorbed. Thus the scattering contrast in bright field looks like absorption.

Non-crystalline materials, such as many biological samples and polymers, scatter diffusely in the TEM and the resulting contrast is *mass thickness contrast*. The image brightness depends on the projected mass thickness (thickness × electron density) of the specimen, the incident electron energy and the objective aperture angle. A smaller objective aperture or lower beam voltage will increase the contrast. When a material is observed under standard conditions, the fraction of the incident beam that is transmitted can be used to measure the specimen thickness, t using the relation:

$$I(\text{sample})/I(\text{hole}) = \exp(-Bt).$$

The constant B is determined by observing films of known thickness under the same standard conditions [7, 41, 42].

If the sample is crystalline, the scattered intensity depends very strongly on the orientation of the crystals and on their thickness. In bright field, a thin crystal will appear dark when it is correctly oriented for diffraction. If the crystal is not perfectly flat, the contours of correct orientation will appear as dark lines, called bend contours. Variation of intensity such as this in crystalline specimens is called *crystallographic* or *diffraction contrast*. Many types of defects in crystals cause localized distortion of the crystal lattice. These defects change the crystal orientation locally and so cause variations in the crystallographic contrast. Detailed information on defects can be obtained by comparing the images produced by different scattered beams to theoretical predictions [7, 43].

If the objects in a bright field image scatter only weakly, as many polymers do, the intensity level will be high but the contrast will be low. In a dark field image the background intensity is low, so the contrast level is high. The dark field images of amorphous samples in the TEM will be particularly dim because the objective aperture can collect only a small part of the scattered intensity. The dark field arrangement in the TEM is more efficient for crystalline samples, where the scattered intensity is concentrated into a few regions of the back focal plane of the objective (Figs 2.5 and 3.13).

3.1.4.2 *Phase and interference contrast*

Any specimen will cause a phase change in the wavefront passing through it. The specimen has an optical thickness nt producing an optical path difference $\Delta = (n-1)t$ and a phase retardation $(2\pi/\lambda)(n-1)t$. To make the phase shift affect the image intensity, the retarded wave must be caused to interfere with another wave. The other wave must be coherent with that passing through the specimen.

Components that cause interference, analogous to those in phase contrast optical microscopy (Section 2.2.3) have been tried in the transmission electron microscope but they are generally impractical [7]. Phase contrast is actually induced in the TEM by imperfect focus. Defocus causes waves from neighboring regions to overlap in the image and interfere. Spherical aberration makes even the best focus imperfect, so phase contrast is always present in the TEM [7, 44]. If the TEM sample is thin and not strongly diffracting, all visible structure is due to phase contrast. This contrast will go through a minimum at the best focus.

Deliberate defocusing enhances phase contrast at lower magnifications [45] but it must be used with caution. If there is only random structure in the specimen, defocus, deliberate or accidental, may induce clearly visible structure unrelated to the specimen—artifacts. Thomas [46] discussed this in detail for polymer microscopy, quoting TEM studies of molecular shape [47], amorphous structure [48] and microphase separation [49] which were dominated by phase contrast artifacts [50, 51]. With care, phase contrast imaging can be successfully applied to polymer systems [52, 53]. A simple rule to avoid misinterpretation is to take an optical transform of the image and consider only structures larger than those corresponding to the innermost ring spacing. Alternatively, a through focus series or a calculation of the MTF will distinguish the specimen structure from the artifacts.

Phase contrast at high resolution produces a lattice image, if diffracted beams from a crystal are allowed to pass through a large objective aperture. The application of this technique to polymers will be described in Section 6.4.

3.1.5 Illumination systems

The illumination system must collect flux from the source and direct sufficient intensity onto the required field of view. The intensity needed at the specimen will increase with the square of the magnification as the area viewed decreases. Typical characteristics of sources are given in Table 3.4, which shows that the small size of the electron sources gives them a very high brightness.

Table 3.4 Characteristics of illumination sources

Source	Optical tungsten filament	Electron tungsten filament*	Field emission source*
Output current (μA)	–	50	5
Output power (W)	25	5	0.5
Diameter	5 mm	25 μm	10 nm
Intensity (W cm^{-2})	100	8×10^5	5×10^{11}
Intensity (particles s^{-1} cm^{-2})	3×10^{20}	5×10^{19}	3×10^{25}
Brightness (W cm^{-2} ster^{-1})	50	10^{11}	10^{16}

* Typical values at 100 kV

3.1.5.1 *Optical microscope*

The illuminating system of the optical microscope is described in detail by Hartley [54]. At high powers a wide cone of rays is needed to fill the aperture of the objective lens. The first condenser lens is weak and collects light for the strong substage condenser that does the demagnifying. The illumination in optical microscopes has been the subject of much confusing discussion about whether the filament itself should be focused onto the specimen (critical illumination). With modern sources this gives an unevenly lit field of view.

The Kohler system shown in Fig. 3.7 gives an evenly lit field (as long as the first condenser is corrected for spherical aberration). It uses the back focal plane of the first condenser lens as the object which is imaged by the substage condenser into the specimen plane. The field diaphragm, at this back focal plane, controls the illuminated field. The aperture at the focal plane of the second (substage) condenser controls the divergence angle of the illuminated rays. When the system is properly adjusted, the two apertures act completely independently. The field diaphragm is closed to match the illuminated area to the viewed area, reducing glare. The second aperture is usually set to fill two-thirds of the objective back focal plane, to balance resolution and contrast. A large illuminated field is needed for a low magnification image, and then the substage condenser must be changed to a lower power. Most bright field condensers have a top lens that can be removed for low power work. There may be a supplementary field lens that should also be removed. Under these conditions, the substage aperture controls the illuminated area (Fig. 3.7B).

3.1.5.2 *Transmission electron microscope*

In the TEM the strongly excited first condenser forms a demagnified image of the filament (actually the cross-over of the electron gun is the effective source). Since beam divergence increases as the spot size falls, much of the flux

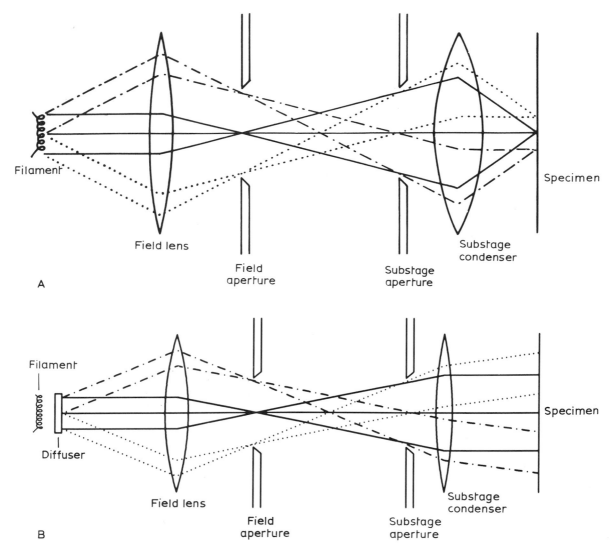

Fig. 3.7 Kohler illumination (A); the field lens produces an image of the filament at the substage aperture. Each part of the specimen receives light from the whole filament. For low magnification operation (B), the substage condenser is weakened to illuminate a larger area. Each part of the specimen now receives light from only a part of the filament, so a diffuser is inserted, as shown, to even the illumination.

is not accepted by the second condenser. The highest intensity at the object and the largest divergence angle are both obtained when the second condenser produces a focused image of the source on the specimen. Since the illumination is then very uneven, in normal operation the second condenser is defocused (under or over focused) to 'spread the beam'. The second condenser aperture then controls the illuminated area.

Increasing the power of the first condenser lens reduces the intensity and divergence of the illumination by reducing the size of the demagnified source. Altering the power of the second condenser changes the area illuminated, and changing the diameter of the condenser aperture does the same in large steps. This arrangement corresponds closely to the Kohler illumination for optical microscopy described above, except that the first and second condenser lenses have exchanged their functions. The correspondence continues at low magnifications in the TEM, where to get a large illuminated area the first condenser lens must be substantially weakened, or turned off. This is the equivalent of removing the top lens of the substage condenser in the optical microscope.

3.2 IMAGING BY SCANNING ELECTRON BEAM

3.2.1 Scanning optics

The basic purpose of the lenses in a scanning electron microscope is to form a focused beam spot at the specimen. By simple geometry, the diameter of the focused spot would be source diameter × demagnification. The normal tungsten filament or lanthanum hexaboride (Lab$_6$) source has a diameter of about 10–50 μm and three condenser lenses are normally used. The effective source size from a field emission gun is only 10 nm, so only one stage of demagnification is required. As in the previous section, the effects of diffraction and aberrations will increase the spot size. The divergence angle of the beam increases as the source is demagnified, so the electrons are at the largest off-axis angles within the final condenser. This is then the critical lens, as the objective lens is in a microscope without scanning. The limiting aperture for the system is within this lens. The resolution of the instrument depends on the interaction volume, not always on the spot diameter (Section 2.3).

The STEM/CTEM is complicated, as the same lenses are used for two tasks. A special objective lens used in STEM/CTEM can be highly excited, that is, it can have large magnetic fields, or more technically, $k^2 > 3$ [7]. The part of the field before the specimen then acts as a third condenser. The remainder of the field continues to act as a magnifying lens, reducing the divergence of the probe and allowing the transfer of electrons to

the transmission detector, which is located below the fluorescent screen. Here the condenser/objective is critical in both modes of operation; for STEM mode it will be called the final condenser.

Spot size d increases with the aberrations of the final condenser, the wavelength of the electrons, and the divergence angle, α, of the probe. Reducing the divergence gives a lower beam current, which may make the resolution worse because of noise if there is a limit to the image recording time. A brighter source reduces this problem, but low beam currents may be needed for beam-sensitive specimens.

For a given beam current i and beam energy E, the probe diameter d is given by:

$$d = \left[\left(\frac{i}{B\alpha} \right)^2 + \left(\frac{\delta E C_c \alpha}{E} \right)^2 + \left(\frac{1.83}{E\alpha^2} \right) + \frac{C_s^2 \alpha^6}{4} \right]^{1/2}$$

where B is the gun brightness;
  δE the energy spread in the beam in eV;
  α the beam divergence angle at the focused spot, diameter d;
  C_c the chromatic aberration coefficient of the final condenser lens;
  C_s the spherical aberration coefficient of the final condenser lens.

The first term is related to noise in the image, the second to chromatic aberration, and the last two are the diffraction and spherical aberration limits to resolution that were met with in Section 3.1.2.3 for the TEM. The smallest possible probe diameter d_{min} can be calculated by finding the value of α that minimizes this expression [11, 55]. This is a lower limit to the size of the resolved detail. If chromatic aberration is neglected and the beam current is very small, d_{min} approaches the calculated minimum size of the resolved detail in TEM, $(C_s \lambda^3)^{1/4}$. Very high resolution can only be obtained when C_s and λ are as small as possible. Since C_s and the focal length are of the same magnitude this requires a short focal length final condenser. The sample must therefore be inside the lens, just as in the TEM, and this is how the highest resolution SEMs and the STEM operate. In the normal SEM a final condenser of longer focal length allows the specimen to be outside the lens.

Chromatic aberration gives a 'disc of least confusion' of $\alpha(\delta E/E)C_c$. C_c, the chromatic aberration coefficient, is similar in magnitude to the focal length. The energy spread in the electron beam δE is fixed at 1–2 eV for a tungsten filament. Chromatic aberration becomes important in the SEM because of the long focal length and low beam voltage. It is particularly important for beam voltages of 1–3 kV, as used for the inspection of radiation sensitive semiconductor devices and polymer coatings [56–58]. It may then be necessary to use other electron sources, since δE is 0.5–2 eV for Lab$_6$ and only 0.2–0.4 eV for a field emission gun operated with its source at room temperature [7, 58]. This allows resolution to be maintained, as shown in Table 3.5.

The values of probe size in Table 3.5 show how the FESEM maintains good resolution to 5 keV, and the HRSEM all the way to 1 keV. (Remember that these are lower to resolution; real specimens may give images that are worse.) The values were calculated with reasonable parameters, but individual instruments may be considerably different. For example $\delta E = 0.4$ eV chosen for the FEG is a high value for a 'cold' FEG, but low for guns which operate at 1800 K. Manufacturers provide graphs showing what is to be expected from their SEM for a given final aperture, accelerating voltage and working distance.

As when imaging with lenses, the depth of field is the resolution divided by the divergence angle α. At a resolution of 10 nm and $\alpha = 5$ mrad, the depth of field is 2 μm. Geometrically, the divergence of the probe in the SEM is controlled by the working distance S and the final aperture diameter D; $\alpha = D/2S$. Typical values for these parameters are 0.2 mm and 20 mm, giving $\alpha = 5$ mrad, very close to the theoretical optimum. The rays passing through the final lens are also deviated by the scanning coils. This angle is small when the field of view is small at high magnifications, and at low magnifications the loss of resolution is not important.

3.2.2 Beam – specimen interactions

3.2.2.1 *Backscattered electrons*

A certain fraction of high energy electrons striking a bulk specimen escape from it with high energy. This fraction increases with atomic number and is independent of incident beam voltage in the usual SEM range of 5–30 kV. The incident electrons lose energy continuously as they pass through the material, so that backscattered electrons of energy very close to the incident energy come only from where the beam strikes the surface. However, for carbon, because most of the electrons in the beam penetrate quite deeply, the most probable energy for backscattered electrons is only half the primary energy. These backscattered electrons are produced in a

Table 3.5 Typical minimum probe size, d_{min}, at a beam current of 1 pA for various SEMs at different beam voltages

	Conventional SEM Tungsten filament δE 1.5 eV	FESEM FEG δE 0.4 eV	HRSEM
Lens focal length	20 mm	20 mm	1.5 mm
Beam voltage (kV)	d_{min} including chromatic aberration (nm)		
20	5.7	2.1	0.8
5	13	5.8	1.6
1	39	20	5.5

significant fraction, perhaps the top one third, of the total interaction volume. They can therefore come from several micrometers deep when the beam voltage is 10 kV or more [11, 55] and leave the surface of the sample over a wide area. A thin coating of heavy metal such as gold on a sample of low atomic number will have little effect on the backscattering yield or resolution. Operation at low voltage (Section 6.3.2) will reduce the depth and the area from which the backscattered electrons come [58].

3.2.2.2 *Secondary electrons*

The energy spectrum of electrons emitted by a specimen in an electron beam has two maxima. One is at high energy where most of the backscattered electrons are and the other, in the low energy region, contains the secondary electrons. The number emitted divided by the number of incident electrons is the secondary emission coefficient. This is not very sensitive to the atomic number of the sample, rising from < 0.1 for atomic low number materials to 0.2 for gold at a beam voltage of 20 kV [59]. A thin gold coating on a polymer specimen will increase the yield of secondaries without degrading their resolution.

Secondary emission depends very strongly on the surface topography (Fig. 2.6). If the beam falls on a tilted surface or onto a peak, more of the interaction volume is near to the surface, so more secondaries will be produced. If the beam falls into a valley or pit, less of the interaction volume will be near the surface, and there will be fewer secondaries. The usual detector for secondary electrons is the Everhart–Thornley detector. It operates with a bias of +250 volts so that all the low energy electrons in the specimen chamber are attracted to it. There is therefore no contrast due to secondaries between regions of the specimen facing towards the detector and those facing away, although the edges will be well defined. The optical analog is of a surface lit by diffuse light. This detector produces good images by collecting some backscattered electrons as well as the secondaries (Fig. 3.8). The result is analogous

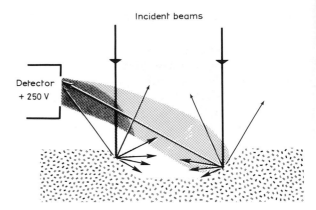

Fig. 3.8 Shaded regions are secondary electron trajectories, drawn to the detector by the +250 V bias. Arrows indicate the backscattered electrons, which are mostly in the forward direction. Very few will reach the detector from the surface facing away from it, so surfaces facing the detector will be brighter.

to directional lighting plus diffuse lighting to fill the shadows – just what a photographer would use in a studio portrait. Since we are accustomed to light shining from above, SEM micrographs taken with this detector system should be oriented with the detector position at the top for viewing.

The dependence of the secondary emission coefficient on the primary beam energy is important. The coefficient increases at very low energies, passing through one at about 1 keV. There is a maximum and then it passes through one again at about 3 keV and continues to fall [58, 60, 61]. In this region, the sample is stable against charging in the beam and does not need a conductive coating (see Section 6.3.1).

3.2.2.3 *X-rays*

When high energy electrons bombard a solid, x-rays are produced, some at specific energies characteristic of the elements present in the solid. This process is not efficient, with only one electron in 10^4 or 10^5 producing a detectable characteristic x-ray [62, 63]. The efficiency falls further for low atomic number elements. This makes x-rays a poor signal to use for simple imaging, so they are used to obtain analytical information.

The electrons cannot produce the characteristic x-rays if they do not have sufficient energy. Thus the x-rays do not come from the whole interaction volume, but from the region where the electron has not lost too much energy. The size of this region will depend on the beam energy, the x-ray energy and the composition of the material. Problems of this sort make quantitative microanalysis difficult (see refs 23–26, 119–122 in Chapter 2).

3.2.2.4 *Interactions in the STEM*

The basic difference between the SEM and the STEM is in the interaction volume, (Section 2.3) which is much smaller in the STEM where very thin specimens are used. High energy electrons can penetrate several micrometers into a thick sample and are scattered in all directions. A beam, however well focused at the surface, will therefore spread out into the bulk. If the specimen is 0.1 μm thick or less, most of the beam passes through without any spreading. The x-ray resolution can therefore be 1000 times better in the STEM than in the SEM [62]. A very large incident beam flux is needed to produce a statistically significant analysis on the STEM because of the low interaction volume and the low efficiency of x-ray production. Analysis of a single spot or line is much more common than 2-D x-ray imaging or mapping because of lack of signal. Even so, radiation sensitive polymers cannot normally stand the irradiation required for such high resolution microanalysis.

For secondary electrons the STEM acts as a high performance SEM. A very thin sample has no low resolution background due to secondaries produced by backscattered electrons, but secondaries come from both surfaces. This means that topographic information from both sides of the specimen is superimposed on the image, making interpretation more complicated. If the specimen is thicker, but still transmits the electrons in a spread beam, there will be a large low resolution signal from the back surface of the specimen. In this case it would be better to mount the specimen on a solid substrate to make it non-transparent.

Backscattering from a thin sample will also have higher resolution and lower efficiency than that from a thick sample. Scattering of all sorts is strongly peaked in the forward direction, so a much bigger signal produced by the same sort of interactions appears below the specimen. The STEM uses forward rather than backward scattered electrons; these electrons have passed through the specimen, and their interactions with it are the same as those in the TEM. The STEM can use electrons scattered to larger angles, and electrons that have lost significant amounts of energy to form images. This is because there is no objective lens, and therefore no spherical or chromatic aberration to consider. Other differences between STEM and conventional TEM are described in Section 2.4.1.

3.2.3 Image formation

In the SEM the brightness and contrast of each image are routinely optimized as an elementary adjustment. Typically a constant background signal is subtracted, and the amplification is increased to set the minimum and maximum signal levels to the dynamic range of the instrument display or storage system. This increases the contrast of any feature, and the noise. The visibility limit of 0.05 contrast described in Section 3.1.2 therefore does not apply to the original signal. When all the image structure is made visible, a detail must have one twentieth of the maximum image contrast. It is common practice to increase the contrast of the detail of interest even at the cost of lost high or low intensity level structure. Here, visibility of a resolved detail is controlled by noise.

For a perfect detector system, the noise is due to the random arrival of the detected particles, usually electrons or x-ray photons. It is $N^{1/2}$ if N particles arrive, so in a detector chain the noise depends on the link where the smallest number of particles is involved. Let this number be fN where N is the number of electrons striking the specimen. In X-ray imaging the smallest number is the number of X-rays arriving at the detector, $f = 10^{-4}$ or less; in BF secondary electron imaging $f \simeq 1$. Consider an image feature of intrinsic contrast C. The signal is CfN and the noise is $(fN)^{1/2}$ so the signal to noise ratio $k = C(fN)^{1/2}$. The probability p that a detail is visible through the noise is $k/(1 + k)$. The visibility limit is usually taken to be an 85% probability that the detail can be seen, so that $k \geq 5$. N depends on the area of the detail, the scan rate and the beam current. $Cf^{1/2}$ depends on the detail and the imaging mode. Take as an example a single pixel detail in a 500×500 image formed at TV scan rates (25 or 30 frames s^{-1}) with a beam current of 0.1 nA. $N \approx 140$, and if $f = 1$, then for $k \geq 5$, $C \geq 0.4$. This limit is unaffected by an increase of contrast in image processing, as the noise is also increased. In the x-ray case, $f = 10^{-4}$ so even at $C = 1$, N must be 250,000 and the minimum size of a visible detail is about 45×45 pixels. X-ray data may be collected at video scan rates, but in practice any x-ray image is obtained by integrating the signal over a much longer time.

From the analysis it seems that there are two possible solutions, to increase the beam current or reduce the scan rate. Neither of these may be possible. At any resolution the beam current is limited by the brightness of the electron source, and the scan time is limited by practical concerns including the patience of the operator. It is difficult to adjust the focus or astigmatism of a slowly scanning image. For polymers there is also the real possibility that radiation damage of the specimen limits the total number of electrons that can be allowed to strike a given specimen area.

3.3 IMAGING BY SCANNING A SOLID PROBE

Information on image formation and resolution in the SPM will be found in Chapter 6. It is not very useful to try to fit SPM imaging theory into the same mold used for imaging with lenses. The resolution depends on the area of interaction between tip and specimen. The tip itself, then, is a main contribution to the limit to resolution, equivalent in general terms to the modulation transfer function of a lensed microscope. For high resolution operation, the tip function depends on the last few atoms of the tip, and this is not well characterized. Even in the best regulated SPMs the image resolution and appearance may change significantly in some arbitrary way during operation because the atomic structure of the tip has changed. Detailed imaging theory is not worth applying except to images of stable structures that are repeatedly seen in the SPM, and preferably images that can be related to observations made using other methods. At lower resolutions image interpretation is simpler, but as with the SEM, detailed accurate metrology requires attention to the details of how edges are imaged, and how the instrument is to be calibrated. Well-characterized tips are again required, but at lower resolution this concerns the complete tip shape rather than the last few atoms, and the larger size scales are stable.

One major advantage of the SPM for polymer samples, shared by the low resolution optical microscope, is the absence of radiation damage. This is a very serious limitation for electron microscopy (see Section 3.5). However, it is not to be thought that the microscopes cannot influence the specimen structure. For soft and weak polymers, it is easy to imagine that a sharp AFM tip in contact with the surface will deform it and move material around as it scans across the surface. The force on the sample must be kept low if this is to be avoided. It seems that the equivalent to low electron dose operation in the TEM is low force operation in the AFM. Operation at low force may cause lower resolution, if the probe is kept in non-contact mode, or more difficult operation. Recently it has been shown that normal operation of an AFM is enough to wipe off atomic steps in a crystal of calcite, as they were only visible when the forces were kept unusually low [64]. Calcite is not a strong solid, but polymer surfaces are likely to be as easily damaged. High resolution images of polymers that do not describe the low forces used may be of damaged surfaces.

Forces are easily measured in the AFM, but as the STM is used to deliberately move atoms around on surfaces in nano-fabrication, so it too can apply large enough forces to the specimen to cause plastic deformation. For a metal coated polymer imaged at lower resolution there should be no problem. If the polymer is a non-conductor adsorbed on a conducting surface, large forces may be generated by the strong inhomogeneous electric fields around the tip. Reliable images are formed when the non-conductor sticks well to the substrate.

3.4 POLARIZING MICROSCOPY

Polarizing microscopy involves the interaction of materials that have anisotropic optical properties with polarized light. The full theory of this interaction is complex. Fortunately, good results in the polarizing microscope are often obtainable without much theory. However, lack of basic understanding can cause difficulties as soon as one goes beyond routine operations. There is a large range of texts, at different levels, with information on the interaction of light with anisotropic materials [1, 2]. For the expert there is Born and Wolf [5] and even more complete, with a very clear presentation, Ramachandran and Ramaseshan [65]. There are several texts on polarizing microscopy [23, 24, 66, 67]. Hartshorne and Stuart [23] contains a section on polymers, much practical information, and also covers theoretical tools, including the important Poincaré sphere construction.

3.4.1 Polarized light

The basic fact that allows light to be polarized and makes its description complicated is that light is a transverse vibration. It consists of an electric field **E** and a magnetic field, at right angles to each other and both perpendicular to the direction of propagation of the light ray. The three directions would form the edges of a cube. The electric field interacts strongly with materials, so its direction and amplitude (or that of the

electric displacement **D**) are used to describe the light.

At an instant in time, the amplitude and direction of the electric field define a point on a plane perpendicular to the direction of the light ray. To an observer looking towards the source, this point would trace out a curve as the field varied. If the curve is simple and repetitive, the light is *polarized* and the form of the curve defines the state of polarization. If the curve is irregular and chaotic, the light is *unpolarized*.

When the electric field oscillates in amplitude but has a fixed direction, the curve traced out is a straight line and the field remains in one plane. This is called the plane of polarization [68], and the light is *linear* or *plane polarized*. When the electric field is of constant amplitude but changes its direction, the point traces out a circle and the light is called *circularly polarized*. If both amplitude and direction change in a regular way, the curve traced out is an ellipse, and the light is *elliptically polarized*. This is the most general polarized state possible (Fig. 3.9).

Any state of polarization can be considered as a combination of (or can be decomposed into) two perpendicularly plane polarized waves with different amplitudes and a specific phase difference. Conversely, adding two such waves can produce any polarization state. If the sum is to be constant over time the two waves must be coherent, so that the phase relation between them remains the same. A state of polarization can also be considered as a combination of a right and left circularly polarized wave. If two such waves have the same amplitude they add up to produce a plane polarized wave. The direction of the plane of polarization is controlled by the phase difference of the two circularly polarized waves.

3.4.2 Anisotropic materials

Birefringent samples have a refractive index that depends on the direction of the electric field in the light. They may be single crystals, or polycrystalline or amorphous polymers with oriented regions [23, 69]. Materials containing

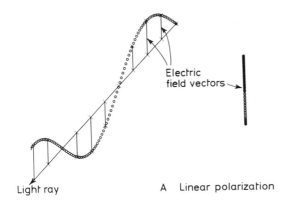

A   Linear polarization

Electric
field vectors

Light ray

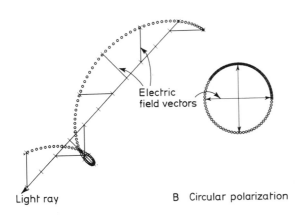

Electric
field vectors

Light ray   B   Circular polarization

many oriented interfaces have *form* birefringence [23]. Light waves passing through a birefringent material can be considered as divided into two perpendicularly plane polarized waves, polarized in the principal directions. The material has different refractive indices, n_1 and n_2, for the two plane polarized waves. As they pass through the material there is a relative *retardation* $R = (n_1 - n_2)t$, and a relative phase shift $\delta = (2\pi/\lambda)(n_1 - n_2)t$. Polarized light passing through a birefringent sample will usually have its polarization state changed. A complicating factor is that in birefringent materials the electric displacement **D** is not generally parallel to the electric field **E**. **E** is perpendicular to the light ray but **D** is tangent to the wavefront (Fig. 3.10). Therefore one must be very careful when talking of the 'direction of the light'.

The optical properties of a birefringent material are displayed by a surface called the *index ellipsoid* or *indicatrix*. It is constructed by moving away from the origin in every direction, and putting a point at a certain radial distance. The distance from the origin is proportional to the refractive index of light that has its electric displacement **D** in that direction. This is *not* the

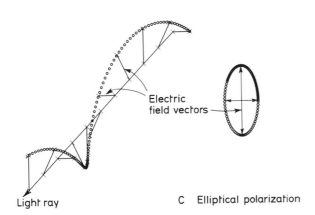

Electric
field vectors

Light ray   C   Elliptical polarization

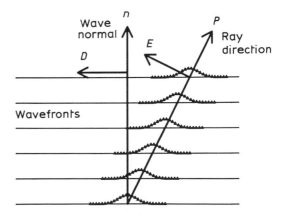

Wave
normal

D

E

n

P

Ray
direction

Wavefronts

Fig. 3.9 Schematic diagram of the electric field vector showing possible states of polarized light. With linear or plane polarized light (A) the electric field remains in a plane. In circularly polarized light (B) the electric field changes direction, but has constant magnitude. In elliptically polarized light (C), the electric field changes in magnitude and direction, tracing out an ellipse.

Fig. 3.10 Parallel light waves and a single light ray in a birefringent substance. The dotted bumps on the waves are an indication of where the intensity of the light is a maximum. The electric field **E** is perpendicular to the ray direction **P**, but does not generally lie along the wavefronts.

refractive index of light that is travelling in that direction. Cross sections of this surface are generally ellipses. The index ellipsoid of an isotropic material would be a sphere with circular cross sections.

There are two classes of birefringent materials, called *uniaxial* and *biaxial*, which have one and two *optic axes*, respectively. A birefringent material appears to be isotropic when a plane light wave passes through it along an optic axis. If the refractive index of a uniaxial material is a maximum when the plane of polarization contains the optic axis it has *positive birefringence*. If it is a minimum under this condition it has *negative birefringence*. Note that this means that positive materials have a minimum of refractive index when they appear isotropic. Uniaxial materials have optical properties with an axis of symmetry, so that the index ellipsoid is an ellipsoid of revolution, and the cross-section perpendicular to the axis of revolution is a circle. The optic axis is the axis of symmetry (Fig. 3.11). When there is no symmetry axis, the index ellipsoid is defined by three axes, the maximum and minimum refractive index direction, and the direction perpendicular to both of these. In the plane containing the maximum and minimum values there must be two directions where the index has the same intermediate value as it does along the third axis. There are therefore two cross-sections that are circles, each containing one of these directions and the third axis. The two optic axes are perpendicular to these cross-sections (Fig. 3.12) and the material is biaxial.

The magnitude of the birefringence of a substance is the difference between the maximum and minimum possible refractive indices, which are observed when the light wave direction is perpendicular to the optic axis or axes. This is *not* the same as the birefringence of a sample of the same material as it is viewed. The magnitude of this sample birefringence is the difference between the two refractive indices for the plane polarized waves that pass through it in the directions of observation, $(n_1 - n_2)$ or Δn. The sample birefringence is normally determined by measuring retardation R and dividing by thick-

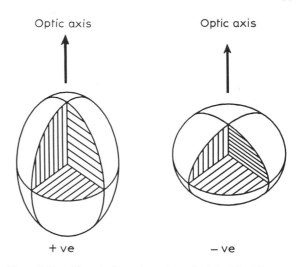

Fig. 3.11 The indicatrix of uniaxial birefringent materials, positive and negative. Positive materials have one principal refractive index greater than the other two. The optic axis is the axis of symmetry of the ellipsoid.

ness, t. A sample may also have a different sign of birefringence than the substance that it is made of. If there is a distinct direction to the sample, such as the long direction of a fiber or the radial direction of a spherulite, the sample is said to have positive birefringence if the refractive index

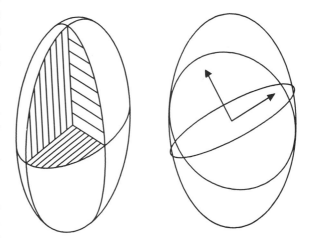

Fig. 3.12 The indicatrix of a biaxial birefringent material. All the three principal refractive indices are different. Two of the cross sections of this ellipsoid are circles. The two optic axes of the material are perpendicular to these two circular cross sections.

is greater when the plane of polarization includes this direction. The crystal form of a polymer often has a positive birefringence. There is a maximum refractive index when the plane of polarization contains the molecular chain axis, and smaller refractive indices in all other directions where the bonds are weaker. However, in the spherulite the molecular chain axes are commonly perpendicular to the radius, so that the spherulite has a negative birefringence.

Another figure often used to display the same information as in the index ellipsoid is the wave surface, or ray velocity surface. This is obtained by plotting all the points where light rays would arrive at a given time after leaving the origin. The advanced references [5, 23, 65] have details on how to use the ray and the wave velocity surface if required.

3.4.3 Polarizing microscopy

If there is no specimen and no residual strain in the optical glass of the lenses, the field of view in crossed polars should be completely dark. A birefringent specimen will split the incident plane polarized light into two components plane polarized along its principal directions. One component will be retarded relative to the other. In general the altered polarization state leaving the specimen will have some component transmitted by the analyzer, so the birefringent material appears bright. If one of its principal planes coincides with the incident plane of polarization, the wave polarized in the other plane will have zero amplitude. Its retardation will then not affect the state of polarization. Thus the specimen will be dark or *extinguish* at four positions 90° apart when it is rotated. It will be brightest at 45° from extinction, when the two waves excited in the specimen have equal amplitude. The standard orientation for measuring retardation is 45° from extinction.

We have already mentioned two reasons why an anisotropic material may be dark between crossed polars:

(1) the sample may be oriented so that it is being

viewed down an optic axis, and so appear isotropic; and

(2) the sample may be in an extinction position.

There is a third possibility in monochromatic light; the sample retardation $R = \Delta nt$ can be an exact multiple of λ so that the phase change is $2m\pi$, where m is an integer. The incident plane polarized light is then reformed, and not transmitted by the analyzer. In white light, some colors will be transmitted and others will not, so the specimen becomes brightly colored. If the birefringent specimen is wedge shaped and viewed with white light in crossed polars, it will show a sequence of polarization colors.

The Michel–Levy chart of these polarization colors [67, 69, 70] can be used to estimate the retardation of a specimen, within the range 200–1600 nm. For example, first order red is at 575 nm, third order green is at 1250 nm. At higher retardations the color is pale pink or pale green due to overlap of the orders. For example, 1700 nm retardation is 4λ for blue and 3λ for yellow. At very low retardations the colors are grays. Complementary colors which appear on a bright field when the analyzer is set parallel to the polarizer can be used to confirm the source of color effects. They are truly complementary only if the specimen is set to the 45° position. A material which is colored or has a birefringence dependent on the wavelength will have a different polarization color sequence from the standard. These colors will be useless for determining retardation.

Retardation is measured more accurately with compensators, with sample and compensator set at $\pm 45°$ between crossed polars. This alignment is critical, as the superposition of birefringent objects at arbitrary orientations produces extremely complicated effects [70, 71]. The compensator is adjusted until the specimen appears dark. Then the retardation of sample and compensator are equal and opposite, $R(\text{net}) = 0$. If the polarization color change on inserting a compensator shows an increase in retardation, then $R(\text{net}) > R(\text{specimen})$, and the compensator is adding its effect to that of the specimen. The specimen must be rotated by 90° so that the effects are subtracted as required.

The most useful and commonly available compensators with fixed retardation are:

(1) *quarter wave plate*, R = 150 nm= $\lambda/4$ (yellow light);
(2) *first order red plate*, R = 575 nm= λ (yellow light).

The first order red plate is also known as the sensitive tint plate. Its main use is in determining the sign of the birefringence of objects of small retardation, ϵ, which have a grey color in crossed polars. With the first order plate in place the net retardation is $(575 + \epsilon)$ nm, which is second order blue, or $(575 - \epsilon)$ nm, which is first order yellow. The eye is very sensitive to color changes in this range, so a very small value of ϵ has a visible effect.

In a spherulite the crystals grow radially outward from a center, so that all orientations are present. In crossed polars the spherulite is bright except at the four perpendicular radial directions where the crystals are in the extinction position. Figure 3.13 in colour section shows high density polyethylene spherulites in a thin melt cast film, under crossed polars and with a first order red plate in place. This material produces spherulites that are unusually perfect, and also have circumferential dark bands. The bands are regions of apparent isotropy where the optic axis is perpendicular to the specimen plane. With the first order red plate placed so that its slow direction is at +45° (top right to bottom left of the image) the spherulites are clearly blue in the upper left and lower right quadrants, and yellow in the other two. Blue means that the retardations of red plate and specimen are added, so that the larger refractive index ('slow') is perpendicular to the radius and the spherulite is negative. Figure 3.14 in colour section shows similar images of polycaprolactone, which gives reasonably large but more realistically irregular spherulites. Note that some show colors in crossed polars. The colors that arise when the first order red plate is added do not mean anything unless the sample has only a small retardation and appears gray in

crossed polars. The spherulites in this figure do not look at all radially symmetric, but a comparison of the two images shows that they are also negatively birefringent.

Crossed circular polars can be obtained with two quarter wave plates. The polars are set to 0° and 90°. One quarter wave plate is placed in the normal compensator slot just below the analyzer with its slow direction at ±45°. An identical plate is placed above the polarizer and below the specimen with its slow direction perpendicular to that of the first plate. The first $\lambda/4$ plate at 45° produces circularly polarized light and the second turns it back to the original plane polarized state. A birefringent sample will give the same image as in crossed linear polars, but without the effects of extinction. The image is the same for all rotations of the specimen stage. The quarter wave plate is also used to determine the sign of birefringence, by its effect on the polarization color of the object.

Commercially available compensators with variable retardation are of the following types.

(1) The *quartz wedge* with range of retardation 2000 nm.
(2) The *Babinet* compensator is two quartz wedges of opposite sign mounted one above the other. One is fixed, the other movable with a micrometer. Range is 2000 nm.
(3) The *Berek* (or *Ehringhaus*) rotary compensator is a disc of uniaxial material cut on the basal plane (normal to the optic axis) and mounted on a ring which can be tilted about a horizontal axis. The retardation increases with tilt. Range is 2000 nm if quartz, over 50 000 nm if calcite.
(4) The *Elliptic* compensator is an accurate $\lambda/10$ or $\lambda/30$ plate that can be rotated about a vertical axis. Range is 50 nm or 20 nm.
(5) The *Senarmont* compensator has two parts, an accurate $\lambda/4$ plate inserted above the specimen in the parallel 0° position and a rotating calibrated analyzer. Range is 150 nm.

The quartz wedge is very useful for rapid qualitative observations of the sign of birefringence. Some wedges are calibrated in retardation, but quantitative measurements are best made with the named devices in the list. No compensator can deal with a sample retardation outside its range, but if the range is too large, accuracy is low. Thus the retardation of the specimen largely determines which compensator to be used. The principles of the Babinet and Berek devices are straightforward. The Senarmont and elliptic compensators act indirectly, and the Poincaré sphere construction is needed for their explanation [23, 65–67]. With accurately calibrated equipment, the elliptic compensator can measure retardations of less than 0.1 nm [72].

Accurate compensation requires monochromatic light because of dispersion in the sample and compensator, but in monochromatic light $R(net) = 0$ cannot be distinguished from $R(net) = \lambda$. Examination in white light is required first, as in white light normally only $R(net) = 0$ is black. Dispersion alters the polarization color sequence, and makes compensation in white light inexact. If the birefringence changes significantly with λ, as it does in several polymers, the band of $R(net) = 0$ is not black. There may be a black band at another retardation [23, 73]. Determining birefringence by the standard method of judging compensation by a black band in white light will then give the wrong answer. The black fringe will 'jump' one position every 6λ of retardation if the sample is PET and the compensator is quartz [73].

Three solutions to this problem have been given by Hartshorne [73].

(1) Use a compensator made from a material similar to that of the sample.
(2) Determine the dispersion of birefringence by measurement on a thin sample over a range of wavelengths, then calculate the effect of dispersion on polarization color.
(3) Prepare a wedge at the edge of the sample, and follow the zero net retardation fringe to the thicker part of the sample as compensation is increased.

One can also measure the principal refractive indices directly and take the difference, but this is not very accurate.

3.5 RADIATION EFFECTS

3.5.1 Effect of radiation on polymers

Some photo-sensitive polymers are affected by all visible light, so that optical microscopy is not possible without chemically changing the specimen. Others are sensitive only to blue light, so the use of 'safelight' colors, red or yellow, permits optical observation without damage to the specimen. However, these are unusual problems, largely confined to lithographic materials. The major concern here is with the effect of radiation on polymers in the electron microscope, for a high energy electron beam alters all organic materials to some extent. Hughes [74] is a basic introduction to radiation chemistry. There are several reviews of radiation damage of organic materials in microscopy [75–77]; reviews of polymer microscopy also deal with the subject [46, 78].

Electron micrographs of polymers cannot be interpreted usefully without considering the possibility that radiation damage has changed the image in some way. It is common to see considerable changes in the image during observation of polymers in the electron microscope, due to radiation damage. Absence of easily visible changes does not prove that there is no damage.

The high energy electrons in the beam of the TEM or SEM interact with electrons in the material as they pass through. They typically transfer tens of electron volts of energy to an electron at the site of the interaction. X-rays or γ-rays act in the same way, locally depositing enough energy to break many chemical bonds. High energy electrons can also interact with atomic nuclei in the sample, knocking them out of position [79–81]. This is not an important damage mechanism in organic materials.

The energy given to the excited electron allows it to leave its atom. The charged defect left behind in organic materials is a positive molecular ion in an excited state, which may dissociate into free radicals [82]. If the ejected electron has enough kinetic energy, it will collide with other electrons near the site and ionize nearby atoms, producing a high concentration of reactive species in a small volume. Most excited species will very rapidly recombine, in 10^{-9} to 10^{-8} s. Many of these recombination events will reform the original local chemical structure and dissipate the absorbed energy as heat. But some will form new structures, breaking chemical bonds and forming others. If the material is initially crystalline, defects will form and eventually it will become amorphous.

The fast processes and intermediate states in radiation chemistry, [83, 84] are certainly not required knowledge for polymer microscopists. The purpose of giving this short description of the details is to show that even in a simple system the processes of radiation damage may be extremely complicated. Mechanisms of radiation induced chemical change of organic molecules in the solid state include energy transfer, diffusion and free radical reactions [82, 85, 86]. Study of the behavior of many different compounds has enabled good estimates to be made of the relative radiation stability and the radiolytic products of different chemical groups. When organic molecules are irradiated, specific bonds or types of bonds are likely to be disrupted. These are not always the least stable energetically; for example in *n*-paraffins, the C–H bonds break much more often than the C–C bonds.

Table 3.6 indicates which bonds break in typical polymers, with associated 'G values' for the specific products of the reaction. The number of units of the product P formed for every 100 eV of radiation absorbed by the specimen is $G(P)$. $G(P)$ depends on the exact chemistry of the molecule, the physical state and the temperature of the material. It may also depend on the type of radiation and the dose rate. If radiation breaks a bond in a polymer that is part of the main chain (m in Table 3.6) it will undergo *degradation* or *scission* ($G(S)$) to products of low molecular weight. If the bond that breaks is part of a side group (s in Table 3.6) of the polymer, a small fragment will be lost. This free valence may bond to another chain, so the two chains become joined by *cross-linking* ($G(X)$). The product has a higher molecular weight than the starting material.

Aromatic compounds are much less sensitive to radiation than aliphatic ones. This has been explained in terms of delocalized excitations, where the excitation energy is spread out around the aromatic rings. This is less likely to cause bond breakage than when the energy is localized on one bond [87]. A phenyl group in a compound can reduce the sensitivity of other chemical groups over 1 nm away [88], so even a few aromatic groups can make a material more radiation resistant. The presence of small amounts of oxygen can have a large effect on the radiolytic yields, due to the formation of peroxides.

G values generally rise at high temperature, are lower in the solid than the liquid state, and lower still if the solid is crystalline. This can be understood in general terms by considering the constraining effect of the surrounding molecules. In a crystal all the surroundings are fixed, and trap the excited atoms very close to their original positions. They are therefore likely to reform the original chemical structure, and the yield of changes will be lower. At high temperature the atoms are on average further away from their equilibrium positions, and less likely to reform bonds as before. This behavior is often called the 'cage effect', as the surroundings act as a cage for the excited species. The opposite effect can also occur, if a bond is sterically hindered. Thus a fully substituted carbon atom is likely to break one of its C–C bonds on irradiation. The substituent groups can get in the way of excited species and so limit recombination.

This general description of the behavior of organic compounds applies to organic polymers when they are irradiated. The radiation chemistry of polymers is described in detail in several texts [82, 86, 89–91].

Polymers are generally divided into two main groups, those that cross-link and those that degrade, for although many polymers do both, one of these processes generally dominates. The group that they belong to depends primarily on the chemical structure of the polymer. For example polyethers will degrade because the C–O–C linkage is easily broken by irradiation.

Table 3.6 Bond breaking in typical polymers

Polymer	Formula unit	G values and bonds broken
Polypropylene	$$\begin{array}{c} H \\ | s \\ -CH_2 \underset{m}{=\!=} C - \\ | \\ CH_3 \end{array}$$	$G(H_2) = 2.8$, $G(S) = 0.2$, $G(X) = 0.16$ s, m, s
Poly(vinyl chloride)	$$\begin{array}{cc} Cl & H \\ s1| & |s2 \\ -CH - & CH - \end{array}$$	$G(HCl) = 13$, $G(X) = 0.1$ s1,2, s2
Polyethylene	$$\begin{array}{cc} H & H \\ s| & |s \\ -CH \underset{m}{=\!=} & CH - \end{array}$$	$G(H_2) = 3.7$, $G(S) = 0.2$, $G(X) = 1.0$ s, m, s
Polyoxymethylene	$$\begin{array}{c} H \\ |s \\ -O \underset{m}{=\!=} CH - \end{array}$$	$G(H_2) = 1.7$, $G(S) = 11$, $G(X) = 6$ s, m, s,m
PMMA	$$\begin{array}{c} MeAcO \\ |s \\ -CH_2 \underset{m}{=\!=} CH - \\ | \\ CH_3 \end{array}$$	$G(\text{ester}) = 2.5$, $G(S) = 3.5$ s, m
Polystyrene	$$\begin{array}{cc} H & H \\ s| & |s \\ -C \underset{m}{=\!=} & C - \\ s| & | \\ H & Ph \end{array}$$	$G(H_2) = 0.03$, $G(S) = 0.01$, $G(X) = 0.03$ s, m, s
PTFE	$-CF_2 \underset{m}{=\!=} CF_2 -$	$G(S) = 0.3$ m
PDMS	$$\begin{array}{c} H \xrightarrow{s2} CH_2 \\ |s1 \\ -O - Si - \\ |s1 \\ H \xrightarrow[s2]{} CH_2 \end{array}$$	$G(H_2, CH_4) = 3.0$, $G(X) = 3.0$ s1,2, s1,2
Polyisobutylene	$$\begin{array}{c} CH_3 \\ |s \\ -CH_2 \underset{m}{=\!=} C - \\ |s \\ CH_3 \end{array}$$	$G(H_2, CH_4) = 2.1$, $G(S) = 4.0$ s, m
Polybutadiene	$-CH_2 \underset{s}{=\!=} CH \underset{m}{=\!\!=\!\!=} CH - CH_2 -$	$G(C{=}C) = -12$, $G(X) = 2.0$ m2, m2

The specimen in the electron microscope receives very large doses of radiation, so that degradation does not stop with a polymer of reduced molecular weight. It continues until the fragments are small enough to volatilize in the vacuum of the microscope, and leave the specimen. In an extreme case there will be a loss of all the material irradiated in a thin specimen, leaving a hole.

Continued cross-linking in the other group of polymers produces an infusible, insoluble, brittle solid [92] of high carbon content. When both processes occur, there will be significant loss of mass, and the residue will be this cross-linked char. There may be some residue even from a sample supposed to degrade completely, because it changes chemically so much during irradiation. More generally, the results of radiation chemistry experiments at low doses, such as *G* values, cannot always be applied to electron microscopy.

3.5.2 Radiation doses and specimen heating

It has been remarked above that electron microscopy produces a large radiation dose in the specimen. The standard unit of radiation dosage used from 1959 until recently, was the *rad*. This is defined as the absorption of 100 erg of energy from ionizing radiation per gram of material irradiated. This unit of *absorbed dose* is used because usually the chemical effect of radiation depends only on the absorbed energy per unit mass. It is independent of the type of radiation or its rate of application.

As the erg and gram belong to the old CGS system of units, the rad has been replaced by the *gray*, Gy. The gray is a SI unit, defined as the absorption of 1 joule of ionizing radiation per kilogram of material, so that 1 Gy = 100 rad. Absorbed doses may also be quoted in units of 100 eV absorbed energy per gram. This seems strange, and gives a very large number. However, this unit multiplied by the *G* value gives the concentration of radiation products in the sample.

Uniform radiation fields can be measured absolutely in such absorbed energy units by calorimetry, but it would be very difficult to measure the energy absorbed in the tiny illuminated region of a TEM specimen. The case of the SEM is even more complicated, as the dose will vary strongly with depth. To avoid these problems, the dose in the electron microscope is normally defined as an *incident dose* or electron flux, in coulombs per square meter (C m^{-2}), or electrons per square nm, (e nm^{-2}), where 1 e nm^{-2} = 0.16 C m^{-2}.

A picoammeter measuring the current *I* nA collected by the fluorescent screen of area *A* cm^2, easily gives the incident dose in the TEM. If the magnification is *M* in thousands and the exposure time is *s* seconds, the dose is 10 *s I M^2*/(*wA*) C m^{-2}, where *w* is the collection efficiency of the screen, typically 0.6–0.8 for 100 keV electrons [93, 94]. The current should be measured with the specimen removed from the beam, so that no electrons are scattered and all are collected. The factor *w* may be determined by inserting a Faraday cup (which has a collection efficiency *w* = 1) and comparing the collected currents. In the SEM a specimen current detector will give the incident dose as 0.1 *s I M^2*/*w* C m^{-2}, assuming a display screen of 10 × 10 cm. *w* will depend on the nature of the sample and the beam voltage.

The conversion factor between incident electron dose in C m^{-2} (or e nm^{-2}) and absorbed dose in rad or Gy, depends on the rate of energy loss of the incident 100 keV electrons. This has been calculated to be 400 to 450 eV μm^{-1} for a material that is largely carbon and has a specific gravity of 1, (a model material for organic polymers) [7, 14–16, 95]. If the specimen is very thin, a significant fraction of the energy transferred to it is lost by the escape of secondary electrons. Secondaries escape from a 10 nm surface layer, so this is the thickness where the effect is important. An approximate calculation for 100 keV incident electrons indicates that at 10 nm thickness, nearly half the energy is lost and 250 eV μm^{-1} is absorbed in the sample [16].

Using the figure of 400 eV μm^{-1} for a single 100 keV electron in the TEM, 1 coulomb deposits 400 J μm^{-1}, so 1 C m^{-2} deposits 400 J in 0.001 kg and is therefore equivalent to 400 kGy (40 Mrad). Similarly, 1 e nm^{-2} is equivalent to 64 kGy or 6.4 Mrad. Remembering that the lethal dose for humans is only 6 Gy (600 rad), these already sound like extremely high doses. For high resolution studies of stable materials, an incident flux of 1 A cm^{-2} is common, which translates into 4 GGy s^{-1} or 4×10^{11} rad s^{-1}.

This energy deposition rate, 4 GGy s^{-1}, corresponds to a temperature increase of ten million degrees per second for a thermally isolated material of specific heat 0.4 J g^{-1} K^{-1}. Irradiation of a large piece of material at a dose rate of 1 GGy s^{-1} would result in its vaporization in a fraction of a second. Apart from transmission electron microscopes, only nearby nuclear explosions and devices that give short pulses of irradiation for the study of fast radiation processes [96] produce such high dose rates. It is important to realize that continuous irradiation in the electron microscope, even at such dose rates, does not have to cause a large rise in the temperature of the specimen. This is because a very small object like the illuminated area in a TEM has a large surface area per unit volume, and so it is efficiently cooled by thermal conduction into the rest of the specimen.

There have been many calculations of the equilibrium temperature rise of a specimen in the beam of a TEM, usually for the ideal case of perfect thermal contact with the surroundings [7, 97, 98]. The temperature rise is approximately proportional to the beam current. For a polymer film firmly mounted on a 200 mesh grid and irradiated with 100 kV electrons, the rise is 1–3 K nA^{-1} [7, 16]. Thus a focused spot 2 μm in diameter with the beam current density used for high resolution, 1 A cm^{-2}, has a total current of 3×10^{-8} A and a temperature rise of 30–90°C.

Polymer microscopy normally uses a beam current ten or a hundred times less than that used in the example above, so the temperature rise is small. Temperature measurement to confirm this is difficult, but TEM observation of low melting point (40–50°C) crystals is convincing [99]. On the other hand, poor thermal contact with the support or a restricted thermal path, as when an

unsupported needle crystal or fiber is illuminated, can produce very high temperatures [100].

Heating effects are more difficult to calculate for scanning microscopes. In the STEM the mean dose rate will be similar to that in the TEM, for a similar magnification, but the focused spot sweeping over the specimen will give short pulses of even higher dose rates. SEM produces smaller temperature increases than TEM because energy is deposited in a thin surface layer and can be conducted away into the depth of the sample [101–103]. Even with beam currents in the microamp range it is not possible to melt the surface of polymer blocks with an SEM. Irradiation of fibers of low thermal conductivity changes this geometry and could produce high temperatures in the SEM.

The complete loss of energy and lateral spreading of incident beam electrons in the interaction volume in the SEM makes a proper conversion of C m^{-2} into Gy almost impossible. The incident dose J C m^{-2}, and the beam voltage E volts, may be combined with the penetration depth p μm to give an estimate of mean absorbed dose as 1000 JE/p Gy.

3.5.3 Effects of radiation damage on the image

3.5.3.1 *Mass loss*

Every bond breaking process produces low molecular weight fragments, and in thin samples these will rapidly diffuse to the surface and evaporate. Bubbles may form at high dose rates in thick specimens when volatile products are trapped, and this is often incorrectly taken to mean that the specimen temperature is high. The mass of the specimen thus decreases during observation; polymers that degrade will lose a lot of mass, and those that cross-link will lose less. In TEM the mass thickness falls and the bright field transmitted intensity increases. The increase of intensity with dose has often been measured [104, 105]. Calibration with films of known thickness allows the mass loss to be calculated if the film composition does not change too drastically. In the SEM the unirradiated bulk of the sample always acts to stabilize the form of the irradiated surface. Mass loss may cause a uniform depression of the sample surface or holes and cracks may appear.

Changes of chemical structure and loss of light elements during electron beam irradiation can be detected with electron energy loss spectroscopy [106–108]. Loss of heavier elements from organic material has been studied by x-ray microanalysis, mostly for its effect on the accuracy of elemental analysis in biological samples [109, 110]. Direct chemical analysis of thin polymer films of large area irradiated with 75 keV electrons at low dose rates has also been used [15, 74]. These results show that the chemical composition and the mass both change for the first 0.1 to 1 C cm^{-2} and then stabilize. The final composition contains more carbon and nitrogen than the initial material and less oxygen, hydrogen and halides. The approach to a steady state makes sense, as groups susceptible to scission and removal are removed during irradiation leaving a more stable material behind.

In a favorable case, the final composition may not be much different from the starting composition. Polystyrene, of elemental composition (CH)$_x$ loses only 15–20% of its total mass before stabilizing at a composition near (CH$_{0.8}$)$_x$ This allows useful quantitative measurement of mass thickness even after long periods of irradiation [42]. In contrast, poly-(oxymethylene), (–CH$_2$–O–) retains only 15% of its mass, and the images after heavy irradiation are normally useless.

Generally mass loss destroys the sample to a greater or lesser extent, but it can be regarded positively, as an etching process using the electron beam, and put to use in a few cases. If filler particles are the objects of study, mass loss in the beam can be a convenient method of removing the matrix, though the state of agglomeration may be disrupted. A more sophisticated idea is to use the differential mass loss of two polymers in a blend to increase the visible distinction between the two phases. Even in a homopolymer the cracks and voids that appear can sometimes be related to the microstructure, making it more visible. For example, cracks form

in molded poly(oxymethylene) samples during irradiation in the SEM [111] which follow spherulite radii and show up the oriented skin.

3.5.3.2 *Loss of crystallinity*

Most chemical changes, and particularly cross-linking, will alter the inter-chain spacing of polymer molecules. In amorphous polymers this alteration can be observed as a change in the electron diffraction pattern [112]. It has no special effect on the image, which is more affected by mass loss and chemical changes [113]. If the polymer is initially crystalline it is a different story. The random insertion of chemical changes ruins the regularity of the chains and the crystallinity is destroyed. In some polymers this is a slow process, but most quickly become amorphous under normal viewing conditions in the TEM. The original crystalline electron diffraction pattern may contain spots, arcs or sharp rings depending on the perfection of orientation of the sample. Whatever the form of the original pattern, radiation damage will transform it into diffuse rings. The local molecular orientation may persist to very high doses, and so may differences in density between regions originally crystalline and regions originally amorphous.

The dose required to change the diffraction pattern into diffuse rings, J_a, has often been used to determine the radiation sensitivity of materials in the microscope [7, 14, 114, 115]. J_a is a good general indication of sensitivity and its measurement has been used to determine the effects of beam voltage, specimen temperature and other variables on damage rate. J_a will not always be appropriate; if the project is to study chemical composition then measurement of changes in EELS or x-ray spectra should be used. If high resolution of the crystalline structure is required, J_a is an over-estimate, and the decay of the relevant diffraction spots is more appropriate [116].

The change in the diffraction pattern takes place in two ways, seen most clearly when single crystals are used to give an initial sharp spot pattern (Fig. 3.15). In some polymers the sharp

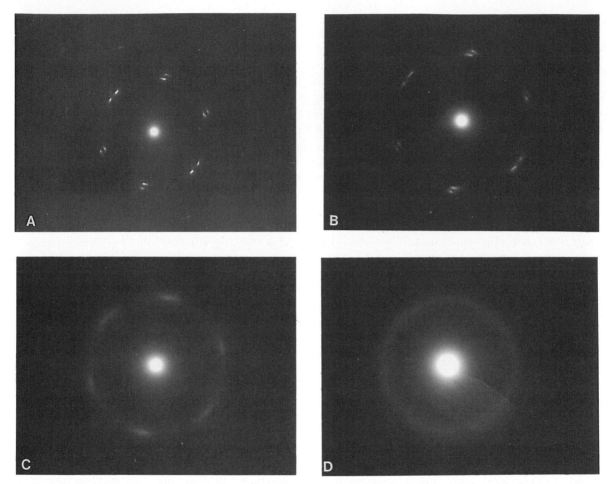

Fig. 3.15 Sequence of electron diffraction patterns from a polyethylene crystal at 100 kV, showing how the sharp spots fade and spread so the final result is a ring pattern. The crystals have become completely amorphous because of radiation damage. The doses are (A) 35–27 C m^{-2}; (B) 53–55.5 C m^{-2}; (C) 70.5–74 C m^{-2}; (D) 123–130 C m^{-2}.

spots simply fall in intensity without changing their position or width and disappear into a diffuse background. In others the spots spread out and shift their position as they become less intense [114, 115]. Poly(oxymethylene) belongs to the first class and polyethylene belongs to the second. From these examples it seems to be simple: materials that degrade lose mass but are otherwise unchanged, so their diffraction patterns lose intensity. Crystals in materials that crosslink are distorted and so the diffraction spacings and peak breadth change. This is generally true, but is not the full story [14, 117].

Loss of crystallinity causes all diffraction contrast features in the TEM image to fade away. Moire fringes, lattice fringes, bend contours and the like will all lose contrast during irradiation [118]. Features that depend on orientation such as bend contours or dislocation strain field images will become smeared out, as directions in imperfect or very small crystals are less well defined (the reciprocal lattice spot increases in size). During irradiation, new contrast features – radiation artifacts – can appear temporarily and then fade with the rest.

Some of these artifacts are simple to explain as

due to the loss of definition of orientation. If a crystal is slightly misoriented and does not diffract, it can diffract after some degree of damage. This could also lead to a temporary increase of contrast for diffraction contrast structures initially present. Other new features, often the more important, are due to inhomogeneous damage processes or motion of the crystals. For example, polyethylene crystals become covered with fine lines or speckles in dark field which mask real features [46, 78].

3.5.3.3 *Dimensional changes*

Large dimensional changes can be induced by radiation that distorts the object and changes the image permanently. This is a serious effect in polymer microscopy because it can be accidentally overlooked. Loss of mass is difficult to miss, and the transient crystallographic contrast details require the operator to be concerned about radiation damage. But the distortion may be instantaneous under the viewing conditions normally used, and the result is a stable and maybe misleading image. If the beam intensity is reduced the changes can be seen, but now recorded images are smeared and useless because the specimen is moving. One must go to very much lower doses to catch the specimen before distortion starts. In some cases this is not worth the effort, for the distorted image can have a high contrast that is in some way related to the original microstructure [119, 120]. Thus as long as the fact of the distortion is recognized, and high resolution details are not pursued, much useful information can be obtained from the 'damaged' specimen.

The distortion of crystalline regions has been described in greatest detail for polyethylene. Large scale changes in the shape of thin films in the TEM and SEM have been related to changes in shape of lamellar crystals due to cross-linking [121, 122]. Similar effects are seen in isotactic polystyrene, natural rubber and nylon, all of them cross-linking materials [123–125]. When thick specimens are used in the SEM, the underlying material prevents large scale distortions.

Changes are seen, similar to those in the TEM, but on a fine scale, less than the penetration depth of the electrons. In spherulitic polyethylene this causes initially smooth surfaces to become full of fine structure-related detail [122, 126–128].

3.5.4 Noise limited resolution

As described in Section 3.2.3, when a finite number of electrons N are used to form an image, there will be a noise limited resolution. Consider a radiation dose of J e nm^{-2} incident on a feature of size d nm, contrast C. If f is the fraction of the incident electrons that contribute to the image, there will be $N = d^2 Jf$ electrons in the image of that feature. The statistical noise is $d(Jf)^{1/2}$ and the signal is $C(d^2 Jf)$. If the signal to noise ratio k is required to be five or more, then $k = Cd(Jf)^{1/2}$, $d = k/C(Jf)^{1/2} \geq 5/C(Jf)^{1/2}$.

For high resolution J must be large, but if the feature is radiation sensitive, J cannot exceed the dose required to destroy it. J_a, the dose required to make the material amorphous, will be the upper limit for all crystallographic contrast features. If the image contrast simply decays without artifacts, distortion or drift, an exact optimum incident dose can be calculated. As these conditions are rarely if ever met, it is much better to try a sequence of low magnification images at low and increasing doses. The maximum usable dose is certainly no more than that where radiation induced changes visibly affect the image features of interest. A study of polyethylene single crystals showed that the artifacts swamped the real structures at 300 e nm^{-2}, 60% of the dose that would be estimated from the decay of diffracted intensities [129].

The best resolution will be obtained when $Cf^{1/2}$ is a maximum, and $Cf^{1/2}$ depends on the imaging mode as well as on the specimen feature. Much of the calculation of resolution limits for radiation sensitive materials becomes a calculation of $Cf^{1/2}$. Calculations have been made assuming that the feature of interest is an isolated fragment of a polyethylene crystal. Reasonable estimates of C and f led to a resolution limit of 5 nm for both bright and dark field in the TEM at 100 keV [129]. Similar values have been obtained for biologically important molecules [130]. Lattice images can be obtained with higher resolution, see Chapter 6.

If the TEM image is recorded on film with a resolution of 100 μm, limited by the focusing screen, the magnification needs to be 20,000 $\times$ or more to see 5 nm structure in the object. For a structure like the polyethylene crystal that can only take 300 e nm^{-2}, the incident electron dose at the film can be no more than $(300f/20,000^2)$ e nm$^{-2} = 0.75f$ e μm^{-2}. The film normally used to record TEM images is very efficient but requires an electron flux of about 1 e μm^{-2} to form an image of optical density one [131]. Thus when f is close to one, as it is in bright field imaging of thin PE crystals, the match of properties is quite good. For small f, as in dark field, the film will be seriously underexposed. Therefore either a more sensitive film must be selected, or more of the resolution capability of normal film must be used. More sensitive films are usually fine-grained x-ray film such as DuPont lo-dose mammography® film, which requires only about 0.2 e μm^{-2} for an optical density of one. The resolution of the faster film is lower, but still better than the usual focusing limit. If the full resolution of the standard film, 20 μm or better, can be used a magnification of only 4000 $\times$ is required for 5 nm image resolution. The flux on the film is then $12f$ e μm^{-2} and dark field images with $f = 0.08$ can be recorded.

A similar calculation for scanning electron microscope images has the same basic equation, $d_{\min} = 5/C(Jf)^{1/2}$. Arguments for the superiority of STEM over CTEM in the imaging of radiation sensitive materials relates to the different values of C and f expected. These considerations are by no means simple [132] and depend on the type of object being imaged. If the probe size is set to be the same as the radiation limited resolution, then the number of electrons in each image point is $Jd^2 = 25/C^2f$. The resolution on the final screen is 10 cm$/1000 = 100$ μm so that the required magnification is $10^5/d$, where d is in nanometers.

These calculations assume that all the electrons

containing useful information form a single image. If m images of the same area are required, the number of electrons available for each falls to J/m. The resolution becomes $5m^{1/2}/C(Jf)^{1/2} = dm^{1/2}$ and the required magnification falls by the same factor. Thus 16 dark field images of polyethylene crystals could be taken at a magnification of 1000, with an ideal resolution of 20 nm [133].

In normal microscope operation, a large flux is used to focus the image and adjust the specimen position. This would be enough to destroy many sensitive samples, and must be avoided. An image intensifier is no more efficient at detecting electrons than photographic film, so it cannot improve the resolution of the recorded image. It does produce a visible image at a very low dose rate. An image intensifier or the equivalent TV imaging device therefore allows focusing and alignment at low doses without straining one's eyes over a dim fluorescent screen in a well-darkened room. Another procedure is to use the beam deflecting system in the TEM to focus and align on a nearby area and then switch the illumination back to record the image [134–136]. This low dose operation is now much easier to set up with computer control of the microscope alignment, and is offered as an option by microscope manufacturers. A very small irradiation may be used to check that an area of interest lies in the area to be imaged, or it may be recorded 'blind'. The latter approaches the ideal of using the maximum number of electrons to form the image, at the cost, perhaps, of many attempts before a useful micrograph is recorded.

3.5.5 Image processing

Resolution gets worse if the available electrons have to be shared by several images; on the other hand, the resolution is improved if the dose received by a large area is combined to form a small image. This is a routine procedure when the image has a regularly repeating structure, such as lattice fringes in a crystal. It has also been done for irregular repeating structures (single biological macromolecules) [137], but for synthetic polymers, only regular structures can be handled at present.

Consider the case when the lattice structure is imaged with incident dose J, but hidden by noise, and an area containing $r \times r$ repeating units is in the image. Adding all the r^2 unit images will give a signal r^2 greater than that for a single unit image, but the noise will largely cancel. The random background will increase by r^2, but the noise will increase only by r and the signal to noise ratio will increase by a factor r. Thus the ratio required in the original image is only $5/r$ and the noise limited resolution becomes $d(r) = 5/rC(Jf)^{1/2} = d_{min}/r$. In principle a very large number of repeating units could be added to give resolution at the limit of the instrument. This would only be possible if the crystal had the same orientation over the whole area, and the image could record the large number of units at high resolution.

The individual images could be added directly, in 'real space', by shifting and adding the digital stored image. The difficulty here is in knowing what shifts to apply to the noisy images. It is much easier in practice to add the images in frequency or reciprocal space. This was originally done by creating and modifying optical diffraction patterns of the image [138], but it is now done by computation of Fourier transforms of the images. The 2-D Fourier transform of an image, like its optical diffraction pattern, is a map of the frequencies present in it. A weak and noisy image of a lattice will produce a map that contains weak but sharp spots corresponding to the regular structure, and diffuse intensity from the noise. The spots are visible even when the structure is not, because they are localized in a few areas. The diffuse intensity is suppressed by multiplying the transform with a filter function that is one at the spots, and zero well away from the spots. An inverse Fourier transform then recreates an image that contains all the signal, and a reduced amount of noise.

If the filter function cuts off very close to the spots, then the reformed image contains only the periodic information, with any local defects suppressed along with the noise. This gives a clean picture of the lattice structure, but in most polymer systems the lattice is well known, and some level of localized information is the purpose of microscopy. If the lattice contains defects such as dislocations, these can also appear in the reconstructed image, but their exact appearance and position can be affected by the choice of filtering aperture and imaged area [139, 140]. In the extreme case, it is easy to show that filtering of pure noise can produce good 'lattice' images. The technique must be treated with caution.

The Fourier transform of an image is frequently used in high resolution TEM to determine the defocus, astigmatism and true resolution [13]. Polymer microscopists should realize that they too are also often working under conditions where simple interpretation is dangerous, because of noise or deliberate large defocus. Image interpretation can often benefit from a check of this type, even when the absolute spatial resolution is much worse than the instrumental limit. If the exact defocus and aberration of an image are determined, the image from an assumed structure can be calculated and compared to the experimental image. This image simulation is a standard procedure in HREM [13, 141], and again it may be used for polymer images at lower resolution.

REFERENCES

1. W. T. Welford, *Optics* (Oxford University Press, Oxford, 1988).
2. F. A. Jenkins and H. E. White, *Fundamentals of Optics*, 4th Edn (McGraw Hill, New York, 1976).
3. L. C. Martin, *The Theory of the Microscope* (Blackie, London, 1966).
4. S. G. Lipson and H. Lipson, *Optical Physics* (Cambridge University Press, Cambridge, 1969).
5. M. Born and E. Wolf, *Principles of Optics*, 6th Edn (Pergamon, Oxford, 1980).
6. M. Spencer, *Fundamentals of Light Microscopy* (Cambridge University Press, Cambridge, 1982).
7. L. Reimer, *Transmission Electron Microscopy, Physics of Image Formation and Microanalysis*, 2nd Edn (Springer, Berlin, 1989).
8. R. D. Heidenreich, *Fundamentals of Transmission*

Microscopy (Wiley, New York, 1964).

9. C. E. Hall, *Introduction to Electron Microscopy*, 2nd Edn (McGraw Hill, New York, 1966).

10. J. J. De Palma and E. M. Long, *J. Opt. Soc. Amer.* **52** (1962) 328.

11. J. I. Goldstein, D. E. Newbury, P. Echlin, D. C. Joy, A. D. Romig Jr., C. E. Lyman, C. Fiori and E. Lifshin, *Scanning Electron Microscopy and X-ray Microanalysis*, 2nd Edn (Plenum, New York, 1992).

12. K. J. Hanszen, in *Advances in Optical and Electron Microscopy*, Vol. 7, edited by R. Barer and V. E. Cosslett (Academic Press, London, 1971).

13. P. Buseck, J. Cowley, L. Eyring, Eds, *High-Resolution Transmission Electron Microscopy and Associated Techniques* (Oxford University Press, Oxford, 1988).

14. H. Von Orth and E. W. Fischer, *Makromol. Chem.* **88** (1965) 188.

15. K. Stenn and G. F. Bahr, *J. Ultrastruct. Res.* **31** (1970) 526.

16. D. T. Grubb, *J. Mater. Sci.* **9** (1974) 1715.

17. I. G. Voigt-Martin, *Advances in Polym. Sci.* **67** (1985) 196.

18. J. Buerger, *Elementary Crystallography* (Wiley, New York, 1963).

19. B. K. Vainshtein, *Modern Crystallography I*, Springer Series Solid-State Science, Vol. 15 (Springer Verlag, Berlin, 1981).

20. International Union of Crystallography, *Fundamentals of Crystallography*, edited by C. Giacovazzo (Oxford University Press, Oxford, New York, 1992).

21. C. Hammond, *Introduction to Crystallography*, Royal Microscopical Society Microscopy Handbook Number 19 (Oxford University Press, Oxford, 1992).

22. A. G. Jackson, *Handbook of Crystallography for Electron Microscopists and Others* (Springer Verlag, New York, 1991).

23. N. H. Hartshorne and A. Stuart, *Crystals and the Polarizing Microscope*, 4th Edn (Elsevier, New York, 1970).

24. P. Gay, *An Introduction to Crystal Optics* (Longmans, London, 1967).

25. L. H. Schwartz and J. B. Cohen, *Diffraction from Materials* (Springer Verlag, Berlin, 1987).

26. J. M. Schultz, *Diffraction for Materials Scientists* (Englewood Cliffs, New York, Prentice Hall, 1982).

27. J. M. Cowley, *Diffraction Physics*, 2nd Edn (North Holland, Amsterdam, 1981).

28. International Union of Crystallography, *Electron Diffraction Techniques*, edited by J. M. Cowley (Oxford University Press, Oxford, New York, 1992–1993).

29. J. C. H. Spence and J. M. Zuo, *Electron Microdiffraction* (Plenum Press, New York, 1992).

30. B. Wunderlich, *Macromolecular Physics*, Vol. 1 (Academic Press, New York, 1973).

31. P. Allen and M. Bevis, *Proc. Roy. Soc.* **A341** (1974) 75.

32. A. J. Lovinger and H. D. Keith, *Macromolecules* **12** (1979) 919.

33. R. H. Geiss, G. B. Street, W. Volksen and J. Economy, *IBM J. Res. Develop.* **27** (1983) 321.

34. W. Claffey, K. Gardner, J. Blackwell, J. Lando and P. H. Geil, *Phil. Mag.* **30** (1974) 1223.

35. F. Brisse and R. H. Marchessault, in *Fiber Diffraction Methods*, edited by A. D. French and K. H. Gardner. ACS Symposium Series, 141 (ACS, Washington DC, 1980).

36. D. L. Dorset, B. Moss, J. C. Wittmann and B. Lotz, *Proc. Nat. Acad. Sci. USA* **81** (1984) 1913.

37. J. R. Fryer and D. L. Dorset, Eds, *Electron Crystallography of Organic Molecules* (Kluwer Academic Publishers, Dordrecht, Boston, 1991).

38. D. L. Dorset, *Macromolecules* **24** (1991) 1175.

39. L. E. Alexander, *X-ray Diffraction Methods in Polymer Science* (Wiley Interscience, New York, 1969).

40. M. Kakudo and N. Kasai, *X-ray Diffraction by Polymers* (Elsevier, Amsterdam, 1972).

41. C. E. Hall, *J. Appl. Phys.* **22** (1951) 655.

42. B. D. Lauterwasser and E. J. Kramer, *Phil. Mag.* **A39** (1979) 469.

43. P. B. Hirsch, A. Howie, R. B. Nicholson, D. W. Pashley and M. J. Whelan, *Electron Microscopy of Thin Crystals* (Butterworths, London, 1965).

44. O. Scherzer, *J. Appl. Phys.* **20** (1949) 20.

45. J. Petermann and H. Gleiter, *Phil. Mag.* **31** (1975) 929.

46. E. L. Thomas, in *Structure of Crystalline Polymers*, edited by I. H. Hall (Applied Science, London, 1984).

47. Th. G. F. Schoon and R. Kretschmer, *Koll. Z. u Z. Polym.* **211** (1966) 53.

48. G. S. Y. Yeh, *Crit. Rev. Macromol. Chem.* **1** (1972) 173.

49. C. L. Marx, J. A. Koutsky and S. L. Cooper, *J. Polym. Sci. Lett.* **8** (1971) 167.

50. H. P. Zingsheim and L. Bachmann, *Koll. Z. u Z. Polym.* **246** (1971) 36.

51. D. R. Uhlmann, A. L. Renninger, G. Kritchevsky

and J. Van der Sande, *J. Macromol. Sci. Phys.* **B12** (1976) 153.

52. D. J. Johnson and D. Crawford, *J. Microscopy* **98** (1973) 313.

53. D. L. Handlin and E. L. Thomas, *Macromolecules* **16** (1983) 1514.

54. W. G. Hartley, *Proc. Roy. Microsc. Soc.* **9** (1974) 167.

55. L. Reimer, *Scanning Electron Microscopy, Physics of Image Formation and Microanalysis* (Springer Verlag, Berlin, 1985).

56. J. Pawley, *J. Microsc.* **136** (1984) 45.

57. D. W. Tuggle, J. Z. Li and L. W. Swanson, *J. Microsc.* **140** (1985) 293 (part of a special issue on low voltage SEM, *J. Microsc.* **140** (1985) 282–349).

58. L. Reimer, *Image Formation in Low-Voltage Scanning Electron Microscopy* (SPIE, Bellingham, 1993).

59. D. B. Wittry, in *X-ray Optics and Microanalysis, 4th Intl. Cong.* edited by R. Castaing, P. Deschamps and J. Philibert (Herman, Paris, 1966).

60. P. H. Dawson, *J. Appl. Phys.* **37** (1966) 3644.

61. D. C. Joy, *Scanning* **11** (1989) 1.

62. J. J. Hren, J. I. Goldstein and D. C. Joy, *Introduction to Analytical Microscopy* (Plenum, New York, 1979).

63. D. E. Newbury, D. C. Joy, P. Echlin, C. Fiori and J. I. Goldstein, *Advanced Scanning Electron Microscopy and X-ray Microanalysis* (Plenum, New York, 1986).

64. F. Ohnesorge and G. Binnig, *Science* **260** (1993) 1451.

65. G. N. Ramachandran and S. Ramaseshan, *Crystal Optics*, in Handbuch der Physik, Vol. 25/1, edited by S. Flugge (Springer Verlag, Berlin, 1961).

66. C. W. Mason, *Handbook of Chemical Microscopy*, 4th Edn (Wiley, New York, 1983).

67. N. H. Hartshorne and A. Stuart, *Practical Optical Crystallography*, 2nd Edn (Elsevier, New York, 1969).

68. The plane of polarization was originally defined by convention to be the plane that we now know to contain the magnetic field. To avoid confusion when using early references, advanced texts tend to use the term 'plane of vibration' to describe the plane containing the electric field.

69. T. G. Rochow and E. G. Rochow, *Resinography* (Plenum, New York, 1976).

70. W. A. Schurcliff and S. S. Ballard, *Polarized Light* (Van Nostrand, New York, 1964).

71. H. D. Mallon, *J. Microsc.* **19** (1969) 107.

72. M. M. Swann and J. M. Mitchison, *J. Exper. Biol.* **27** (1950) 226.

73. N. H. Hartshorne, *Science Progress* **50** (1962) 11.

74. G. Hughes, *Radiation Chemistry* (Clarendon Press, Oxford, 1973).

75. G. F. Bahr, F. B. Johnson and E. Zeitler, *Lab. Invest.* **14** (1965) 1115 (also available in *Quantitative Electron Microscopy*, edited by G. F. Bahr and E. Zeitler (Williams and Wilkins, Baltimore, 1965)).

76. B. M. Siegel and D. R. Beaman, Eds, *Physical Aspects of Electron Microscopy and Microbeam Analysis* (Wiley, New York, 1975).

77. M. S. Isaacson, in *Principles and Techniques of Electron Microscopy*, Vol. 7, edited by M. Hayat. (Van Nostrand and Reinhold, New York, 1977).

78. D. T. Grubb, in *Developments in Crystalline Polymers*, edited by D. C. Bassett (Applied Science, London and New York, 1982).

79. G. J. Dienes and G. H. Vineyard, *Radiation Effects in Solids*, Ch. 3 (Interscience, New York, 1957).

80. V. E. Cosslett, in *Electron Microscopy and Analysis 1979*, edited by T. Mulvey (Institute of Physics, London, 1979).

81. L. W. Hobbs, *Ultramicroscopy* **23** (1987) 339.

82. M. Dole, Ed., *The Radiation Chemistry of Macromolecules* (Academic Press, New York, 1973).

83. P. Ausloos, Ed., *Fundamental Processes in Radiation Chemistry* (Interscience, New York, 1964).

84. G. E. Adams, E. M. Fielden and B. O. Michael, Eds, *Fast Processes in Radiation Chemistry* (Wiley, New York, 1975).

85. R. O. Bolt and J. G. Carroll, Eds, *Radiation Effects on Organic Materials* (Academic Press, New York, 1963).

86. R. L. Clough and S. W. Shalaby, Eds, *Radiation Effects on Polymers*, ACS Symposium Series, 475 (American Chemical Society, Washington, DC 1991).

87. B. Pullman and A. Pullman, *Quantum Biochemistry* (Interscience, New York, 1963).

88. P. Alexander and A. Charlesby, *Nature* **173** (1954) 578.

89. A. Charlesby, *Atomic Radiation and Polymers* (Pergamon, Oxford, 1960).

90. A. Chapiro, *Radiation Chemistry of Polymeric Systems* (Interscience, New York, 1962).

91. S. Okamura, *Recent Trends in Radiation Polymer Chemistry*, Advances in Polymer Science, Vol. 105 (Springer Verlag, Berlin, 1993).

92. A. Charlesby, *Proc. Roy. Soc.* **A215** (1952) 817.
93. D. T. Grubb, *J. Phys. E; Sci. Instrum.* **4** (1971) 222.
94. L. H. Bolz, D. H. Reneker and K. W. Yee, *J. Phys. E; Sci. Instrum.* **5** (1972) 1037.
95. K. Kiho and P. Ingram, *Makromol. Chem.* **118** (1968) 45.
96. F. M. Charbonnier, J. P. Barbour and J. L. Brewster, in *Fast Processes in Radiation Chemistry*, edited by G. E. Adams, E. M. Fielden and B. O. Michael (Wiley, New York, 1975).
97. B. Gale and K. F. Hale, *Br. J. Appl. Phys.* **12** (1961) 115.
98. J. Ling, *Br. J. Appl. Phys.* **18** (1967) 991.
99. A. Keller, *Phil. Mag.* **6** (1961) 63.
100. K. Kanaya, *J. Electr. Micr. Japan* **4** (1956) 1.
101. L. G. Pittaway, *Br. J. Appl. Phys.* **15** (1964) 967.
102. Y. Talmon and E. L. Thomas, *J. Microsc.* **111** (1977) 151.
103. H. Kohl, H. Rose and H. Schnabl, *Optik* **58** (1981) 11.
104. V. E. Cosslett, in *Proc. Europ. Conf. on Electron Microscopy*, edited by A. L. Houwink and B. J. Spit (Nederlandse Vereniging voor Electronenmikroskopie, Delft, 1960).
105. R. Freemen and K. R. Leonard, *J. Microsc.* **122** (1981) 275.
106. M. S. Isaacson, *Ultramicroscopy* **4** (1979) 193.
107. R. F. Egerton, *Ultramicroscopy* **5** (1980) 521.
108. R. F. Egerton, *J. Microsc.* **126** (1982) 96.
109. T. A. Hall and B. L. Gupta, *J. Microsc.* **100** (1974) 177.
110. H. Shuman, A. V. Somlyo and P. Somlyo, *Ultramicroscopy* **1** (1976) 317.
111. J. W. Heavens, A. Keller, J. M. Pope and D. M. Rowell, *J. Mater. Sci.* **5** (1970) 53.
112. G. S. Y. Yeh, *J. Macromol. Sci. Phys.* **B6** (1972) 451.
113. D. Vesely, *Ultramicroscopy* **14** (1984) 279.
114. K. Kobayashi and K. Sakaoku, *Lab. Invest.* **14** (1965) 1097.
115. D. T. Grubb and G. W. Groves, *Phil. Mag.* **24** (1971) 815.
116. D. L. Dorset and F. Zemlin, *Ultramicroscopy* **17** (1985) 229.
117. G. Ungar, D. T. Grubb and A. Keller, *Polymer* **19** (1980) 1284.
118. The decay of crystallinity is equivalent to an increase in the extinction distance, and this can result in a change from dynamic to kinematic diffraction conditions, with rise or fall of intensity. Since most polymer crystals are much thinner than an extinction distance to begin with, the result is usually a continuous decline of diffracted intensity.
119. F. P. Price, *J. Polym. Sci.* **37** (1959) 71.
120. J. Dlugosz and A. Keller, *J. Appl. Phys.* **39** (1968) 5776.
121. D. T. Grubb, A. Keller and G. W. Groves, *J. Mater. Sci.* **7** (1972) 131.
122. D. T. Grubb and A. Keller, *J. Mater. Sci.* **7** (1972) 822.
123. E. H. Andrews, *Proc. Roy. Soc.* **A270** (1962) 232.
124. C. G. Cannon and P. H. Harris, *J. Macromolec. Sci. Phys.* **B3** (1969) 357.
125. J. Dlugosz, D. T. Grubb, A. Keller and M. B. Rhodes, *J. Mater. Sci.* **7** (1972) 142.
126. J. E. Breedon, J. F. Jackson, M. J. Marcinkowski and M. E. Taylor, *J. Mater. Sci.* **8** (1973) 1071.
127. D. Fotheringham and B. Paker, *J. Mater. Sci.* **11** (1976) 979.
128. S. Bandyopadhy and H. R. Brown, *Polymer* **19** (1978) 589.
129. E. L. Thomas and D. G. Ast, *Polymer* **15** (1974) 37.
130. R. M. Glaeser, *J. Ultrastruct. Res.* **36** (1976) 466.
131. G. C. Farnell and R. B. Flint, *J. Microsc.* **97** (1973) 271.
132. D. L. Misell, *J. Phys. D; Appl. Phys.* **10** (1977) 1085.
133. J. R. White, *Polymer* **16** (1975) 157.
134. R. C. Williams and H. W. Fischer, *J. Mol. Biol.* **52** (1970) 121.
135. K. H. Herrmann, J. Menadue and H. T. Pearce-Percy, in *Electron Microscopy 1976*, edited by D. G. Brandon (Tal International, Jerusalem, 1976).
136. Y. Fujiyoshi, T. Kobayashi, K. Ishizuka, N. Uyeda, Y. Ishida and Y. Harada, *Ultramicroscopy* **5** (1980) 459.
137. M. Kessel, J. Frank and W. Goldfarb, in *Electron Microscopy at Molecular Dimensions*, edited by W. Baumeister and W. Vogell (Springer Verlag, Berlin, 1980).
138. G. Harburn, C. A. Taylor and T. R. Walberry, *Atlas of Optical Transforms* (Cornell University Press, Ithaca, 1975).
139. C. A. Taylor and J. K. Ranniko, *J. Microsc.* **100** (1974) 307.
140. D. C. Martin, K. R. Shaffer and E. L. Thomas, in *Electron Crystallography of Organic Molecules*, edited by J. R. Fryer and D. L. Dorset. NATO ASI Series, p. 129 (Kluwer Academic Publishers, Dordrecht, Boston, 1991).
141. J. C. H. Spence, *Experimental High Resolution Transmission Electron Microscopy*, 2nd Edn (Oxford University Press, Oxford, 1988).

CHAPTER FOUR

Specimen preparation methods

4.1 SIMPLE PREPARATION METHODS

Specimen preparation ranges from direct and simple methods to complex, time consuming and even frustrating ones. Fortunately, there are a number of simple methods which are quite adequate for some materials. For example, many particulate materials may be handled by the simple methods. This section covers a wide range of these more simple and generally direct methods which are described in broad subsections: optical microscopy, scanning electron microscopy (SEM) and transmission electron microscopy (TEM) preparations. It must be emphasized that quick observation of most materials by a combination of a simple microscopy technique and direct preparation methods is often helpful in shedding light on the problem. This aids determination of the best approach to a solution. In many cases there is no one *correct* approach, but there may well be approaches that can save time, if they are conducted early in the study. Tradenames of products used in specimen preparation are mentioned in the text and, unless otherwise stated, these are standard materials available from the suppliers (Appendix V). Specific microscopes are not mentioned but microscope vendors are listed in Appendix VI.

4.1.1 Optical preparations

The single most important preparation instrument in the microscopy laboratory is the stereo binocular microscope. These instruments are inexpensive and readily available. Materials may be observed in either transmitted or reflected light, and the result often provides insights into the problem. Even rather large parts may be examined as part of the important first step in choosing the area of a sample to be analyzed.

Transparent specimens, generally those less than 100 μm thick, may be directly mounted onto standard glass microscope slides with cover slips, for transmitted light observation. For magnifications less than 100 times, this preparation might be sufficient. For higher magnification examination, suitable mounting media are usually required in order to reduce surface reflections. Immersion oils of specific refractive indices, or special mounting media, such as Permount (available from Fisher Sci.), may be chosen to provide contrast between the material and the mountant, in the case of particles, whereas matching refractive index oils may be used with fibers and films to permit observation of internal structures. Fibers, particles, small strips of films and membranes can be prepared in this manner for optical study. Simple reflected light microscopy of large and irregularly shaped specimens can be aided by pressing the underside of the specimen into modeling clay, to provide an even top surface.

A direct method of preparing fluids is to use a cavity slide to permit a known fluid thickness to be examined optically. A crystal suspension may be examined in this way [1, 2]. Solutions or solid materials may be placed in a cavity slide or onto a slide with a cover slip (under an inert or dry atmosphere, if needed).

4.1.2 SEM preparations

A major advantage of the SEM for surface observations is that sample preparation is generally simple. In the simplest case, the material to be examined, chosen carefully from a larger sample, is placed on double sided sticky tape on a specimen stub. In order to maintain contact with the stub for conduction, the tape covers only part of the specimen stub so the sample is placed partly on the tape and partly on the stub. Conductive paints, such as silver or carbon suspensions, are used to attach the specimen to the sample holder. Such paints can also be dabbed onto the tape or the base of the specimen to provide contact with the stub. Care must be taken not to abuse the material during handling. Such simple preparation methods are very successful for fibers, films, membranes and even rather large plastic parts. Preparation is not complete as polymers generally require a conductive coating (Section 4.7.3) to be applied for imaging. A simple method for minimizing charging problems is the use of antistatic sprays such as Duron. Sikorski *et al.* [3] sprayed and soaked materials in Duron. Sprayed droplets of this material can be observed above 1000×.

Hearle *et al.* [4] described several specimen holders that are useful for simple preparations of fibers and fabrics where the sample must just be attached to the holder. Each SEM manufacturer and the various EM supply laboratories provide different microscope stubs and modifications can readily be made in a machine shop.

4.1.3 TEM preparations

Samples for transmission electron microscopy must fit onto a specimen support known as a grid or screen. This grid is a metal mesh screen, generally 2–3 mm in diameter, which fits into the specimen holder of the microscope. Grids come in a variety of mesh sizes and shapes. Sizes in general use range from 50 to 400 mesh, that is from 50 to 400 holes per inch. The grid mesh used is typically square; but for some materials, slotted screens, rectangles, hexagons, or single holes might be preferred. Grids are made of copper, for the most part, but beryllium, gold, polymer and nickel grids are used for various applications, for their chemical resistance and for x-ray analysis. Microtomed sections, which contain polymer embedded in a resin, are usually directly supported on the grid. Very small sections or sections which break apart may require a support to hold them onto the grid. Particles, crystals, emulsions and other fine materials are placed on an electron transparent support film on the TEM grid. The preparation of such support films will be described below.

Specimen preparation for TEM generally involes the formation of a thin film of the material less than 100 nm thick. The methods used for this preparation depend upon the nature of the polymer and its physical form. In the case of thick or bulk specimens, microtomy is generally used. In the case of solutions, powders or particulates, simpler methods can provide a thin, dispersed form of the material. Three types of simple preparations will be described later in this section: dispersion, disintegration and film casting. The more complex methods such as microtomy, replication, etching and staining will be described in other sections of this chapter.

4.1.3.1 *Specimen support films*

Plastic, carbon and metal films (Section 4.7) are used as specimen supports on TEM grids. There are two plastic support materials in use: collodion, 0.5% solution of nitrocellulose in amyl acetate; and formvar, 0.25% polyvinyl formal in ethylene dichloride. These polymers are available as powders, solutions or prepared films on TEM grids. Formvar films, especially holey ones, are used as substrates for the formation of holey carbon films. Today, collodion is not used too often, as it is not as stable in the electron beam as formvar or carbon films.

Premixed solutions of formvar and collodion are recommended for high quality film supports as the powders take several days to dissolve and are less uniform. Collodion is generally film cast on a water surface. A large petri dish filled with distilled water is allowed to settle and collodion is dropped onto the water surface and allowed to dry. The first film cast is used to clean the water surface. The films are placed onto the grids by any one of several methods. In one method, the grids are placed upside down on the film surface and are lifted up with a glass microscope slide placed on their surface and then scooped down into the water and up with the grids on the glass, under the film. The film is allowed to dry and a razor is used to separate the film around the grids. In another method, the grids are placed on a mesh screen below the water surface, and the water is slowly removed allowing the film to settle on the grid surfaces. A sintered glass filter apparatus is useful in this case.

Formvar solutions are placed in any tall glass container that has a lid. Clean glass slides, pretreated with a detergent to aid release of the cast film, are dipped into the solution and then quickly lifted up above the solution, covered and permitted to dry slowly in the solvent vapor. To free the film, the glass slide is scored around the perimeter with a razor, scalpel or needle. The film may then be floated onto a water surface and the grids placed on it and picked up, as for collodion. Alternately, the slide can be scored into about 3 mm squares. Either form of film is slipped into the water by placing the slide at an angle to the surface and slowly immersing it beneath the surface. The small squares of film are picked up individually by placing the grids under the water surface and scooping them up. Coated grids are dried on filter paper and carbon coated to increase their beam stability. Breathing on the slide prior to immersion in the water bath aids release of the film from the glass.

Support films generally used for microscopy above $50,000\times$ magnification are either perforated, holey, carbon coated plastic, or holey carbon films for highest resolution. In the case of holey plastic supports, the presence of moisture

or some other immiscible liquid in the solution used for coating will provide holes in the films. A fuller discussion on this topic is found in the specimen preparation text by Goodhew [5]. In our experience, moisture from the air or from one's breath on the slide causes the formation of holes and, in fact, it may be more difficult to prepare continuous films. These perforated supports provide an area of the specimen supported by the film with other areas suspended over the holes.

4.1.3.2 *Dispersions*

Solutions or suspensions of polymers can be dispersed by several methods including atomizing or spraying, or the fluid might simply be allowed to spread over the support film on the TEM grid. The grids are placed on filter paper to remove the excess material. Preparation of polymer materials suspended in oils requires removal of the oil prior to application of a conductive coating. A droplet of the material in the oil is permitted to spread over the support film. The grids are placed on a fine mesh metal screen in a petri dish containing a solvent at the level just below the screen. The solvent is replenished until the oily deposit is removed. Metal shadowing (Section 4.7.2.2), evaporated at an angle of 30–60° to the specimen, is generally employed to add contrast to the material prior to examination.

Dilute solutions of some high molecular weight polymers can be dispersed directly onto carbon films to provide macromolecules collapsed into spherical shapes, which can be observed in the microscope. The size of the particle is related to the molecular weight. Hobbs [6] described a process for the examination of single molecules of rubbers or glassy polymers which involved the addition of a nonsolvent to aid collapse and the spraying of a fine mist onto a carbon coated TEM grid. Dissolving polyethylene in a 0.005% *n*-hexadecane solution was shown to be successful [7]. Richardson [8] showed that very diluted solutions of polystyrene in *n*-butanol were satisfactory for such studies.

Dilute latex emulsions, with a glass transition above room temperature, can be atomized or sprayed using a fine pipette attached to a standard can of Freon gas. Difficulties arise if the latex is not dilute enough, as clumps of particles are then observed. With low glass transition latexes, cryogenic or chemical hardening methods must be used to ensure the mechanical stability of the polymer particles.

Powders are suspended in a fluid, mixed and dropped onto the plastic or carbon coated support grid. If the powders do not disperse by this method, the powder can be rolled onto a plastic support film on a glass slide, and cut, scored, floated off and picked up. Another variation is that powders may be mixed with the collodion or formvar and a film cast with the material held in the support film.

4.1.3.3 *Single crystal formation*

Single crystals are readily formed by precipitation during cooling of dilute solutions of polymers, such as polyethylene (PE) from xylene. PE single crystals can be prepared for TEM by placing a drop of the crystal suspension, in a solvent for the polymer, on a carbon coated grid and shadowing to enhance contrast. Bassett and Keller [9] used this method to show the effect of temperature on PE morphology. An example of a TEM micrograph of PE single crystals is shown in Fig. 4.1. Bassett *et al.* [10] used this technique to prepare polyoxymethylene in bromobenzene and poly(4-methylpentene-1) in xylene. Single crystals of linear PE were grown isothermally in xylene by self seeding, as shown by Blundell and Keller [11]. Single crystals of poly(tetramethylene adipate) were prepared from dilute solutions in ethanol, *n*-propanol, *n*-butanol and other solvents [12]. In a typical preparation, the solution was maintained at 80°C for 30 min and transferred to an oven at two degrees above the clouding temperature and held 16 h before cooling. A drop of the turbid solution was placed on a carbon coated grid and shadowed. In this study, two different forms of crystals, identified

by electron diffraction, were produced depending on the solvent used.

More recently, Chang *et al.* [13] prepared single crystals of high density polyethylene (HDPE) by casting a xylene suspension on carbon coated glass slides and evaporating the solvent. Carbon platinum shadowed specimens (Section 4.7.2.2) were studied to determine the nature of the chlorination reaction. Single crystals of poly-(*p*-xylylene) have been observed directly in a high voltage TEM [14–16]. Crystals were obtained by dissolving Parylene-N film in chloronaphthalene by heating at 254°C and holding the 0.05–0.1% solution at 210°C overnight, followed by cooling to room temperature. Individual chains in an unknown form of the crystal have been imaged [14] and the structures analyzed [15]. Direct observations have been made of dislocations in poly(*p*-xylylene) [16].

Although early TEM studies of polymer single crystals [17] provided much of the fundamental knowledge of polymers, study of solution grown single crystals does not directly lead to an understanding of the more complex bulk polymer textures, and thus these methods are not commonly used in the industrial laboratory.

4.1.3.4 *Disintegration*

The use of an ultrasonic bath to disintegrate cellulose into fibrils observable in the TEM was described in 1950 [18]. Even earlier papers were referenced, where either mechanical or ultrasonic vibrations were used to provide thin TEM specimens. Morehead [18] cut samples into short lengths and treated them in water in an ultrasonic bath for about 20 min. Hearle and Simmens [19] described the disintegration of fibers using a blender or an ultrasonic disintegrator.

More recently, high modulus fibers, such as aromatic polyamides, that are difficult to prepare by other methods, were fragmented into fibrillar fine structures [20] by high wattage ultrasonic irradiation in water. This is extremely useful for producing very finely divided material for transmission methods such as lattice imaging and electron diffraction. The major drawback is

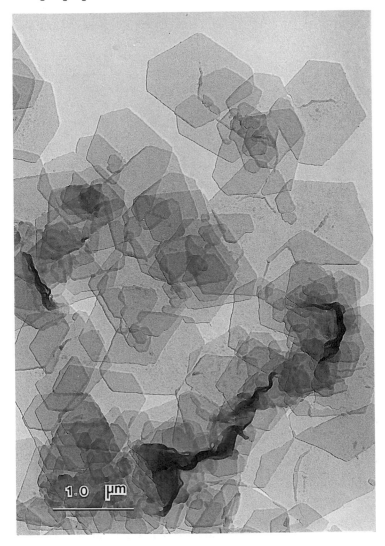

Fig. 4.1 TEM micrograph of poly-ethylene single crystals produced by self seeding from a 0.05% solution of Sclair 2907 in xylene at 83°C.

that the position of the sample in the original fiber is unknown. The choice of liquid used in the sonication preparation is also very important. The effect of the liquid depends on the boiling point, surface tension and polymer interaction, but ethanol, water or ethanol/water mixtures are the best choices. In addition, for best results, the materials must be mechanically broken down prior to sonication. Materials often require 30 min or more to fibrillate and the liquid must be kept cool. The sonicator can be pulsed on and off, or the bath cooled directly.

4.1.3.5 *Casting thin films*

Thin films can be cast from solutions of crystalline or noncrystalline polymers where the film thickness is controlled by the solution concentration. Polyethylene films have been cast from boiling dilute solutions in xylene [21]. In most cases, a glass slide is dipped into the solution and suspended to dry in the vapors of the solvent to permit slow and uniform drying. Roche *et al.* [22] prepared poly(butylene terephthalate) (PBT) in a 1% solution in hexafluoroisopropanol (HFIP) by depositing drops onto a glass slide placed on an incline in a beaker of the solvent, to limit fast drying. Thicker regions on the bottom of the slide were used for optical study while the thinner upper regions were used in the TEM. Samples were used for scanning transmission electron microscopy (STEM) and electron diffraction in order to define the nature of the spherulites.

Vadimsky [23] described a useful method for the preparation of thin films from the melt or solution. The method involves the evaporation of carbon onto freshly cleaved mica or fractured NaCl crystal substrates. The thin polymer film is cast onto the carbon coated substrate or the substrate is dipped into a polymer solution. After solvent evaporation, the film is scored and removed from the glass by floating it onto a water surface. The specimen can also be deposited directly onto the substrate followed by carbon coating. Geil [24, 25], in a variation of this method, deformed PE single crystals by deposition on a Mylar substrate and drawing it before carbon coating and TEM examination. Windle and coworkers [26–27] prepared thin films by shearing molten polymers on freshly cleaved rock salt with a sharp razor blade. The films, of thermotropic copolyesters (Section 5.6), were quickly quenched, the rock salt dissolved, and the films annealed to assess crystallite growth.

Thin films may also be formed by casting on a liquid surface to allow easy removal. A range of liquids have been used, such as glycerol, orthophosphoric acid and mercury. Grubb and Keller [28] placed a few drops of orthophosphoric acid or glycerol on a microscope slide, on a hot bench, and subsequently a drop of the PE slurry was allowed to fall on this hot surface. The polymer melted as the solvent evaporated, and the film was solidified by cooling. The film was floated onto a water surface and picked up on TEM grids. These self supported thin films were found to be spherulitic by optical observation and large areas of the film were thin enough for TEM [29]. In a study by Chu and Wilkes [30], films of poly(ethylene terephthalate) (PET) were cast from trifluoroacetic acid and the spherulitic films

were studied by polarized light microscopy, SEM and light scattering.

Howell and Reneker [31] prepared thin polymer films from solutions spread on water for HREM studies. PEEK in *a*-chloronaphthalene, PS in benzene, PE in hot xylene, for instance, were collected on carbon films on grids that had been thinned in an oxygen glow discharge plasma to produce holes or cracks in the carbon. The polymer films were then stained in RuO_4 vapor for about 15 min (Section 4.4.6) to add contrast to the molecular scale features observed by HREM. Highest resolution was achieved by focusing with a large objective aperture; the lattice planes in the particles of graphite were imaged to ensure highest resolution. Thin layers were also formed for HREM [31] by trapping polymer solutions between freshly cleaved sheets of mica separated by a wedge. The wedged mica sheets were placed in a solution of PE in boiling xylene, the wedge removed and the mica sheets clamped together, withdrawn from the solution, cooled and the solvent evaporated. The mica sheets were separated, shadowed with platinum–carbon, coated with evaporated carbon (Section 4.7.2), floated off the mica and picked up onto grids for HREM imaging.

Cast films exhibit a range of morphologies due to the effect of solvents, substrates and orientation. In the case of block copolymers the choice of solvent is quite important to the final structure. Spherulites, a common textural structure observed in crystalline polymers, are formed in many industrial processes where the polymer is melted prior to forming the article of interest. Films produced by these industrial processes differ from films used in model studies as the former are usually not thin in the microscopic sense. In true thin films, the spherulitic texture is two dimensional, whereas in these thicker film materials the spherulites are three dimensional.

4.1.3.6 *Drawing thin films*

Thin cast films may be drawn manually at room temperature or above. The film to be drawn is transferred to a glass slide coated with an inactive liquid, for example glycerol or silicone oil. Needles or blunted razor blades are pressed into the film and drawn apart. Small regions of drawn film suitable for TEM are produced, but there is no control over the draw ratio or the draw rate. Petermann and Gohil [32] developed a method for forming highly oriented, ultra-thin films of crystalline thermoplastics under more controlled conditions. The first step is to make a small quantity of dilute solution (0.3–1%) of the polymer. A thin film of molten polymer is then made by placing a drop of the solution onto a hot glass and allowing the solvent to evaporate. If the polymer dissolves only at elevated temperatures, the finely divided suspension which forms on cooling the solution can be used, as the polymer re-dissolves as the drop heats up. The hot glass plate is kept at a temperature where quiescent crystallization is very slow. For HDPE this is about 125°C. An enclosure may be used to limit temperature fluctuations. A glass rod coated with the polymer is touched to the melt film and slowly drawn away, at a few cm s^{-1}. The local orientation causes immediate crystallization, producing a solid drawn film which can be as thin as 20 nm. A schematic diagram of the drawing of a film is shown in Fig. 4.2 [32]. The thin drawn film is collected on a clean glass microscope slide, cut into small pieces, floated off onto a clean water surface and mounted on grids for the TEM. The melt film remaining on the glass plate may be used directly in the OM after cooling. The drawing zone is limited to a region of about the same size as the melt film thickness, about 1 μm, so a draw rate of only 1 cm s^{-1} corresponds to a very high strain rate, 10^4 s^{-1}. A longitudinal flow gradient of this magnitude is close to the conditions that are used in the industrial processes of high speed spinning.

The method has been used primarily for the study of HDPE (linear polyethylene), but it also works for other crystalline polyolefins. The oriented films that are formed are smooth surfaced and uniform in structure. They contain stacks of parallel lamellae on edge, and the lamellae extend through the entire film. This makes the interpretation of transmission images particularly simple, as there is no overlapping of adjacent structural regions. For the same reason, defocus imaging [33–35] is very straightforward and enhances the visibility of the crystals in the film (see Fig. 5.20). The HDPE specimens produced by this method are thin and uniform, with a well defined crystal orientation and arrangement. Preparation is comparatively easy and needs no special equipment beyond a hot plate. Therefore they make excellent training aids, for a microscopist who is learning how to take electron diffraction patterns or dark field micrographs from radiation sensitive polymers.

4.1.4 SPM preparations

A number of scanning probe microscopy (SPM) techniques have been developed and recently applied to investigations of polymer materials. From a specimen preparation point of view the major difference between the scanning tunneling microscope (STM) and atomic force microscope (AFM) is that STM requires the sample to be conducting, whereas a non-conducting material can be imaged by AFM. Methods used for the production of fine grained conductive coatings for SEM have been used to provide conducting

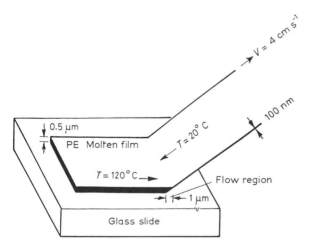

Fig. 4.2 The method of producing a highly oriented melt drawn film is shown in this schematic. (From Petermann and Gohil [32]; reproduced with permission.)

surfaces for STM (Section 4.7.3). Many of the methods used for SEM and TEM may be applied to SPM imaging, but as with these conventional techniques the potential of artifacts is ever present and from new and possibly unexpected sources in these high resolution techniques. Substrates that are most often used for SPM are silicon, highly oriented pyrolitic graphite (HOPG) and mica, which have all been used for decades for TEM and SEM as they are all flat and easy to prepare. HOPG and mica are freshly cleaved to form new, clean and flat surfaces. Flatter substrates and samples are required for AFM while coating with a conducting layer is usually required for STM, unless the sample is very thin, as is likely for Langmuir–Blodgett (LB) films. Albrecht *et al.* [36] imaged LB films using both STM and AFM on samples prepared in three ways. Films were prepared by (a) raising a horizontally held graphite substrate from the bottom of the LB trough, (b) by lowering the horizontally held graphite to bring it into contact with the spread film and (c) by dipping the graphite vertically into the water and then lifting it vertically out [36]. The authors suggested the SPM imaging reveals molecular scale orientation and morphology.

Fibrillated liquid crystalline polymer fragments [53] were suspended in water or a water–ethanol mixture and dropped onto HOPG and thin peels and sections were also mounted on silicon and HOPG for FESEM, STM [37–38] and AFM [39]. These studies were not at atomic scale resolution, and thus there was no observable effect of the underlying substrate. Early work in this field, such as reviewed by Leggett *et al.* [40], has shown that a number of features of HOPG surfaces, such as cracks and steps, can produce image features through thin polymer and biopolymer layers. Even worse, some graphite features and artifacts resemble molecules and are easily misinterpreted [40]. The recurrent theme in this book, the need to conduct complementary imaging, especially when using new imaging techniques, thus continues to be of utmost importance for the use of SPM techniques.

Applications of STM and AFM for polymers are now numerous, and a few examples will be mentioned here and in Chapter 5. PS films formed by spin coating solutions onto silicon were studied by Stange *et al.* [41] after carbon coating for STM which showed isolated molecules. Suzuki *et al.* [42] also studied spin cast films, in this case films of polyimide precursors on indium–tin–oxide (ITO) coated glass, followed by imidization at elevated temperatures. STM images of gold coated films showed the details of the film surface; the authors also imaged the ITO substrate and measured the grain structure for comparison with the polymer. Reneker *et al.* [43] studied PE samples prepared from a solution by the method they also used for HREM [31] (Section 4.1.3.5). Once the solution of PE was trapped between the mica sheets the device was removed from solution, cooled, the solvent evaporated and the sheets, separated and shadowed as for HREM, were also imaged with an STM operated in air. These authors did careful work by both HREM and STM that showed the same lateral resolution by both techniques but much better vertical resolution by STM.

The feasibility of imaging polymer crystal samples was shown by Piner *et al.* [44] who conducted STM imaging of gold and chromium coated PE lamellae. They found that thick regions of crystals were more difficult to image than thinner regions on edges. Magonov *et al.* [45] used AFM to study cold extruded PE with atomic resolution and reported a fibrillar structure with highly oriented molecules was present. Fritzsche *et al.* [46] studied polyethersulfone ultrafiltration membranes by drying the samples by a solvent exchange procedure for SEM, using isopropanol and hexane and by observing the wet membranes directly by AFM. They were able to detect and measure pores by both methods although neither technique could resolve the surface pore structures. The porosity seen in AFM and SEM differs, as expected, due to the drying and metal coating for SEM versus observation of the virgin membrane via AFM. In AFM vertical profile measurements yield the pore diameters.

4.2 POLISHING

The topic of polishing may appear out of place in a text on polymers and may seem more appropriate in a discussion of metals, ceramics or rock specimens. After experience in a geology laboratory the idea of grinding and polishing appeared worth a try. The methods work remarkably well with a range of materials, including hard to section polymers and composites containing glass or carbon fibers. Variations in the polishing method are used to provide thick specimens with a polished surface for reflected light or SEM and thin specimens for transmitted light. In the former category are specimens that must be polished flat for microanalysis. There is little published on the polishing of polymers, although such methods have been used for over a decade in the Hoechst Celanese laboratory [47]. One review by Holik *et al.* [48] gives a description of the adaptation of metallographic and petrographic preparation techniques to the preparation of polymers for light microscopy. A general description of metallographic polishing techniques is given by Samuels [49] while Goodhew [5] provides a general text on preparations in materials science which includes this topic. Machined semicrystalline polymer and short glass fiber composite surfaces were shown in the SEM to have details produced by machining [50] which showed the need for polishing.

4.2.1 Polishing artifacts

Polishing specimens can produce artifacts at each stage of the preparation. Cutting the specimen and final polishing are the steps with the greatest potential for artifacts. The initial cutting can cause the formation of microcracks which can be misinterpreted as voids or cracks in the specimen. In order to limit such cracking, the specimen should be rough cut several inches from the surface of interest and then cut more carefully, with a water cooled diamond saw, near that region. Grinding should be done with gentle pressure on the mount and should never be done dry. Each grit should be used only until the scratches from the previous grit are removed. Cleaning of the specimen is also important as larger grits or polishes on the specimen can contaminate the fine polishing cloth. A 30 s ultrasonic rinse in water followed by drying off with a can of Freon gas is adequate to remove the grit. However, the gas can should not be turned upside down as this will result in a frozen and ruined specimen polish. Finally, undercutting of the specimen, that is preferential polishing of the embedding resin compared to the specimen, is another problem that must be overcome when developing a polishing method.

4.2.2 Polishing specimen surfaces

Standard metallographic methods [5] are used for grinding and polishing. Large samples are precut with a saw to fit into 1–2 inch diameter metal or rubber molds. Small samples are mounted on a jig, such as a piece of cardboard, to hold them in the proper orientation and then they are placed in the molds. Epoxy is mixed, poured into the mold and cured to provide support. Epoxide resins and hardeners are acceptable for most polymer and composite materials. Fast cure resins are difficult to use as the exothermic reaction creates large bubbles that hinder polishing. A diamond saw is used to precut cured specimens, and wet grinding is done with sandpapers ranging from 120 to 600 grit. As the specimen is ground on each of the successively finer papers, it is turned 90° so that the new scatches are clearly visible compared to the coarser scratches being replaced. Water flowing over the grinding papers keeps the specimen from heating up and also carries the debris away. Polishing is conducted using graded alumina suspensions in water (1, 0.3 and 0.05 μm), chromium oxide slurries or diamond paste suspended in oils. Polishing is done on a rotating wheel with a layer of fluid on the surface of the polishing cloth. A benchtop ultrasonic cleaning unit is used to remove the polishing media from the specimen. The quality of the polish can be checked in the optical microscope.

4.2.2.1 *Polishing surfaces*

Bartosiewicz and Mencik [51] polished nylon and polypropylene prior to etching to reveal the polymer structure. The specimen surface was wet ground with 320 and 400 grit silicon carbide paper and 600 grit aluminum oxide paper. Polishing was done using a 15 μm aluminum oxide slurry on medium high nap felt cloth followed by use of 1 and 0.5 μm aluminum oxide slurries. Final polishing was with a 0.05 μm chromium oxide slurry on Microcloth (Buehler). In our laboratory [47] a separate wheel is used for each polish grit size to limit contamination. Polishing is conducted with a firmly held specimen, lightly pressed down on the moving wheel and constantly moving on the surface of the polish cloth. The polish cloth is chosen for the specific sample, as certain cloths limit undercutting of hard fiber containing polymers. Polishing should be done in short steps of up to a few minutes, followed by cleaning, drying, examination and then more polishing. Ultrasonic cleaning is important for cleaning the specimen between polishing steps. Polishing the specimen too long is a common cause of undercutting, i.e., the softer embedding media, or the polymer in a fiber filled composite, is polished away more rapidly than the fiber, resulting in a three dimensional surface. Brittle fracture of the fibers can occur if the specimens are polished for too long, or by too heavy a hand, or if ultrasonic cleaning is too extensive. If the specimen is pulled from the hands during polishing or grinding, or if undercutting is present, the specimen must be reground or, for deeper problems, it can be sawn to provide a new surface.

Typical specimens that can be polished and observed optically include composites of polymers and hard fillers, tough polymers, molded parts and extrudates. The distribution of the fillers, such as glass or graphite fibers or mineral particles, can be observed by reflected light. The shape of the fiber cross section, the presence of voids and the overall structure of the material can be observed by this method. Figures 4.3 and

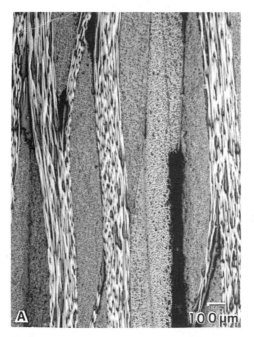 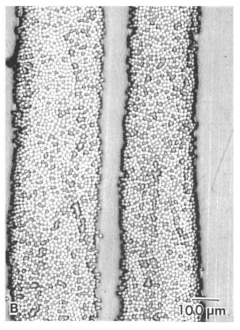

Fig. 4.3 Reflected light micrographs of polished composite specimens show the carbon fibers (white) and their orientation within the polymer matrix. Black regions are voids. In (A) various layers are oriented normal to one another while in (B) two uniaxially fiber tows are shown.

Fig. 4.4 Reflected light micrographs of polished longitudinal sections of a polymer extrudate show the flow pattern at low magnification (A). Differential interference contrast (DIC) of a similar specimen (B) shows this flow pattern in greater detail.

4.4 show two applications of polishing polymer materials. A carbon fiber composite, in an epoxy matrix, is shown in a reflected light micrograph (Fig. 4.3A). The small, round, white regions are fiber cross sections whereas the longer white regions are fibers oriented parallel to the polished face. The black regions are voids in the original composite. At a higher magnification another composite material (Fig. 4.3B) exhibits uniform packing of the fibers in the polymer matrix. The flow pattern of the polymer is seen in a longitudinal polished section of a polymer extrudate (Fig. 4.4A) and reflected differential interference contrast (DIC) more clearly reveals the surface detail (Fig. 4.4B).

Specimens for OM or SEM can be prepared by polishing followed by etching, either chemically or with plasma or ion beams (Section 4.5) to bring out relief or detail. In the case of crystalline polymers, etching brings out detail relating to the crystalline structures present. Guild and Ralph [52] studied cross sections of glass fibers following embedding in polyester resin. Sections cut perpendicular to the fiber direction were mounted in cold mounting resin, ground and diamond polished. Etching in hydrofluoric acid (40%) for 5–10 s revealed the glass fibers more clearly.

4.2.2.2 *Polishing thin sections*

Thin sections (about 2–40 μm thick) provide ideal specimens for study using optical microscopy. However, some materials are too tough, brittle, or hard to be sectioned by microtomy. In geology thin sections of rocks are commonly made by polishing techniques and this method also works well for polymers that cannot be microtomed because they are too hard at room temperature.

Samples are first cut, embedded in a support resin, ground and wet polished on one face. The sample is glued, polished face down, to a glass microscope slide with fast cure epoxy or cyanoacrylate glue. The cured sample is cut, ground and polished on the opposite side until it is thin enough to transmit light. Care must be taken, as mentioned earlier (Section 4.2.1) regarding cutting of the specimen and limiting polishing artifacts. The thin sectioning method has an added problem in that the thinner the specimen, the better the optical image but also the greater likelihood of the specimen pulling out of the mount. The suggestion here is caution; polish for short times, such as 30–60 s, clean and examine the specimen before proceeding.

Applications of the polished thin sectioning method [48] include extrudate cross sections, glass fiber reinforced polycarbonate and rubber toughened nylon. Recently, the fine structure of bulk thermotropic liquid crystalline polymers (Section 5.6.2) has been observed in thin specimens prepared by the polishing method [53]. This method was required as microtomy was not possible for these tough polymers.

4.3 MICROTOMY

Microtomy, or sectioning, methods involve the preparation of thin slices of material for microscopy observation. Generally, *microtomy* refers to sectioning for observation in an optical microscope by transmitted light. Microtomed sections are cut with steel or glass knives to about 1– 40 μm thickness. *Ultramicrotomy* methods involve the preparation of ultrathin sections of material for observation in an electron microscope. Ultramicrotome sections are cut with glass or diamond knives in ultramicrotomes to about 30– 100 nm thick. If both OM and TEM are to be conducted on the same specimen the TEM preparation method can be utilized for both. Optical observation provides an overview of the specimen whereas TEM sections can be studied at higher magnifications.

Ultramicrotomy is very commonly used in the preparation of biological specimens for electron microscopy. The materials must be carefully fixed, stained and embedded prior to sectioning. The aim is to provide sections with visible fine structure that represents the original material. Stains (Section 4.4), such as osmium tetroxide and chlorosulfonic acid, are used to enhance contrast by increasing the electron density in specific structures in the material. Combined staining and sectioning methods have found great utility in the preparation of polymer materials. Polymers are generally easier to prepare, prior to sectioning, but are much more difficult to section than biological materials. Fixation and staining methods may be needed as often the low atomic number of the material does not provide much contrast for electron microscopy. Some polymers are too soft to be sectioned at room temperature and these must be

hardened either chemically or by cooling below room temperature during microtomy. This latter method, *cryosectioning*, is an increasingly popular biological specimen preparation method for unfixed materials, and it is also applied to polymers.

4.3.1 Peelback of fibers/films for SEM

Synthetic fibers were among the earliest polymers produced for commercial applications. Morphological studies were undertaken in order to understand structure-property relations and to improve properties. During the 1950s the SEM was not available. Although microtomy was already being used by Scott and Fergerson [54] and others, little could be seen by direct observation of thin sections in the TEM. This resulted in a need for other preparation methods. One of the most important of these methods, currently used in modified form, is the *peelback* method or longitudinal splitting of fibers [55]. The early peelback technique was used to provide surfaces for replication or thin films, both for the TEM. Today the method is used to prepare fibers and films for SEM study.

Scott [55] described the peelback method as a split section or cleavage technique by what he called orientation and cleavage plane splitting. The idea is to open a fiber (or film) with minimal disruption and provide a section or replica for TEM imaging. Orientation splitting involves the cutting of a fiber with a razor blade, at an oblique angle, halfway into the fiber, followed by a cut along or parallel to the fiber axis. This second cut below the first one is peeled back with forceps to provide a thin section aligned with the longitudinal axis. In cleavage plane splitting, a razor is used to make an oblique cut halfway through the fiber and then the polymer is peeled back along the fiber axis with forceps. The split works its way along the fiber to the outer surface where it then continues along the surface skin. Additional splitting of the same fiber provides layers aligned along the molecular fiber axis. Fibers prepared by peelback methods are now generally observed in the SEM where the internal texture and fracture

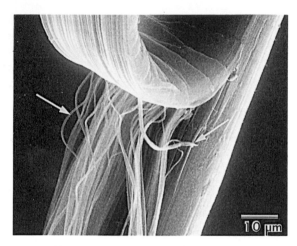

Fig. 4.5 SEM micrograph of a polymer fiber prepared by the peelback method reveals the internal fibrillar texture (arrow).

mechanism can be clearly imaged. An example of a peeled back PET fiber, showing the fibrillar texture, is seen in an SEM image in Fig. 4.5.

4.3.2 Sections for OM

Specimens for transmitted light microscopy must be transparent; as the practical resolution of the optical microscope is about 0.5–1 μm, specimens about 1–40 μm thick must be prepared. Very simple hand held devices using a razor blade may be sufficient or complex preparations involving several steps prior to microtomy may be required. For example, alignment of fibers for sectioning involves mounting the fibers in a specific orientation, placing them in a capsule or mold, adding appropriate embedding resins, curing, and trimming before sectioning. Much of the preparation is common to both optical and TEM sectioning so the reader should be aware that both sections of this chapter should be read for a complete discussion. Since fibers are among the most difficult specimens to cut they are used as an example in the following sections.

4.3.2.1 *Microtomes*

Sectioning is accomplished using several types of manual and mechanical techniques. Hand held

and home made forms of sectioning devices with razor blades are commonly found in any textile fiber laboratory. Stoves' book on *Fibre Microscopy* [56] describes several of these methods, including a hand held metal plate, the Hardy microtome and other mechanical sectioning devices. Such instruments for cutting sections for optical microscopy were in use as early as the eighteenth century [57].

The simplest and most useful of these devices for fibers, films, yarns and fabrics is a modified metal plate, as shown in Fig. 4.6. Sections cut using such a plate tend to be ragged around the edges, nonuniform in thickness and often damaged and distorted. On the other hand, the method is simple, and it can provide rapid specimen observation. The plate is 3×1 inch in size, with slots from the edge to the center, ending in small holes, about 1 mm across. The specimen is combined with a colored filler yarn and forced into the slot. A razor blade is used to cut both top and bottom surfaces, the plate is

Fig. 4.6 A metal plate used to provide rapid cross sections of fibers, yarns, and fabrics. The plate is the same size as a 1×3 inch glass microscope slide.

placed on a glass microscope slide and a drop of immersion oil and a cover slip are placed on the top surface before viewing in transmitted light.

Rotary and sledge microtomes are used with steel knives to provide sections for observation in the optical microscope. These mechanical systems provide consistent thin sections in the range of 5–40 μm thickness. Thinner sections are cut using glass knives which provide good sections 1–4 μm thick.

4.3.2.2 *Specimen mounting*

Specimens for microtomy must either be cut to fit in the chuck of the microtome and be self supporting, or they must be embedded in a supporting medium. Pieces of molded or extruded plastics are trimmed to fit the chuck using small saws and razor blades. In the case of films and fibers, however, the samples must be embedded in a resin for both support and orientation. The steps for mounting specimens will be described here. A more complete description of embedding media will be found in Section 4.3.4.

One method for mounting samples for sectioning is as follows.

(1) A glass microscope slide is put on top of a piece of graph paper to aid alignment. Fibers are placed lengthwise on the glass slide and secured at the ends with double sided tape.
(2) Fibers can be consolidated by tying them at intervals with a cord or yarn, preferably of a different color than the specimen.
(3) Samples are supported with cardboard frames, cut in a 'C' or oval donut shape, placed under the fibers and glued to the frame with Elmer's white glue or Eastman 910 fast drying glue.
(4) Fibers are mounted for cross sections as shown in Fig. 4.7A. For longitudinal sections,

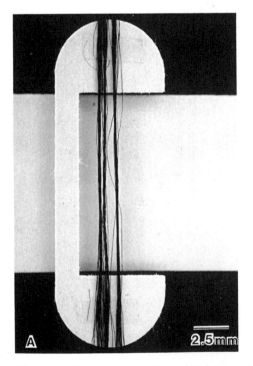
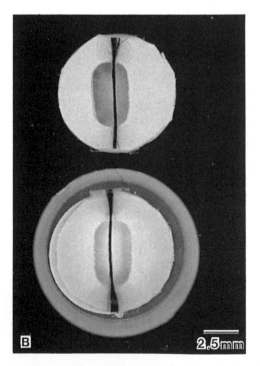

Fig. 4.7 Fibers are shown mounted on a "C" shaped frame (A), punched out of light cardboard, in order to orient them for cross sections. For longitudinal sections fibers are mounted perpendicular to the direction shown, but on two "C" frames, or they are mounted on a frame that fits into an embedding capsule (B).

two frames can be glued together and slipped under the fibers so they are parallel to the short axis of the frame, or they are mounted across a donut shaped frame (Fig. 4.7B top) and placed in the top of the capsule, face down (Fig. 4.7B bottom).

(5) The Elmer's glue is hardened in a 60–70°C oven for about 15 min and the frame placed in a capsule or mold.

4.3.2.3 *Specimen embedding*

Embedding media are chosen, mixed, added to the capsule and allowed to cure, according to the details supplied by the manufacturer. In our laboratory, one or two embedding media are in general use while a wider range are on the shelf, in case they are needed. In this way, different types of materials may be embedded at once and the person performing the microtomy can become familiar with the sectioning characteristics of those specific resins. Paraffin, long used for embedding biological and polymer materials, will not be described in any detail as its use is not reliable: cutting is poor at room temperature, it is messy to use and maintaining the hot paraffin is time consuming. Room temperature curing epoxies work well for optical sectioning of most dry polymer materials. Wet materials are more compatible with water soluble resins as materials need not be totally dehydrated prior to embedding. Processing is rapid and high quality sections can be obtained using a glass knife. Where optical studies will be followed by TEM the EM embedding media are best used for both studies. Media used for TEM will be described in the next section.

A variety of molds are available to encapsulate samples for embedding. BEEM type capsules are useful for small pieces of materials that can be placed in the tip which then provides pre-trimmed specimen blocks. Gelatin capsules are rounded in shape and thus must be trimmed, but they have the advantage that they can be removed by soaking in water. Other shapes are good for specific sample forms. Flat embedding molds are excellent for films and membranes,

especially when orientation of the specimen is important. A range of these capsules and molds should be kept in the laboratory.

4.3.2.4 *Sectioning*

For trimming, the cured block is placed in a vise and cut with a jeweler's saw. Fine trimming with a razor blade can be done under a stereomicroscope to provide a square or trapezoidal block face shape (Fig. 4.8). The trimmed sample is secured in the chuck of the microtome and sectioned to the desired thickness. A common problem is that sections tend to curl due to distortions caused by the sectioning process. Curled sections can be flattened by manually uncoiling them under a stereomicroscope and placing them in immersion oil between a glass slide and a cover slip. The sections can also be relaxed by heating with an infrared lamp or by flotation on hot glycerol.

Typical polymer materials studied in optical thin section include: extrudates or molded parts, such as semicrystalline polyoxymethylene; multiphase polymers, such as rubber toughened nylon; filled polymers, such as carbon black filled nylon; fibers, such as polyester and rayon; and films which are too thick to transmit light. Two examples are of nylon imaged in polarized

Fig. 4.8 The photograph shows the trapezoid shaped face of a trimmed block ready to be sectioned.

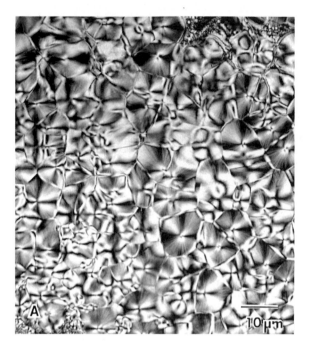 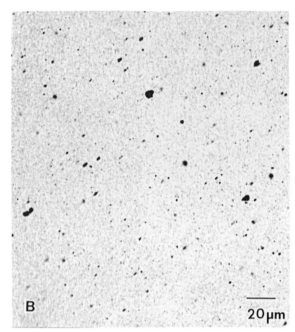

Fig. 4.9 Two examples of microtomed sections viewed in the optical microscope are shown: (A) a section of a nylon pellet, in polarized light, reveals a coarse spherulitic texture; and (B) a section from a black, molded nylon part, in a bright field micrograph, shows the size and distribution of the carbon black filler.

light (Fig. 4.9A), which reveals its spherulitic texture, and imaged in bright field (Fig. 4.9B), which reveals the dispersed carbon black particles.

4.3.3 Sections for SEM

The SEM is used to study the surfaces of polymer materials, such as plastics, fibers, yarns, membranes, films and composites. The outer surfaces can be studied or bulk SEM studies can be performed on samples that have been fractured (Section 4.8) or sectioned. The flat surface remaining after sectioning can be used as a sample for SEM, particularly for x-ray analysis. A conductive coating (Section 4.7.3) is generally applied to the surfaces of interest for SEM.

Many of the methods used to provide cross sections for the SEM are modifications of optical techniques, such as the Hardy microtome [58] and a method [59] using an adaptation of the Shirley plate (Section 4.3.2). An advantage of using paraffin wax embedding and sectioning for this is that the wax can be removed from the sections with xylene. Kershaw and Lewis [60] described a similar method with wax embedding of textile fabrics or yarns that provided a section of the intact assemblage. Epoxy resin sections of biological tissue were prepared for the SEM by removal of the epoxy using a concentrated solution of sodium methoxide [61].

Hearle *et al.* [4] described several special holders that are modifications of the metal plate method. Specimen holders which have holes for packing materials such as fibers, yarns, or fabrics for razor blade cutting are useful methods which involve the insertion of a looped thread through the hole and the placement of the material in the loop. This material is then pulled through the hole and sectioned with a razor blade. Modifications to the stubs on any SEM can be made readily in a machine shop.

4.3.4 Ultrathin sectioning

Ultramicrotomy is an obvious example of a preparation method developed by biologists and now in general use by polymer microscopists. In the early 1950s Porter and Blum [62], Sjostrand [63] and Haanstra [57] were constructing microtomes to provide sections, about 0.1 μm thick (or less), for TEM study. Today, ultramicrotomes are available from companies such as LKB and Reichert that permit ultrathin sections of polymers to be obtained on a routine basis.

Sectioning is one of the most widely used methods in the preparation of polymers for electron microscopy. Microtomy permits the observation of the actual structure in a bulk material which is not possible by methods such as thin film casting or surface replication. Polymers (in common with biological tissues) require care in handling, embedding in resins for support and addition of stains to enhance contrast in the TEM.

A general method will be described for the preparation and sectioning of polymers, pointing out those areas of difficulty and citing some of the excellent references available. The aim of the preparation is sections that truly represent the original bulk structure and which are thin, flat, not deformed, have contrast and can be used to form images that can be interpreted. Experience is required to obtain such sections of polymer specimens consistently. Fortunately, the tedious fixation and dehydration steps required for biological materials are not required for most polymers. The steps involved in sample preparation for ultramicrotomy include: (1) fixation/ staining, (2) drying (if needed), (3) specimen mounting (Section 4.3.2), (4) embedding and curing, (5) trimming and sectioning and (6) carbon, metal or polymer coatings.

Staining is often required in order to increase the electron scattering of polymers selectively and, thus, aid contrast and resolution of details. Staining is performed either on small pieces of the material before embedding or by post-staining the sections themselves, or both. This topic is so vast and important to the polymer microscopist that it will be treated separately (Section 4.4).

4.3.4.1 *Specimen drying*

Most polymer materials are dry, or can be air dried without structural change, while others, such as some membranes and emulsions, are wet and require drying prior to embedding. Air drying may destroy or distort structures due to the deleterious effects of surface tension. A method that has been used in the case of wet, porous membranes [64] is to cut strips of the membrane in a jar containing water and to transfer the strips directly into the embedding resin in a disposable weighing pan. This sample is placed in a vacuum for several hours at room temperature to remove the water, which is replaced by resin. After vacuum impregnation, the membrane samples are placed in flat embedding molds in fresh resin. Other methods of drying might include critical point drying, freeze drying or chemical drying (Section 4.9).

4.3.4.2 *Embedding media*

There are many embedding media available, almost too many for the novice to choose one over another for a given specimen. Glauert [65] is an excellent reference that includes a description of the various media and the reasons for their use. There are three media in general use: epoxy resins, polyester resins and methacrylates. The ease of sectioning and stability in the high vacuum and the electron beam are important. A hardness match of the polymer and embedding media is also important to good sectioning. Variations can be made in the recipe for each medium in order to change its hardness. One of the reasons for the widespread use of epoxies is the fact that they are the most stable in the electron beam, whereas the methacrylates are the least stable. Availability and good results have been our experience with the epoxies. Epon 812 (a Shell resin) was replaced in the early 1980s with equivalent epoxies. They are used in most applications except where the greater potential infiltration of low viscosity resins is required. The embedding media should be cured using the times and temperatures recommended by the supplier.

Recently, biologists have suggested the use of microwave techniques for more rapid embedding [66–70], and recipes have been developed for a reduction in curing from 48 h to 15 min or less at a power of 700 W in a 2540 MHz microwave oven [67].

4.3.4.3 *Block trimming*

Specimen preparation methods, trimming blocks and sectioning were discussed earlier (Section 4.3.2) and are covered in detail for biological materials by Reid [71]. A factor to consider for TEM sectioning is the orientation of the specimen in the trimmed block, which is important to section quality and structural interpretation, especially of anisotropic specimens. Longitudinal sections of fibers and films should be aligned so that the long axis of the material is *not* parallel or perpendicular to the knife edge. The specimen should be oriented at an oblique angle to the bottom of the trapezoid or perpendicular to it for best sectioning. Trimming a block to a specific shape and size is more critical for TEM than for OM. The cured specimen block is cut close to the specimen plane with a jeweler's saw or a razor blade. The pretrimmed block in a microtome chuck is placed in a holder built to fit in the stage of a stereomicroscope. The top surface is cut or faced and then the sides of the block are cut at an oblique angle to provide a trapezoidal shaped block face. Where possible, the epoxy is all trimmed away; where this is not possible, a minimum sized block face with little epoxy is prepared. Smaller block faces are better for TEM. The bottom of the section, the part that will be cut by the knife first, should be the longer side of the trapezoid.

4.3.4.4 *Ultramicrotomy*

Glass knives are used to face the block up to the specimen and for final trimming either in the ultramicrotome itself or in a microtome made for that purpose, such as the LKB Pyramitome. Ultrathin sections are cut with either a glass or a diamond knife in the ultramicrotome (ultratome). Glass knives have the advantage of being inexpensive, easy to make and sharper than razor blades. The disadvantage of glass knives is that they do not remain sharp. Glass knives provide good sections of samples that are softer than glass, such as unfilled nylon, polyoxymethylene (POM) and PE. In the case of very hard materials that might crack a diamond knife, such as polymer composites, or very high modulus materials, it is better to use an expendable glass knife. Diamond knives are sharp, can be resharpened and can cut very thin sections, but they are very expensive, and they can chip or crack. Old diamond knives are quite useful for cutting hard materials that require diamond but might chip or crack a good knife. In practice, proficiency should be demonstrated with a glass knife prior to attempting diamond knife sectioning. Finally, the nature of the trough and the trough fluid, used to float the sections so they can be picked up on grids, must be considered. Diamond knives are supplied with a trough that is simply filled with water during room temperature sectioning. Glass knives are made as they are used and various sticky tapes may be used to produce troughs.

Sectioning problems, collection and trouble-shooting are all described by Reid [71]. Perseverance is the key word, for with polymers sectioning is tedious and hours are often spent before usable sections are obtained. A longitudinal section of a multiphase polymer is shown in the micrograph in Fig. 4.10 to have dense particles aligned with the fiber axis although no staining was used in this case.

4.3.5 Ultrathin cryosectioning

Cryomicrotomy and cryoultramicrotomy are sectioning methods performed at low temperatures to produce thin or ultrathin sections, respectively, of polymers too soft for room

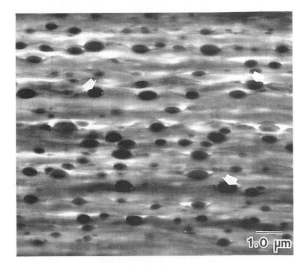

Fig. 4.10 An ultrathin section of a multiphase blend reveals the size and distribution of the dispersed phase particles (arrows).

temperature sectioning. The methods are not routine, but they appear to be gaining in popularity recently as a direct result of the ease of use of new commercial instruments. Zierold [72] provides an excellent discussion of the optimal conditions for the preparation of ultrathin cryosections of biological materials. In this work, the mean ice crystal size was shown to be critical, and the section quality was seen to increase as the ice crystal size decreased. Polymers that have a glass transition temperature below room temperature are soft and can be hardened by cooling, prior to sectioning. Sectioning in the -20 to $-40°C$ range is fairly straightforward as liquids may still be used to sections off the knife. At lower temperatures, such as $-120°C$, sectioning is more difficult as a dry knife must be used. Dry sections are difficult to pick up without damage. Ionizing antistatic devices are used to inhibit sections flying away due to static charge. Special instrumentation is required to control the temperature of the knife and the specimen accurately, as the section thickness is affected by thermal expansion and contraction. The cryomicrotome must also limit frost buildup on the specimen, as frost is a problem in sectioning at low temperatures.

4.3.5.1 *Literature reviews*

Chappius and Robblee [73] used solid carbon dioxide and alcohol as a coolant while Leigh-Dugmore [74], Hess and Ford [75], Andrews and Stubbs [76] and Andrews [77] used liquid nitrogen to cut about 50 nm thick sections of rubber. Andrews *et al.* [78] examined sections of several semicrystalline polymers produced by cryoultramicrotomy at $-150°C$ using a modified microtome. Ethylene glycol was used to wet the knife edge, but no fluid was used to pick up the sections. Dlugosz and Keller [79] produced sections of spherulitic PE by cryoultramicrotomy and studied the beam induced band structures.

Early results with cryomicrotomes were described by Cobbold and Mendelson [80]. Polyurethane elastomer, a blend of crystalline and noncrystalline polymers, showed spherulitic textures after sectioning at about $-70°C$. Injection molded polypropylene (PP) was also sectioned at about $-70°C$, while polytetrafluoroethylene (PTFE) was sectioned at much lower temperatures. The authors concluded that the technique, though difficult, had potential. Extruded styrene–butadiene–styrene (SBS) copolymer was prepared by cryosectioning with a diamond knife in liquid air at -85 to $-115°C$, followed by osmium tetroxide vapor staining for one hour [81]. This method revealed the alternating sequence of the polystyrene and polybutadiene lamellae. Odell *et al.* [82] prepared extruded triblock copolymer by first chemically hardening the polybutadiene, with osmium tetroxide, followed by cryoultramicrotomy to produce 30 nm thick sections which showed fine structure details. Parallel polystyrene rods were observed in the SBS copolymer. Ultramicrotomy and selective staining with osmium tetroxide was also used in the preparation of a binary blend of PP and thermoplastic rubber [83].

Fridman *et al.* [84] sectioned polyester based polyurethanes using glass knives with a $45°$ cutting angle and propanol at $-90°C$ as the trough fluid for floating off the cryosections. The trough (or boat) was filled with water and allowed to freeze, forming a shelf onto which a drop of propanol or isopentane was placed. The alcohol lubricates the knife edge permitting sections to be brushed onto a TEM grid. The structure and morphology of segmented polyurethanes was also shown by TEM observation of samples prepared by cryoultramicrotomy [85]. A glass knife was used in the range of −75 to −65°C. A dry knife and a wet knife, with propanol as trough fluid, were used for the preparation of sections less than 100 nm thick. Sitte [86] described a microtome developed specifically for cryoultramicrotomy. Zierold [72] reviewed the status of cryomicrotomy for biological specimens where he showed that the method is useful for the determination of ultrastructure [87] and for elemental distribution [88, 89]. Prefreezing with liquid propane cooled liquid nitrogen, controlled specimen and knife temperature and use of a discharge device to limit static charge were conditions that had to be optimized for successful sectioning. The theoretical and practical applications of low temperature techniques, including cryoultramicrotomy, have been presented in a comprehensive text by Echlin [90].

4.3.5.2 *Summary*

Cryoultramicrotomy is not a routine method of specimen preparation for polymer specimens. However, new instruments are now available that permit control of the operating variables. Cryosectioning has several advantages: specimen embedding is not required, which limits the potential of chemical reaction; soft polymers can be sectioned, which may not be possible at room temperature; and hardening is not performed by a chemical reaction. Disadvantages include: it is time consuming; special equipment is required to control knife and specimen temperature; static charge affects picking up of sections; and frost buildup limits the method. This method is used more often in polymer microscopy than in the past, as it is valuable for materials science studies.

4.4 STAINING

4.4.1 Introduction

In transmission electron microscopy, image contrast is the result of variations in electron density among the structures present. Unfortunately, most polymers in common with biological materials are composed of low atomic number elements, and thus they exhibit little variation in electron density. In addition, the production of very thin specimens, by microtomy (Section 4.3), for example, is difficult. TEM micrographs of multiphase polymers often do not provide enough contrast to image the phases clearly. Methods which have proven useful in contrast enhancement include shadowing with a heavy metal in a vacuum evaporator and staining, generally by the addition of heavy atoms to specific structures. Shadowing is a topic of a later section of this chapter (Section 4.7.2.2) and staining is the topic of the present discussion.

Staining, in the most general terms, involves the incorporation of electron dense atoms into the polymer, in order to increase the density and thus enhance contrast. In this general sense of the term, staining will refer to either the chemical or the physical incorporation of the heavy atom. Biological materials often have sites, or specific structures, that react with various stains to yield contrast and thus information about these structures. Polymers are quite another story. Although there is little differential contrast in polymers and thus there is the need for stains, there is no single source book of the various stains matched with their reactive polymer groups. Several authors [35, 91, 92] have provided reviews which include some of the stains actively used for polymers.

Most of the stains applied to polymers are positive stains. In positive staining, the region of interest is stained dark by either a chemical interaction or by selective physical absorption. Chemical reactions are preferred as stains that are only physically absorbed (such as iodine) may be removed in the vacuum of an electron microscope. In microstructural work, staining may occur chemically after the staining agent penetrates some regions because of a higher diffusion rate. In negative staining the shape of small particles mounted on smooth substrates is shown by staining the regions surrounding the particles rather than the particles themselves. Such staining methods are often applied to latex or emulsion materials. Many of the stains used for polymers have been adapted from biological methods [93–95].

4.4.1.1 *Literature review*

Staining of textile fibers, with high atomic number elements, has been employed since the mid-1950s. Boylston and Rollins [96] reviewed the use of osmium tetroxide, iodine, phosphotungstic acid and uranium salts, stains which were initially used by biologists. Early studies generally involved investigation of the fine structure of cellulosic textile fibers. Maertens *et al.* [97] used osmium tetroxide and silver nitrate to stain cellulose, and Hess and coworkers [98] deposited iodine in natural cellulose fibers. Phosphotungstic acid was used as a negative stain [99] to show that cellulose is composed of elementary fibrils, about 3.5 nm in diameter, organized into microfibrils about 10–30 nm. Kassenbeck and Hagege [100], among others, reported thin sectioning studies of these natural fibers. Iodine was described [96] as a stain for nylon 6 fibers; phosphotungstic acid was proposed as a stain for nylon, polypropylene and Terylene; and osmium tetroxide was advocated as a stain for polyester and polybutadiene. Hagege *et al.* [101] described an interesting method of inclusion of stainable unsaturated polymers within the fibrillar framework of cellulosic fibers.

Walters and Keyte [102] first observed dispersed particles in blends of rubber polymers by phase contrast optical microscopy. Marsh *et al.* [103] studied elastomer blends by both optical phase contrast and TEM. Electron microscopy was applied to study blends of natural rubber, styrene–butadiene rubber (SBR), *cis*-polybutadiene (PB) and chlorobutyl rubber [104]. It became obvious that both hardening of the rubber and staining were necessary for producing sections with contrast for TEM. Today, the most common methods of observing multiphase polymers are by phase contrast OM of thin sections, TEM of stained ultrathin sections and SEM of etched or fractured surfaces.

Staining permits TEM observation of the dispersed phases in multiphase blends. Osmium tetroxide is the most commonly used stain for this application, while other stains have more limited application. Detailed fine structure of polymers is also made visible by staining. For example, chlorosulfonic acid staining enhances the lamellar texture of PE [105]. There are cases where a stain has been associated with a specific functional group of polymers. A specific stain for nylon, developed by Reimschuessel and Prevorsek [106], showed the sizes of the macrofibrils and microfibrils of nylon. Fibers were immersed in 10% aqueous solution of $SnCl_2$ for 10 min at 100°C, rinsed, placed in NH_4OH solution to convert the tin chloride to insoluble SnO and then embedded for ultrathin sectioning.

Staining of polymers can be conducted either before or after sectioning. The sample is cut into small blocks, about 1–3 mm across, and immersed in the stain solution or exposed to the vapor. Materials can be embedded and the blocks faced and then stained, especially when the stain diffuses into the polymer slowly. This method permits the sectioning and collection of the near surface material which is the most thoroughly stained. If sections can be cut prior to staining then they are stained either in the vapor, immersed in the solution, or placed on the surface of a stain droplet.

4.4.2 Osmium tetroxide

Multiphase polymers containing an unsaturated rubber phase form the largest single group of polymers studied by microscopy. This is due in no small measure to their ability to be stained with osmium tetroxide. The staining and hardening of rubber phases with osmium tetroxide was introduced by Andrews and Stubbs [76] and

Andrews [107], who stained unsaturated synthetic rubbers, and then further developed by Kato [108–110], to show the morphology of rubber modified plastics and unsaturated latex particles. The polybutadiene in ABS polymers is not apparent in unstained cross sections in the TEM, but staining results in contrast enhancement due to increased density of the unsaturated phase. Latex particles flatten and aggregate upon drying and early attempts at hardening, such as by bromination, were not considered successful. Thirty years after its first application, the method of osmium tetroxide staining is still widely and successfully applied to unsaturated rubbers and latexes.

Osmium tetroxide reacts with the carbon–carbon double bonds in unsaturated rubber phases enhancing the contrast in TEM by the increased electron scattering of the heavy metal in the rubber compared to the unstained matrix. The reaction is very important as it both fixes and stains the polymer. This *fixation*, as it is termed in biology, is a chemical crosslinking, or bridging, of the rubber which causes hardening and increased density. Staining is also known to take place by selective absorption in both semicrystalline and amorphous polymers. The reaction is slow, often taking days to weeks, when staining a block of material in an aqueous solution. Solvents are often added in order to increase diffusion of the stain. The high vapor pressure of OsO_4 is beneficial, making vapor staining of sections viable; however, this vapor pressure, combined with the toxicity of the stain, makes it very dangerous to use, and appropriate care must be taken to handle this material in a hood with good ventilation as otherwise harm could come to the eyes and mucous membrane (see suppliers for safety data sheets).

4.4.2.1 *Literature review*

Some examples of osmium tetroxide staining are worthwhile to discuss as they describe staining methods for specific polymers. Osmium tetroxide vapor was used to stain and harden (3 days) a thin film of a two phase blend containing crystalline polychloroprene [111]. The spherulitic texture was observed, likely by a combination of staining due to the unsaturation present and due to differential absorption by the crystalline and amorphous regions in the spherulites as the crystalline lamellae are unstained. Vapor staining with osmium is a general method for staining thin, melt crystallized polymer films.

Osmium tetroxide staining and ultrathin cryosectioning [82] of SBS copolymers showed structures which contain the styrene phase in cylindrical form arranged in a regular hexagonal macrolattice. A combination of prestaining, cooling with liquid air to harden the polymer and subsequent vapor staining of the sections with OsO_4 (24 h) revealed the now classical structure. Molair and Keskkula [112], Kato [110] and Matsuo *et al.* [113] have shown staining of high impact polystyrene (HIPS). Fridman and Thomas [114] used OsO_4 to reveal the structure of crystalline polyurethanes. The unsaturated, hard segment in the poly(propylene oxide)/4,4'-diphenylmethane diisocyanate/butenediol (PPO/MDI/BEDO) polyurethane was preferentially stained permitting visualization of the microstructure.

4.4.2.2 *Preferential absorption*

Preferential absorption of OsO$_4$ has been shown [115] to reveal spherulites in semicrystalline PET. Stefan and Williams' [116] work on ABS-polycarbonate blends also showed contrast by selective absorption. The dark SAN polymer, in this latter study, contains the osmium stained rubber particles while the polycarbonate was not stained. Niimoni *et al.* [117] found that there is often enough phase contrast in stained copolymers which have different degrees of unsaturation or functional groups like –OH, –O–, or –NH$_2$, as they each vary in reactivity with the stain. A specially constructed pressure bomb was developed by Edwards and Phillips [118] in order to terminate crystallization and fix polymers with OsO$_4$ at elevated pressure. This method has permitted determination of lamellar growth rates and the observation of developments in crystalline morphology.

An example of the staining of semicrystalline PET is shown in Fig. 4.11. Unoriented polyester chip was melt crystallized at 235°C for 2–3 h and subsequently stained by immersion in 4% aqueous osmium tetroxide. The amorphous regions in the spherulites appear to have enhanced electron density, due to the stain. The polarized light micrograph (Fig. 4.11A) shows the overall texture of the spherulites, while the higher magnification TEM micrograph of an ultrathin section (Fig. 4.11B) shows the detailed spherulitic texture where the amorphous regions exhibit enhanced electron density.

4.4.2.3 *Two step reactions*

Several two step reactions have extended the range of OsO$_4$ staining to materials that cannot be stained directly. Alkaline saponification at boiling temperature followed by reaction of the

Fig. 4.11 Unoriented, melt crystallized polyester stained for seven days by immersion in 4% osmium tetroxide exhibits a spherulitic texture in polarized light (A) and TEM (B).

hydroxyl groups with OsO$_4$ was used to study poly(vinyl chloride)/ethylene-vinyl acetate systems [119]. Riew and Smith [120] exposed samples of rubber modified epoxy resins to OsO$_4$ dissolved in tetrahydrofuran (THF). The THF diffuses into the epoxy, speeding the reaction with the rubber. Hobbs [6] recommended a concentration of 10% or less. Aqueous formaldehyde was used in the staining of polyamides [121]. Thin films of melt crystallized polyamide samples were placed in a mixture of equal parts of 30% aqueous formaldehyde and 1% OsO$_4$ for three days. The osmium tetroxide is reduced by the formaldehyde and reacts selectively, resulting in better definition of the polyamide structure. Kanig [122] developed a stain for butyl acrylate rubber by treatment with hydrazine or hydroxylamine and post-staining with OsO$_4$, which works for polymers containing acid and ester groups. The Os is deposited after the ester is reduced by the hydrazine. Hutchins [123] used solvent assisted osmium staining to characterize butylacrylate and ethylene propylenediene in a SAN matrix. Faced microtomy specimens were soaked in 1,7 octadiene, to introduce C=C sites and then stained in 1% aqueous OsO$_4$ at 60°C for microtomy and TEM.

Fleischer *et al.* [124] developed a two stage method involving selective dehydrochlorination, to form double bonds, followed by reaction with osmium tetroxide. Siegmann and Hiltner [125] applied this method to image chlorinated PE in an impact modified poly(vinyl chloride). Small blocks were cut and immersed in a bicyclic amine, 1,8-diazabicyclo[5.4.0]undec-7-ene, for 2 days at 0°C. Samples were rinsed with 2 N HCl and stained in aqueous osmium tetroxide for 14 days prior to ultrathin sectioning. The dispersed phase particles were observed showing that the morphology changed from discrete particles to a

network structure as the impact modifier was increased. Wegner *et al.* [126] developed a method of staining polyesters and segmented copolymers having ester bonds by reacting thin solution cast films with allylamine, which is absorbed to the boundary between crystalline and amorphous regions.

4.4.2.4 *General method*

Osmium tetroxide is available in small ampoules either as crystals, ready to dissolve in water, or as premixed solutions. The ampoules are generally prescored and require scoring to cut open and pour into glass containers. Staining prior to embedding can enhance contrast and harden the material. Penetration is rather poor and days to weeks may be needed to stain a specimen by immersion in a 1–2% aqueous solution. Embedded and faced specimen blocks can also be stained by immersion. Crystals can be placed in the bottom of a tube and the specimens or sections on grids placed above them for vapor staining. To speed the reaction, the tube used for vapor staining is placed in a beaker of water on a hot plate. Vapor staining at 50°C requires about 8 h for a bulk specimen or 1–2 h for sections. Diluted emulsion or latex particles are dropped onto coated grids and stained over 1% aqueous osmium tetroxide in a closed vessel for about 30 min. Stain times are dependent upon the form of the specimen, the mechanism of reaction, the degree of unsaturation and the temperature used.

4.4.2.5 *Inclusion methods*

Polymers containing unsaturated rubber and semicrystalline polymers are often effectively stained using osmium tetroxide. What about materials that do not show such differential staining? Two examples will be described where reactive (unsaturated) materials are included into the polymer to provide reaction sites. Inclusion of a stainable unsaturated polymer was shown for cellulosics [101] and synthetic fibers [127]. The initial work was focused on improvement of the properties of cellulosics by inclusion of an elastomer between the microfibrils. OsO₄ staining revealed that a lamellar sheet structure was present. Marfels and Kassenbeck [127] used a similar method with polyester and nylon fibers.

An outline of the isoprene inclusion method is as follows.

(1) Treat the fibers overnight in a 1% solution of benzoyl peroxide catalyst, in freshly distilled isoprene.
(2) Rinse the fibers in fresh isoprene.
(3) Suspend the fibers in a metal autoclave filled with 5–20 ml distilled isoprene. Seal and heat to 90°C for 6–8 h at about 2–3 kg/cm² pressure.
(4) Dry and embed the fibers for ultrathin sectioning.
(5) Stain sections in OsO₄ vapor for one hour at 50°C.

Application of this method to polyester fibers [128] is shown in the TEM micrographs in Fig. 4.12. Figure 4.12A shows a section of an untreated fiber with dense titanium dioxide particles present. A section of a treated and stained fiber (Fig. 4.12B) shows long, thin, dense structures, along the fiber axis that may be voids. Dense, stained isoprene regions are observed in teardrop shaped voids adjacent to delustrant particles. This method is quite useful for observation of void sizes and shapes which relate to factors such as dyeability.

The second reactive inclusion method was developed [129] for microporous membranes. Stretched polypropylene, Celgard 2500 (trademark, Hoechst Celanese Corp.), shows little fine structure after ultrathin sectioning and examination in the TEM (Fig. 4.13A), although SEM study clearly reveals a surface pore structure. In order to enhance contrast, the membrane was treated with an unsaturated surfactant followed by osmium tetroxide staining and ultrathin sectioning. A 1–2% solution of polyoxyethylene allyl ether, Brij 97 (available from ICI Americas Inc.), in 50/50 methanol/water solution was used to

 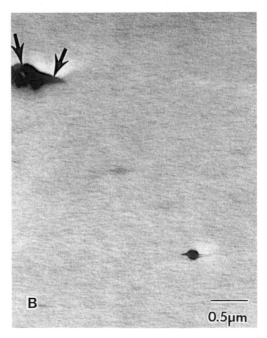

Fig. 4.12 Longitudinal sections of polyester fibers shown by TEM (A) do not have any interesting morphology, unstained, except for particles of TiO₂ (arrows). Impregnation with isoprene followed by OsO₄ staining results in electron dense longitudinal striations and elongated voids (arrows) adjacent to the particles (B).

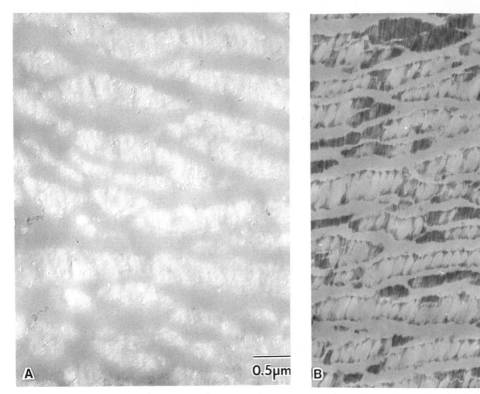

Fig. 4.13 TEM micrographs of a sectioned microporous membrane show little detail of the structure due to the low electron scattering of the polymer (A). Treatment with an unsaturated surfactant followed by osmium staining results in sections with enhanced contrast which permits assessment of the microporous structure (B).

treat the membrane for about 10 min. Strips of the treated membrane were placed in a test tube containing OsO_4 crystals, the tube was sealed with a cork and placed in a beaker of water at 50°C for about 8 h in a well ventilated hood. Figure 4.13B shows that the pores in the membrane have been coated with the surfactant and stained, revealing the microporous structure. Higher magnification micrographs of similar membranes are shown in Section 5.2. This method has general utility for revealing the structure of porous polymer materials.

4.4.2.6 *Staining for SEM*

Rubber particle morphology has also been shown by Keskkula and Traylor [130] who removed the rubber from impact polystyrene (IPS) and ABS polymers and examined their surface topography

in the SEM. The method involved reaction of a small piece of polymer with 1% OsO_4 solution in cyclohexane on a steam bath. The polystyrene was separated from the hardened particles by washing with isopropanol. Comparison with other methods shows this to be interesting for the study of particle size, shape and deformation. Some caution is in order, however, as comparison of diameter measurements made by electron microscopy of stained diblock copolymers of polybutadiene spheres in a polystyrene matrix and small angle neutron scattering have shown some discrepancy [131].

4.4.3 Ebonite

The study of phase size and compatibility requires that the different phases be observed, distinct from one another, and that there is minimum distortion in the polymers. However, multiphase polymers often cannot be stained or sectioned uniformly. There are composite structures that are combinations of soft rubbers, coatings and oriented fibers which cannot be stained with a single staining agent and the sections may be deformed or distorted, limiting both observation and interpretation. The ebonite method was developed by Smith and Andries [132] to stain and uniformly harden polymers.

4.4.3.1 *Literature review*

Smith and Andries [132] modified a method using molten sulfur to transform rubber to ebonite that could be polished for surface examination. A small block of the sample was held in a molten mixture of sulfenamide accelerator/zinc stearate/sulfur in a 90/5/5 weight ratio at 120°C for 8 h. A control method these authors used was cryosectioning with subsequent OsO$_4$ staining. Blends of styrene–butadiene rubber (SBR) with chlorobutyl rubber (CB) or *cis*-PB, prepared by the two methods, revealed that the overall structures were the same, but the ebonite method was the simpler of the two. With ebonite, both phases stained, but the SBR appeared to have more contrast. With newer and supposedly easier to use cryomicrotomes, it may be that the ebonite method has less current utility. It is certainly not as popular as OsO$_4$ even though the materials are less toxic to use. Three diblock copolymers of 1,4-polybutadiene and *cis*-1,4 polyisoprene and blends of these copolymers with the corresponding homopolymers were prepared by the ebonite method [133], thereby providing useful observations of the dispersed phases.

4.4.3.2 *General method and examples*

A modified ebonite method was developed [134] to study the interfaces associated with polymer tire cords. Tire cords composed of PET, rayon or nylon fibers are generally bonded to rubber with a resorcinol–formaldehyde–latex (RFL) adhesive. The nature of the interfaces is of interest in tire cord studies. OsO$_4$ may be used to stain and harden the RFL, but the soft rubber is not affected by this treatment, and, in fact, it forms a barrier to stain penetration. The ebonite reaction hardens the rubber and hardens and stains the RFL while maintaining the geometrical integrity of the composite.

The reaction medium consists of molten sulfur/accelerator (*N,N*-dicyclohexyl-2-benzothiazolylsulfenamide)/zinc stearate in the weight ratio 90/5/5. Small pieces of the cord are cut from the tire carefully trimming some of the rubber but leaving a thin, undisturbed surface layer. Eight hours are required for the reaction at 120°C. Samples are removed, scraped off and placed in a 120°C oven to remove the excess ebonite. The treated cords are embedded and sectioned with a diamond knife to reveal the interfaces that relate to the adhesive coating and the tire cord adhesion. The fiber–RFL–rubber interfaces are all observed in the tire cord cross section (Fig. 4.14). A more complete description of this type of study is described in Chapter 5.

In summary, the ebonite method can be used routinely for polymer blends or composite specimens, such as tire cords, where hardening,

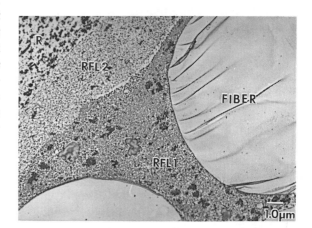

Fig. 4.14 TEM micrographs of sectioned industrial tire cords are prepared by the ebonite method to enhance the contrast of the various structures and to harden the adhesive and the rubber for sectioning. Fiber cross sections, two adhesive layers (RFL) and the rubber (R) are shown by TEM.

penetration and staining are required. Although sulfur is a poison that must be used in a hood, it is easier to use and less toxic than OsO$_4$. The method is less time consuming. The resolution possible with the ebonite method is quite good, enabling the imaging of the fine structure of the latex adhesive, which is only possible with uniform staining and ultrathin sections about 50–60 nm thick.

4.4.4 Chlorosulfonic acid

An important staining technique was developed by Kanig [135] for the enhanced contrast of polyethylene, a material which has been a model compound for fundamental polymer studies. PE crystals cannot be sectioned, nor are they stable in the electron beam, due to radiation damage. The chlorosulfonation procedure crosslinks, stabilizes and stains the amorphous material in crystalline polyolefins, permitting ultrathin sectioning and stable EM observation. Chlorosulfonic acid diffuses selectively into the amorphous material in the semicrystalline polymer, increasing the density of the amorphous zone compared to the crystalline material. The treatment stains the surfaces of the lamellae primarily due to incorporation of chlorine and sulfur. Treatment with a salt solution results in a reaction with the polar groups, and metal ions are deposited resulting in increased electron density. Post-staining with uranyl acetate intensifies and stabilizes the contrast. Lamellar structures are revealed in PE that are now known to be typical of semicrystalline polymers.

4.4.4.1 *Literature review*

Chlorosulfonic acid was used to show the lamellar structure in both linear and branched PE and to study the effects of drawing and annealing [122, 136, 137]. Dlugosz *et al.* [138] examined cryosections of drawn, rolled and annealed, bulk oriented PE which clearly showed a lamellar texture. Lamellae were shown to have two preferred orientations and to be arranged in

stacks perpendicular to the draw direction. Experiments using chlorosulfonation on other olefins and polyesters were shown to result in their dissolving in the acid rather than staining. Hodge and Bassett [139] prepared and observed the lamellar texture in bulk PE documenting an evaluation of the staining method, mechanism, application and limitations using PE in the chain extended form. The reaction time was shown to vary depending on the diffusion channels open. Voigt-Martin *et al.* [140] prepared chlorosulfonic acid stained cryosections [141] of melt crystallized linear PE and compared the morphology with TEM of replicas and light scattering. Results showed lamellar crystallites in the entire molecular weight range studied. Kanig [142] described the lamellar crystallization of polypropylene from the melt using this method. Staining was also applied (120°C) to the crystallization of PE [142, 143]. The chlorosulfonation method has been applied to bulk polymers and high modulus fibers [144–146]. Smook *et al.* [146] studied the fracture process of ultrahigh strength PE fibers and used the method in a unique application to show the nature of the kink bands present. Highly oriented ultrahigh molecular weight PE fibers were shown to be preferentially attacked at these kink bands when exposed to the acid for 45 min at 80°C.

Schaper *et al.* [147] developed structure–property relations for surface grown PE fibers before and after zone drawing by studying ultrathin sections stained with chlorosulfonic acid. Computer processing was used to reveal detail in HREM imaging which showed the microfibrillar superstructure of the fibers. Additionally, shish kebab structures were shown by three methods, chlorosulfonic acid staining, gold decoration (Section 4.7.5) and after permanganate etching (Section 4.5.3). SEM showed fiber buckling and kinking which is another effect of the highly oriented fibrillar structure [147]. Kalnins *et al.* [148] studied PE and PP treated with chlorosulfonation and collected data for the penetration depth versus treatment time using optical microscopy. The penetration time was found to be constant with time in homogeneous, bulk samples of PE. Penetration was affected by thermal oxidation and a decrease in crystallinity.

4.4.4.2 *General method and examples*

A general method for staining PE with chlorosulfonic acid is as follows.

(1) Treat the sample with chlorosulfonic acid for 6–9 h at 60°C.
(2) Wash the stained sample in concentrated sulfuric acid and then in water.
(3) Dry and embed the sample in epoxy resin.
(4) Cut ultrathin sections with a diamond knife.
(5) Post-stain the sections in 0.7% aqueous uranyl acetate (3 h).

The stained lamellar texture is shown (Fig. 4.15) to result from the chlorosulfonation of linear PE crystallized isothermally from the melt [105]. Although large parts of the section show no detail, some regions do contain the parallel dark lines shown in this figure. These lines are a few hundred Ångströms apart, and they appear and

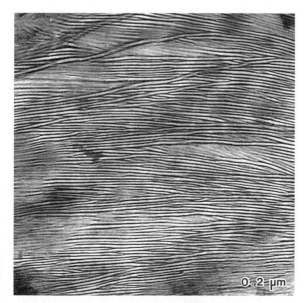

0.2 µm

Fig. 4.15 TEM micrograph of a chlorosulfonic acid stained linear polyethylene crystallized isothermally from the melt reveals the electron dense interlamellar surfaces typical of polyethylene.

disappear as the section is tilted in the microscope. The interlamellar surfaces are electron dense, showing that they are stained. It should be remembered that uranyl acetate alone can also be used for staining polymers.

4.4.5 Phosphotungstic acid

Phosphotungstic acid was first used for biological staining of structures about 1945. Hayat [149] described that early work and what is known about the mechanism of staining. Phosphotungstic acid (PTA) is an anionic stain with a high molecular weight (3313.5), which imparts high density to the stained material. There is no agreement, apparently, among biologists as to the interaction of this stain with organic materials, although it is known to stain proteins. Two interpretations are the formation of a complex in aqueous solution and ionic precipitation. In any case, the specificity of staining, at least in biological tissue, appears related to the pH of the solution. This is due to the fact that the PTA molecule is unstable and degrades when the pH is higher than 1.5.

4.4.5.1 *Literature review*

Phosphotungstic acid staining was used to show the fine structure in nylon 6 fibers [150] by soaking the fibers in 9–11% aqueous salt solutions and staining with 1.5 and 4.8% PTA. Longitudinal periodicities were shown for the stained fibers by TEM. However, the lamellae showed a change in size, depending on the concentration. At 4.8% PTA the lamellae were about 7 nm. These authors reported [151] that 9–11% HCl treatment bound 15% of the weight of PTA in the unoriented or amorphous regions of nylon. TEM and electron diffraction showed a range of periodicities in nylon from the ordered, oriented and unoriented regions.

Spit [152, 153] showed detailed spherulitic structures in solvent cast nylon 6 and nylon 6,6 films cast from formic acid onto water and stained with 2% PTA. Spherulitic textures with white and dark bands or lines were observed.

The dark bands contained the electron dense stain while the white, crystalline areas did not absorb the stain, enhancing the contrast in the 5 nm wide lamellae. The fine structures of nylon 6,6 in bulk and fiber form were shown to have fibrils about 10 nm wide in spherulites [154] following PTA staining and thin sectioning. Elongated spherulites were observed oriented to the fiber axis in undrawn fibers. The fine structures were observed due to the absorption of the phosphotungstic acid into the amorphous regions. According to Boylston and Rollins [96], PTA was used to reveal the fine structure of polyesters [155, 156]. Polyoxymethylene was also stained with PTA [157].

The staining of polypropylene was not as straightforward as that of the polyamides and polyesters. Hock [158] developed a method for staining melt crystallized PP and showed it to be composed of spherulites containing lamellae. The melt crystallized polymer was boiled for various times in 70% HNO_3 at 120°C. After 4 h the reaction had leveled off and the polymer exhibited no further measurable change. Small chips of the oxidized polymer were reacted in 5% aqueous PTA for three days at room temperature. Functional groups that polyamides have in common with proteins might be the reason why they can be stained with PTA. However, the direct staining of olefins by PTA would not be expected. In the case of oxidized PP staining apparently occurred where the folded chains had been cut by the acid and the $-NO_2$ and $-COOH$ groups were attached to the remaining short molecules. Staining clearly revealed the interlamellar regions adjacent to the unstained crystalline lamellae, and lamellar thicknesses were measured as a function of the crystallization temperature.

4.4.5.2 *General methods and examples*

PTA is known to react with monomer epoxy resins which extract the stain [93], precluding its use prior to epoxy embedding. Therefore, PTA stained material is usually either embedded in glycol methacrylate or polyester resins, or sec-tions are post-stained in cured epoxies. PTA penetration is slow and about 100 μm penetration into a block of material can be expected. Sections on grids are immersed into the solution or placed on a droplet. Pretreatment in absolute ethanol increases exposure to the stain. Martinez-Salazar and Cannon [159] reported a new method of PTA staining of nylon 6 and nylon 6,6 using 2% PTA and 2% benzyl alcohol. Thin films on a specimen grid were floated on a drop of the mixed solution for 10 min and then washed in water several times.

PTA reacts with surface functional groups such as hydroxyl, carboxyl and amines [160] as a positive stain. Shaffer *et al.* [161] used PTA for negative and positive staining which they found useful in enhancing the contrast in TEM imaging of latexes, such as poly(butyl acrylate) and poly(ethyl acrylate). One drop of latex was added to 1 ml of *fresh* 2% PTA to stain deformable or low glass transition temperature material. A drop of the stained latex was placed on a carbon–formvar coated, stainless steel grid. After removing excess fluid with filter paper, the specimen was placed in a TEM cold stage and frozen. Figure 4.16 shows TEM micrographs of a latex, observed using a cold stage, with and without staining [162]. The unstained latex shows (Fig. 4.16A) discrete particles that are somewhat aggregated. After PTA staining, the particles could be imaged more clearly (Fig. 4.16B). Core–shell latex particles, such as electron dense polystyrene, can be observed by this method. Sections of materials can also be stained; for example, poly(vinyl acetate) sections were stained with 5% PTA for 30 min [161].

4.4.6 Ruthenium tetroxide

Ruthenium tetroxide is known to be a stronger oxidizing agent than osmium tetroxide and supposedly superior for staining rubber [163]. It is known that ruthenium tetroxide will oxidize aromatic rings yielding either mono- or dicarboxylic acids [164]. Although osmium tetroxide is used in more polymer staining experiments than any other reagent, there are materials which

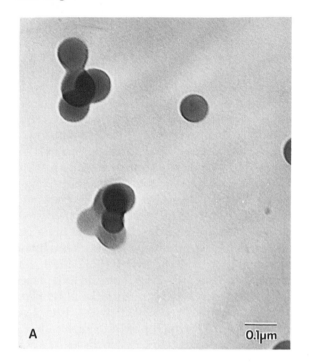 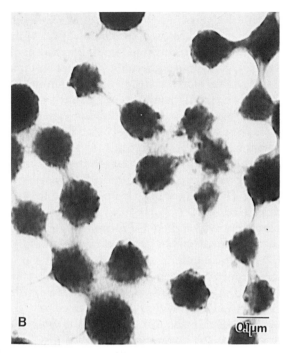

Fig. 4.16 TEM micrographs of latex particles are shown of images taken in a microscope with a cold stage (A) and following both staining with PTA and using the cold stage (B).

it cannot stain, including saturated polymers, and it also exhibits slow and relatively poor diffusion.

4.4.6.1 *Literature review*

Two groups independently introduced the use of RuO_4 as a staining agent. Vitali and Montani [163] have shown the staining of latex and resin materials. Polybutadiene latexes, treated in OsO_4 and RuO_4 vapor, for comparison, showed no significant differences. ABS resins were treated by both reagents and the TEM thin sections showed they both reacted with the unsaturated rubber. Treatment of a saturated acrylonitrile–styrene–acrylate (ASA) resin resulted in no staining with OsO_4 whereas the saturated rubber phase was hardened and stained by 1% RuO_4 treatment. The reaction is attributed to a cross-linking mechanism on the ester groups of the acrylate phase.

Trent *et al.* [165–167] introduced ruthenium tetroxide stain for the electron microscopy of polystyrene/poly(methyl methacrylate) (PS/PMMA), ABS and nylon 11. Films were cast from toluene onto glass slides and sections were picked up on TEM grids. Staining was conducted by mounting the grids on a glass slide suspended over a 0.3% aqueous solution of RuO_4 (30 min) at room temperature. The polystyrene was stained in PS/PMMA.

RuO_4 stains polymers containing ether, alcohol, aromatic or amine moieties [166], and no reaction was observed for PMMA, PVC, PVF_2 and PAN. HDPE, isotactic PP and atactic PP were said to be lightly stained. Thin films cast from 1% solutions were stained in RuO_4 vapor for 120 min. HIPS was cast from toluene; ABS was cast from ethyl acetate; nylon 11 was cast from 50/50 phenol/formic acid; and PS was cast from toluene. The combined use of OsO_4 and RuO_4 staining revealed interesting detail in ABS. The spherulitic texture of nylon 11 was observed after a thin film was exposed to RuO_4 vapor. Morel and Grubb [168] conducted staining experiments on amorphous and spherulitic films of isotactic PS by vapor staining from a 1% solution for 5 min. The lamellar morphology of melt crystallized PS was shown by the staining experiment with the added benefit of stability against radiation damage. In contrast to OsO_4, RuO_4 appears to stain the surface layers and react more with the lamellar surfaces than the amorphous regions. The method has also been used to stain crazes [168, 169].

Frochling and Pijpers [170] used RuO_4 to stain the amorphous rubber phase in impact modified polyamide containing poly(propylene glycol) as added rubber. A blend sample was cryoultrathin sectioned at $-60°C$, a block copolymer thin film was cast and both were vapor stained in 0.5% aqueous solution of RuO_4 for 30 min or in the vapor of a 2% aqueous solution of OsO_4 for 10 min. The OsO_4 staining did not reveal contrast but the RuO_4 clearly did stain the polyether. Ohlsson and Tornell [171] used RuO_4 vapor to stain blends of PP and SBS. The SBS used was Kraton 1101 from Shell, a triblock thermoplastic elastomer with polystyrene endblocks and a central block of polybutadiene. The solid polymers were cooled in liquid nitrogen, cut in a microtome and vapor stained for 3 to 24 h for backscattered electron imaging in an SEM and for sectioning for TEM imaging. Selective etching with xylene for 24 h dissolved the thermoplastic elastomer for SEM imaging. Ruthenium in hypochlorite was used as a stain for TEM of ternary blends of PP, EPDM and HDPE prepared by two different process methods [172]. Samples were trimmed for microtomy, treated for 16 h with a 2% solution of ruthenium trichloride (see below) [176], sectioned and examined by TEM. This method worked very well and showed the lamellar structure of the PP matrix and the dispersed HDPE phase as the amorphous regions were stained.

Hobbs *et al.* [173–174] have studied the morphology and toughening mechanisms of toughened blends of poly(butylene terephtha-late) (PBT) and BPA polycarbonate using a combination of staining and etching methods to prepare samples for SEM and TEM. Use of several methods is an aid to the detection of artifacts that can occur with both staining and etching preparation methods. Both osmium and ruthenium tetroxide were employed for selective staining. The impact modifier reacted with OsO_4, as it oxidizes double bonds. Immersion in a 1% solution in hexane for 30 min was sufficient. The PC absorbed RuO_4, and the PBT did not. Copper grids with thin sections on them were glued to a glass slide and suspended above the RuO_4 solution in a stoppered bottle for 30 min. The RuO_4 solution was prepared by the first method described in the next section [175]. SEM was conducted on samples crystallized in a hot stage and on cross sections faced with the ultramicrotome after etching the PC with diethylene triamine (DETA), which had little effect on the PBT. This thorough work of Hobbs *et al.* [173–174], showed that the various combinations of stains and etching were effective in revealing the morphology of these blends. The PBT was seen to be the continuous phase, with the core–shell impact modifier isolated in islands in the PC. The toughening mechanisms were determined by correlation of morphology and physical properties.

4.4.6.2 *General method and discussion*

The general method for ruthenium tetroxide staining is to stain sections over a *fresh* 1% solution for about 5–30 min. This is a problem as ruthenium tetroxide solutions are quite unstable, though Trent *et al.* [166] have frozen solutions in sealed glass containers for periods up to six months and report that ruthenium tetroxide can be prepared by oxidation of hydrated ruthenium dioxide using sodium periodate (available from Morton Thiokol, Inc., Alfa Products). The reaction, shown in the following equation [177]:

$$2NaIO_4 + RuO_2 \xrightarrow[H_2O]{1°C} RuO_4 + 2NaIO_3$$

$$\quad 4g \qquad 0.6g \quad 100\,ml$$

is complete in 3–4 h [175]. Trent [175] and Montezinos *et al.* [176] reported this instability and suggested alternative reactions for preparation. Montezinos *et al.* [176] discussed the preparation of RuO_4 by oxidizing ruthenium dioxide or trichloride with sodium metaperiodate and extracting the tetroxide with chloroform. However, it seems this is not simple or fast, whereas oxidation of ruthenium trichloride with sodium hypochlorite [178] is a viable alternative:

$$2\,RuCl_3 \cdot 3H_2O + 8\,NaClO \xrightarrow{\;H_2O\;}$$

$$0.2\,g \qquad\qquad 10\,ml \qquad 100\,ml$$
$$5\%\ aq$$

$$2\,RuO_4 + 8\,NaCl + 3\,Cl_2 + 3\,H_2O$$

This latter, one step procedure was applied [176] to PE films and blends of PE and PP with elastomers. Treatment times of 30 min to 3 h were followed by drying and embedding in Spurr epoxy resin for ultramicrotomy. Care must be taken as ruthenium tetroxide is volatile and toxic although little is known regarding health hazards. The use of materials safety data sheets supplied with these chemicals is recommended. The expected reaction of unsaturated chains with ruthenium tetroxide is given by:

$$-CH=CH- + RuO_4 \longrightarrow -C\underset{O}{\overset{H}{\diagdown}}$$

$$+ CH_2=CH-$$
$$+ RuO_2$$

4.4.6.3 *Examples*

Recently, Wood [179] showed the benefit of microwave stimulated heavy metal staining with RuO_4 of conventional and ultrahigh molecular weight polyethylene, segmented block copolymers, ethylene–propylene copolymers and polystyrene. Microwave methods have been used for fixation and staining of biological samples [69–70, 179] but they have not been reported for staining polymers, although they are used for

digestion and ashing. The staining agent used was RuO_4 vapor generated from ruthenium trichloride [176] in an industrial microwave oven installed in a chemical fume hood. A staining vessel with sections on grids suspended over the solution was transferred to an oven operated at 17% power for 10 to 30 min. Changes in staining were attributed to heating the aqueous mixture, enhanced reaction rate and improved interactions of the water and polymers [179].

Polymers with carbon black were not used due to potential intense heating. Wood [179] showed interfacial staining at the boundary of polyethylene domains dispersed in HIPS was enhanced relative to room temperature staining (Fig. 4.17). The polyethylene-rich regions also contain small unstained domains of crosslinked polyethylene as well as two other ethylene copolymers (Fig. 4.17A). A comparison of TEM images of occluded PS particles in the HIPS prepared using microwave (Fig. 4.17B) versus conventional staining (Fig. 4.17C) shows greater contrast at the HIPS/rubber interface, which may contain some grafted polymer, in the case of microwave staining. The effect of two different stains is shown in Fig. 4.17D of a pigmented Noryl blend of poly(phenylene oxide) PPO and HIPS [180]. The pigment particles are best seen in an unstained section (Fig. 4.17E) whereas OsO_4 reveals the butadiene rich structure in HIPS (Fig. 4.17E), while regions rich in PS are seen best by RuO_4 (Fig. 4.17F).

The study of core–shell morphologies is very important to the field of emulsion polymers and also to toughened polymer blends. Shaffer *et al.* [181–183] are well known for their work in this field and have developed many methods for analyzing morphology. Methods used include negative staining with phosphotungstic acid [181] (Section 4.4.5), OsO_4 [181] (Section 4.4.3) and RuO_4 [182]. A few drops of the latex mixture are combined with a few drops of 2% aqueous solution of uranyl acetate, which acts as a negative stain. A drop of this mixture is deposited on a formvar coated stainless steel grid. The grid is then exposed to RuO_4 vapors to differentiate the phases in the core and shell in

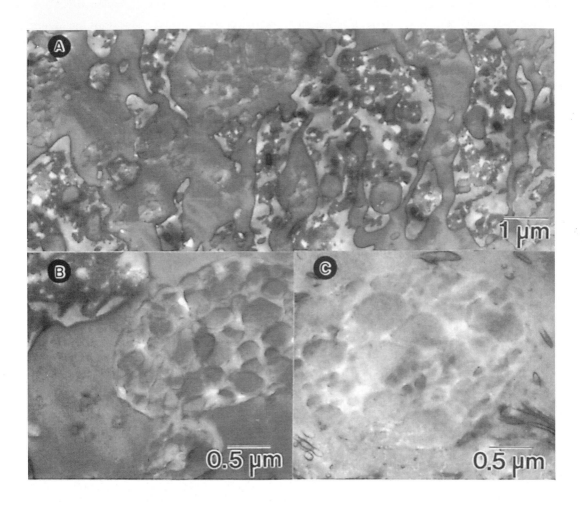

Fig. 4.17   TEM images [179] of microtomed section of HIPS/PE/ethylene copolymer blend stained with RuO$_4$ in a microwave oven, showing selective staining at the HIPS/PE interface (A). Higher magnification of HIPS region of material (B); HIPS prepared by conventional RuO$_4$ staining (C); rubber phase appears white. The effect of two different stains in a pigmented blend of PPO and HIPS is shown [180]: the pigment particles are best seen in an unstained section (D) whereas OsO$_4$ reveals the butadiene rich structure in HIPS (E), while regions rich in PS are seen best by RuO$_4$ (F). (From Wood [179, 180], reproduced with permission.)

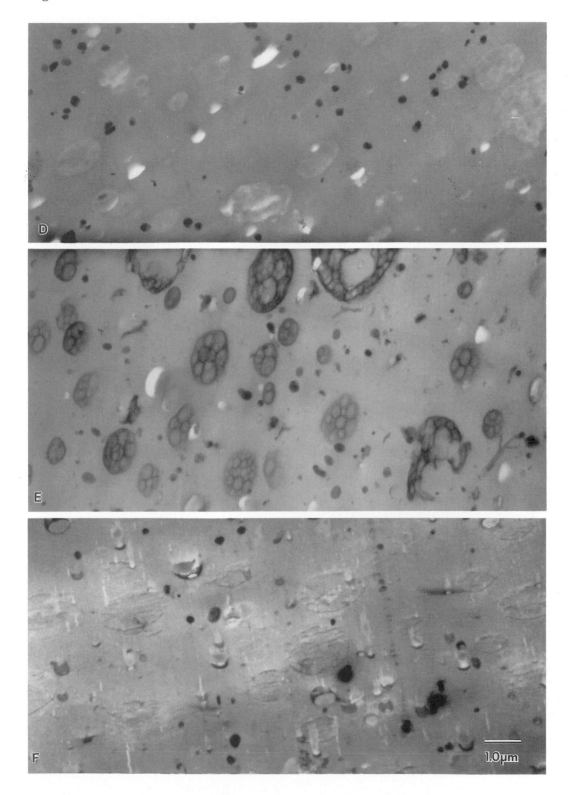

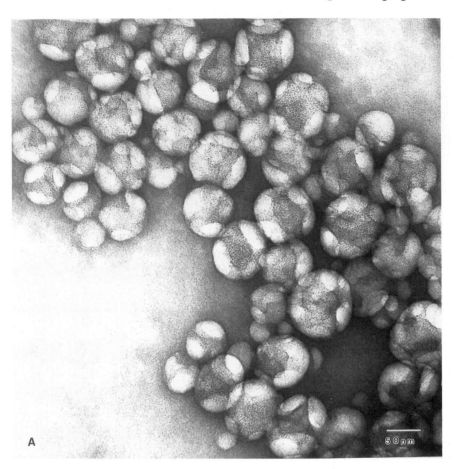

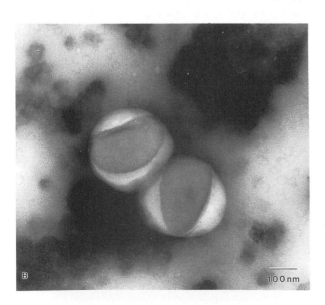

Fig. 4.18 TEM images of latex particles prepared by staining with OsO_4, RuO_4, and phosphotungstic acid [183]. TEM image (A) shows latex particles with dark phases, the core of PBA–PB stained with OsO_4 and RuO_4, and negative staining due to the PTA. 'Lumps' of PMMA can be seen on the core. TEM image (B) shows a core–shell latex of PS with a PBA shell, prepared by the same method without the OsO_4; the RuO_4 stained the PS core and this is seen against a lighter PBA incomplete shell. (From O. L. Shaffer, unpublished [183]).

SAN, HIPS, ABS, ASA and nylon 11; HDPE and PP are only lightly stained by this method.

4.4.7 Silver sulfide

Microporous structures in biological and polymer materials are often difficult to observe, even in the TEM, as there is limited differential electron scattering and thus limited contrast between the embedding resin and the specimen. The electron contrast can be increased by the addition of a high atomic number element, such as silver, atomic number 47, to porous regions. This was shown earlier for osmium tetroxide (Section 4.4.2.3). The silver sulfide method has been described for use in both natural and synthetic fibers. Sotton [184] studied wool, regenerated cellulose, poly(vinyl alcohol), acrylics and polyesters. Hydrogen sulfide was liquefied and the samples treated in an autoclave under pressure prior to treatment with a silver nitrate (O.1 N) solution which precipitated silver sulfide in the pores. Silver nitrate can be used with water, acetone or ethylene glycol as solvents.

Hagege, Jarrin and Sotton [185] have used a similar silver sulfide insertion technique in the study of aramid fibers. The fibers were treated according to the method of Sotton [184] (described above) and also by the method of Hagege *et al.* [101], using isoprene but excluding the osmium tetroxide staining. Sections examined in the TEM showed details of the void microstructure. Dobb *et al.* [186] also used this method to study the aramid structure. The silver sulfide insertion technique appears to have potential in providing contrast between the polymer and microvoids in some fibers.

This method continues to be used by the Leeds group. Dobb *et al.* (for instance, [187]) used the silver sulfide method to prepare new aramid fibers for TEM in order to evaluate the nature of the microvoids and to determine their influence on the tensile and compressive behavior of the fibers. The modification of the earlier method is the TEM image [182]. Figure 4.18 shows TEM images of latex particles prepared by staining with OsO_4, RuO_4 and phosphotungstic acid [183]. A latex with a core of poly(butylacrylate–butadiene) (PBA–PB) and a shell of PMMA was diluted in distilled water and stained by adding three drops of the 2% aqueous OsO_4 to the diluted dispersion. This dispersion was then diluted in 2% aqueous phosphotungstic acid (PTA) and a drop placed on a carbon coated formvar stainless steel TEM grid. The grid was then placed on a glass slide and stained in RuO_4 vapor for 10–20 min and air dried. Fig. 4.18A shows latex particles with dark phases, the core of PBA–PB stained with OsO_4 and RuO_4, and negative staining due to the PTA [183]. 'Lumps' of PMMA can be seen on the core. Fig. 4.18B is of a core–shell latex of PS with a PBA shell, prepared by the same method without the OsO_4. The RuO_4 stained the PS core and this is seen against a lighter PBA incomplete shell [183].

In summary, RuO_4 is a new oxidizing agent that appears somewhat similar to OsO_4 in the staining of unsaturated phases. Some saturated polymers may be stained with this reagent, by vapor phase reaction of sections for short times (30 min). Although the reactive moieties appear to include ethers, alcohols, aromatics and amines, it must be remembered that ruthenium tetroxide is known to oxidize aromatic rings and cleave double bonds or rings rather than bonding to them. Polymers that may be stained include PS, described below.

(1) Treat the specimen with gaseous hydrogen sulfide at a pressure of 1380 kPa at 20°C for 16 h; wash in alcohol.
(2) Immerse the specimen in a 5% aqueous solution of silver nitrate, at 20°C for 3–4 h [188].
(3) Embed the specimen in a low viscosity resin (e.g. Spurr resin).
(4) Ultrathin section with a diamond knife.

4.4.8 Mercuric trifluoroacetate

The structural elucidation of multiphase polymers is a continuing problem as, at times, the few commonly used staining reagents are not satisfactory. Osmium tetroxide has the widest range of applicability, but there are polymers that cannot be differentiated by this stain. Mercuric trifluoroacetate has been described for the staining of several polymers [189]: polystyrene, poly(2,6-dimethyl-1,4-phenylene oxide) and saturated styrene–butadiene–styrene block copolymers (Kraton G, trademark, Shell Oil Company).

The reaction of mercuric trifluoroacetate is said to take place by electrophilic substitution and is activated by electron donating groups, such as oxygen, which are attached to the ring. Apparently, bisphenol A polycarbonate/polyethylene blends can also be stained with reagent but must be diluted with water and used for shorter times (20 min) to prevent crumbling of the sample [189]. This staining reagent is very toxic and it must be handled in a hood, with proper protective clothing, although little is known about this specific substance. The slow rate of diffusion limits penetration of the reagent into the polymer. Long exposure times are limited by acid attack which results in swelling and disruption of the specimen surface [190]. In addition, the major disadvantage of the method is that there is no hardening of the polymers, making room temperature ultramicrotomy difficult.

4.4.8.1 *General method*

Trimmed and faced blocks are stained by immersion in a 10% solution of HgO in trifluoroacetic acid for 10–60 min. Samples are washed in a dilute solution of trifluoroacetic acid followed by distilled water. Poly(phenylene oxide) (PPO) appears to have a higher mercury uptake than polystyrene in a bonded laminate of the two films [6]. Blends made by coextrusion of Kraton G and PPO show dispersed particles in a matrix. Kraton G is the lighter contrast polymer as it seems to take up less of the stain than the PPO. In summary, mercuric trifluoroacetate staining has been shown for several polymers where the dispersed phase particles are differentiated by this stain. The method has limited application.

Table 4.1 Polymer functional groups and stains

Polymers	Stains
Unsaturated hydrocarbons, alcohols, ethers, amines	Osmium tetroxide
Acids or esters	(a) Hydrazine (b) Osmium tetroxide
Unsaturated rubber (resorcinol–formaldehyde–latex)	Ebonite
Saturated hydrocarbons (PE and PP)	Chlorosulfonic acid/uranyl acetate
Amides, esters and PP	Phosphotungstic acid
Ethers, alcohols, aromatics, amines, rubber, bisphenol A and styrene	Ruthenium tetroxide
Esters, aromatic polyamides	Silver sulfide
Acids, esters	Uranyl acetate

4.4.9 Iodine

Early treatment of materials with iodine revealed structures in nylon 6 filaments [191, 192]. The longitudinal periodicity of nylon was shown by the differential absorption of the iodine in the crystalline and noncrystalline regions. Hess and Mahl [98] treated poly(vinyl alcohol) fibers with 5% iodine which revealed a long period (16 nm) which relates to the absorption of iodine in the noncrystalline or amorphous regions. Hess *et al.* [193] detected structures of the order of tens of nanometers in cellulose treated with iodine. The fibers were soaked in 12% iodine following homogenization to break them up into fibrils.

Peterlin *et al.* [157] described the use of iodine vapor, at 60°C (24 h) to reveal the structure in PE fibrils drawn from solution grown single crystals. Andrews *et al.* [78] extended this method to prestaining a block of material prior to ultrathin sectioning. They showed the contrast enhancement of PE spherulites by action of the iodine. However, the iodine dissipates upon standing in air and is known to vaporize in the vacuum of the electron microscope. The chlorosulfonic acid method (Section 4.4.4) has replaced iodine staining of polyethylene.

The structure of PET fibers was described [194] using the iodine sorption method. Iodine sorption of the fibers was shown as a function of the test temperature, or the heat history of the process, whereas polyester fiber structures are a function of the textile fiber process. This structure was monitored by the study of the change in sorption of iodine, said to be related to the amorphous, free area and the dyeability of the fiber.

Oil based finishes on fiber or yarn surfaces, such as PET, have been stained with iodine [195]. The yarn samples were placed overnight in an airtight glass chamber containing iodine crystals. The treated yarns were assessed by using low magnification (250×) on an ordinary research microscope and using a stereo binocular microscope for finish uniformity highlighted by the adsorption of iodine.

There are few current applications of the iodine staining or sorption method found in the literature. Even in early studies, the destaining of polymers at room temperature and the volatilization in the vacuum of the electron microscope showed limitations of the method. Overall, there appears to be simple physical adsorption of the iodine, preferentially in the amorphous regions of semicrystalline polymers.

Table 4.2 Specific functional groups, examples and stains

Functional group	Examples	Stains
−CH−CH−	Saturated hydrocarbons (PE, PP) (HDPE)	Chlorosulfonic acid Phosphotungstic acid Ruthenium tetroxide
−C=C−	Unsaturated hydrocarbons (Polybutadiene, rubber	Osmium tetroxide Ebonite Ruthenium tetroxide
−OH, −COH	Alcohols, aldehydes (Polyvinyl alcohol)	Osmium tetroxide Ruthenium tetroxide Silver sulfide
−O−	Ethers	Osmium tetroxide Ruthenium tetroxide
−NH₂	Amines	Osmium tetroxide Ruthenium tetroxide
−COOH	Acids	Hydrazine, then Osmium tetroxide
−COOR	Esters (butyl acrylate) (polyesters) (ethylene−vinyl acetate)	Hydrazine, then Osmium tetroxide Phosphotungstic acid Silver sulfide Methanolic NaOH
−CONH₂ −CONH−	Amides (nylon)	Phosphotungstic acid Tin chloride
Aromatics	Aromatics Aromatic polyamides Polyphenylene oxide	Ruthenium tetroxide Silver sulfide Mercury trifluoroacetate
Bisphenol A based epoxies	Epoxy resin	Ruthenium tetroxide

Table 4.3 Osmium tetroxide staining

(a) Multiphase polymers stained by OsO_4

Acrylonitrile−butadiene−styrene
Acrylonitrile−styrene−acrylate
Styrene−butadiene−styrene
High impact polystyrene
Impact poly(vinyl chloride)

Copolymers of 1,4-polybutadiene and *cis*-1,4-polyisoprene

Blends with unsaturated rubber, isoprene, or isoprene included in fibers, e.g. PET or nylon

Polyoxyethylene allyl included in membranes

(b) Polymers requiring pretreatment prior to OsO_4

Polymer	Pretreatment
Acids, esters	Hydrazine
Epoxy thermosets	Tetrahydrofuran
Ethylene−vinyl acetate copolymers	Alkaline saponification
Chlorinated PE	Bicyclic amine
Polyesters	Allyl amine

4.4.10 Summary

Staining of polymers is an important part of sample preparation for electron microscopy as it provides the enhanced contrast required to image the structures. There are few staining techniques that work for a range of polymers and there are several stains with limited applicability. For many polymers there is no proven stain, and thus preparative treatments must be found by experimentation. This summary will provide a listing (Tables 4.1–4.3) of those polymers which have been shown to be stained by the various reagents described in this section. In addition, several stains that are applicable for only a few polymers are also listed here, even though they were not fully described.

4.5 ETCHING

4.5.1 Introduction

Etching is another preparation method which potentially enhances the information available by microscopy. Samples are etched and then replicas taken or a conductive coating is applied to the etched surface. There are three general categories of etching: dissolution, chemical attack and bombardment with charged particles as in plasma and ion beam etching. Dissolution implies the removal of whole molecules of a material as it dissolves. Solvent extraction with xylene was used for the study of PE [196] although the dissolution method in general is not recommended due to the artifacts which can result from swelling or reprecipitation. In chemical etching there is a chemical attack of molecules and removal of fragments from the specimen. Acid treatments which selectively oxidize one phase present in a multiphase material, aiding contrast between the various phases, are a form of chemical etching. Charged species activated by high voltages (ion etching) or in a radiofrequency plasma [197] are also used to etch away material differentially, in this case the surface of the specimen. Plasmas and ion beams are employed to bombard, or sputter and remove surface atoms

and molecules. Etching will be considered in three parts here: (1) plasma and ion etching, (2) chemical etching and (3) acid etching.

4.5.2 Plasma and ion etching

A major difficulty in plasma and ion etching is that textures, such as steps, cones and holes, can be produced which are artifacts and which do not reveal the microstructure. Of all the methods of specimen preparation, etching is the most prone to such artifacts and thus image interpretation is very difficult. Etching preparations are useful for comparison with structures formed during other specimen preparation processes, especially microtomy. Such complementary studies are essential to the determination of the true polymer structure.

Etching is commonly used today to thin, or ion mill, metals and ceramics for TEM and to produce SEM specimens. Barber [198] applied ion bombardment to prepare thin foils of non-metals for TEM. The theory and practice of sputtering metal specimens has been described [199–205]. Artifact or cone formation has been commonly observed in metals [201, 206] and a roughening of the surface and the production of conical protrusions has been observed [207, 208]. Rough surfaces are known to be enhanced and surface cones produced, if specimens are not flat and rotated during sputtering [209]. Also, 'ripples' and dune structures have been observed [210, 211]. Factors affecting ion etching of materials have been discussed [212]. Barnet and Norr [213] used an oxidizing plasma to etch embedded carbon fibers for SEM by a method that has been applied to polymer fibers. With polymers, artifacts may be formed and the factors affecting the etching process must be understood before etching studies can be undertaken successfully.

4.5.2.1 Literature review

Etching has been used to reveal structures in polymer fibers [214, 215], polymer blends [216, 217] and in bulk polymers, such as PET [218] and PE [219]. Periodic structures were seen to develop on argon plasma etched, oriented materials, such as nylon and PET, thought to be characteristic of the fold period [220]. Such structures were also shown for ion etched (argon) fibers [221]. Fine scale structures were observed for etched polyamide [222], PET [19, 223] and aramid fibers [222] which showed ripple structures transverse to the fiber axis.

Goodhew [224–227] discussed the formation of large scale structures (0.1–1 μm) perpendicular to the drawn fiber axis upon ion etching carbon fibers. He attributed these structures to be derived from the original surface striations. Such structures were not formed on glass fibers which were used as an unoriented, amorphous control. Optimization of the ion etching method was achieved [228] by first acid etching the carbon fibers to remove the surface striations. Transverse striations were still observed on the ion etched fibers, although no striations were formed on glass fiber controls. In general, authors using controls and optimum conditions for both plasma and ion etching are able to relate the structures observed to the original crystalline structure of the material.

Blakey and Alfy [229] used oxygen plasma etching (5–30 min) to reveal the nature of delustrant particles in polymer fibers. Friedrich *et al.* [230] applied plasma etching to HDPE where differential etching resulted from greater etching of the amorphous regions compared to the crystalline regions. These authors [231] used oxygen plasma etching to reveal the nature of polymers blended with glass fibers. PBT and poly(vinyl acetate) with glass fibers were ion etched (5 kV, 5–30 min) [6] to degrade polymer blocks selectively after sectioning. Kojima and Satake [232] ion etched PP, HDPE and propylene–ethylene block copolymers and revealed lateral structures resulting from preferential etching of the amorphous regions. Gupta *et al.* [233] plasma etched fractured cured epoxy resins with argon to reveal their heterogeneous nature. Woods and Ward [234] studied the oxygen gas plasma treatment of high modulus PE fibers in order to determine possible mechanisms which result in improvements in fiber–resin adhesion.

The mechanisms are surface oxidation, cross-linking and surface etching that could be related to improvement in interlaminar shear strength of the composites. SEM of the plasma treated fibers showed the presence of micro-surface cracks after relatively short plasma treatment times. A comparison of ion beam and radiofrequency plasma etching for biological ultrastructure is quite useful [235]. Clearly, artifacts may be formed during the etching process and such effects must be correctly interpreted for the method to have any utility. Finally, Mijovic and Koutsky [236] and Hemsley [237] have reviewed polymer etching.

4.5.2.2 *General method and examples*

There is no one general method that is applicable for either ion or plasma etching polymers. Etch times are generally short, on the order of 5–30 min, as it is important not to leave a residue on the sample surface. Rotation of the specimen and the target and cooling of the ion guns are factors that minimize potential artifact formation. In addition, the smoother the original surface texture, the fewer artifacts are expected. In our experience, plasma etching is much less complicated and there is less chance of artifacts than in ion etching.

Controlled etching experiments were designed, keeping in mind that etching could provide some insights into polymer fine structure *if* great care was taken. Etching, as conducted in a low temperature RF plasma asher (LTA), is generally used to oxidize organic matter. Such devices provide a nondirectional plasma which minimizes artifact formation. Oxygen and argon were used in these experiments, at 162°C, from 5 to 30 min. Temperatures below the crystalline melting temperature should be used. Etching time is affected by the surface area of the specimen. Ion etching was conducted using the Ion Tech microsputter gun (Section 4.5.2) with two water cooled ion guns, at 7 kV, for 10–30 min. The guns are directed at the region of interest as the specimens are rotated. A major benefit of the system used was the diffusion pumped vacuum system that permitted slow etching with a fast pump rate. This results in reduced ash deposition and limits the formation of coarse structures that result from restructuring of the material.

The etching study was conducted to complement ultrathin sectioning of high modulus oriented fibers [238]. Representative results of the plasma etching experiment are shown in the secondary electron images (SEI) (Fig. 4.19). Glass fibers and amorphous polyester film were used as controls. Oriented PET and aramid fibers were etched for 15 min in argon in order to evaluate the effect of the treatment on oriented crystalline

and liquid crystalline materials, respectively. Glass fiber surfaces (Fig. 4.19A) show a very fine, disordered texture whereas amorphous PET shows that delustrant particles are present, but no detailed surface textures are revealed (Fig. 4.19B). PET fibers (Fig. 4.19C) show delustrant particles, but a striated texture is observed normal to the fiber axis as it is in the etched aramid (Fig. 4.19D) where a large particle of unknown origin was seen. Oxygen etching is shown for a glass fiber surface (Fig. 4.19E) and an aramid fiber (Fig. 4.19F) for comparison with argon etching. The glass fiber surface is mottled, but it has no structure, whereas the aramid fiber

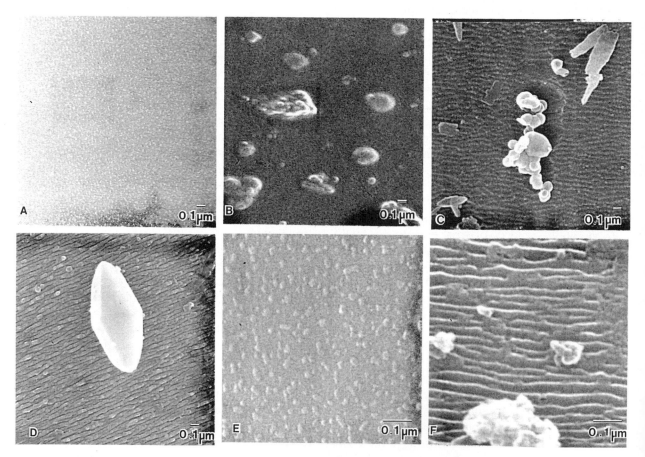

Fig. 4.19 SEI images show the result of plasma etching with argon: (A) a glass fiber surface shows no ordered details; (B) amorphous PET film with particles but no order; (C) an oriented crystalline PET fiber surface with lateral striations, and (D) an aramid surface also with lateral striations. The effect of plasma etching with oxygen is shown for a glass fiber surface (E) which reveals no detail and (F) an aramid fiber which exhibits a lateral striated texture.

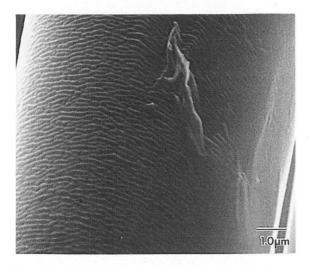

Fig. 4.20 SEI shows the effect of argon ion etching on an aramid fiber surface where lateral striations are observed normal to the fiber axis.

has a striated texture normal to the fiber axis. This striated texture, observed in semicrystalline and liquid crystalline materials [239], is similar for both argon and oxygen etching. An aramid fiber etched with the Ion Tech ion beam micro-sputter gun is shown in a representative SE image in Fig. 4.20. The structures observed are similar to those formed by plasma etching but there is a tendency to some directionality in the structures depending upon sample orientation with respect to the ion guns.

4.5.2.3 *Summary*

Factors known to limit artifacts in etched polymers are:

(1) plasma etching appears less directional than ion etching;

(2) argon generally results in cleaner surfaces than oxygen;

(3) slow etching and fast pump rates provide cleaner surfaces;

(4) specimen rotation and cooling limit heating effects;

(5) ion gun cooling and low current also limit heating effects; and

(6) low angle of incidence of two ion guns minimizes artifacts.

In most etched, oriented semicrystalline polymers striated, or lateral, structures are observed normal to the draw direction. These structures, termed ripples, striations or corrugations, appear to fall into two size ranges: 5–20 nm and 100 nm or greater. The fine textures often correspond to the SAXS fold period (the lamellar crystal periodicity); these are most likely related to the material microstructure. The larger textures are more likely to be artifacts of the method. Experiments have shown that the striations are observed in oriented, semicrystalline materials, strongly suggesting that there is preferential etching of the amorphous regions resulting in a texture associated with the original microstructure of the polymer. There is no consensus on interpretation of the striations or just what these textures imply regarding liquid crystalline polymers. These examples of the etching method were used specifically because the textures are obvious and have been observed in many other laboratories: as yet, there is a range of opinion regarding their interpretation.

Different specimen types yield a range of results upon ion or plasma etching. Multiphase polymers generally etch differentially, enhancing the contrast. Melt crystallized polymers can be etched to reveal the spherulites. Surface protuberances and particulate fillers can and do form cones or ridges when etched. Oriented semicrystalline polymers, on the other hand, appear to be the most controversial with respect to the resulting surface textures. Clearly, in such cases the specimen should be prepared by other methods for comparison, and control experiments are essential. There are problems in the industrial laboratory that can be solved, in part, by microscopy of surfaces prepared by etching techniques: however, these are far fewer than those addressed by other specimen preparation methods.

4.5.3 Solvent and chemical etching

Solvent and chemical etching is conducted for the same reason as ion and plasma etching, that is as a complementary technique in the determination of microstructure. Chemical etching has been arbitrarily divided into two sections for the purpose of this discussion. Acid etching will be dealt with in a separate section. Early studies involved etching to reveal the interior of polymers for replica formation.

4.5.3.1 *Literature review*

Peck and Kaye [240] immersed cellulose acetate specimens in acetone, at −50°C, and then flooded the surfaces with cold absolute alcohol, followed by replication which showed the skin, orientation, voids and pigment. Reding and Walter [21] etched PE with hot carbon tetrachloride (high density PE), benzene (low density PE) or toluene, which removed the amorphous material. Bailey [241] used a rapid xylene etch to reveal spherulites in PE and PP, while Li and Kargin [242] etched with benzene. A method was developed for obtaining PE for TEM. Melt crystallized PE, backed with carbon and gelatin, was exposed to xylene and dissolved to a thin layer and examined directly by TEM [243, 244].

Isopropanol vapor was used to dissolve the matrix in polymer blends [245]. Williams and Hudson [246] etched microtomed blocks of high impact polystyrene so that the rubber particles protruded from the matrix. Later, Kesskula and Traylor [130] removed rubber particles from HIPS and ABS polymers by dissolving the matrix in a cyclohexane solution of osmium tetroxide and extracting the dispersed phase for SEM. Olefin particles were removed from impact modified nylon and polyester [6]. Selective etching of the polycarbonate phase with triethylamine in a mixture with styrene–acrylonitrile copolymer (SAN) revealed the nature of the blend [247].

Amine etching was used to reveal the structure of PET as early as 1959 [55] when PET fibers were etched with *n*-propylamine for replica formation. Methylamine was also used [248], although the selectivity of the reagent was questioned. Tucker and Murray [249] etched PET filaments with 42% aqueous solutions of *n*-propylamine at 30°C. Apparently, the first step in the reaction is the removal of the fiber skin and then crazing. A periodic arrangement of discs was revealed at about 10 μm intervals along the fiber axis, joined to a central core [250–252]. Warwicker [253] showed disc forms in PET swollen in dichloroacetic acid. If aminolysis is to be used to study PET, then the excellent review of the reagents, reactions and earliest studies by Sweet and Bell [254] should be evaluated. These authors used primary amines to degrade PET selectively by removing the less ordered regions. In some cases the aminolysis was shown to degrade and crystallize PET, which limits the use of the method for morphological studies. Forty percent aqueous methylamine did not cause such solvent induced crystallization.

An etching technique was developed to study the structure of crystalline polymers, such as nylon 6, nylon 6,6 and PP [51]. Aromatic and chlorinated hydrocarbons were used to etch polished surfaces of ground and polished plastic parts. Grinding and polishing removed the surface detail with a series of polishing media down to a 0.05 μm alumina slurry. Choice of specific etchant was made so that it would be a poor solvent for the plastic so as to suppress swelling and dissolution of the crystalline phase. Results were evaluated using reflected light and Nomarski differential interference contrast optical microscopy.

A new etching method has been developed for the observation of the internal crystalline structure of water-soluble poly(ethylene oxide) using sodium ethoxide in ethanol. Bu *et al.* [255] found the optimal etching condition is to use 21 wt % NaC_2H_5 in ethanol for 10 min at c. 298°K with frequent agitation. The sample must be washed in absolute ethanol for 10 min and then dried prior to making one stage replicas by shadowing with heavy metal, backing with carbon and then removing the polymer with water. Blends of thermoplastic polyurethane elastomer and ABS resin were observed by SEM for the morphology of the blends following etching with methyl ethyl ketone (MEK) [256]. Samples showed variations in structures depending on the etch time with 3 h proving adequate to etch these blends whereas serious artifacts were observed at 4 h. This points to the need for carefully controlled experiments.

Hobbs *et al.* [173, 257] determined the morphology and deformation behavior for toughened blends of PBT, PC and PPE using complementary staining and TEM as well as etching for SEM techniques on the same specimens. Samples for SEM were faced with an ultramicrotome and etched by brief immersion in diethylene triamine, a selective etchant for PC. This etching method was said to be preferable to solvent or plasma etchants which attack both components to various degrees and can obscure fine details. Zeronian *et al.* [258–259] have studied the surface modification of polyester by alkali-metal hydroxides and amines to change the properties of the polyester fabrics, such as drape and hydro-philicity. Weight loss occurs rapidly with time in methanolic sodium hydroxide at 21°C, whereas it is much slower for treatment with 10% aqueous sodium hydroxide at 60°C. The reaction is thought to occur at the surface of PET fibers with chain scission products removed into the solution. In the SEM, surface pitting is obvious as is a decrease in the fiber diameter, with larger pits being observed for fibers with titanium dioxide delustrant or other inert material [258–259]. The reader interested in the use of these techniques to modify the properties of the polyester fibers is referred to the review of this topic [258].

4.5.4 Acid etching

Oxidizing acids have been used to etch chemically resistant semicrystalline polymers, such as polyolefins, as the acid preferentially diffuses into and attacks the amorphous regions. After slight etching the crystalline lamellae remain proud of the surface and details of the morphology may be seen in the SEM after metal coating, or in the TEM after shadowing and replication. Large scale features, such as spherulites, may be visible by reflected light microscopy.

4.5.4.1 *Literature review*

The general historical trend has been to start with very severe etchants and move to weaker ones which more reliably show fine scale structures. The progression of acids with time has been as follows: nitric, chromic, permanganate, permanganate/sulfuric, permanganate/sulfuric/phosphoric acids. Palmer and Cobbold [260] first etched bulk samples of melt crystallized PE with fuming nitric acid (95%) at 80°C and observed the lamellar morphology. Hock [261] used boiling nitric acid (70%) to reveal the microstructure of PP. Hock [262] saw spherulites in the oxidized polymer and, upon sonication, also observed lamellar fragments. Kusumoto and Haga [263] treated nylon 6,6 with 18% nitric acid at 60°C and showed it was more easily oxidized than PE.

A major problem with nitric acid etching is that it is too strong, etching not only the amorphous surface regions but also the bulk. Chromic acid etching was employed to alleviate this problem. Armond and Atkinson [264] treated PP with fuming nitric acid and then with chromic acid to reveal the cracks and fractures of bulk annealed PP. Bucknall *et al.* [265] studied polymer blends following etching with chromic and phosphoric acids. Microtomed blocks provided flat, smooth surfaces for etching of HIPS and blends with PPO. The solution contained 400 ml sulfuric acid, 130 ml phosphoric acid, 125 ml water and 20 g chromic acid. Reaction for 5 min at 70°C resulted in the removal of the unsaturated rubber, permitting assessment of the dispersed phase size and shape. Bucknall and Drinkwater [266] developed a method for ABS blends by fracturing in liquid nitrogen and etching, at 40°C for 5 min, in concentrated aqueous chromic acid solution. Briggs *et al.* [267] used chromic acid etching on blends with polyolefins, and Bubeck and Baker [268] applied this method to etching PE. Finally, epoxy resins have also been etched with chromic acid [269] and sulfuric acid [270].

4.5.4.2 *Permanganate etching*

Permanganic acid, a weaker acid, has been found to remove selectively the amorphous regions of polyolefins. It has been applied by Olley *et al.* [271] and others [140, 141, 272] as a complementary method to chlorosulfonation. The original permanganate etching method involved the following steps.

(1) Add 7% potassium permanganate to concentrated sulfuric acid (not the other way around).
(2) Treat the specimen for 15 min at 60°C, or for 15–60 min at room temperature [271].
(3) Etching is stopped by treatment in 5 cm³ reagent, cooling to 0°C, decanting, then washing successively in cold dilute sulfuric acid, hydrogen peroxide, water and acetone [273].

This method can be used to reveal lamellar detail in surfaces of PE, isotactic PP and isotactic poly(4-methylpentene-1). Addition of orthophosphoric acid to the reagent decreases the presence of artifacts in drawn linear PE and in blends [272]. Figure 4.21 [274] shows the types of structures observed by permanganate etching followed by replication and TEM. Strobl *et al.* [141] ion etched for 30 min prior to a permanganate/crown ether treatment and followed that by a sulfuric acid and hydrogen peroxide wash and carbon–platinum replication. Naylor and Phillips [275] developed a method, used in the production of the micrograph in Fig. 4.21, with 2% w/w potassium permanganate in concentrated sulfuric acid. This method shows details of the structure with minimum artifacts. They also [275] adapted the permanganate acid etching method for SEM evaluation by a study of etching as a function of time, temperature and concentration for a series of polyethylenes. They identified artifacts which limit study, but which can be minimized by treatment in an ultrasonic bath.

Bassett and Olley [276] studied the lamellar morphology of isotactic PP which involves treatment of a glass knife microtomed specimen with 0.7% w/v solution of potassium permanganate in 2:1 concentrated sulfuric acid/dry phosphoric acid for 15 min, with shaking. Nomarski differential interference contrast optics were used to judge the etched surfaces. Two stage replicas were prepared for TEM by metal shadowing at the first stage.

Fig. 4.21 Portion of a banded spherulite of linear polyethylene, also showing a spherulite boundary, as revealed by one hour etching using 2% w/w potassium permanganate in concentrated sulfuric acid as described by Naylor and Phillips [275]. The micrograph is of a chromium shadowed carbon replica studied in the TEM where each disk-like object in the bands is composed of lamellae nucleated at that point by a screw dislocation. (From Phillips and Philpot [274]; used with permission.)

As predicted in the first edition, there has continued to be progress in the interim in further development and utilization of the permanganate method for a variety of polymers. It is important to understand that the correct 'recipe' for etching is very dependent on your own composition of the polymer and on its morphology. Care must be taken to identify the appropriate reference and then do controlled experiments on your own materials. Bashir *et al.* [277] for instance conducted a comparative study of etching techniques for the electron microscopy of a melt processed plug or extrudate of PE using permanganic etch and chlorosulphonic acid treatment. In the case of these melt extruded PE fibers they found the permanganic etch required 2 h at 60°C as compared to 15 min for spherulitic PE. The time for the chlorosulphonic treatment was one day at 70°C. The etched samples were then replicated and compared to thin sections of the chlorosulphonic treated samples. TEM of samples from both preparation methods showed interlocked shish-kebab structures. More recent study of compacted PE fibers, by Olley, Bassett, Hine and Ward [278] showed the microstructure of aggregates of parallel high modulus fibers which had good mechanical properties. The preparation method used was as follows: 1% w/v potassium permanganate dissolved by continuous stirring in an acidic mixture of 10 volumes sulfuric acid, 4 volumes orthophosphoric acid, 1 volume water, with the sample shaken for 2 h in the mixture [278]. Recovery of the etched specimens was done by adding a small quantity of hydrogen peroxide to the chilled acid mixture instead of it being applied as a second step. The standard two stage replica process was used with cellulose acetate shadowed with tantalum tungsten followed by deposition of a carbon film and extraction of the replica (Section 4.6.2).

A modification of the procedure has also been developed for PEEK. Olley, Bassett and Blundell [279] found that a 2% w/v solution of potassium permanganate in a mixture of 4 volumes of orthophosphoric acid and 1 volume of water was effective. Samples were etched at room temperature for 50 min, then etching was stopped by adding the reagent to twice the volume of hydrogen peroxide solution followed by washing and replication. The method clearly revealed the spherulitic detail in PEEK [279].

Sutton and Vaughan [280] described a new method of preparation using permanganic etching of specimens which are difficult to handle for geometrical reasons. Polypyrrole *p*-toluenesulfonate films embedded in acrylic or epoxy resins are etched with a number of different reagents

Table 4.4 Etchants for polymers and multiphase polymers

(a) Polymers

Polymer	Etchant
PE	Hot carbon tetrachloride, benzene or toluene
PE, PP	Xylene or benzene
Melt crystallized PE	95% fuming nitric acid (80°C)
PE, isotactic PP	Chromic acid or 7% permanganate/conc. sulfuric acid, 15 min at 60°C or 15–60 min at room temperature or add orthophosphoric acid
PET	42% *n*-propylamine, 1 h, room temperature or *o*-chlorophenol or methylamine
Nylon 6, 6,6	Aromatic and chlorinated hydrocarbons
Cellulose acetate	Acetone at −50°C, then cold ethanol
Polycarbonate	Triethylamine, chloroform vapor
Polyoxymethylene	Iodobenzene and HFIP

(b) Multiphase polymers

Polymer	Etchant
Matrix in HIPS	Cyclohexane in osmium tetroxide
HIPS	100 ml sulfuric acid, 30 ml phosphoric acid, 30 ml water, 5 g chromic acid
ABS–rubber phase	10 M chromic acid, 5 min, 40°C
ABS–rubber phase slower on SAN matrix	Sulfuric, chromic acids and water for 5 min at 70°C
PU/ABS or PU/SAN (polyester based)	2% aqueous osmium tetroxide for 24 h, liquid nitrogen fracture, then use following:
SAN	Methyl ethyl ketone, 4 h
Polyurethane	Tetrahydrofuran vapor, 1 h or dimethylformamide
PMMA in SAN/PMMA	Chain scission in electron beam
Poly(vinyl methyl ether)	Chain scission in electron beam

and etching rates determined for various polymers. Procedures were developed to etch the polymer and also to attack the embedding media. The method has the advantage for handling thin films and fibers, but caution must be used when using a range of etchants in the interpretation of the resulting images.

Lemmon, Hanna and Windle [281] appear to be among the first to use potassium permanganate to etch thermotropic liquid crystalline polymers (TLCPs) (Section 5.6). Thin films formed from the melt on glass substrates were etched for a few minutes in a solution of potassium permanganate in sulfuric and phosphoric acids. Discrete entities, corresponding to the 'non-periodic layer crystallites' are observed in the SEM of an etched film of a very low molecular weight polymer. Ford *et al.* [282] also claim to be among the first to report on the application of permanganic etching to TLCPs. In this work the etchant was prepared by dissolving potassium permanganate (10 mg to 1 ml) in a $2:1:1$ mixture of orthophosphoric acid, sulfuric acid and water for 45 min at ambient temperature, followed by washing as mentioned earlier in this section. These authors dried the specimens and coated them with gold for examination by OM and SEM. The use of gold coating implies that the work was not conducted at very high resolution. Bedford and Windle [283] studied shear induced textures in thermotropic LCPs also using the method of Bassett and coworkers [272]. Hanna *et al.* [284] from the same laboratory further discussed the dimensions of crystallites in the TLCPs, using two specimen preparation methods. A melt shearing process was used to prepare thin films for TEM (see Section 5.6) and samples for SEM were prepared by treating thin films on glass substrates with 2 wt % potassium permanganate in orthophosphoric acid, followed by washing and metal coating. Anwer and Windle [285] used the same method for study of TLCPs following magnetic alignment, and fracturing for imaging in the SEM. The etching was observed to reveal crystallites which lie normal to the local chain axis. Hudson and Lovinger [286] used a similar technique [272] to investigate the morphology of a similar TLCP. They used 0.5% potassium permanganate with a solvent ratio of $2:2:1$ of water, phosphoric acid and sulfuric acid for about 2 h at 30°C. They used a two stage replica process to prepare samples for TEM which showed the fine fibrillar structure of these polymers.

It is now known that a wide array of polymers can be etched using potassium permanganate [273] although some care must be taken to limit the effect of artifacts. The list includes linear and branched PE, PP, PS, poly(4-methylpentene-1), poly(butene-1), PVF2, PEEK, PET and various copolymers such as EPDM terpolymers [273]. More recent work has shown that even liquid crystalline polymers can be etched by a variation of this method. Controls and complementary microscopy are essential to ensure that the experimentalist is not led astray imaging artifacts, hills and valleys or missing fine structure, lost in the wash baths.

4.5.5 Summary

Table 4.4 includes functional groups and polymers and their respective etchants. Chemical etching, such as with solvents and acids, and ion and plasma etching are conducted in order to reveal selectively structures in polymers that may not be observed directly. In all these methods, interpretation of the structures formed can be more difficult than specimen preparation. Accordingly, the etching methods are best used to complement other methods, such as microtomy, fractography and staining. Controls are essential to any experiment of this type, but, with care, the structures of semicrystalline polymers and polymer blends may be observed.

4.6 REPLICATION

Replication is one of the oldest methods used for the production of thin TEM specimens. The procedure was first introduced by Bradley [287, 288], and it is well documented in texts on specimen preparation [289, 290]. Replicas have the surface characteristics or topography of the original specimen. Replicas are used in optical, scanning, or transmission electron microscopes. Reflected light microscopy of polymer surfaces is often difficult because glare from the surface limits visible detail, although metal coatings reduce the glare. The method is useful for beam sensitive materials, materials which must be kept intact, are too large for direct observation or are curved and thus limit imaging. However, direct SEM has reduced the need for replication in recent years. High resolution secondary electron imaging and FESEM has matched the resolution available by TEM replication. Accordingly, replicas are not commonly used in the industrial research laboratory.

Replica methods most often used for TEM imaging include single or direct replicas, double or two stage replicas and extraction replicas. Direct replicas have the highest resolution but are the most difficult to prepare. Double or two stage replicas are easiest to prepare, but their preparation is quite time consuming. Extraction replicas provide a thin layer of the specimen attached to the replica, permitting analytical study. Application of conductive coatings and shadowing, an integral aspect of the replication process, will be described in the next section. Specimens for replication are often pretreated in order to reveal the internal or bulk structures. Examples of such treatments include etching with solvents, chemicals, ions or plasmas. The overriding disadvantages of replication are that they are time consuming to prepare and the interpretation of the resulting images is difficult, at best.

4.6.1 Simple replicas

4.6.1.1 *Replicas for OM*

Several simple replica methods for optical microscope examination will be described. Hemsley [237] described two methods for preparing replicas: application of a 5% PS solution in benzene (or xylene) or a solution of gelatin in water. These are dried, stripped from the surface and metal shadowed or coated to enhance the detail. A silastic replica method is in general use

for replication of internal surfaces where either surface topography or critical size measurements are required. Dow Corning Silastic, such as RTV silicone rubber, with silastic E curing agent (Dow Corning, Midland, MI), can be used as the replicating medium by dropping or brushing it onto the specimen. The silastic ingredients are measured according to the manufacturer's instructions. If the part has a fine opening, the part itself can be attached to the stopper of a vacuum flask and a vacuum applied while adding the silastic mixture. The replica is allowed to cure overnight and the silastic is carefully removed and metal coated to enhance surface detail. Examination is generally by reflected optical microscopy.

4.6.1.2 *Replicas for SEM*

Several methods have been specifically developed for the preparation of replicas for the SEM. Peck [291] developed a method to determine the nature and location of volatile and nonvolatile smoke deposits on cigarette filters. Eastman 910 adhesive, methyl-2-cyanoacrylate monomer liquid, was heated and vaporized adjacent to the specimen as the *in situ* replicating material. Eastman 910 apparently works well because it polymerizes rapidly in a water environment. However, as a result of the reaction in water, best results are obtained when producing the replica in a dry nitrogen atmosphere.

Oliver and Mason [292] have used replica methods to assess the effect of surface roughness on the spreading of liquids and to measure contact angles. For stationary studies, small beads of poly(methyl methacrylate) (PMMA) were melted and the molten droplets spread and solidified on the surfaces. Dynamic studies involved polymer melts mounted on a remotely controlled hot stage stub in the SEM and the experiments were video recorded.

Plastic and silicone replicas for the SEM were compared in a study of large specimens which could not be destroyed. Orthopedic implants, made of HDPE, are large and studies of wear are needed even though the implant will see continued use. Robbins and Pugh [293] compared various replication media for such studies. The methods they used to form negative replicas of an HDPE knee prosthesis appear general and will be described. Collodion (2–4%), in amyl acetate, was applied and allowed to spread to a thin layer and dry for about 20 min. The replica was removed using sticky tape. and attached to the SEM stub. Robbins and Pugh [293] also used a silicone replication medium, Xantopren Blue (a silicone substance manufactured by Unitek for dental impressions). A mixture of the silicone base and the hardener was applied to the surface in a thin layer and allowed to dry for 5–10 min. The replica was removed with forceps, trimmed to size and glued to the SEM stub with a drop of the silicone mixture.

4.6.1.3 *Methods and examples*

The replication methods described have a basic limitation in that they are all negative impressions of the specimen. This is a major factor in image interpretation. Wood [294] developed a method to provide positive replicas of polymer fiber surfaces. Polymer fibers often have surface coatings, such as finishes in textile fibers, that aid handling. Finish evaluation is difficult as these materials might vaporize in the SEM. However, reflected light microscopy, at high magnifications, reveals only a portion of the curved surface at any one level of focus. Therefore, the depth of focus of the SEM is required.

The general replica method for SEM of fibers or other polymers is as follows.

(1) Place the specimen onto and into thick tape to permit removal after replication.
(2) Pour a mixture of a silicone and a cure agent over the fibers, permit it to cure and peel it off. The replicating mixture can be Dow Corning RTV 3112 Encapsulant and Catalyst F (fast cure) [294], or Dow Corning RTV silicone rubber with silastic curing agent, or Xantopren (Unitek) [293].
(3) Clean the specimen in an ultrasonic bath.
(4) Drop a low viscosity embedding resin, such

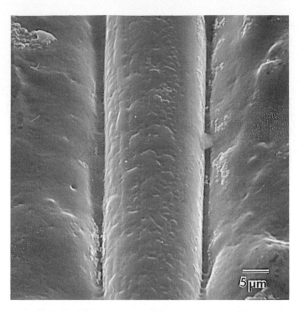

Fig. 4.22 SEM of a fiber surface replica shows details of the surface finish which is not possible in cases where the finish vaporizes in the vacuum of the SEM. (From Wood [294]; unpublished.)

as Epotek 301 (Epoxy Technology Inc., Billerica, MA) into the molded silicone (negative) replica. Turn the mold over onto an SEM stub and allow to cure overnight.
(5) Remove the silicone from the resin replica and metal coat the resulting positive replica for the SEM.

An SEM image of such a fiber replica is shown in Fig. 4.22. The rough fiber surface texture is associated with the finish. Gordon [295] reviewed the artifacts associated with silicone elastomer surface replicas, especially bubbling, and suggests that these methods must be used with great care.

4.6.2 Replication for TEM

4.6.2.1 *Direct replicas*

Direct or single stage replicas have the best possible resolution, are the fastest and, unfortunately, are the most difficult to prepare. The method involves the deposition of the replicating media and its removal or dissolution of the polymer. Materials used for direct replication include polymers, evaporated carbon films or metal oxides. Carbon is widely used for replication, especially if the polymer specimen to be replicated can be readily removed or dissolved.

Carbon replicas [288] are formed by the evaporation of a thin layer of carbon in a vacuum evaporator. Metal shadowing, at an angle of 20–45° to the specimen surface, is performed while the specimen is in the evaporator. The highest resolution direct replica material is carbon–platinum (C–Pt). Following evaporation, the thin replica is stripped from the substrate by etching, dissolving or some other method. In some cases, the carbon replica film can be scored and floated off onto a water surface and then picked up onto TEM grids. Carbon replicas do not usually just simply float off the substrate and treatment is required for their removal. Evaporation of a wetting release agent, such as Victawet, prior to carbon evaporation aids replica stripping although it can affect resolution.

The removal of direct carbon replicas is dependent upon the polymer. Boiling xylene vapor was used to remove drawn PE from replicas [296] in work on drawn polymer morphology. Hobbs and Pratt [297] described a direct carbon replica method for replication of a PBT impact fracture surface by evaporation of platinum at 20° and PBT removal in hexafluoroisopropanol (HFIP). Latex film coalescence in poly(vinyl acrylate) homopolymer and vinyl acrylic copolymer latexes was studied using direct replicas [298]. As the latex films have a low glass transition temperature, they were cooled by liquid nitrogen to about −150°C in the vacuum evaporator and shadowed with Pt/Pd at 45° followed by deposition of a carbon support film at 90° to the specimen surface. The latex films were dissolved in methyl acetate/methanol. TEM micrographs of the latex films show the difference between films aged for various times (Section 5.5.2).

Polymers are used to aid removal of thin replicas when the specimen cannot be dissolved.

Polymers that are used include poly(acrylic acid) (PAA), gelatin, formvar and collodion. Peterlin and Sakaoku [299] and O'Leary and Geil [300] described the deposition of PAA onto specimens to aid removal of replicas. In the former work, C–Pt replicas of drawn PE were backed by a 5% aqueous solution of PAA, dried and stripped off the specimen. The PAA was dissolved in water. The advantage of using PAA or gelatin is that the replicas are left floating on the water surface and they can be readily picked up on a TEM grid. Similarly, Balik and Hopfinger [301] used collodion to strip a C–Pt replica from POM on mica. The polymer solution, in iodobenzene, was crystallized directly onto mica and the collodion was extracted with amyl acetate. A variety of methods [302] are used to wash away the backing plastic. Carbon replicas, with the plastic film, are placed on the TEM grid, on filter paper in a dish, and a volatile solvent is introduced onto the paper. The addition of solvent into the covered dish is often successful in dissolving the plastic while the carbon replica settles onto the grid. Alternatively, the solvent can be allowed to drip over the grid on a mesh screen, or a sophisticated extraction apparatus can be used to permit the solvent vapor to extract the plastic material.

An example of direct or single stage carbon replicas is shown in Fig. 4.23 of a grooved polycarbonate disc [303]. The polycarbonate disc was shadowed with Au–Pd at a shallow angle (c. 30°) and then a thin carbon layer was deposited in a vacuum evaporator. The disc was dissolved using methylene chloride and the replicas were placed on copper grids for TEM evaluation. In this particular case the replicas were made by shadowing from opposite directions, normal to the grooves to illustrate the asymmetry of the groove structure (Fig. 4.23), as one side has a steeper slope than the other. The TEM study was complementary (see Section 5.3.2) to other studies in TEM, of thin sectioned substrates, cut perpendicular to the grooves; SEM of an IBS coated (Section 4.7.3) disc showed details and these were compared to FESEM images of uncoated discs.

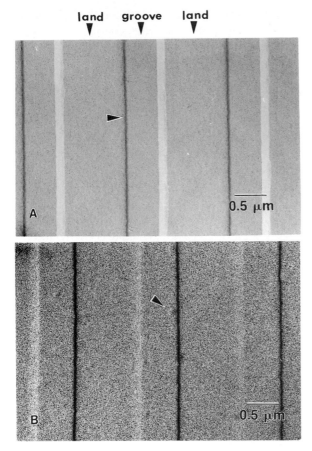

land   groove   land

0.5 μm

A

B

0.5 μm

Fig. 4.23 TEM micrographs of replicas from a polycarbonate disc, made by shadowing from opposite directions normal to the grooves to illustrate the asymmetry of the grooves. In the top photo the shadowing direction is radially inward while for the bottom photo the shadowing direction is radially outward [303]. The inner edge (arrow in top photo) was found to have a much steeper slope than the outer edge (arrow in bottom photo). (From Baro *et al.* [303], reproduced with permission.)

4.6.2.2 *Two stage replicas*

Two stage, or double, replicas provide positive impressions of the specimen surface, although they require more time to prepare. The steps involved in formation of a two stage replica are:

(1) form the replica;
(2) strip the replica from the specimen;
(3) shadow cast in a vacuum evaporator;

(4) carbon coat in a vacuum evaporator;
(5) place the replica on grids and dissolve or extract the plastic.

Replicating films or solutions may be used to form a first stage replica. The replica is stripped from the specimen, using tweezers or double sided sticky tape. Breathing on the replica is another method to get it to release, generally followed by floating it off onto a water surface. In some cases a second thick layer of the same plastic is applied in order to provide a backing. Unfortunately, the thicker films aid stripping at the expense of longer drying times.

Examples of replicating plastics (along with their solvent) are cellulose acetate (acetone), gelatin (water or dilute sodium hydroxide), acrylic resin (acetone) and poly(vinyl alcohol) (water). Solutions of 2% formvar in chloroform or dioxane and 1–4% collodion in amyl acetate are appropriate for rough surfaces where tapes and films are not easily removed. Washing and extractions are similar to those described above for one stage replicas. A major problem can be the incomplete dissolution of the plastic replica which interferes with TEM imaging.

Replicas of fibers and yarns are difficult to strip off due to their size and shape. Fibers can be prepared for replication by semiembedding them first in a resin or gelatin [304–306]. The fibers are replicated with metal and carbon and then stripped. Scott [55] used the peelback method to prepare fibers for internal or bulk study. Replicas were formed by coating the peeled fibers with 10 nm of chromium at a 30° angle to the specimen. Fibers were placed, chromium side down, onto a 3% solution of PAA in water and allowed to dry before removing the fiber and leaving the replica on the PAA. Scott coated the replica with polystyrene in CCl_4, removed the PAA with water and the styrene with CCl_4.

A commonly employed method for the preparation of positive, two stage replicas is the modified method of Scott [55] described by Geil [17] using PAA as the first stage replica. This general method is as follows.

(1) Place the fiber or film on a slide and shadow

with a metal (such as chromium) at 30° to the specimen.

(2) Place 2–3 drops of 3% PAA in water on the shadowed specimen and allow to dry overnight, or place the specimen onto a drop of the PAA [307].

(3) Peel the PAA/Cr replica from the specimen and turn PAA side down on the glass for carbon evaporation onto the Cr.

(4) Float the replica onto distilled water for 4–6 h to dissolve the PAA and then pick up on TEM grids.

In the modified method [307] the specimen is peeled away leaving the Cr face up on the PAA ready for carbon evaporation. After carbon coating, the entire slide is placed in water to surround the PAA, leaving the Cr/C replica to float. Hudson and Lovinger [286] used a concentrated aqueous solution of PAA, cast on the polymer, dried, detached, shadowed with Pt and coated with carbon. Olley *et al.* [278] made an impression of etched PE fibers in softened cellulose acetate, shadowed with tantalum tungsten, followed by deposition of carbon and extraction of the replica. Figure 4.24 is a TEM micrograph of a replicated experimental, stretched polypropylene film made by this method with chromium as the shadowing metal. The replica shows the elliptically shaped pores formed by stretching the PP. Unstretched lamellae are seen in rows between the rows of pores. The replica method can produce excellent results; however, it is very time consuming as the drying and dissolution of the PAA extend the preparation over several days.

In summary, there are many plastic materials available to prepare two stage replicas, although these methods are quite time consuming. The plastics that dry in less than 30 min (e.g. formvar and collodion) are the most useful as many replicas may be made of the surfaces under study. Although the PAA method is the most commonly used, it has the decided disadvantage of taking over 24 h to dry and taking additional hours to dissolve after replication. Thick replicating tapes also take longer and do not reproduce

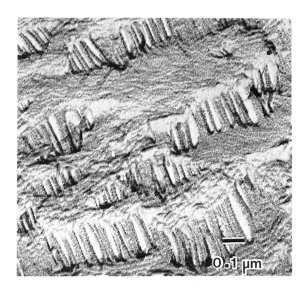

Fig. 4.24 TEM micrograph of a two stage replica made of the surface of an experimental stretched polypropylene film. The replica was shadowed with chromium at an angle of 30°. Slit-like voids are seen which are formed in rows or channels separated by oriented fibrils.

topography as accurately as thinner tapes. If the rapid methods cannot be used, plan on a week-long effort and use PAA.

4.6.2.3 *Extraction replicas*

Extraction or detachment replicas provide a replica of the surface along with some thin fragments of the actual specimen. The advantage is that the surface topography is recorded and analytical experiments, such as electron diffraction, may also be conducted. The method is complementary to microtomy or ultrasonication. Methods for such replication are similar to those described above for single and double stage replicas. The samples are either small, such as powders or crystals, or they are pretreated to encourage fibrillation or splitting. Fibers prepared by the Scott peelback method, for instance, when replicated generally provide some of the fibrils on the replica. Much work on single crystals has been conducted by this method where the *in situ* replicas, the crystals themselves, are present in the final preparation. Similar methods can be used for bulk crystallized polyethylene.

Bassett and Keller [308] studied the *in situ* shape of PE crystals by extraction replication. A PE solution was evaporated directly onto viscous poly(vinyl alcohol) (PVA) in water, and the PVA was allowed to dry. The crystals were shadowed and carbon coated, the PVA was dissolved and the crystals were picked up on carbon coated support grids. Solution grown PE single crystals on Mylar were elongated and replicated for TEM [24, 25] and the crystals examined directly. The crystals were removed by being stripped with PAA, metal shadowed, backed with carbon and the PAA dissolved. Petermann and Gleiter [309] deformed carbon coated grids by 5% in a miniature tensile device, forming cracks about 1–5 μm wide, prior to the evaporation of a xylene suspension of PE crystals onto the grids. The samples were stretched a second time to deform the areas where the original cracks were formed so that the fibrils extending across the broken carbon film could be imaged.

Jones and Geil [310] investigated deformation and micronecking in polymer single crystals. The preparation method involved the deformation of the crystals on Mylar film and subsequent metal shadowing. The crystals were stripped from the film using aqueous PAA neutralized with ammonia. The PAA/crystals were lifted off the Mylar, carbon coated and floated on water. The sample remaining consisted of the carbon/crystal/metal which was stable in the TEM. Micronecks bridged by fibrils pulled from the crystals were observed, for example, in deformed single crystals of polyoxymethylene.

Structural studies were conducted on PE and PTFE (polytetrafluoroethylene) fibrils extracted by coating with C–Pt and stripping with a backing layer. In this case [299, 300] the backing layer was 5% PAA in water. After dissolution of the PAA, fibrils were left for TEM observation. Thin shreds or layers of rubber were torn off blends of natural rubber, polyisoprene, SBR and neoprene using gelatin for the extraction replica [311].

4.7 CONDUCTIVE COATINGS

4.7.1 Coating devices

There are a range of devices available for the deposition of conductive coatings. The device commonly in use for preparation of TEM replicas and carbon support films (*c.* 1–50 nm thick) is the vacuum evaporator. Sputter coaters have more recently become popular for coating SEM specimens, to provide a conductive layer about 1–100 nm thick, which emits secondary electrons producing SEM images. There are a wide range of sputter coaters that provide various means of protecting the specimen from heating during the deposition. Sophisticated devices for deposition of fine grain coatings include electron beam, ion beam microsputter and magnetron sputtering systems. All these methods have in common the production of a thin conductive film essential to the study of polymers by scanning electron microscopy.

4.7.1.1 *Vacuum evaporators*

Vacuum evaporators have been in use for several decades for the thermal evaporation of materials, such as metals, onto a specimen to provide a conductive layer and dissipate charge during electron microscopy. Holland [312] is an early authority on the topic of vacuum deposition. Typically, a 12 inch diameter bell jar is fitted onto a vacuum system which includes a rotary pump and diffusion pump. Electrodes are fitted onto the baseplate of the evaporator and connected to a transformer. These electrodes are used for attachment of the metals and the carbon rods for evaporation. An important accessory is a liquid nitrogen trap which is fitted above the diffusion pump to trap oil vapors and keep the vacuum clean, which is essential for carbon coating.

The time needed for vacuum evaporation depends upon the state and speed of the vacuum system, but the actual coating time is usually about 2 min after 30 min pumping to obtain an appropriate vacuum. Preparation of metal wires for evaporation is very straightforward. The length of the wire, its thickness and the distance to the specimen determine the coating thickness. Metal wires are typically wrapped around a 'V' shaped filament of tungsten wire attached to the electrodes in the vacuum chamber and evaporated from the heated tungsten filament. The filaments can be purchased or bent manually from tungsten wire. Specimen rotation devices and shutters are available that limit specimen heating. Potential disadvantages of evaporative coatings include nonuniformity, heat damage to the specimen, oil contamination and a grain size resolvable by modern instruments.

4.7.1.2 *Sputter coaters*

In sputter coating the metal (gold–palladium or platinum) on a target is dislodged by inert ions and directed onto the specimen. Sputter coating was described by Echlin and Hyde [313] and Echlin [314]. In simple units, the small chamber is evacuated with a rotary vacuum pump, the system is flushed several times with argon and the argon flow is controlled during the sputter coating as argon ions bombard the metal target and the metal is deposited on the specimen. Diode systems have generally been replaced by triode systems, as the sample is the anode in the diode systems, which causes deleterious sample heating. A comparison of the diode, triode and vacuum evaporative systems was described by Ingram *et al.* [315]. A major problem with sputter coaters is heating of the specimen. Panayi *et al.* [316] developed a so called 'cool' sputter system, which has been used, in part, in the design of the Polaron E5100 coater. Robards *et al.* [317] sputtered gold onto frozen specimens using a permanent magnet to confine the plasma and, thus, limited heating effects.

4.7.1.3 *High resolution coating devices*

There are a variety of other coating methods. Three of these are the electron beam (E beam) technique, Penning sputtering and saddle field ion beam sputtering. The methods require more time and effort than the techniques described, but they result in finer grained coatings that are needed for high resolution SE imaging and for TEM. Ion beam sputtering will be described, as an example of these systems, in the section on coatings for SEM.

4.7.2 Coatings for TEM

Replication is a major preparation using conductive coatings for TEM. Carbon is used as a support in two stage replicas, or as a direct replica, but it does not impart any electron contrast. The specimen is generally metal shadowed at an oblique angle, in order to highlight the surface topography, enhancing the electron contrast, followed by evaporation of a carbon support film. Carbon support films are used in place of plastic supports for high resolution imaging. Polymer sections often require a light (1–5 nm) carbon coating in order to limit charging and protect the specimen from beam instabilities. This coating helps the situation, although beam damage and radiation damage still occur.

4.7.2.1 *Carbon coatings*

Carbon films are used as specimen supports, for replicas and as coatings for ultrathin sections. Carbon evaporation is difficult to perform as a high current is needed, and the carbon rod often moves or breaks. Carbon rods are prepared by sharpening one end to a short cylinder about 4 mm long. The other rod is flattened at an oblique angle and the rods are set so that the cylinder faces the specimen while making contact with the flat rod. One method of carbon coating [318] is rapid application of current. In our experience, best results are obtained by applying the current smoothly, in about 20–30 s, trying not to shock the rods into losing contact. A shutter is used between the specimen and the carbon rods to limit heat exposure.

Carbon support films are evaporated directly onto freshly cleaved mica, NaCl, plastic coated grids or plastic coatings on glass slides. There is difficulty in stripping the carbon film from the support. With mica and NaCl, the films are floated off onto water and picked up on TEM grids. The carbon is left on plastic support films, unless very high resolution is required. Carbon films about 1–3 nm thick enhance beam stability of ultrathin sections while about 10–20 nm may be required for replicas. The color of the carbon film can be used to judge the thickness. A drop of diffusion pump oil is placed on a glass slide on top of a piece of white filter paper to aid thickness determination. Quartz monitors are available for film thickness measurements.

4.7.2.2 *Shadowing*

Shadowing [290] is used to enhance specimen contrast by increasing the electron scattering. Replicas are metal shadowed prior to carbon coating, in order to accentuate surface topography by the application of a heavy metal in front of raised areas while no metal is deposited on the other side. Shadowing at known angles, generally 20–45°, is useful for the measurement of particle heights. Fine detail is accentuated by smaller shadowing angles while larger angles are used for larger structures. Metals generally used for shadowing include Au–Pd (60/40), Pt–Pd (80/20), gold, tungsten oxide, chromium, platinum and C–Pt. Platinum is the most difficult metal to evaporate, but it provides the finest grained shadow. Chromium is quite coarse and useful for very low magnification studies. Aluminium is used for optical microscopy.

4.7.3 Coatings for SEM and STM

SEM and STM samples are coated to provide an electrically conductive layer, to suppress surface charges, to minimize radiation damage and to increase electron emission. The coating is intended to be a thin *in situ* replica of the specimen surface. The thickness and texture of the coating must be minimized for smaller textures and higher resolution. Coating techniques for SEM have been reviewed [319], although higher

resolution microscopes are now available that require finer grained coatings as does STM.

4.7.3.1 *Sputter or evaporative coating*

There are several types of coating devices to choose from for the preparation of specimens for SEM, including sputter coaters and vacuum evaporators, described above. Sputter coaters have some definite advantages: short preparation time (5–10 min), multiple specimens and uniform coatings on rough textured specimens. Special stages fitted onto the baseplate of an evaporator provide for sample rotation for thorough coating. Overall, finer grained and thinner conductive films are obtained with the vacuum evaporator, but it is not recommended for heat sensitive or rough textured specimens. When magnifications under 5000× are required, the sputter coater is the preparation method of choice. As the texture of the specimen decreases in size and the required resolution increases, there are definite differences observed in the method of application of conductive coatings, metal type and thickness.

There are no resolvable variations among different metal coatings sputtered to thicknesses of 20 nm or more, when examined in microscopes that only resolve 10–15 nm. However, this is certainly not the case in the FESEM where resolutions can be less than 4 nm. Thick gold coatings tend to be granular, cracked and nonuniform and they may be resolved in the SEM, whereas Au–Pd and platinum are less likely to be resolved. Reviews of conductive coating methods [320, 321] are useful for an overall understanding of this topic. Braten [322] showed that diode sputtered gold gave a granular and cracked appearance to the surface, and recommended vacuum evaporation of Au–Pd. For highest resolution dedicated SEM imaging, a thickness of about 5 nm (or less) of sputtered platinum is recommended. This has also been suggested by other authors [323–326].

A study of various coating devices was conducted several years ago to identify the best conductive coating for high resolution SEI taken using a TEM (with a scanning attachment) with a resolution of 3 nm [325]. The results of that study are even more relevant today as the FESEM also has potentially similar resolution. A stretched polypropylene membrane was used in this study as there are fine and coarse textures and the film is extremely beam and heat sensitive. Coatings were made of similar thickness (5 nm) and metal (Au–Pd) in a range of devices. SE images taken in a late 1970s vintage dedicated SEM show that diode sputtering results in a cracked metal surface (Fig. 4.25A), whereas ion beam sputtering results in no cracking (Fig. 4.25B). SE images taken in an analytical electron microscope (AEM) show that evaporative metal coatings result in a granular texture (Fig. 4.25C) compared to an ion beam sputtered sample (Fig. 4.25D).

A mid-1980s dedicated SEM, with 4 nm resolution, was found to resolve even thin platinum

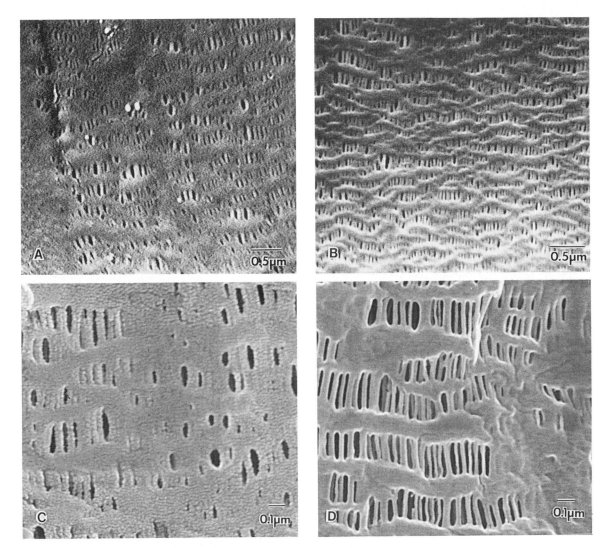

Fig. 4.25 Micrographs of a commercial stretched polypropylene membrane (Celgard 2400) prepared for microscopy by different metal coating devices are shown by SEI in a dedicated SEM (A, B) and in an AEM (C, D): (A) Au–Pd sputter coating, (B) ion beam sputtered with Au (same magnification as in (A)), (C) Au–Pd evaporation and (D) ion beam sputtering with Au (same magnification as (C)).

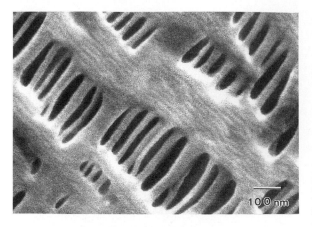

Fig. 4.26 FESEM image of Celgard 2500 taken at 5 kV with *c.* 3 nm Pt coated by ion beam sputtering.

sputter coatings that had not been resolved earlier. Contrary to earlier assumptions, these coatings were even resolved at low magnifications. What is implied by the statement that coatings are 'resolved' is simply that a granular texture not associated with the specimen is observed. This texture is an *artifact* if it is misinterpreted as being related to the specimen. Figure 4.26 is a high resolution SE image of an ion beam sputtered membrane where the fine texture is resolved.

4.7.3.2 *Ion beam sputter coating*

The saddle field ion gun was first used for traditional ion thinning of materials, such as metals [327]. The ion beam sputtering (IBS) technique uses a fine, collimated and water cooled ion source that sputters metal from a target onto the specimen surface. Argon is directed into the ion gun, and the specimen is neither heated nor bombarded by the plasma as is the case in ordinary sputter coating. The Ion Tech EM-Microsputter system with a saddle field ion gun is used to form thin metal films. This device is used on a vacuum station with a diffusion pump and a liquid nitrogen trap to limit contamination. The geometry is such that the ion source impinges on the rotating target at an angle of about 35–45 °. The specimen is placed

at a 90° angle to the target and out of range of the ion source (Fig. 4.27). In a more complete description of this technique [328] the metal grain sizes were shown to be about 1 nm by TEM cross sections.

Clay and Peace [329] compared evaporative, sputtering and ion beam sputter coatings and showed that IBS produces a fine grained, uniform film of gold. In a comprehensive review of advances in metal coatings Echlin [330] described several factors important to the production of high resolution coatings. The success of ion beam sputtering can be attributed to a number of factors: slow deposition rate, clean environment and the fact that the specimen is not exposed to high energy electrons. Overall, the system provides thinner films with smaller grain sizes [331] compared to most other coating methods. Detailed studies have compared the grain sizes of ion beam sputtered chromium and iridium [332]. Grains are smaller if the deposition chamber has a liquid nitrogen trap attached. Similar results can be expected for coatings prepared by Penning sputtering [333] or by electron beam coating [330]. Unfortunately, these devices have some drawbacks that limit their use. Specimens generally must be fairly flat for coating. In addition,

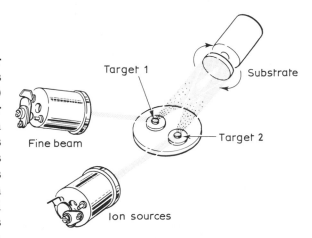

Fig. 4.27 The geometry in the ion beam sputtering unit is shown. Two water cooled ion guns form a cone of ions/atoms of argon which are focused on the metal target. The sputtered metal is then focused onto the rotating specimen.

the apparatus is more expensive and the coating times are longer than by any of the other systems described. These factors notwithstanding, ion beam sputtering is one coating method of choice for high resolution metal coatings. Thus, SE imaging is now possible at 60 kV for up to 150,000× with minimum beam damage and beam heating [129]. Recently, magnetron sputtering [334] has also been used to produce fine grain metal coatings with 1–2 nm of Pt deposited on a cold substrate. Ion beam and magnetron sputtering both can provide fine grain size conductive coatings required for high resolution FESEM and STM imaging.

In summary, compromises must be made in the choice of conductive coating techniques, depending on resolution. Issues to consider are: capital cost, coating deposition time, ease of use and the resolution required for observation.

4.7.3.3 *Carbon coating*

Chemical imaging and x-ray analysis modes in the SEM require that samples be conductive, and yet the information of interest cannot be masked by a heavy metal coating. Carbon coatings are applied to such specimens in order to dissipate the charge, although carbon does not provide much electron emission, and often metal coatings must be applied for imaging. Carbon coatings are also used to provide a continuous coating prior to metal coating. Production of carbon coatings has been described above (Section 4.7.2.1).

4.7.4 Artifacts

It is appropriate to discuss the potential 'imaging effects', or *artifacts*, in specimen preparation. For this discussion, the term 'artifacts' will be reserved for those textures that are an effect of the preparation method and are incorrectly attributed to the specimen. The polymer microscopist has to deal with such preparation effects daily and, thus, should be fully aware of both the causes of such effects and the appearance of the resulting structures. How can the microscopist know when the structures observed are an adverse preparation effect or a real structure of the material? Experience is the best teacher, but there are some effects of metal coating, charging and beam damage that are generally known to cause artifacts.

4.7.4.1 *Charging*

Polymers are generally insulators, and poor conductive coatings will cause a charge buildup in the specimen. Charging is a reversible process associated with a high negative charge on the specimen which can cause bright spots in the image. Other effects, such as the specimen moving in the beam, image shift, poor signal output and 'snowy' images are all due to charging. Charging may be decreased by carbon coating prior to metal coating, thicker metal coatings, better contact between the specimen holder and the specimen and low accelerating voltage SEM operation. The top of the specimen surface may be connected to the specimen stub by silver paint or a carbon paste in the case of x-ray specimens. Typical charging effects are shown in the micrographs in Fig. 4.28.

Charging continues to cause image effects even with the use of field emission electron guns and low voltages. As the specimen discharges there is unstable imaging. At high beam energies, 20 kV for example, the total electron yield is less than one. The specimen emits fewer electrons than it receives and charges negatively. At lower beam energies the incident electrons do not penetrate so far into the specimen. Electrons, particularly the low energy secondary electrons, can escape the specimen more easily and the electron yield increases. An area that has charged negatively has a lower effective incident beam energy, a higher electron yield, and thus appears bright. At some low incident energy $E2$, the total electron yield becomes unity and there is a dynamic charge balance. If the beam voltage is reduced to $E2$, an uncoated nonconducting polymer sample may, in principle, be imaged with no charging effects.

Joy [335–338] has provided careful experimental studies of the effect of imaging at low voltages

Fig. 4.28 SEM micrographs show a few of the spurious imaging effects which result from specimen charging. The banding shown in (A) and what appears as movement of the specimens are also due to charge buildup. The edges of the fiber are bright due to a bright edge effect (B).

Table 4.5 Measured values of $E2$ for a range of polymers [338]

Source	Polymer	Measured $E2$ (kV)
Hoechst Celanese	Celcon	1.2
Hoechst Celanese	PBT	1.1
Dow	nylon 6	1.2
Exxon	HDPE	1.5
Exxon	EVOH	1.4
Exxon	LDPE	1.5
Exxon	PC	1.3
Exxon	PS	0.9
Goodyear	PET	0.9
Rohm & Haas	PMMA	1.6

specimen at some energy $E > E2$ the charge of the specimen will be negative as more electrons are received by the specimen than it emits. When viewing at $E < E2$ there will be a positive charge. The SEM should be set at the lowest usable operating voltage and the following test should be done to determine where the setting should be ($E2$). The test is as follows.

(1) Scan the area at low magnification.
(2) Go to higher magnification and count to five.
(3) Return to the original magnification.
(4) If the area in the center of the image is bright there is negative charging and the beam energy, $E > E2$.
(5) If the area in the center of the image is dark there is positive charging and the beam energy, $E < E2$.

If the sample is charging positively then increase the kV before operating. If there is negative charging, then the kV is too high; the specimen may be tilted and then tested again.

The beam voltage that gives charge balance depends on the exact nature of the specimen, and charging will produce some effects on the image even at low voltage. The charges build up slowly on the specimen, so low beam current and rapid scanning give less charging, but more noise. A digital framestore gives a less convenient method for integrating over many rapid scans to remove this noise. Heating the specimen to 50–200°C or putting it on a high atomic number conducting

in the SEM and the control of charging relating to many materials, including polymers. Calculated and measured values of $E2$ for a range of polymers are in the range 0.9 to 1.6 kV [336, 337] (see Table 4.5). The sample can be tilted to increase $E2$ to a higher voltage. He also developed a simple test for determining $E2$ for a specimen during viewing. When viewing the substrate, may reduce charging by increasing leakage currents, but neither is as reliable as a light coating of metal.

Recently, Butler, Joy and Bradley [338] showed unexpected and unusual charging effects using a field emission LVSEM with an immersion type objective lens. For strongly insulating samples such as polymers they showed that even working at $E2$ some charging effects were observed. Charging has been observed as a function of magnification; at high magnifications the sample might charge positively whereas at lower magnifications the sample might charge negatively. Charging was also observed to be worse when the beam is exactly focused on the specimen, while slightly out of focus the charging is less. Positive and negative charging can also be observed within the same image. These authors [338] have termed these effects 'dynamic charging' and it is quite clear that more work will be done in this area. In our laboratory [339], after several years imaging in a field emission SEM, the practice to limit charging effects is to lightly coat the specimens with 1–3 nm Pt by the ion beam sputtering technique (Section 4.7.2) and to image at 5 kV. The addition of a metal coating increases the apparent $E2$ and the charging effects and artifacts are limited by this preparation. This method makes it easier and faster to focus and to collect images, either by traditional methods, or by digital scanning, at a range of magnifications.

4.7.4.2 *Beam damage*

The effect of beam damage must be distinguished from that of conductive coatings as cracking and buckling of the surface may be observed in both cases. Examples of irreversible beam damage effects are shown in Figs 4.29 and 4.30. Beam damage causes different changes in different polymers. For instance, PMMA is very beam sensitive with changes thought to be caused by chain scission, whereas polycarbonate is thought to crosslink in the beam and is, thus, less sensitive [342]. Radiation damage in POM is rapid and is due to chain scission [343]. In the

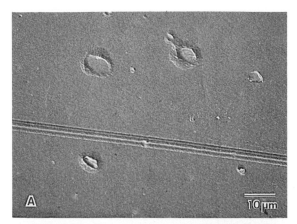

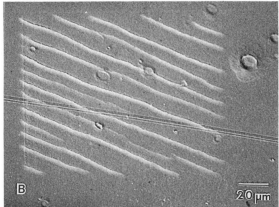

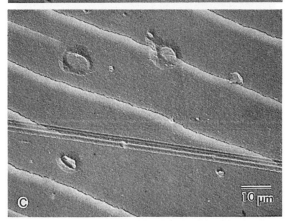

Fig. 4.29 The effect of beam damage is shown here: (A) SEM of the surface of molded polyoxymethylene taken first, while (B) and (C) were taken afterwards. The diagonal lines were caused by the electron beam, but they could be misinterpreted, perhaps as a cracked metal coating or a specimen defect, if only (C) is examined.

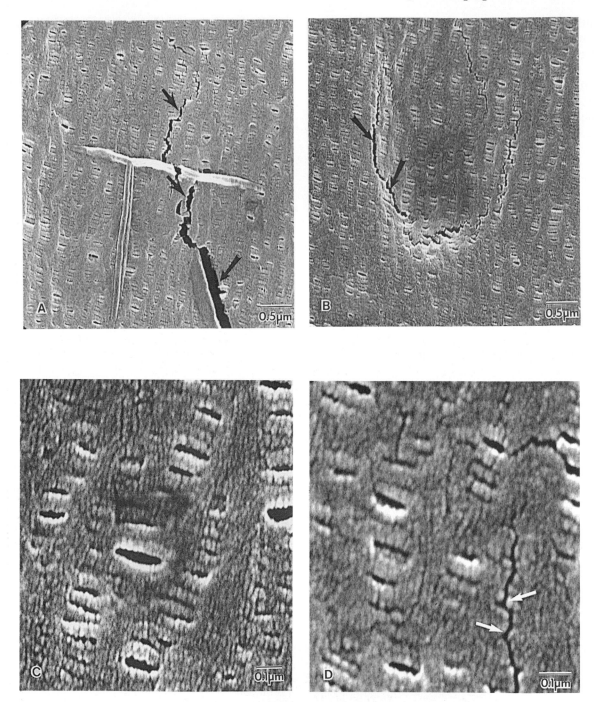

Fig. 4.30 Even thick metal coatings do not always protect the specimen from beam damage. SE images of a microporous membrane surface exhibit a granular, or pebbly, texture with fine cracks due to resolution of the evaporated Au–Pd coating. In addition, beam damage is seen as cracks (arrows) in the surface and a collapsed region in (B) which was examined at higher magnification prior to taking this micrograph.

case of resolved metal coatings, the morphological effects are observed all over the specimen, and there is no additional charging in these cracked areas. In cracked or buckled textures resulting from beam damage, the textures and charging increase with time, and fresh areas do not exhibit these textures. Resolution of metal grain textures is more likely by FESEM or HRSEM and should not be misinterpreted.

A good example of a cracked, beam damaged specimen is the surface of a POM molding, shown by SEM in Fig. 4.29. The molded surface has a production related scratch mark and several particulate 'pock' marks (Fig. 4.29A). Seconds after taking this micrograph the region has the appearance shown in Fig. 4.29C. The diagonal lines could be misinterpreted as being cracks from a thick metal coating or as being due to the surface morphology of the specimen. However, reducing the magnification (Fig. 4.29B) reveals the pattern in the picture frame (rectangle) that was exposed to the electron beam for the higher magnification micrograph. Another example of the effect of beam damage that might be mistaken for too thick a metal coating is shown in Fig. 4.30 of a membrane surface prepared for SEM by evaporation of Au–Pd.

The metal granularity has been described earlier: however, the large cracks are due to beam damage.

4.7.4.3 *Low voltage FESEM imaging artifacts*

The higher brightness of field emission guns allows operation at lower beam voltages, and under these conditions there are new challenges in imaging and interpretation. A major effect of low voltage imaging is simply that it is more sensitive to surface detail. This can be an advantage of low voltage operation, but it means that the images are much more sensitive to surface contamination. An excellent example is shown in Fig. 4.31A of the surface of a grooved polycarbonate optical disc, imaged using FESEM at 1 kV. Several effects can be noted. First, the picture frame contrast, from focusing on the central region at higher magnification, is very obvious and clearly the SEM was contaminated as shown by the dark region. The picture frame is also darker at the left side of the image due to the rastering of the beam. Figure 4.31B shows a mottled, contaminated surface even at 3 kV. A fine, linear pattern which runs horizontally is also observed in the images, especially in Fig.

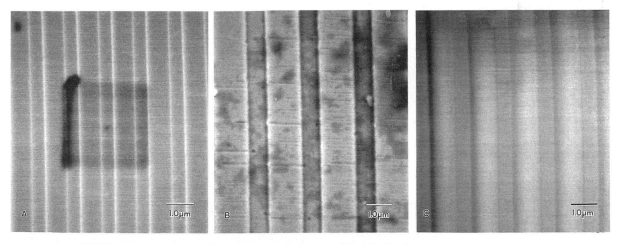

Fig. 4.31 FESEM image of a polycarbonate grooved disk at 1 kV: (A) clearly shows contamination in the 'picture frame' caused by focusing at higher magnification, (B) shows mottled surface contamination, and (C) shows a horizontal pattern due to emission noise from the gun. (From M. Jamieson, unpublished [339].)

4.31C [339], that is due to field emission noise, suggesting the microscope and gun settings were not optimized. Again, use of fine metal coatings and 3–5 kV limits the observation of contamination and field emission noise which can detract from resolving specimen details.

Combined low voltage and field emission SEM have been used to study microporous membranes [340–341], shown earlier in this chapter. Figure 4.32 shows some of the potential artifacts that have been imaged and are examples for consideration when imaging at very high magnifications and thus irradiating very small sample volumes [341]. In Fig. 4.32A there is a thin layer within each of the pores in the membrane that occurred over time in the microscope and was not present in the first images of this area. This is a contamination effect. Figure 4.32B and C show less obvious artifacts. The fibrils that separate the pores in this membrane are not usually filled in nor are they collapsed on one another, except during longer times in the microscope. This might not be obvious at higher voltages but at 1 kV the surface detail is very clear and does not represent the original membrane (see Fig. 4.26). These are just some of the artifacts that are possible using lower voltages. Higher resolutions are possible but the interpretation of the resulting images remains one of the key issues in these imaging techniques.

4.7.5 Gold decoration

Gold decoration, described by Bassett [344], is a method of highlighting very fine surface steps. A very light coating of gold, about 0.3–1 nm, is evaporated onto the specimen surface followed by carbon evaporation. The metal nucleates along the edges of the steps on crystal specimens. Thin materials are examined directly with the gold–carbon film. The carbon film with the gold may be stripped from thicker specimens. Polymer thin films and single crystals have been explored using the gold decoration technique.

Studies using gold decoration have been performed in several laboratories [345–349]. Thin solution cast films of nylon were studied by Spit

[346], while Krueger and Yeh [347] studied stirred PE solutions which formed shish kebabs. Kojima and Magill [350] studied the morphology of spherulites of block copolymers by gold decorating sections. Shimamura [351] applied

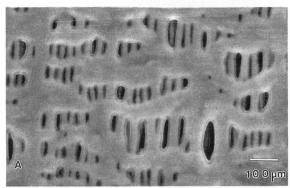

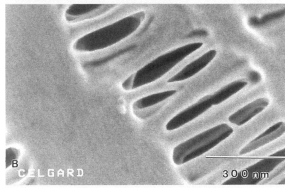

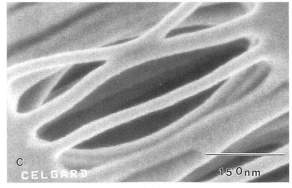

Fig. 4.32 FESEM of a Celgard microporous membrane shows (A) that the pores are filling in due to contamination, (B) fibrils are joining together and one pore region is filled, and (C) two fibrils are formed into an X pattern due to time in the electron beam. (From M. Jamieson, unpublished [341].)

this method to the study of the internal structure of PE fibers.

4.8 YIELDING AND FRACTURE

4.8.1 Fractography

Microscopy is used to study the plastic flow and fracture of a range of biological, ceramic and polymer materials. In polymer microscopy, the most common technique is the SEM of fracture surfaces. Traditional fractography studies were conducted by optical microscopy techniques, but as most fracture surfaces are very rough, the superior depth of field ·of the SEM makes it the better choice even at low magnifications. Optical microscopy is still used for some studies; for example, internal flaws (cracks, crazes and shear bands) can be seen in transparent materials and so can stress whitening, which occurs in the plastic zone of some polymers. The fine structure of fracture surfaces used to be studied by TEM of replicas, but the modern SEM has sufficient resolution for this purpose. Thus, fracture itself is now rarely studied in the TEM, but it is used extensively for the study of local plastic deformation in shear zones or crazes which are the precursors or early stages of fracture in glassy and semicrystalline polymers. Crazes have been studied by both OM and SEM, but TEM is required for resolution of the fine structure.

4.8.1.1 *Fracture types*

The plastic flow and fracture that is to be studied in the microscope may come from one of three different sources: (1) deformation of the sample in a standard mechanical testing device, such as a tensile (Instron) tester, impact (Charpy or Izod) machine or tear tester; (2) deformation or fracture of a microsample by special miniature testing devices, either before or during observation in the microscope; (3) yielding or fracture of a real product during either service or testing. Numerous studies have been carried out with materials from these sources. An extensive discussion of polymer fractography, including the topic of

crazing, is found in books by Kausch [352, 353]. Basic texts on fracture, such as by Liebowitz [354], discuss fracture mechanics and fractography. In this section, discussion will focus on the techniques required to prepare fractured or deformed specimens for microscope observation.

In the first case, mechanical deformation is part of the sample preparation, where the sample is formed by testing used to obtain mechanical data. The purpose of conducting microscopy on the fractured sample is normally to determine the mode and propagation of failure and to correlate the mechanical data and sample microstructure. SEM of the fracture surfaces provides these observations for comparison with standards of similar materials [355]. The fracture could simply be used to provide access to an internal surface of the polymer for microstructural investigation. In this case, it must be remembered that the fracture surface is *not* a random section through the material, but one where the fracture required the least energy.

As an example of such studies, Bhowmick, Basu and De [356] studied tensile, tear, abrasion and flexing failure modes of a nitrile rubber vulcanizate. Tear testing showed that the addition of particulate fillers caused strengthening. In the tensile case, a flow initiated fracture showed typical fracture morphology. The dispersed phase morphology and adhesion of the polymers was shown to relate to the region studied in the SEM. Observations of a fracture surface must be conducted carefully, relating the morphology to the nature of the fracture. For example, different regions of impact fracture surfaces in poly(butylene terephthalate) show variations in dispersed phase morphology [6].

In the second case, the mechanical deformation method will not be standard, but suitable samples will be designed to fit the microscope. The *in situ* deformation of polymers is almost always conducted in the SEM at low magnification and is most often used for fibers and fabrics. Other microdeformation methods are used to prepare thin films for TEM observations, to model fiber structure or to investigate crazing. Since the mechanical deformation technique is controlled by microscopy, it is appropriate to describe it in this chapter.

In the third case, where the failure and fracture occur outside the laboratory, microscopy is often used in a forensic manner to determine the failure mode and its point of initiation. Microanalysis may be used to pinpoint a specific cause and assign responsibility for the failure. Here the special features of specimen preparation relate to the nature of the work and optical photography may be required to document the relation of the specific SEM samples to the original object.

4.8.2 Fracture: standard physical testing

Fractured bulk polymers and composites require only coating with a conductive layer (Section 4.7.3) before observation in the SEM, although some composite fracture surfaces are so rough as to make deposition of a thin continuous conductive film very difficult. High resolution is rarely required in these materials so the common solution is to use a thick coating, and often carbon is evaporated followed by metal coating. Fibers, particularly textile fibers and thin films, have such a small cross sectional area that the main difficulty is in handling the broken sample.

4.8.2.1 *Fiber fractography*

Hearle and Cross [357] broke thermoplastic fibers at normal rates of extension in an Instron and examined them in the SEM. They developed a special stub for mounting the fiber ends. A step was cut from the circular stub and an elliptical hole cut in the remaining stub and a screw placed in the cut step. This provides a space in the center of the stub for fibers to be mounted on double sided sticky tape. The screw is used to attach the cut step portion to the remainder of the stub.

Studies have been conducted by breaking polymer fibers on devices, such as an Instron and then selecting fiber ends for mounting and SEM examination. This method is time consuming but many more well characterized fibers can be evaluated by breaking fibers in the microscope. Thus, the method of choice is the one

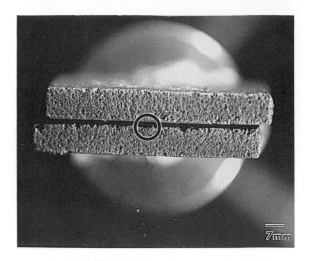

Fig. 4.33 Photograph of matched fiber ends mounted between two carbon blocks to insert in the SEM for imaging of the fracture surface and x-ray microanalysis of the locus of failure.

where controlled failure occurs outside the microscope. Fibers fractured either by standard testing or by deformation in the SEM are prepared by the following method. The fibers are placed onto double sided sticky tape on a rectangular piece of a carbon stub and attached to another stub (Fig. 4.33). Mounting fibers is best done under a stereo binocular microscope. Only a few fibers can be studied by this method in a reasonable time; however, morphological analysis of the fracture aids determination of the nature and cause of failure.

SEM images of a matched pair of tensile failed PET fibers are shown in Fig. 4.34. A classical slow fracture zone, or *mirror*, is seen adjacent to the locus of failure. A typical ridged or *hackle* morphology is exhibited as the crack propagates and accelerates away from the failure locus. In this study, an inorganic residue from the polymer process was shown to be the cause of failure [358]. The value of such a fractography is that the flaws causing failure can be determined and this information can be used to modify the process, and improve mechanical properties. There are many good reviews of the polymer fracture process [352, 359–362].

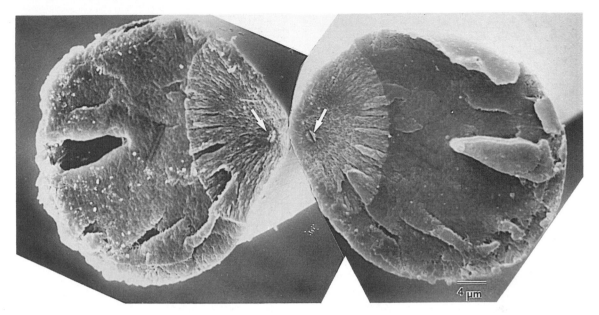

Fig. 4.34 SEM images of a matched fractured PET fiber show a defect at the locus of failure (arrows). The region surrounding the locus of failure, the mirror, is the slow fracture zone. As the fracture accelerates across the fiber, ridges, or hackles, are formed in the outer fracture surface.

4.8.2.2 *Fracture of plastics*

The fracture process and morphology of thermoplastics, glassy thermoplastics, and thermosetting resins can all be studied by fracture methods. In one example, two samples of a nylon polymer exhibited very different elongation properties, although they were processed similarly. SEM study showed that the specimen with poor elongation properties exhibited brittle fracture morphology (Fig. 4.35A and B), due to a contaminant, whereas the specimen with higher elongation properties exhibited ductile fracture morphology (Fig. 4.35C–F).

Multiphase polymers are observed by fractography for evaluation of the rubber or the dispersed phase size, shape and morphology. Bucknall [363] described the control of the structure during blend manufacture and the resulting effects on the properties including the microscopy characterization. Reed [364] described the impact performance of polymers. Evaluation of fractured materials resulting from

various physical property tests is conducted in the SEM where the fracture mechanism and the cause of failure are related to the properties. A detailed study [365] of impact toughened polyamides showed mechanisms of failure. Ductile fracture surfaces of water conditioned nylon 6 showed extensive yielding and shear bands. Stress whitened regions beneath the ductile fracture surfaces were cryogenically dissected both transverse and longitudinally with respect to the ductile crack growth direction. Cavitation was observed around the rubbery particles. Representative SEM images (Fig. 4.36) show a range of different multiphase polymers in notched Izod impact fractured specimens. A polymer with large, nonuniform, dispersed phase particles, not well adhered to the matrix, is shown in Fig. 4.36A. A much finer dispersed phase is shown in Fig. 4.36B where both particles and holes from particle pullouts are observed. Smaller particles are not as obvious in Fig. 4.36C, although the dispersed phase accounts for 15% of the specimen. Finally, Fig. 4.36D does not clearly

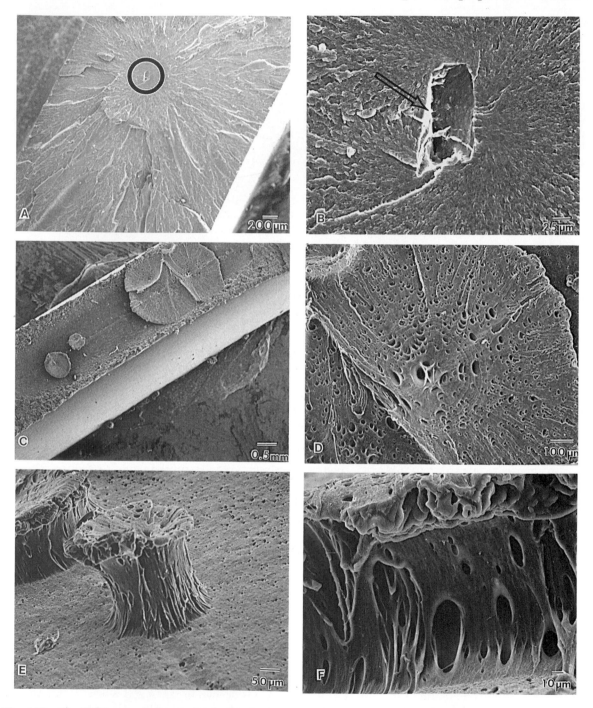

Fig. 4.35 The SEM was used to compare the fracture morphology of two semicrystalline specimens that were processed similarly and yet produced different elongation properties. The nylon specimen with lower properties (A, B) has a brittle fracture morphology with a rectangular shaped particle at the locus of failure (B). The specimen with higher elongation properties (C–F) exhibited ductile failure.

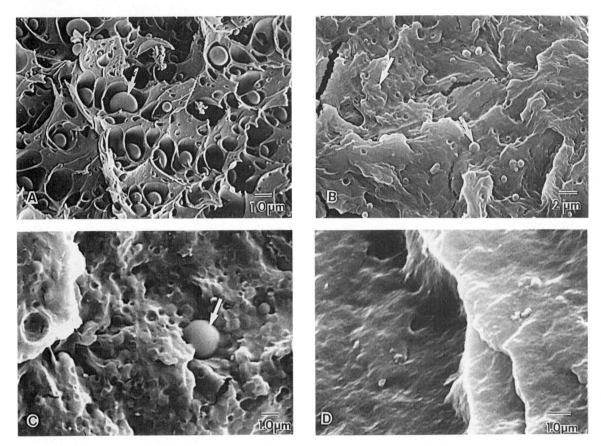

Fig. 4.36 SEM images of representative fractured, dispersed phase polymer specimens: (A) has obvious dispersed phase particles in a ductile matrix that have poor adhesion; (B) has much finer dispersed phase particles; (C) has particles and holes where particles pulled out and are on the other fracture face – the adhesion is much better in this case compared to (A); (D) has barely distinguishable particles, although over 20% elastomer is present.

reveal the elastomer. In this case, the SEM does not provide observation of the size and distribution of the dispersed phase, and such information must be provided by some other microscopy technique.

4.8.2.3 *Composite fractures*

Studies of the relation of structure to the mechanical behavior of fiber reinforced composites are commonly encountered as these materials find application in both structural and nonstructural applications. The fracture behavior

of composites is affected by many variables, including the nature of the fiber and matrix, the fiber–matrix bond, fiber orientation, stacking angle and sequence, void level, loading and the environment. As a result of the breadth of the variables Mulville and Wolock [366] stated that generalizations relating to fracture behavior are difficult to make. Some possible damage modes in composites, however, are matrix cracking, fiber–matrix interfacial bond failure, fiber breakage, void growth and delamination. SEM examination of the failure zone may reveal all of these failure modes with no clear evidence of the initiation site. However, the SEM does provide some insights into the nature of the failed surface. Figure 4.37 shows the nature of two such composite fracture surfaces. The clean surface of the glass fibers, shown by SEM (Fig. 4.37A and B), results from poor bonding of the fibers to the matrix and failure at the fiber–matrix interface. In another case (Fig. 4.37C and D), the fibers have resin on their surfaces and the failure appears to have taken place in the matrix rather than at the fiber–resin interface.

The identification, characterization and quantification of fracture modes of graphite fiber resin composites in a specific deformation (tensile testing) mode have been determined using the SEM [367] where three different modes of fracture were identified for off-axis fiber–resin composites. Kline and Chang [368] investigated the fracture surface features associated with various failure tests, including tension, compression and tension fatigue. Flexural fatigue loading of graphite epoxy composites results in matrix cleavage, hackle formation and wear failure features [369].

Polymer matrices are also commonly reinforced with mineral fillers or fibers, such as calcium carbonate, talc, wollastonite, clay and mica [370]. SEM images of fracture surfaces show the wetting behavior or adhesion of the filler by the polymer matrix. Figure 4.38 shows secondary electron image (A) and backscattered electron image (B) micrographs of a mineral filler in a matrix of a commercial polymer. SEM does not reveal the nature of the filler in the matrix whereas BEI does reveal the mineral filler due to atomic number contrast. BE imaging is important in the observation of mixed polymer and inorganic materials, enhancing the contrast between these materials. Kubat and Stromvall [371] studied the reinforcement of polypropylene and polyamide 6 with mineral fibers. Finally, polymer fibers are also used to increase the fracture toughness of polymer composites. Hamer and Woodhams [372] showed that the enhanced impact strength of fractured polypropylene containing chopped PET fibers was due to fibers which were not well dispersed.

4.8.3 *In situ* deformation

Deformation experiments can be conducted within the microscope in order to assess *in situ* the nature of the change in structure as a function of the specific deformation process. Although there is little space in the specimen area of the TEM, stages can be accommodated. The chamber of the SEM is quite large and it can accommodate a variety of stages, including hot, cold and wet stages, tensile and straining stages. Hot stages and deformation stages are also commonly applied to optical microscopy experiments. The types of structure information that are monitored in such *in situ* experiments are the crystal or phase structure in TEM, by electron diffraction and dark field imaging, and the topography of the fracture surface in the SEM. The chemical composition of the specimen can also be monitored by x-ray microanalysis techniques in the electron microscope. Optical microscopy studies are excellent for the identification of phase structures and phase changes as a function of temperature.

A major problem in deformation experiments with polymers in the SEM is that a conductive coating is normally required for such imaging. Where such a coating cannot be applied, due to the formation of new surfaces during deformation, other methods are applied to increase conduction or decrease charge buildup in the specimen. Charge neutralizer devices use low energy ion sources to neutralize the negative surface charge by irradiation with a flux of positively charged ions [373]. Antistatic sprays can either be sprayed on the material or the fibers dipped in it for low magnification studies (less than 1000×). McKee and Beattie [374] described a stage for the SEM where they deformed fibers, yarns and fabrics after spraying with either Duron antistatic spray or Triton X-100. Triton X-100 is a commercial surfactant (trademark of Rohm and Haas, Philadelphia, PA), a nonionic oxyphenoxyethanol. Charge neutralization and antistatic sprays have in common usefulness only at low magnifications.

The SEM permits the observation of *in situ*

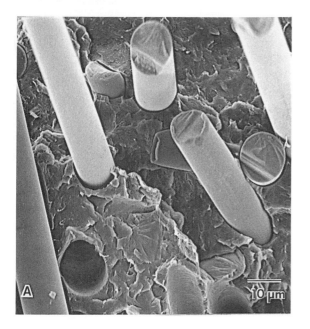 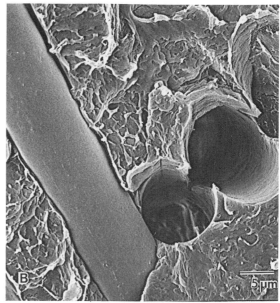

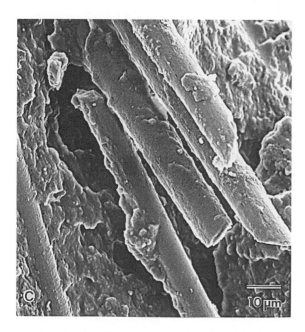 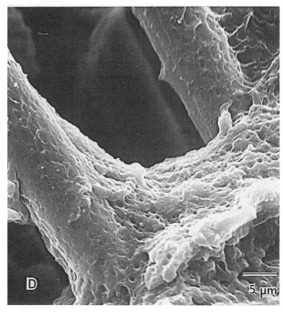

Fig. 4.37 SEM images of two glass filled composites with different adhesion properties are shown: (A) is a composite with poor adhesion, as shown by the clean glass fiber surfaces resulting from failure at the fiber–resin interface; (B) has holes where fibers with poor adhesion pulled out of the matrix; (C) shows failure in the matrix and thus there is resin on the fiber surfaces; also seen magnified in (D).

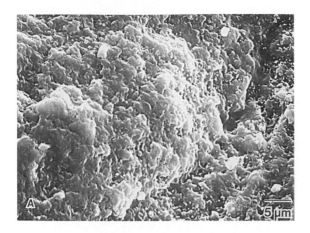

Fig. 4.38 Secondary electron image of a mineral filled polymer composite (A) does not reveal the nature of the dispersed filler particles. Atomic number contrast in the backscattered electron image (B) clearly shows the mineral filler (brightness increases with atomic number).

deformation and fracture of small specimens, such as fibers and films, deformed using specially built tensile and straining stages within the SEM [375]. Fibers and woven and nonwoven fabrics have been deformed under tension [4] and have aided understanding of the failure mechanism. Recent experiments of this type have deformed the specimens from both ends so that the central region remains stationary [376, 377]. Mao *et al.* [378] have developed a useful method where a thermoset resin photoresist layer was spread over the specimen surface, and the specimen was exposed to strong light, except in regions of a special grid which left a pattern on the surface. Once the specimen was deformed the displacement could be measured using the change in the pattern. Craze growth at the crack front in polypropylene was studied in the SEM and backscattered electron images obtained using a poor vacuum without a conductive coating [376, 379]. A common method to limit specimen charging during *in situ* deformation studies is to use low accelerating voltages, generally less than 1 kV.

An example of a study conducted using a tensile stage in the SEM is the evaluation of the ductile failure of poly(vinyl chloride). Smith *et al.* [380] stamped dumbbell shaped pieces of polymer from 1 mm thick sheets and extended them to a neck in an Instron tester. The prestrained pieces were then strained in the SEM. Low accelerating voltage was used for imaging of the uncoated specimens. These experiments showed that, after neck formation, fracture occurs by crack propagation from a flaw or cavity within the surface craze.

The *in situ* deformation of amorphous polymers by shear deformation and craze growth has been observed in optical microscope studies by Donald and Kramer [381]. Grids with thin films of various polymers and polymer blends were prepared on copper grids which were strained in air on a strain frame held in an optical microscope. The films were precracked in an electron microscope by a method more fully described by Lauterwasser and Kramer [382]. Many crazing studies are evaluated by *in situ* methods, and optical microscopy plays a major role in providing an overview of the deformation structure. Crazing studies will be more fully explored in the next section.

4.8.4 Crazing

Crazing is the first stage of fracture in many glassy polymers and also in blends and semicrystalline polymers where there is a glassy matrix. Crazing is a localized tensile yielding process that produces thin sheets of deformed, 'crazed', material with the sheets perpendicular to the principal stress axis. Within a craze the continuous phase is void and there are fine (*c.* 10 nm) fibrils of oriented polymer parallel to the stress axis that extend across the craze. Craze material might have an average density of only 20% of that of the bulk polymer (i.e. if 80% of the craze is void) but the craze can still support a significant load because of the oriented fibrils. The load bearing capacity of crazes can be seen directly in some tensile test samples of transparent polymers where crazes visible to the eye extend right across the specimen while it retains most of its strength [353].

Eventually, crazes break down to form cracks, and when the cracks grow to critical size the sample fails. Although crazes lead to failure in this way they can be useful, since if many crazes are produced before failure occurs, energy is absorbed by the material as the local yielding takes place. The impact strength of modified blends is due to the large number of crazes formed. Even when there is no macroscopic indication of crazing, microscopy may show local yielding, by crazing or shear deformation, taking place just in advance of a crack tip. This ductile zone controls the fracture toughness of the polymer just as it does in metals. Examples of polymers showing this effect are HIPS that has undergone fatigue cracking [383], polycarbonate [381] and lightly crosslinked epoxy resins. Kramer [384] described the formation and breakdown of fibrils quite thoroughly. The TEM is required to see the fine structure of crazes and their relation to second phase particles in blends. Crazing and fracture have been reviewed by Kambour [385], Rabinowitz and Beardmore [386], Kramer [387] and Kausch [352, 353].

4.8.4.1 *Preparation for TEM*

Preparation of crazed polymers for TEM is quite difficult. First, the whole specimen must be stressed to failure, resulting in crazes which are weak and full of voids. Worse yet, the craze structure is unstable in the absence of applied stress. Relaxation of the oriented fibrils and of the elastic deformation of the bulk material tends to close up the craze, destroying its original structure. Sectioning a bulk sample containing crazes is therefore very likely to destroy the information that is being sought. If thin sections or solvent cast thin films are deformed, to avoid the problem of sectioning crazed material, the craze structure may not be the same as that in the bulk material, since the surfaces of a thin film have more freedom. One can imagine a thin film necking down to an even thinner film on drawing, but this could not happen if the same material was inside a thick specimen, without creating large voids. In the terms of mechanical testing, the thin films are deformed in plane stress, while inside the bulk such sheets of material are deformed in plane strain.

Microscopists have either attempted to stabilize the crazes in bulk samples by infiltration with supporting materials before sectioning, or they have used thin films in the TEM and a range of techniques including diffraction to try to ensure that the thin film structures are representative. Kambour [388, 389] stabilized crazes in polycarbonate by impregnation with silver nitrate for 4–10 days. Selected area electron diffraction showed that the material deposited was metallic silver. Later, Kambour and Holik [390] used liquid sulfur to impregnate and reinforce crazes in PPO. The crazes were formed in ethanol and then were filled with formamide at 125°C for 1 h to keep the voids open. Specimens were transferred to liquid sulfur, at the same temperature, for 24 h. Carbon coating the sections supported the structure and permitted sublimation of the sulfur. This method is useful in cases where the material is not changed during treatment. Kambour [385] outlined the requirements for a craze 'infusant' material: (1) the craze should be completely filled with a liquid, below the temperature where the craze loses its strength; (2) this material should be a solid at the temperature for microtomy; and (3) this material should have higher electron density than the polymer, if it remains in the craze.

Beahan *et al.* [391] compared crazes in thin

films of polystyrene with fractured bulk specimens. They found the morphology of crazes to be the same in both cases. They used a full range of microscopy preparation methods including thin sectioning of precrazed bulk material impregnated to reinforce the structure and limit damage, straining of sections cut from bulk material, crazing solvent cast thin films and straining a section of a thin film directly on a TEM grid. The article serves as an excellent review of preparation methods.

Kramer *et al.* [392] developed a new TEM method to observe and measure deformation and fracture mechanisms at planar interfaces which should have general applicability and so will be described in some detail. The samples studied were PS and PVP homopolymers as an immiscible polymer pair and a PS–PVP block copolymer as a compatibilizer to reinforce the interface between the PS and PVP. PS and PVP were fabricated by compression molding. A thin film of the block copolymer was spin cast from toluene solutions on the PVP slab. The coated PVP molding was joined to the PS molding, annealed and allowed to cool to permit diffusion at the interface. The sandwich was cut with a diamond saw in slabs for fracture toughness measurements and then cut into smaller pieces for TEM preparation. These small pieces were embedded in epoxy resin and cured at room temperature for 2 h, microtomed with a glass knife perpendicular to the interfacial plane to obtain thin films, *c.* 0.5–1.0 μm at a knife angle of 45° and clearance angle of 9°. The films were placed on ductile Cu grids with 1 mm squares previously coated with the block copolymer, so that the interface was aligned perpendicular to the straining direction. The film was bonded to the grid by exposure to vapor of the solvent and then dried at 50°C for 12 h in vacuum. The grid was strained in tension at a constant strain rate of 4×10^{-4} s^{-1} with a servocontrolled motor drive at room temperature. The film was then exposed to iodine vapor at room temperature for 6–12 h to stain the PVP phase prior to TEM observation. Kramer *et al.* [393] also developed a technique for the study of crazing in glassy block copolymers.

They prepared thin films *c.* 0.5 μm thick by spin casting diblock and triblock copolymers of PS and PVP from benzene onto a rock salt substrate [393]. The dry films were exposed to benzene vapor for 24 h and the glassy films were floated off the rock salt onto a water bath surface where they were picked up onto a grid coated with a thin film of PS. The film adhered to the grid after a brief exposure to the solvent. An SEM beam was used to 'burn' a thin slit in the material and cracks 50–100 μm long by 10 μm wide were introduced in the center of each grid square. Grids were deformed as described above and examined by OM to locate areas of interest for TEM study and measurement of craze fibril extension ratios [382].

4.8.4.2 *Deformation methods*

Even if it can be assumed that crazes formed in thin films are representative of bulk behavior, there has been a problem that the mechanical deformation itself has been poorly controlled. Deformation by hand held tweezers or by exposing strained films to solvent is a poorly controlled experiment (e.g. ref. 394) which cannot give well defined mechanical histories. This has made it difficult to relate the observed structure to mechanical properties. Lauterwasser and Kramer [382] solved this problem by bonding solvent cast films of polystyrene to a ductile copper grid (1 mm mesh) and drawing the grid in a micrometer screw driven tensile stage to a well defined strain. As the deformation of the grid is completely plastic, it can be removed from the tensile stage, and individual 1 mm squares of interest can be cut out for study in the TEM without relaxing the load on the polymer film.

A quantitative analysis of craze shape and mass thickness contrast within the craze allowed Lauterwasser and Kramer [382] to derive the stress profile existing along a polystyrene craze. Kramer and his coworkers have extended this study to many other polymers, relating the mean density of craze material to entanglement density in the polymer glass and to toughness [395] without a basic change of preparation technique.

Clearly, if a rubber modified blend contains particles larger than the film thickness (about 1 μm), the film cannot be representative of the bulk. For the study of craze tips and craze growth, a double tilting stage was used to obtain stereo pairs of the craze tip [396].

The microstructure observed for thick films shows fibrils, about 4–10 nm in diameter for polystyrene, in agreement with SAXS measurements on the crazes in the bulk polymer. Very thin films of polystyrene (100 nm) show modification in the craze structure as there is no plastic restraint normal to the film [397]. Deformation zones have also been studied in polycarbonate, polystyrene–acrylonitrile and other polymers [398]. Crazes in thermosets can be studied in thin films spun onto NaCl substrates which can be washed away when the film has been cured. Mass thickness measurements are difficult to make in radiation sensitive materials; that is why most TEM work has been done on polystyrene and least on PMMA. After developing the techniques described above for TEM Donald and Kramer [398] applied similar methods in optical microscopy to study radiation sensitive materials and the kinetics and growth of deformation zones. Thin films were strained on grids *in situ* in a reflecting OM. Change of interference color, which depends on the film thickness, was a very sensitive method for observing film deformation.

A study of a PS–PB block copolymer showed variation in craze behavior as a result of rubber particles added to modify the otherwise brittle, glassy polymers. Such copolymers were studied under the high strain of physical laboratory testing where the polybutadiene in the copolymer was stained with osmium tetroxide prior to microtomy [399]. The brittle behavior of the glassy polymers was shown by TEM and STEM to be modified by the rubber particles which provide toughening by control of the craze behavior. In a study of the craze behavior in isotactic polystyrene [169], films of polystyrene were drawn from dichlorobenzene solution and cast onto glass microscope slides, followed by various treatments including isothermal crystallization. TEM specimens were produced, as described earlier [382], strained and examined by OM and TEM. Crazes in amorphous isotactic polystyrene were shown to be similar to crazes formed in atactic polystyrene. Clearly, crazes are important to the understanding of the fracture behavior of both isotropic and oriented glassy polymers and also to the understanding of fracture toughness in toughened polymers. TEM micrographs of a craze are shown in Fig. 4.39.

4.9 FREEZING AND DRYING METHODS

Generally, the methods described here involve special drying of the specimen or some kind of freezing technique. Materials as diverse as emul-

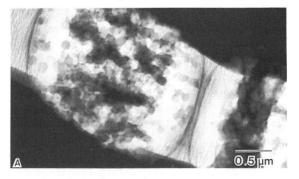 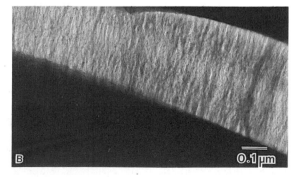

Fig. 4.39 Typical TEM micrographs of crazed thin films showing a deformed rubber particle in HIPS (A) and craze fibrils in unmodified PS (B).

sions, latexes, polymer blends, wet membranes and ductile polymers often require these special methods to provide appropriate specimens for electron microscopy.

In the case of emulsions, latexes, some adhesives and wet membranes, the specimen of interest is wet, generally with water and must be dried prior to electron microscopy. The deleterious effects of air drying result from the stress of surface tension forces. The methods used by biologists [400] to avoid this problem are: (1) the replacement of water with an organic solvent of lower surface tension, (2) freeze drying or (3) critical point drying. Some latexes can be 'fixed' in their original shape by chemical or physical treatment that makes the particles sufficiently rigid to withstand the surface tension forces [401]. Treatments that have been used are bromine [402], osmium tetroxide [109] and high energy irradiation [402]. Unfortunately, many polymers, especially those that are chemically saturated, such as vinyl acetate or acrylates, are unaffected by such treatments. Sectioning films of latexes is also possible, but the individual particles are often not observed by this method. In cases where the specimen is too soft or ductile for routine fractography, low temperature methods, such as freeze fracture and freeze etching, can provide an appropriate specimen for study. Special TEM preparation methods of such materials, by both direct imaging techniques and replication methods, have been described and reviewed by Bachmann and Talmon [403].

4.9.1 Simple freezing methods

There are some simple freezing methods that provide adequate preparation for some polymers. Cold stage microscopy of colloidal suspensions, microemulsions and liquids is possible by fast freezing and examination of the thin, frozen specimen in an EM. Talmon *et al.* [404] developed a rather interesting technique in which a thin sample is trapped between two polyimide films. The liquid layer is about 100 nm thick, while the films are about 40 nm thick. Film selection is quite important as the polyimide is more radia-

tion resistant than traditional support films. The films are formed by dipping glass microscope slides into a 0.75% solution of the prepolymer in *N*-methylpyrrolidone and xylene. The slide is dried for 10 min at 90°C and then cured at 300°C for 3.5 h. The polyimide coated slides are dipped into 12% hydrofluoric acid for 5–19 s and then floated off onto water and picked up onto grids. A drop of the specimen is placed onto one film coated grid and covered with another film coated grid prior to immersion into liquid nitrogen. A transfer module was designed in this study. Examples were shown for benzene plasticized polystyrene latex and a surfactant. Falls *et al.* [405] used this fast freeze cold stage in a study of hexagonal ice. Knowledge of this structure is essential to understand the morphology of frozen fluids.

Manual methods of freeze fracture are often useful in providing specimens for study in the SEM. An example of a 'freeze shattering' method was described by Stoffer and Bone [406] for comparison with microtomy results. Polymers immersed in liquid nitrogen were mechanically shattered with a hammer, mounted, vacuum pumped and sputter coated for observation. This simple method is useful if the materials cannot be sectioned. However, fine structural details are not conclusive when specimens are prepared by such methods.

Finally, multiphase polymers and blends often have such high impact properties that they do not fracture during room temperature impact testing. Prefreezing in liquid nitrogen, after notching, has been used successfully to fracture such polymers in an impact tester for property and structure evaluation. This method is far superior to shattering with any mechanical device which deforms the specimen nonreproducibly. Polymers which are ductile at room temperature and which smear upon fracturing may also be frozen and fractured in this way.

4.9.2 Freeze drying

Of the three methods that are used to diminish the surface tension effects resulting from air drying, the replacement of water with an organic solvent, such as ethanol or amyl acetate [407], is certainly the simplest. However, controlled experiments must be conducted to ensure that the organic solvent has no detrimental effect on the polymer. The second method used to avoid surface tension effects, *freeze drying*, enjoyed great popularity during the 1970s. The method was introduced for biological specimens in 1946 [408] and has been developed and promoted over the years for biological [400, 409, 410] and polymer [401] specimens. Freeze drying permits the sublimation of water as a solid to the gas phase, thus avoiding surface tension effects. The fundamentals of freeze drying have been reviewed by Rowe [411, 412].

Freeze drying involves rapid freezing, sublimation of the frozen water and application of a conductive coating to add stability to the material. The actual freeze drying takes place in a vacuum evaporation unit set up with a special specimen freezing device and a cold trap for trapping of the water molecules during the procedure. There are several points involved in freeze drying any material [413] including: (1) rapid freezing of a thin specimen layer, (2) minimum ice crystal formation and (3) slow but complete sublimation of the ice. Rather than describing those techniques used in biological studies [400] a practical application of this method for polymers will be described.

4.9.2.1 *General method and examples*

Walter and Bryant [414] described a method for freeze drying latex specimens in a home made vacuum system rather than a commercially available device (as was typical of the state of the art at that time). Later, a freeze drying/image analysis method using commercially available equipment was described [413]. Important details of that method included: specimen preparation, placement onto a TEM (or SEM) grid, the hardware for the experiment and the metal coating.

A general method for the preparation of latex for TEM study has been used on film forming latex, important in coatings and adhesives applications. Emulsion particles collapse into a film upon air drying which is typical of latexes with a glass transition below room temperature. Solutions must be very dilute, so as to obtain a monolayer of uniformly frozen particles. Two methods generally used to transfer latex particles to the TEM grid are spraying and placement of a microdroplet onto a plastic coated grid. Freezing must be rapid, so that it occurs before any air drying occurs. Direct spraying or dropping onto a frozen grid is difficult, at best, and often the specimen freezes in the air above the grid. A simple prefreezing table (adapted from Walter [415]) was constructed (Fig. 4.40A) to permit good sampling and rapid freezing. The steel or aluminium table is placed in a Styrofoam vat filled with liquid nitrogen up to the level of the bottom surface (Fig. 4.40A). Drops of solution are placed onto a room temperature grid, and then the specimen holder is quickly placed on the precooled table, so that freezing occurs prior to any air drying. The cover limits air access to the specimens and the possibility of frost formation.

An Edwards evaporator with freeze fracture accessory was used in this experiment, although any commercial freeze etch device can be used. The system must have provision for pumping liquid nitrogen into the specimen holder and temperature sensing and controlling devices. The specimens are quickly transferred from the prefreezing table to the precooled specimen stage in the evaporator at its lowest possible temperature ($-150°C$). Transfer takes place through the port in the stainless steel collar, below the bell jar, under a flow of dry nitrogen which is used to limit ice formation on the frozen specimen surface. Handling of the grids is simplified by placing them on the special lipped specimen holders (Fig. 4.40B) which permit the rapid transfer (i.e. less than 30 s) of several specimens into the vacuum chamber. Samples are freeze dried at -60 to $-80°C$, depending on the specific material, for about 8 h.

Freeze dried specimens are generally shadowed (at about $-150°C$) in the vacuum chamber. Shadowing provides a metal cap that is the

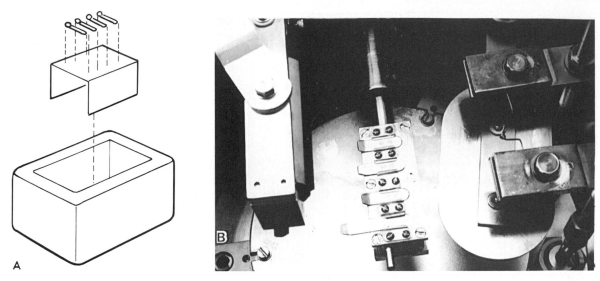

Fig. 4.40 A Styrofoam cooler, used as a prefreezing chamber, is shown in this sketch (A) with an aluminium specimen table and small, rectangular holders for the TEM grids. The cooler lid (not shown) is important for limiting frost formation on the specimen. A photograph of the specimen holders (with lips for easy accessibility) is shown on the cold stage in the vacuum evaporator (B).

shape of the frozen, undistorted particle and limits distortion during examination in the electron microscope [416]. Replication of the latex particles, or macromolecules, can also be done if the specimen is expected to change at room temperature. For SEM specimens, the same procedure is used, but the specimens are dried onto small glass cover slips which are attached to the stub with silver paint. Watanabe *et al.* [417] used a similar approach in their preparation of latex paper coatings. A wet sample was cut, cooled in liquid nitrogen, placed on an SEM mount, held by a metal ring and placed in a vacuum evaporator for freeze drying and metal coating while still cold. The method limited the coalescence of a polystyrene–butadiene latex. Katoh [418] described a method for the preparation of film forming latex particles. The emulsions were diluted to about 0.1–0.01%, dropped onto a hydrophilic glass substrate and the excess was removed with filter paper prior to placement into liquid nitrogen. As in the general method below, the sample, poly(ethyl acrylate), was transferred to the vacuum evaporator, freeze dried and metal coated.

Results of the experiment described are shown in Fig. 4.41. A monodisperse latex of known particle size (Fig. 4.41A) was used both as a control and for calibration of the particle size distribution measurement [419]. A film forming latex is shown following both air drying (Fig. 4.41B) and freeze drying (Fig. 4.41C). Clearly, the flat, film forming, air dried particles are three dimensional following freeze drying.

Thus, the steps in the general specimen preparation are as follows.

(1) Dilute the latex to about 1:1000 with water or until only a slightly milky blue cast is seen in the clear solution.
(2) Place microdroplets of diluted latex on formvar coated grids which are on rectangular, lipped specimen holders. Quickly place holders on the precooled (liquid nitrogen) table and cover.
(3) Transfer specimen grids on the holders to the precooled stage in the vacuum evaporator under flowing nitrogen.
(4) Freeze dry specimens at −60 to −80°C (about 8 h).

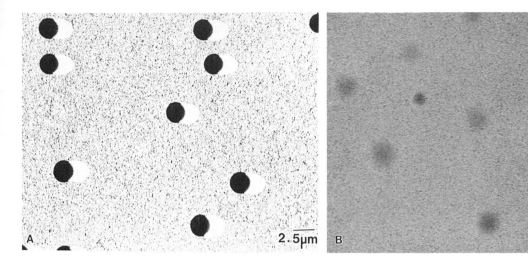

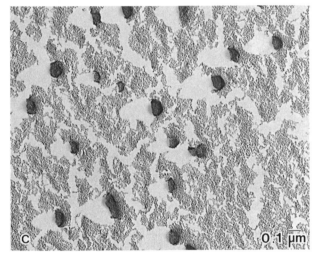

Fig. 4.41 TEM micrographs of freeze dried polystyrene latex (A) used as a control for the experiment are three dimensional and show no deformation, whereas an air dried film forming latex (B) shows barely visible, flat regions that have no shadow. The same latex as in (B) after the freeze drying experiment has shadows (C), showing that the particles are three dimensional.

(5) Lower the temperature to −150°C, and shadow samples with a heavy metal.

4.9.3 Critical point drying

The drying preparation method in greatest use among biologists today is critical point drying (CPD). The method was first described by Anderson [420–422] and then by Hayat and Zirkin [423]. A clear explanation of the method and applications are given by Anderson [422, 424] and Cohen [425], while a somewhat more current review by Cohen [426] is recommended to the interested reader.

In ordinary drying, the liquid in a specimen evaporates, and the resulting surface (interfacial) tension can distort the structure. In critical point drying [425], heating a specimen in a fluid above the critical temperature to above the critical pressure permits the specimen to pass through the 'critical point' (that temperature and pressure where the densities of the liquid and vapor phases are the same and they coexist and thus there is no surface tension). By definition, a gas cannot condense to a liquid at any pressure above the critical temperature. The 'critical pressure' is the minimum pressure required to condense a liquid from the gas phase at just below the critical temperature. Thus, CPD allows the specimen and fluid to be taken directly to a gas phase without experiencing any surface tension effects and resulting distortion. At above the critical temperature, the gas is bled off leaving the specimen dried for study.

Critical point drying is conducted using transitional fluids which go from liquid to gas through the critical point. The critical temperature (more than 300°C) and pressure (above 21 MPa) of water are much too high for it to be used. Unfortunately, this requires the removal of water and its replacement by a transitional liquid. Water is removed and replaced by dehydration fluids which are replaced by the selected transitional fluid. A typical transition sequence is:

(1) a graded water–ethanol (or water–acetone) series;
(2) 100% ethanol (or acetone);
(3) transitional fluid, usually Freon (13 or 16) or CO_2.

Graded series are combinations of fluids that are used to gradually replace the water with the dehydrating fluid, such as water/ethanol: 90/10, 75/25, 50/50, 25/75, 0/100. Freon TF (134) is useful as an intermediate fluid, as it does not have to be fully flushed out of the system when used prior to drying with CO_2. Critical constants for Freon 13 (CCl_3), Freon 116 (CF_3-CF_3), and carbon dioxide (CO_2) are in the range of 20–40°C and 3.75–7.5 MPa. The CPD preparation is conducted in a pressure vessel to control both the temperature and the pressure.

4.9.3.1 *General method and examples*

One example [427] of this method will be briefly described. A wet polymer membrane composed of polybenzimidazole (PBI) becomes brittle and distorted upon air drying, due to surface tension effects. Samples prepared for the SEM by immersion in liquid nitrogen and then hand fractured are distorted and often ductile fracture is observed. Standard critical point drying from carbon dioxide, following dehydration into ethanol, yields a membrane which fractures with no evidence of ductility. Most exciting is the fine structure visible in these fractures. Comparison of SEM micrographs of a membrane prepared by freeze fracturing (Fig. 4.42A) and a membrane dried by CPD (Fig. 4.42B) shows large differences in the structure. The freeze fracture method reveals a ductile, distorted structure, whereas the CPD fracture clearly reveals a monolayer of granular particles in the outer surface, or skin layer, and a more open substructure of similar

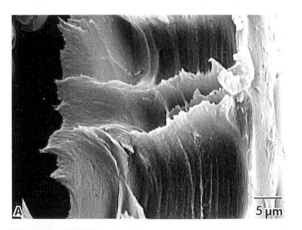 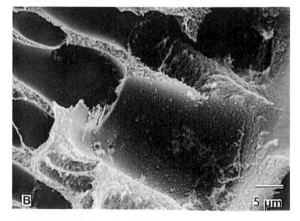

Fig. 4.42 SEM images show a comparison between fracturing of a wet membrane following immersion in liquid nitrogen (A) and following critical point drying (B). Fracturing after freezing results in a deformed ductile failure whereas the fracture following critical point drying shows no deleterious effects of surface tension and the result is a brittle fracture with excellent detail of the internal morphology.

granular particles. These results are consistent with a more complete microstructural study [64].

Finally, for critical point drying of polymers a suitable series of fluids and conditions must be chosen that will not damage the specimen. Carbon dioxide dissolves in epoxies and in polystyrene. Apparently long term high pressure exposure and then a sudden release can turn these polymers into popcorn. Microscopic comparison of a material prepared by a variety of different preparation methods is the best way to uncover any possible artifacts caused by the specimen preparation method.

4.9.4 Freeze fracture-etching

Freeze fracture and freeze etching are distinctly biological techniques that have also been used by the polymer microscopist. Freeze fracture means fast freezing of a specimen followed by fracture with a cold knife in a vacuum chamber to reveal the internal structure of a bulk specimen. Freeze etching is somewhat of a misnomer as this process is the surface freeze drying of the freshly fractured specimen or the sublimation of ice from the frozen surface. Typical conditions are to sublime the surface ice at about $-100°C$ for 1–5 min at a good vacuum. This method has the advantage of exposing the underlying true surface features, by removing about 20 nm of ice, for replication and TEM evaluation or for direct SEM observation of the surface structures.

4.9.4.1 *Biological method*

An excellent review of freeze fracturing is found in a chapter by McNutt [428]. The method has been described for the preparation of biological membranes. An older, improved version of the method was described by Steere [429] and Moor *et al.* [430], while a review of the method and application to membranes was described by Branton [431]. The method involves cementing a 1–2 mm piece of the material onto a copper disc with gum arabic dissolved in 20% glycerin and then transferring the material to liquid Freon 22 (chlorodifluoromethane). The specimen is fractured with a cold knife in a vacuum evaporator,

and it may be etched prior to replication. A replica can be cast on the surface of the hydrated material at low temperature in a vacuum [428] for examination in the TEM; or the replica, or the shadow cast specimen surface, can be examined directly in the SEM.

4.9.4.2 *Literature review*

Several polymer studies have been reported where the specimens were prepared by freeze fracture techniques. A modification of the freeze fracture method was used by Singleton *et al.* [432] in the preparation of plasticized PVC. The sample was notched, cooled and fractured and then immediately replicated with platinum–carbon. Replicas were stripped after warming to room temperature. The authors noted that the preparation was not highly reproducible, perhaps due to nonuniform cooling of large specimens. The results of such a study must be compared with other characterizations for accurate analysis.

The refolding of poly(ethylene oxide) chains in block copolymers was studied by Gervais and Gallot [433] using the freeze fracture replica method. Specimens in the form of viscous gels were quenched from -25 to $-160°C$ by immersion into liquid Freon 22 cooled with liquid nitrogen. The styrene and ethylene oxide block copolymer specimens were fractured and platinum–carbon replicas cast in a vacuum evaporator. The lamellar structure of the system was shown by this method. These same authors [434] used the freeze fracture method for a similar study of block copolymers with an amorphous and a crystallizable block. Lamellar crystalline structures were observed in block copolymers of polystyrene or polybutadiene and poly(ethylene oxide). Freeze fracture and etching was also used to study the structure of PVC/solvent gels [435]. The gels were quenched in solid/liquid Freon 22 $(-160°C)$ and then fractured (at $-150°C$). The gel was replicated following partial solvent sublimation in a Balzers freeze etching system. After warming, the gels were dissolved in tetrahydrofuran.

The freeze fracture method has been used to study the structure of colloidal particles in water–oil mixtures stabilized by polymer emulsifiers. Microemulsions consisting of water, toluene and graft copolymer composed of a polystyrene backbone and a poly(ethylene oxide) graft were deposited onto a small gold plate, quenched in liquid nitrogen in equilibrium with its own solid phase [436]. Replicas of the fractured surfaces were washed with tetrahydrofuran, which showed the micellar structure of the copolymers. A similar method was used for the preparation of polystyrene polymer latexes for TEM study of the size distribution [437]. In this case, the frozen droplet was microtomed, with a cold knife at −100 to −120°C, etched for up to 90 s and then a platinum–carbon replica was prepared. Etching was found to be unnecessary and a potential cause of error. The remaining latex was dissolved away before examination of the replica. Such replicas can reveal the size distribution and structure of the latex particles.

The freeze fracture method has been used to show the structure of several types of polymer materials. In addition to the examples described above, classical microemulsions have been studied [438], while micellar aggregates of copolymers have been shown [439, 440]. Polymer latexes have been prepared using similar methods by Sleytr and Robards [441]. The emphasis in this review was on the plastic deformation observed in the freeze fracture method and in ultrathin frozen sections. Network segregation morphology of aqueous polymer systems has also been described by Purz *et al.* [442], who measured the cooling rate and its effect on the structure of polymer solutions and gels. In general, the freeze fracture technique is quite useful for gels and emulsions.

REFERENCES

1. D. C. Bassett, F. C. Frank and A. Keller, *Phil. Mag.* **8** (1963) 1739.
2. S. Mitsuhashi and A. Keller, *Polymer* **2** (1961) 2.
3. J. Sikorski, J. S. Moss, A. Hepworth and T. Buckley, *J. Sci. Inst. Ser. 2* **1** (1968) 29.
4. J. W. S. Hearle, J. T. Sparrow and P. M. Cross, *The Use of the Scanning Electron Microscope* (Pergamon, Oxford, 1972).
5. P. J. Goodhew, Specimen Preparation in Materials Science, in *Practical Methods in Electron Microscopy*, edited by A. M. Glauert (North Holland – American Elsevier, Amsterdam, 1973).
6. S. Y. Hobbs, *J. Macromol. Sci., Rev. Macromol. Chem.* **C19** (1980) 221.
7. M. Furuta, *J. Polym. Sci., Polym. Phys. Edn.* **14** (1976) 479.
8. M. J. Richardson, *Proc. R. Soc., Lond. A* **279** (1964) 50.
9. D. C. Bassett and A. Keller, *Phil. Mag.* **8**(7) (1962) 1533.
10. D. C. Bassett, F. R. Dammont and R. Salovey, *Polymer* **5** (1964) 579.
11. D. J. Blundell and A. Keller, *J. Macromol. Sci., Phys.* **B2** (1968) 337.
12. R. Minke and J. Blackwell, *J. Macromol. Sci., Phys.* **B18**(2) (1980) 233.
13. B. H. Chang, A. Siegmann and A. Hiltner, *J. Polym. Sci., Polym. Phys. Edn.* **22** (1984) 255.
14. M. Tsuji, S. Isoda, M. Ohara, A. Kawaguchi and K. I. Katayama, *Polymer* **23** (1982) 1568.
15. S. Isoda, M. Tsuji, M. Ohara, A. Kawaguchi and K. I. Katayama, *Polymer* **24** (1983) 1155.
16. S. Isoda, M. Tsuji, M. Ohara, A. Kawaguchi and K. I. Katayama, *Makromol. Chem., Rapid Commun.* **4** (1983) 141.
17. P. H. Geil, *Polymer Single Crystals* (Interscience, New York, 1963).
18. F. F. Morehead, *Text. Res. J.* **8** (1950) 549.
19. J. W. S. Hearle and S. C. Simmens, *Polymer* **14** (1973) 273.
20. M. G. Dobb, D. J. Johnson and B. P. Saville, *J. Polym. Sci., Polym. Symp.* **58** (1977) 237.
21. F. P. Reding and E. R. Walter, *J. Polym. Sci.* **38** (1959) 141.
22. E. J. Roche, R. S. Stein and E. L. Thomas, *J. Polym. Sci., Polym. Phys. Edn.* **18** (1980) 1145.
23. R. G. Vadimsky, in *Methods of Experimental Physics*, Vol. 16B, edited by R. A. Fava (Academic Press, New York, 1980) p. 185.
24. P. H. Geil, *J. Polym. Sci.* **A2** (1964) 3813.
25. H. Kiho, A. Peterlin and P. Geil, *J. Polym. Sci.* **B3** (1965) 157.
26. A. M. Donald and A. H. Windle, *Colloid Polym. Sci.* **261** (1983) 793.
27. R. J. Spontak and A. H. Windle, *J. Polym. Sci: Part B, Polym. Phys.* **30** (1992) 61.

28. D. T. Grubb and A. Keller, *J. Mater. Sci.* **7** (1972) 822.

29. D. T. Grubb, A. Keller and G. W. Groves, *J. Mater. Sci.* **7** (1972) 131.

30. Chuen-mei Chu and G. L. Wilkes, *J. Macromol. Sci., Phys.* **B10**(2) (1974) 231.

31. B. F. Howell and D. H. Reneker, *Mat. Res. Soc. Symp. Proc.* **115** (1988) p. 155.

32. J. Petermann and R. M. Gohil, *J. Mater. Sci.* **14** (1979) 2260.

33. J. Petermann and H. Gleiter, *Phil. Mag.* **28** (1973) 1279.

34. M. J. Miles and J. Petermann, *J. Macromol. Sci., Phys.* **B16** (1979) 1.

35. E. L. Thomas, in *Structure of Crystalline Polymers*, edited by I. H. Hall (Elsevier – Applied Science, London, 1984) p. 79.

36. T. R. Albrecht, M. M. Dovek, C. A. Lang, P. Grutter, C. F. Quate, S. W. J. Kuan, C. W. Frank and R. F. W. Pease, *J. App. Phys.* **64**(3) (1988) 1178.

37. I. H. Musselman, P. E. Russell, R. T. Chen, M. G. Jamieson and L. C. Sawyer, in *Proc. XIIth International Congress for Electron Microscopy*, edited by W. Bailey (San Francisco Press, San Francisco, 1990) p. 866.

38. L. C. Sawyer, R. T. Chen, M. G. Jamieson, I. H. Musselman and P. E. Russell, *J. Mater. Sci.* **28** (1993) 225.

39. I. H. Musselman and P. E. Russell, *Microbeam Anal.* **26** (1991) 377.

40. G. J. Leggett, M. C. Davies, D. E. Jackson, C. J. Roberts and S. J. B. Tendler, *Trends Polym. Sci.* **1**(4) (1993) 115.

41. T. G. Stange, R. Mathews, D. F. Evans and W. A. Hendrickson, *Langmuir* **8** (1992) 920.

42. M. Suzuki, T. Maruno, F. Yamamoto and K. Nagai, *J. Vac. Sci. Technol.* **A8**(1) (1990) 631.

43. D. H. Reneker, J. Schneir, B. Howell and H. Harary, *Polymer Commun.* **31** (1990) 167.

44. R. Piner, R. Reifenberger, C. Martin, E. L. Thomas and R. P. Aparian, *J. Polym. Sci., Part C: Polym. Lett.* **28** (1990) 399.

45. S. N. Magonov, K. Qvarnstrom, V. Elings and H. J. Cantow, *Polymer Bull.* **25** (1991) 689.

46. A. K. Fritsche, A. R. Arevalo, A. F. Connolly, M. D. Moore, V. Elings and C. M. Wu, *J. Appl. Poly. Sci.* **45** (1992) 1945.

47. M. A. Sieminski and L. C. Sawyer, unpublished.

48. A. S. Holik, R. P. Kambour, D. G. Fink and S. Y. Hobbs, *Microstructural Science*, Vol. 7 (Elsevier – North Holland, Amsterdam, 1979).

49. L. E. Samuels, *Metallographic Polishing by Mechanical Methods* (Pitman, Belfast, 1971).

50. B. Haworth, C. S. Hindle, G. J. Sandilands and J. R. White, *Plast. Rubber Proc. Appl.* **2** (1982) 59.

51. L. Bartosiewicz and Z. Mencik, *J. Polym. Sci., Polym. Phys. Edn.* **12** (1974) 1163.

52. F. J. Guild and B. Ralph, *J. Mater. Sci.* **14** (1979) 2555.

53. L. C. Sawyer and M. Jaffe, *J. Mater. Sci.* **21** (1986) 1897.

54. R. G. Scott and A. W. Fergerson, *Text. Res. J.* **26** (1956) 284.

55. R. G. Scott, *ASTM Spec. Tech. Publ.* No. 257 (1959) 121.

56. J. L. Stoves, *Fibre Microscopy* (Van Nostrand, New Jersey, 1958).

57. H. B. Haanstra, *Philips Tech. Rev.* **17** (1955) 178.

58. C. A. Anderson and M. Lipson, *Text. Res. J.* **39** (1969) 88.

59. C. D. Felton, *J. Mater. Sci. Lett.* **6** (1971) 89.

60. A. Kershaw and J. Lewis, *J. Microsc.* **113**(1) (1978) 109.

61. W. L. Steffens, *J. Microsc.* **113**(1) (1978) 95.

62. K. R. Porter and J. Blum, *Anat. Record* **117** (1953) 685.

63. F. S. Sjostrand, *Experientia* **9** (1953) 114.

64. L. C. Sawyer and R. S. Jones, *J. Membrane Sci.* **20** (1984) 147.

65. A. M. Glauert, Fixation, Dehydration and Embedding of Biological Specimens, in *Practical Methods in Electron Microscopy*, edited by A. M. Glauert (North Holland – American Elsevier, Amsterdam, 1975).

66. B. L. Giammara, in *Proc. 43rd Ann. EMSA* (1985) 706.

67. B. L. Giammara and J. Hanker, *Stain Tech.* **61** (1986) 51.

68. B. L. Giammara, *Scanning* **14**(11) (1992) II-60.

69. G. R. Logan and A. M. Dvorak, *Methods of Microwave Fixation for Microscopy* (1993).

70. L. P. Kok and M. E. Boon, *J. Microscopy* **158**(3) (1990) 291.

71. N. Reid, Ultramicrotomy, in *Practical Methods in Electron Microscopy*, edited by A. M. Glauert (North Holland – American Elsevier, Amsterdam, 1975).

72. K. Zierold, *Ultramicroscopy* **14** (1984) 201.

73. M. M. Chappius and L. S. Robblee, *Rubber World* **136** (1957) 391.

74. C. H. Leigh-Dugmore, *Microscopy of Rubber* (Heffer, Cambridge, 1961).

75. W. M. Hess and F. P. Ford, *Rubber Chem. Technol.* **36** (1963) 1175.

76. E. H. Andrews and J. M. Stubbs, *J. R. Microsc. Soc.* **82** (1964) 221.

77. E. H. Andrews, *J. Polym. Sci.* **3** (1965) 353.

78. E. H. Andrews, M. W. Bennett and A. Markham, *J. Polym. Sci.* **A2**(5) (1967) 1235.

79. J. Dlugosz and A. Keller, *J. Appl. Phys.* **39**(12) (1968) 5776.

80. A. J. Cobbold and A. E. Mendelson, *Sci. Tools* **18**(1) (1971) 1.

81. J. Dlugosz, M. J. Folkes and A. Keller, *J. Polym. Sci., Polym. Phys. Edn.* **11** (1973) 929.

82. J. A. Odell, J. Dlugosz and A. Keller, *J. Polym. Sci., Polym. Phys. Edn.* **14** (1976) 861.

83. A. Ghijels, N. Groesbeek and C. W. Yip, *Polymer* **23** (1982) 1913.

84. I. D. Fridman, E. L. Thomas, L. J. Lee and C. W. Macosko, *Polymer* **21** (1980) 393.

85. C. H. Y. Chen, R. M. Briber, E. L. Thomas, M. Xu and W. J. MacKnight, *Polymer* **24** (1983) 1333.

86. H. Sitte, in *Proc. EM Workshop on Polymers*, edited by E. L. Thomas (Danvers, MA, May 1983) p. 6.

87. A. W. McDowall, J. J. Chang, R. Freeman, J. Lepault, C. A. Walter and J. Dubochet, *J. Microsc.* **131** (1983) 1.

88. T. A. Hall, *J. Microsc.* **117** (1979) 145.

89. A. P. Somlyo, A. V. Somlyo, H. Shuman and M. Stewart, *Scanning Electron Microsc.* **11**(2) (1979) 711.

90. P. Echlin, *Low Temperature Microscopy and Analysis* (Plenum Press, New York, 1992).

91. S. Y. Hobbs, in *Plastics Polymer Science and Technology*, edited by M. D. Bayal (Wiley– Interscience, New York, 1982) p. 239.

92. D. T. Grubb, in *Developments in Crystalline Polymers–1*, edited by D. C. Bassett (Applied Science, London, 1982) p. 1.

93. M. A. Hayat, *Positive Staining for Electron Microscopy* (Van Nostrand Reinhold, New York, 1975).

94. M. A. Hayat, *Principles and Techniques of Electron Microscopy–Biological Applications*, 3rd Edn (CRC Press, Boca Raton, 1989).

95. P. R. Lewis and D. P. Knight, Staining Methods for Sectioned Material in *Practical Methods in Electron Microscopy*, edited by A. M. Glauert

(North Holland–American Elsevier, Amsterdam, 1977).

96. E. K. Boylston and M. L. Rollins, *Microscope* **19** (1971) 255.

97. C. Maertens, G. Raes and G. Vandermeerssche, *Proc. of the Stockholm Conf.* (1956) 292.

98. K. Hess and M. Mahl, *Naturwissenschaften* **41** (1954) 86.

99. K. Muhlethaler, *Z. Schweiz. Forst.* **30** (1960) 55.

100. P. Kassenbeck and R. Hagege, *Text. Res. J.* **38** (1968) 196.

101. R. Hagege, P. Kassenbeck, D. Meimoun and A. Parisot, *Text. Res. J.* **39** (1969) 1015.

102. M. H. Walters and D. N. Keyte, *Rubber Chem. Technol.* **38** (1965) 62.

103. P. A. Marsh, A. Voet and L. D. Price, *Rubber Chem. Technol.* **39** (1966) 359.

104. W. M. Hess, C. E. Scott and J. E. Callan, *Rubber Chem. Technol.* **41** (1968) 344.

105. D. T. Grubb and A. Keller, *J. Polym. Sci., Polym. Phys. Edn.* **18** (1980) 207.

106. A. C. Reimschuessel and D. C. Prevorsek, *J. Polym. Sci., Polym. Phys. Edn.* **14** (1976) 485.

107. E. H. Andrews, *Proc. R. Soc. Lond.* **A 227** (1964) 562.

108. K. Kato, *J. Electron Microsc.* **14** (1965) 220.

109. K. Kato, *J. Polym. Sci., Polym. Lett. Edn.* **4** (1966) 35.

110. K. Kato, *Polym. Eng. Sci.* **7** (1967) 38.

111. H. Yoshimoto, *Kolloid Z. Z. Polym.* **236**(2) (1969) 116.

112. G. E. Molair and H. Keskkula, *J. Polym. Sci.* **A14** (1966) 1595.

113. M. Matsuo, J. Sagae and H. Asai, *Polymer* **10** (1967) 79.

114. I. D. Fridman and E. L. Thomas, *Polymer* **21** (1980) 388.

115. N. C. Watkins and D. Hansen, *Text. Res. J.* **38** (1968) 388.

116. D. Stefan and H. L. Williams, *J. Appl. Polym. Sci.* **18** (1974) 1451.

117. M. Niimoni, T. Katsuta and T. Kotani, *J. Appl. Polym. Sci.* **19** (1975) 2919.

118. B. C. Edwards and P. J. Phillips, *Polymer* **15** (1974) 491.

119. Y. Jyo, C. Nozaki and M. Matsuo, *Macromolecules* **4** (1971) 517.

120. C. K. Riew and R. W. Smith, *J. Polym. Sci.* (1971) 2739.

121. G. Weber, D. Kuntze and W. Stix, *Colloid Poly. Sci.* **260** (1982) 956.

122. G. Kanig, *Proc. Colloid Polym. Sci.* **57** (1975) 176.

123. G. A. Hutchins, *Proc. 51st Ann. MSA* (1993) 900.

124. D. Fleischer, E. Fischer and J. Brandrup, *J. Macromol. Phys.* **B14** (1977) 17.

125. A. Siegmann and A. Hiltner, *Polym. Eng. Sci.* **24**(11) (1984) 869.

126. G. Wegner, Li-Lan Zhu and G. Lieser, *Makromol. Chem.* **182** (1981) 231.

127. H. Marfels and P. Kassenbeck, *Chemie. text. Ind.* **27/79** (1977) 788.

128. L. C. Sawyer and S. Garg, unpublished.

129. T. Sarada, L. C. Sawyer and M. Ostler, *J. Membrane Sci.* **15** (1983) 97.

130. H. Keskkula and P. A. Traylor, *Polymer* **19** (1978) 465.

131. C. V. Berney, R. E. Cohen and F. S. Bates, *Polymer* **23** (1982) 1222.

132. R. W. Smith and J. C. Andries, *Rubber Chem. Technol.* **47** (1974) 64.

133. R. E. Cohen and A. R. Ramos, Office of Naval Research, Task No. NR 356–646, Technical Report No. 2, July 6, 1978.

134. G. Gillberg, L. C. Sawyer and A. L. Promislow, *J. Appl. Polym. Sci.* **28** (1983) 3723.

135. G. Kanig, *Kolloid Z. Z. Polym.* **251** (1973) 782.

136. G. Kanig, *Kunststoffe* **64** (1974) 470.

137. D. T. Grubb, J. Dlugosz and A. Keller, *J. Mater. Sci.* **10** (1975) 1826.

138. J. Dlugosz, G. V. Fraser, D. T. Grubb, A. Keller, J. A. Odell and P. L. Goggen, *Polymer* **17** (1976) 471.

139. A. M. Hodge and D. C. Bassett, *J. Mater. Sci.* **12** (1977) 2065.

140. I. G. Voigt-Martin, E. W. Fischer and L. Mandelkern, *J. Polym. Sci., Polym. Phys. Edn.* **18** (1980) 2347.

141. G. Strobl, H. Schneider and I. G. Voigt-Martin, *J. Polym. Sci., Polym. Phys. Edn.* **18** (1980) 1361.

142. G. Kanig, *J. Crystal Growth* **48** (1980) 303.

143. G. Kanig, *Colloid Polym. Sci.* **261** (1983) 373.

144. D. T. Grubb, in *Developments in Electron Microscopy and Analysis*, edited by D. L. Missell, Inst. Phys. Conf. Ser. 36 (Institute of Physics, Bristol, 1977) p. 399.

145. J. A. Odell, D. T. Grubb and A. Keller, *Polymer* **19** (1978) 617.

146. J. Smook, W. Hamersma and A. J. Pennings, *J. Mater. Sci.* **19** (1984) 1359.

147. A. Schaper, D. Zenke, E. Schulz, R. Hirte and M. Taege, *Phys. Stat. Sol.* **116** (1989) 179.

148. M. M. Kalnins, M. T. Conde Brana and U. W. Gedde, *Polymer Testing* **11** (1992) 139.

149. M. A. Hayat, *Positive Staining for Electron Microscopy* (Van Nostrand Reinhold, New York, 1975) p. 47–62.

150. K. Hess, E. Gutter and H. Mahl, *Naturwissenschaften* **46** (1959) 70.

151. K. Hess, E. Gutter and H. Mahl, *Kolloid Z.* **168** (1960) 37.

152. B. J. Spit, in *Proc. Fifth Int. Congr. for Electron Microscopy, Philadelphia*, edited by S. S. Bresse, Jr (Academic Press, New York, 1962) Vol. 1, p. BB7.

153. B. J. Spit, *Faserforsch. Textiltech.* **18** (1967) 161.

154. J. A. Rusnock and D. J. Hansen, *J. Polym. Sci.* **A3** (1965) 647.

155. E. Belavtseva, *Vysokomol. Soed.* **5** (1963) 1847.

156. E. Belavtseva and K. Gumargalieva, *Zadodsk Lab.* **29** (1966) 966 (Chem. Abstr. 59–116650).

157. A. Peterlin, P. Ingram and H. Kiho, *Makromol. Chem.* **86** (1965) 294.

158. C. W. Hock, *J. Polym. Sci.* **A2**(5) (1967) 471.

159. J. Martinez-Salazar and C. G. Cannon, *J. Mater. Sci. Lett.* **3** (1984) 693.

160. C. W. Pease, *J. Ultrastruct. Res.* **15** (1966) 555.

161. O. L. Shaffer, M. S. El-Aasser and J. W. Vanderhoff, in *Proc. 41st Annu. Mtg EMSA, Phoenix*, August 1983, edited by G. W. Bailey (San Francisco Press, California, 1983) p. 30.

162. O. L. Shaffer, unpublished.

163. R. Vitali and E. Montani, *Polymer* **21** (1980) 1220.

164. U. A. Spitzer and D. G. Lee, *J. Org. Chem.* **39** (1974) 2468.

165. J. S. Trent, J. I. Scheinbeim and P. R. Couchman, *J. Polym. Sci., Polym. Lett. Edn.* **19** (1981) 315.

166. J. S. Trent, P. R. Couchman and J. I. Scheinbeim, *Polym. Sci. Technol.* **22** (1983) 205.

167. J. S. Trent, J. I. Scheinbeim and P. R. Couchman, *Macromolecules* **16** (1983) 589.

168. D. E. Morel and D. T. Grubb, *Polym. Commun.* **25** (1984) 68.

169. D. E. Morel and D. T. Grubb, *Polymer* **25** (1984) 417.

170. P. E. Frochling and A. J. Pijpers, *J. Polym. Sci., Part B: Polym. Phys.* **25** (1987) 947.

171. B. Ohlsson and B. Tornell, *J. Appl. Polym. Sci.* **41** (1990) 1189.

172. Y. Tervorrt-Engelen and J. van Gisbergen, *Polym. Comm.* **32** (1991) 261.

173. S. Y. Hobbs, M. E. J. Dekkers and V. H. Watkins, *J. Mater. Sci.* **23** (1988) 1219.

174. M. E. J. Dekkers, S. Y. Hobbs and V. H. Watkins, *J. Mater. Sci.* **23** (1988) 1225.

175. J. S. Trent, *Macromolecules* **17** (1984) 2930.

176. D. Montezinos, B. G. Wells and J. L. Burns, *J. Polym. Sci., Polym. Lett. Edn.* **23** (1985) 421.

177. P. J. Beynon, P. M. Collins, D. Gardiner and W. G. Overend, *Carbohydr. Res.* **6** (1968) 431.

178. S. Wolfe, S. K. Hasan and J. R. Campbell, *Chem. Commun.* (1970) 1420.

179. B. A. Wood, *Proc. 51st MSA* (San Francisco Press, San Francisco, 1993) 898.

180. B. A. Wood, in *Advances in Polymer Blends and Alloys Technology*, Vol. 3 (Technomic Publishing, Lancaster, 1992) p. 24.

181. O. L. Shaffer, M. S. El-Aasser and J. W. Vanderhoff, *Proc. 45th EMSA* (San Francisco Press, San Francisco, 1987) 502.

182. I. Segal, O. L. Shaffer, V. L. Dimonie and M. S. El-Aasser, *Proc. 51st MSA* (San Francisco Press, San Francisco, 1993) 882.

183. O. L. Shaffer, unpublished micrographs.

184. M. Sotton, C. R. *Acad. Sci. Paris* **270B** (1970) 1261.

185. R. Hagege, M. Jarrin and M. Sotton, *J. Microsc.* **115** (1979) 65.

186. M. G. Dobb, D. J. Johnson, A. Majeld and B. P. Saville, *Polymer* **20** (1979) 1289.

187. M. G. Dobb, C. R. Park and R. M. Robson, *J. Mater. Sci.* **27** (1992) 3876.

188. M. Sotton and A. M. Vialard, *Text. Res. J.* **41** (1971) 834.

189. S. Y. Hobbs, V. H. Watkins and R. R. Russell, *J. Polym. Sci., Polym. Phys. Edn.* **18** (1980) 393.

190. S. Y. Hobbs (1985), personal communication.

191. K. Hess and H. Kiessig, *Naturwissenschaften* **31** (1943) 171.

192. K. Hess and H. Kiessig, *Z. Phys. Chem.* **A193** (1944) 196.

193. K. Hess, H. Mahl and E. Gutter, *Kolloid Z.* **155** (1957) 1.

194. J. Gacen, J. Maillo and J. Bordas, *Bull. Sci.* **ITF6** (1977) 167.

195. G. Gillberg, A. Kravas and J. Langley, *J. Microsc.* **138**(1) (1985) RP1.

196. M. M. Winram, D. T. Grubb and A. Keller, *J. Mater. Sci.* **13** (1978) 791.

197. A. Garton, P. Z. Sturgeon, D. J. Carlsson and D. M. Wills, *J. Mater. Sci.* **13** (1978) 2205.

198. D. J. Barber, *J. Mater. Sci.* **5** (1970) 1.

199. H. Z. Fetz, *Physik* **119** (1942) 590.

200. G. J. Wehner, *J. Appl. Phys.* **25** (1954) 270.

201. G. J. Wehner, *J. Appl. Phys.* **30** (1959) 1762.

202. E. Jakopic, in *Proc. Eur. Reg. Conf. on Electron Microscopy*, edited by A. L. Houwink and B. J. Spit (Nederlandse Vereniging voor Electronenmikroskopie, Delft, 1960) p. 559.

203. D. E. Harrison, Jr, N. S. Levy, J. P. Johnson III and H. M. Effron, *J. Appl. Phys.* **39** (1968) 3742.

204. M. J. Nobes, J. S. Collingon and G. Carter, *J. Mater. Sci.* **4** (1969) 730.

205. G. Carter, J. S. Collingon and M. J. Nobes, *J. Mater. Sci.* **6** (1971) 115.

206. R. S. Dhariwal and R. K. Fitch, *J. Mater. Sci.* **12** (1977) 1225.

207. A. D. G. Stewart and M. W. Thompson, *J. Mater. Sci.* **4** (1969) 56.

208. I. H. Wilson and M. W. Kidd, *J. Mater. Sci.* **6** (1971) 1362.

209. P. Sigmund, *J. Mater. Sci.* **8** (1973) 1545.

210. J. P. Ducommun, M. Cantagrel and M. Moulin, *J. Mater. Sci.* **10** (1975) 52.

211. I. S. T. Tsong and D. J. Barber, *J. Mater. Sci.* **7** (1977) 687.

212. B. Salehpoor and P. M. Marquis, *J. Microsc.* **124**(3) (1981) 239.

213. F. R. Barnet and M. K. Norr, *Carbon* **11** (1973) 281.

214. B. J. Spit, in *Proc. Eur. Reg. Conf. on Electron Microscopy*, edited by A. L. Houwink and B. J. Spit (Nederlandse Vereniging voor Electronenmikroskopie, Delft, 1960) p. 564.

215. B. J. Spit, *Polymer* **4** (1962) 109.

216. J. Dlugosz, in *Proc. Fifth Int. Congr. for Electron Microscopy, Philadelphia* (Academic Press, New York, 1962) Vol. I, BB11.

217. L. Moscou, in *Proc. Fifth Int. Congr. for Electron Microscopy, Philadelphia* (Academic Press, New York, 1962) Vol. I, BB5.

218. A. Keller, in *Proc. Fifth Int. Congr. for Electron Microscopy, Philadelphia* (Academic Press, New York, 1962) Vol. I, BB3.

219. J. E. Breedon, J. F. Jackson and M. J. Marcinkowski, *J. Mater. Sci.* **8** (1973) 1071.

220. F. R. Anderson and V. F. Holland, *J. Appl. Phys.* **31** (1960) 1516.

221. N. V. Hien, S. L. Cooper and J. A. Koutsky, *J. Macromol. Sci.-Phys.* **B6**(2) (1972) 343.

222. G. Carter, A. E. Hill and M. J. Nobes, *Vacuum* **29** (1979) 213.

223. M. R. Padhye, N. V. Bhat and P. K. Mittal, *Text. Res. J.* **46** (1976) 502.

224. P. J. Goodhew, *J. Phys. E: Sci. Instrum.* **4** (1971) 392.

225. P. J. Goodhew, *Nature* **235** (1972) 437.
226. P. J. Goodhew, in *Proc. 5th Eur. Congr. on Electron Microscopy*, Vol. 5 (1972) p. 300.
227. P. J. Goodhew, *J. Mater. Sci.* **8** (1973) 581.
228. S. B. Warner, D. R. Uhlmann and L. H. Peebles, *J. Mater. Sci.* **10** (1975) 758.
229. P. R. Blakey and M. O. Alfy, *J. Text. Inst.* (1978) 38.
230. J. Friedrich, J. Gahde and M. Pohl, *Acta Polym.* **31**(5) (1981) 310.
231. J. Friedrich, J. Gahde and M. Pohl, *Acta Polym.* **33**(3) (1982) 209.
232. M. Kojima and H. Satake, *J. Polym. Sci., Polym. Phys. Edn.* **20** (1982) 2153.
233. V. B. Gupta, L. T. Drzaland and Y. L. Chen, in *Proc. 41st Annu. Mtg. EMSA, Phoenix*, August 1983, edited by G. W. Bailey (San Francisco Press, California, 1983) p. 34.
234. D. W. Woods and I. M. Ward, *Surface and Interface Anal.* **20** (1993) 385.
235. R. W. Linton, M. E. Farmer, P. Ingram, J. R. Somner and J. D. Shelburne, *J. Microsc.* **134** (1984) 101.
236. J. S. Mijovic and J. A. Koutsky, *Polym. Plast. Technol. Eng.* **9** (1977) 139.
237. D. Hemsley, Microscopy of Polymer Surfaces, in *Developments in Polymer Characterization*, edited by J. V. Dawkins (Applied Science, London, 1978) p. 245.
238. L. C. Sawyer, *J. Polym. Sci., Polym. Lett. Edn.* **22** (1984) 347.
239. L. S. Li, L. F. Allard and W. C. Bigelow, *J. Macromol. Sci., Phys.* **B22**(2) (1983) 269.
240. V. Peck and W. Kaye, *Text. Res. J.* **4** (1954) 295.
241. G. W. Bailey, *J. Polym. Sci.* **62** (1962) 241.
242. L. S. Li and V. A. Kargin, *Vysokomol. Soed.* **3** (1961) 1102.
243. K. Kubota, *J. Polym. Sci.* **3** (1965) 403.
244. K. Kubota, *J. Polym. Sci.* **5** (1967) 1179.
245. H. Keskkula and P. A. Traylor, *J. Appl. Polym. Sci.* **11** (1967) 2361.
246. R. J. Williams and R. W. A. Hudson, *Polymer* **8** (1967) 643.
247. G. Kaempf, in *9th Int. Congr. on Electron Microscopy*, Vol. I (1978) p. 491.
248. G. Farrow, D. A. S. Ravens and I. M. Ward, *Polymer* **3** (1962) 17.
249. P. Tucker and R. Murray, in *Proc. 33rd Annu. Mtg EMSA*, August 1975, edited by C. J. Arcenaux (Claitors, Baton Rouge, 1975) p. 82.
250. C. Chu and G. Wilkes, *Polym. Prepr.* **13** (1973) 1282.
251. P. Tucker, D. Johnson, M. Dobb and J. Sikorski, *Text. Res. J.* **47** (1977) 29.
252. R. Murray, H. A. Davis and P. Tucker, *J. Appl. Polym. Sci.* **33** (1978) 177.
253. J. O. Warwicker, *J. Appl. Polym. Sci.* **22** (1978) 869.
254. G. E. Sweet and J. P. Bell, *J. Polym. Sci., Polym. Phys. Edn.* **16** (1978) 1935.
255. H. S. Bu, S. Z. D. Cheng and B. Wunderlich, *Polymer* **29** (1988) 1603.
256. R. de J. Santos, J. C. Bruno, M. T. M. B. Silva and R. C. R. Nunes, *Polymer Testing* **12** (1993) 393.
257. M. E. J. Dekkers, S. Y. Hobbs and V. H. Watkins, *Polymer* **32** (1991) 2150.
258. S. H. Zeronian and M. J. Collins, *Textile Inst.* **20** (1989) 1.
259. M. J. Collins, S. H. Zeronian and M. Semmelmeyer, *J. Appl. Poly. Sci.* **42** (1991) 2149.
260. R. P. Palmer and A. J. Cobbold, *Makromol. Chem.* **74** (1964) 174.
261. C. W. Hock, *Polym. Lett.* **3** (1965) 573.
262. C. W. Hock, *J. Polym. Sci.* **4** (1966) 227.
263. N. Kusumoto and Y. Haga, *Rep. Prog. Polym. Phys. Japan XV* (1972) 583.
264. V. J. Armond and J. R. Atkinson, *J. Mater. Sci.* **4** (1969) 509.
265. C. B. Bucknall, I. C. Drinkwater and W. E. Keast, *Polymer* **13** (1972) 115.
266. C. B. Bucknall and I. C. Drinkwater, *Polymer* **15** (1974) 254.
267. D. Briggs, D. M. Brewis and M. B. Kovieczo, *J. Mater. Sci.* **11** (1976) 1270.
268. R. A. Bubeck and H. M. Baker, *Polymer* **23** (1982) 1680.
269. K. Selby and M. O. W. Richardson, *J. Mater. Sci.* **11** (1976) 786.
270. D. J. Boll, R. M. Jensen, L. Cordner and W. D. Bascom, *J. Composite Materials* **24** (1990) 208.
271. D. C. Bassett, *Principles of Polymer Morphology* (Cambridge University Press, Cambridge, 1981) p. 102–14.
272. R. H. Olley and D. C. Bassett, *Polymer* **23** (1982) 1707.
273. D. C. Bassett, in *Developments in Crystalline Polymers – 2*, edited by D. C. Bassett, Chapter 2 (Elsevier Applied Science, London, 1988) p. 67–114.
274. P. J. Phillips and R. J. Philpot, *Polym. Commun.* **27** (1986) 307.
275. K. L. Naylor and P. J. Phillips, *J. Polym. Sci., Polym. Phys. Edn.* **21** (1983) 2011.

276. D. C. Bassett and R. H. Olley, *Polymer* **25** (1984) 935.

277. Z. Bashir, M. J. Hill and A. Keller, *J. Mater. Sci. Lett.* **5** (1986) 876.

278. R. H. Olley, D. C. Bassett, P. J. Hine and I. M. Ward, *J. Mater. Sci.* **28** (1993) 1107.

279. R. H. Olley, D. C. Bassett and D. J. Blundell, *Polymer* **27** (1986) 344.

280. S. J. Sutton and A. S. Vaughan, *J. Mater. Sci.* **28** (1993) 4962.

281. T. J. Lemmon, S. Hanna and A. H. Windle, *Polymer Comm.* **30** (1989) 2.

282. J. R. Ford, D. C. Bassett, G. R. Mitchell and T. G. Ryan, *Mol. Cryst. Liq. Cryst.* **180B** (1990) 233.

283. S. E. Bedford and A. H. Windle, *Polymer* **31** (1990) 616.

284. S. Hanna, T. J. Lemmon, R. J. Spontak and A. H. Windle, *Polymer* **33** (1992) 3.

285. A. Anwer and A. H. Windle, *Polymer* **34** (1993) 3347.

286. S. D. Hudson and A. J. Lovinger, *Polymer* **34** (1993) 1123.

287. D. E. Bradley, *J. Appl. Phys.* **27** (1956) 1399.

288. D. E. Bradley, Replica and Shadowing Techniques, in *Techniques for Electron Microscopy*, edited by D. H. Kay (Blackwell, Oxford, 1965) p. 96.

289. P. J. Goodhew, Specimen Preparation in Materials Science, in *Practical Methods in Electron Microscopy*, edited by A. M. Glauert (North Holland, Amsterdam, 1980).

290. J. H. M. Willison and A. J. Rowe, Replica, Shadowing and Freeze-etching Techniques, in *Practical Methods in Electron Microscopy*, edited by A. M. Glauert (North Holland, Amsterdam, 1980).

291. V. G. Peck, *Appl. Polym. Symp.* **16** (1971) 19.

292. J. F. Oliver and S. G. Mason, *J. Colloid Interface Sci.* **60** (1977) 480.

293. P. Robbins and J. Pugh, *Wear* **50** (1978) 95.

294. E. Wood (1980), private communication.

295. K. D. Gordon, *J. Microsc.* **134**(2) (1984) 183.

296. K. Sakaoku and A. Peterlin, *J. Macromol. Sci.* **B1** (1967) 103.

297. S. Y. Hobbs and C. F. Pratt, *J. Appl. Polym. Sci.* **19** (1975) 1701.

298. B. R. Vijayendran, T. Bone and L. C. Sawyer, *J. Dispersion Sci. Technol.* **3**(1) (1982) 81.

299. A. Peterlin and K. Sakaoku, *J. Appl. Phys.* **38** (1967) 4152.

300. K. O'Leary and P. H. Geil, *J. Appl. Phys.* **38** (1967) 4169.

301. C. M. Balik and A. J. Hopfinger, *Macromolecules* **13** (1980) 999.

302. P. J. Goodhew, Specimen Preparation in Materials Science, in *Practical Methods in Electron Microscopy*, edited by A. M. Glauert (North Holland – American Elsevier, Amsterdam, 1973) p. 144–6.

303. A. M. Baro, L. Vazquez, A. Bartolame, J. Gomez, N. Garcia, H. A. Goldberg, L. C. Sawyer, R. T. Chen, R. S. Kohn, R. Reifenberger, *J. Mater. Sci.* **24** (1989) 1739.

304. J. Dlugosz, *Proc. 1st Eur. Reg. Congr. on Electron Microscopy* (Stockholm, 1956) p. 283.

305. N. Ramanthan, J. Sikorski and H. J. Woods, *Proc. Int. Conf. Electron Microscopy* (London, 1954) p. 482.

306. M. H. Walters and D. N. Keyte, *Trans. Inst. Rubber Ind.* **39** (1962) 40.

307. P. H. Geil, *J. Macromol. Sci., Phys.* **B12** (1976) 173.

308. D. C. Bassett and A. Keller, *Phil. Mag.* **6** (1961) 345.

309. J. Petermann and H. Gleiter, *J. Polym. Sci., Polym. Phys. Edn.* **10** (1972) 2333.

310. J. B. Jones and P. H. Geil, *Makromol. Chem.* **181** (1980) 1551.

311. H. Kiyek and T. G. F. Schoon, *Rubber Chem. Technol.* **40** (1967) 1238.

312. L. Holland, *Vacuum Deposition of Thin Films* (Wiley, New York, 1956).

313. P. Echlin and P. J. W. Hyde, *Scanning Electron Microsc.* **5** (1972) 137.

314. P. Echlin, *Scanning Electron Microsc.* **7** (1974) 1019.

315. P. Ingram, N. Morosoff, L. Pope, F. Allen and C. Tisher, *Scanning Electron Microsc.* **9** (1976) 75.

316. P. N. Panayi, D. S. Cheshire and P. Echlin, *Scanning Electron Microsc.* **10**(1) (1977) 143.

317. A. W. Robards, A. J. Wilson and P. Crosby, *J. Microsc.* **124** (1981) 143.

318. P. J. Goodhew, Specimen Preparation in Materials Science, in *Practical Methods in Electron Microscopy*, edited by A. M. Glauert (North Holland – American Elsevier, Amsterdam, 1973) p. 140.

319. J. I. Goldstein, H. Yakowitz, D. E. Newbury, E. Lifshin, J. W. Colby and J. R. Coleman, *Practical Scanning Electron Microscopy* (Plenum Press, New York, 1975).

320. P. Echlin, *Scanning Electron Microsc.* **8** (1975) 217.

321. P. B. DeNee and E. R. Walker, *Scanning Electron Microsc.* **8** (1975) 225.

322. T. Braten, *J. Microsc.* **113** (1978) 53.
323. I. M. Watt, in *Proc. 9th Int. Electron Microscopy Congr. of Electron Microscopy*, Vol. II, Toronto (1978) p. 94.
324. P. Echlin and G. Kaye, *Scanning Electron Microsc.* **11**(1) (1978) 109.
325. L. C. Sawyer, unpublished.
326. H. S. Slayter, *Scanning Electron Microsc.* **13**(1) (1980) 171.
327. J. Franks, P. R. Stuart and R. B. Withers, *Thin Solid Films* **60** (1979) 231.
328. J. Franks, C. S. Clay and G. W. Peace, *Scanning Electron Microsc.* **13**(1) (1980) 155.
329. C. S. Clay and G. W. Peace, *J. Microsc.* **123** (1981) 25.
330. P. Echlin, *Scanning Electron Microsc.* **14**(1) (1981) 79.
331. B. H. Kemmenoe and G. R. Bullock, *J. Microsc.* **132** (1983) 153.
332. G. F. Cardinale, V. L. Carlino and D. G. Howitt, *Scanning* **15** (1993) 25.
333. K. R. Peters, *Scanning Electron Microsc.* **13**(1) (1980) 143.
334. T. Muller, P. Walther, C. Scheidegger, R. Reichelt, S. Muller and R. Guggenheim, *Scanning Microscopy* **4** (1990) 863.
335. D. C. Joy, *Scanning* **11** (1989) 1.
336. D. C. Joy, notes from the course on SEM, Lehigh University.
337. D. C. Joy, unpublished.
338. J. H. Butler, D. C. Joy and G. F. Bradley, *Proc. 51st MSC* (San Francisco Press, San Francisco, 1993) 870.
339. M. Jamieson, unpublished.
340. L. C. Sawyer and M. Jamieson, in *Proc. 47th EMSA* (San Francisco Press, San Francisco, 1989) 334.
341. M. Jamieson, unpublished.
342. A. Colebrooke and A. H. Windle, in *Scanning Electron Microscopy: Systems and Applications*, edited by W. C. Nixon, Inst. Phys. Conf. Ser. 18 (Institute of Physics, Bristol, 1973) 132.
343. D. T. Grubb and G. W. Groves, *Phil. Mag.* **24** (1971) 190.
344. G. A. Bassett, *Phil. Mag.* **3** (1958) 1042.
345. G. A. Bassett, D. J. Blundell and A. Keller, *J. Macromol. Sci., Phys.* **B1** (1967) 161.
346. B. J. Spit, *J. Macromol. Sci., Phys.* **B2** (1968) 45.
347. D. Krueger and G. S. Yeh, *J. Macromol. Sci., Phys.* **B6** (1972) 431.
348. D. J. Blundell and A. Keller, *J. Macromol. Sci., Phys.* **B7** (1973) 253.
349. J. J. Metois, J. C. Heyraud and C. Finck, *J. Microsc. Spectrosc. Electron.* **4** (1979) 225.
350. M. Kojima and J. H. Magill, *J. Macromol. Sci., Phys.* **B15**(1) (1978) 63.
351. K. Shimamura, *J. Macromol. Sci., Phys.* **B16**(2) (1979) 213.
352. H. H. Kausch, *Polymer Fracture* (Springer-Verlag, Berlin, 1978).
353. H. H. Kausch, Ed., *Crazing in Polymers*, Adv. Polym. Sci. Ser. 52/3 (Springer-Verlag, Berlin, 1983).
354. H. Liebowitz, Ed., Fracture, Vol. VII in *Fracture of Non-Metals and Composites* (Academic Press, New York, 1972).
355. L. Engel, H. Klingele, G. W. Ehrentstein and H. Scherper, *An Atlas of Polymer Damage* (Prentice Hall, New Jersey, 1981).
356. A. K. Bhowmick, S. Basu and S. K. De, *Rubber Chem. Technol.* **53** (1980) 321.
357. J. W. S. Hearle and P. M. Cross, *J. Mater. Sci.* **5** (1970) 507.
358. L. C. Sawyer and M. Jamieson, unpublished.
359. E. H. Andrews, *Adv. Polym. Sci.* **27** (1978) 1.
360. A. Peterlin, *Fracture* **1** (1977) 471.
361. J. G. Williams, *Adv. Polym. Sci.* **27** (1978) 67.
362. A. S. Krausz and H. Eyring, *Deformation Kinetics* (Wiley, New York, 1975) p. 331.
363. C. B. Bucknall, *Toughened Plastics* (Applied Science, London, 1977).
364. P. E. Reed, Impact Performance of Polymers, in *Developments in Polymer Fracture−1*, edited by E. H. Andrews (Applied Science, London, 1979) p. 121.
365. F. Speroni, E. Castoldi, P. Fabbri and T. Casiraghi, *J. Mater. Sci.* **24** (1989) 2165.
366. D. R. Mulville and I. Wolock, Failure of Polymer Composites, in *Developments in Polymer Fracture−1*, edited by E. H. Andrews (Applied Science, London, 1979) p. 263.
367. J. H. Sinclair and C. C. Chamis, *34th Annu. Tech. Conf.* 1979, Reinforced Plastics/Composites Inst., Soc. of the Plastics Ind., Inc. 22A, p. 1.
368. R. A. Kline and F. H. Chang, *J. Composite Mater.* **14** (1980) 315.
369. R. Richard-Frandsen and Y. Naerheim, *J. Composite Mater.* **17** (1983) 105.
370. J. E. Theberge, *Polym. Plast. Technol. Eng.* **16**(1) (1981) 41.
371. J. Kubat and H. E. Stromvall, *Plast. Rubber Process.* (1980) 45.

372. J. W. Hamer and R. T. Woodhams, *Polym. Eng. Sci.* **21** (1981) 603.

373. J. R. White and E. L. Thomas, *Rubber Chem. Technol.* **57** (1984) 457.

374. A. N. McKee and C. L. Beattie III, *Text. Res. J.* **40** (1970), 1006.

375. R. M. Minnini and M. Jamieson, unpublished.

376. S. Y. Hobbs, *Rev. Sci. Instrum.* **53** (1982) 1097.

377. R. H. Hoel and D. J Dingley, *J. Mater. Sci.* **17** (1982) 2990.

378. T. H. Mao, P. W. R. Beaumont and W. C. Nixon, *J. Mater. Sci. Lett.* **2** (1982) 613.

379. A. Chudnovsky, A. Moet, R. J. Bankert and M. T. Takemore, *J. Appl. Phys.* **54** (1983) 5562.

380. K. Smith, M. G. Hall and J. N. Hay, *J. Polym. Sci., Polym. Lett. Edn.* **14** (1976) 751.

381. A. M. Donald and E. J. Kramer, *J. Mater. Sci.* **17** (1982) 1871.

382. B. D. Lauterwasser and E. J. Kramer, *Phil. Mag.* **A39** (1979) 469.

383. J. A. Manson and R. W. Hertzberg, *J. Polym. Sci., Polym. Phys. Edn.* **11** (1973) 2483.

384. E. J. Kramer, *Polym. Eng. Sci.* **24** (1984) 761.

385. R. P. Kambour, *J. Polym. Sci.* **D7** (1973) 1.

386. S. Rabinowitz and P. Beardmore, *CRC Crit. Rev. Macromol. Sci.* **1** (1972) 1.

387. E. J. Kramer, *Developments in Polymer Fracture*, edited by E. H. Andrews (Chapman and Hall, London, 1979).

388. R. P. Kambour, *Polymer* **5** (1964) 143.

389. R. P. Kambour, *J. Appl. Polym. Sci., Appl. Polym. Symp.* **7** (1968) 215.

390. R. P. Kambour and A. S. Holik, *J. Polym. Sci.* **A2**(7) (1969) 1393.

391. P. Beahan, M. Bevis and D. Hull, *Polymer* **14** (1973) 96.

392. J. Washiyama, C. Creton and E. J. Kramer, *Macromolecules* **25** (1992) 4751.

393. C. Creton, E. J. Kramer and G. Hadziioannou, *Colloid Polym. Sci.* **270** (1992) 399.

394. E. L. Thomas and S. J. Israel, *J. Mater. Sci.* **10** (1975) 1603.

395. A. M. Donald and E. J. Kramer, *J. Polym. Sci., Polym. Phys. Edn.* **20** (1982) 899.

396. A. M. Donald and E. J. Kramer, *Phil. Mag.* **A43** (1981) 857.

397. A. M. Donald, T. Chan and E. J. Kramer, *J. Mater. Sci.* **16** (1981) 669.

398. A. M. Donald and E. J. Kramer, *Polymer* **23** (1982) 1183.

399. A. S. Argon, R. E. Cohen, B. Z. Jang and J. B. Vander Sande, *J. Polym. Sci., Polym. Phys. Edn.* **19** (1981) 253.

400. M. A. Hayat, *Principles and Techniques of Electron Microscopy: Biological Applications* (Van Nostrand Reinhold, New York, 1977).

401. E. B. Bradford and J. W. Vanderhoff, *J. Colloid Sci.* **17** (1962) 668.

402. E. B. Bradford and J. W. Vanderhoff, *J. Colloid Sci.* **14** (1959) 543.

403. L. Bachmann and Y. Talmon, *Ultramicroscopy* **14** (1984) 211.

404. Y. Talmon, H. T. Davis, L. E. Scriven and E. L. Thomas, *Rev. Sci. Instrum.* **50**(6) (1979) 698.

405. A. H. Falls, S. T. Wellinghoff, Y. Talmon and E. L. Thomas, *J. Mater. Sci.* **18** (1983) 2752.

406. J. O. Stoffer and T. Bone, *J. Dispersion Sci. Technol.* **1**(4) (1980) 393.

407. A. K. Kleinschmidt, D. Lang, D. Jacherts and R. K. Zahn, *Biochim. Biophys. Acta* **61** (1962) 857.

408. R. W. G. Wyckoff, *Science* **104** (1946) 36.

409. R. C. Williams, *Biochim. Biophys. Acta* **9** (1952) 237.

410. R. C. Williams, *Exp. Cell Res.* **4** (1953) 188.

411. T. W. G. Rowe, *Ann. NY Acad. Sci.* **85** (1960) 641.

412. T. W. G Rowe, *Current Trends in Cryo Biology*, edited by A. V. Smith (Plenum Press, New York, 1970).

413. L. C. Sawyer, B. Strassle and D. J. Palatini, in *Proc. 37th Annu. Mtg EMSA, San Antonio*, August 1979, edited by C. J. Arcenaux (Claitors, Baton Rouge, 1979) p. 620.

414. E. R. Walter and G. H. Bryant, in *Proc. 35th Annu. Mtg EMSA, Boston*, August 1977, edited by C. J. Arcenaux (Claitors, Baton Rouge, 1977) p. 314.

415. E. R. Walter, private communication.

416. S. A. McDonald, C. A. Daniels and J. A. Davidson, *J. Colloid Interface Sci.* **59** (1977) 342.

417. J. Watanabe, G. Seibel and M. Inoue, *J. Polym. Sci., Polym. Lett. Edn.* **22** (1984) 39.

418. M. Katoh, *J. Electron Microsc.* **28** (1979) 197.

419. J. W. Vanderhoff, *J. Macromol. Sci., Chem.* **A7**(3) (1973) 677.

420. T. F. Anderson, *Trans. NY Acad. Sci.* **13** (1951) 130.

421. T. F. Anderson, *C. R. Prem. Congr. Int. Microsc. Electron., Paris* (1953) 567.

422. T. F. Anderson, in *Proc. 3rd Int. Conf. on Electron Microscopy, London*, 1954 (Royal Microscopy Society, London, 1956) p. 122.

423. M. A. Hayat and B. R. Zirkin, *Principles and Techniques of Electron Microscopy: Biological*

Applications, Vol. 3, edited by M. A. Hayat (Van Nostrand Reinhold, New York, 1973) p. 297.

424. T. F. Anderson, *Physical Techniques in Biological Research*, Vol. III (Academic Press, New York, 1966), 2nd Edn., p. 319.

425. A. L. Cohen, in *Scanning Electron Microscopy*, Vol. 1, edited by M. A. Hayat (Van Nostrand, New York, 1974) p. 44.

426. A. L. Cohen, *Scanning Electron Microsc.* **10** (1977) 525.

427. L. C. Sawyer, unpublished.

428. N. S. McNutt, *Dynamic Aspects of Cell Surface Organization* (Elsevier – North Holland Biomedical Press, Amsterdam, 1977) p. 75.

429. R. L. Steere, *J. Biophys. Cytol.* **7** (1957) 167.

430. H. Moor, K. Muhlethaler, H. Waldner and A. Frey-Wyssling, *J. Biophys. Cytol.* **10** (1961) 1.

431. D. Branton, *Proc. Natl. Acad. Sci. USA* **55** (1966) 1048.

432. C. J. Singleton, T. Stephenson, J. Isner, P. H. Geil and E. A. Collins, *J. Macromol. Sci., Phys.* **B14**(1) (1977) 29.

433. M. Gervais and B. Gallot, *Makromol. Chem.* **180** (1979) 2041.

434. M. Gervais and B. Gallot, *Polymer* **22** (1981) 1129.

435. Y. C. Yang and P. H. Geil, *J. Macromol. Sci., Phys.* **B22**(3) (1983) 463.

436. F. Candau, J. Boutillier, F. Tripier and J. C. Wittmann, *Polymer* **20** (1979) 1221.

437. R. Reed and J. R. Barlow, *J. Appl. Polym. Sci.* **15** (1971) 1623.

438. J. Biais, M. Mercier, P. Lalanne, B. Clin, A. M. Bellocq and B. Lemanceau, *C. R. Acad. Sci. Paris* **285** (1977) 213.

439. C. Price and D. Woods, *Eur. Polym. J.* **9** (1973) 827.

440. A. Rameau, P. Marie, F. Tripier and Y. Gallot, *C. R. Acad. Sci. Paris* **286** (1978) 277.

441. U. B. Sleytr and A. W. Robards, *J. Microsc.* **110** (1977) 1.

442. H. J. Purz, M. Schlawne, A. Buchtemann and J. Hartmann, *Acta Histochem. Suppl.* **23** (1981) 89.

CHAPTER FIVE

Polymer applications

5.1 FIBERS

5.1.1 Introduction

Characterization of the microstructure of polymer fibers can provide insights into the fundamental structures present and into the relationship between structure and properties important for applications. Morphological characterization provides information to help understand the effects of processing history on mechanical and other physical properties. Microscopy techniques are used to observe features such as: fiber shape, diameter, structure (crystal size, voids, etc.), molecular orientation, size and distribution of additives, structure of yarn and fabric assemblages and failure mechanisms. These features are directly related to specific mechanical and thermal properties. Emphasis in this section is on assessment of the structure of polymer fibers as it relates to solving problems or evaluating the effect of process modifications. Fibers prepared from liquid crystalline polymers require special techniques and interpretation which will be described later (Section 5.6). It must be emphasized that any study of polymer fibers will be incomplete if the only technique applied is microscopy. X-ray diffraction, thermal analysis (DSC, TGA, heat shrinkage) and spectroscopy (IR, Raman, XPS) are among the many techniques which complement microscopy investigations (Section 7.4).

The polymers used in fibers are linear, so the molecules are a few nanometers across and

several hundred nanometers long. In unoriented materials, the molecules are coiled and folded into loose isotropic spheres. When a fiber is oriented, by drawing for example, the molecular chains become aligned parallel to the fiber axis (uniaxial fiber orientation) and the stiffness and strength improve. In most fibers, the molecules are still coiled and folded, although they are oriented. Only in ultrahigh modulus fibers (Section 5.1.4.2) or in fibers formed from liquid crystalline precursors (Section 5.6.4) are the molecules highly elongated and extended parallel to the fiber axis.

Natural and synthetic textile fibers were among the earliest materials studied by electron microscopy. Guthrie [1] and Stoves [2] described the techniques and applications of fiber microscopy to industrial practice. Somewhat later, evidence was provided for an oriented microfibrillar texture in polymer fibers [3]. X-ray diffraction suggested an arrangement of fine structures about 50 nm long and 5 nm wide in semicrystalline fibers [4, 5]. Peterlin [6, 7] observed the formation of fibrils and microfibrils by the deformation and transformation of spherulites using various microscopy techniques.

A basic element of semicrystalline fibers is the *microfibril* Microfibrils may be bundled into fibrils, about several hundred nanometers thick. A mechanically weak boundary between the fibrils results in fibrillation during deformation. Barham and Keller [8] and Prevorsek *et al.* [9] discussed the microfibrillar model, and the latter authors summarized the effects of fiber structure on textile properties. Microfibrils are known to exist in most fibers and are also known to be present in drawn single crystals, such as single crystal PE mats, crazes, melt extrudates and solid state extrudates. In addition, it is known [9] that larger structures, macrofibrils, are composed of microfibrils and that crystallites, disordered domains and partially extended noncrystalline molecules are present in fibers. Fiber structure and properties for nylon 6 and poly(ethylene terephthalate) (PET) fibers were further elaborated [10–13] using both electron microscopy and small angle scattering. A major point of these studies was evidence supporting the strong lateral interactions between the microfibrils.

Reviews of specimen preparation methods for fiber microscopy and instrumental techniques applied to fibers were published during the early 1970s [14–16]. This section contains applications of microscopy to the understanding of fiber microstructures used in the industrial laboratory for modification of fiber formation processes to improve specific mechanical properties or for problem solving.

Fibers have been produced from a wide range of polymers (Appendix III). A handy listing of common textile fibers is found in the *Textile World Manmade Fiber Chart*, issued each year by Textile World [17]. This comprehensive chart lists the various fiber names, types, optical micrographs of cross sections and longitudinal views, mechanical properties, etc., of about 35 textile fibers.

5.1.2 Textile fibers

5.1.2.1 *Optical microscopy of textile fibers*

The optical microscope is used to study various fiber features, such as: (1) size, (2) cross section (shape), (3) uniformity, (4) molecular orientation and (5) distribution of fillers. Specimen preparation methods include direct observation and sectioning. Fibers are embedded prior to sectioning by microtomy (Section 4.3) or polishing (Section 4.2) methods. Figure 5.1 contains optical

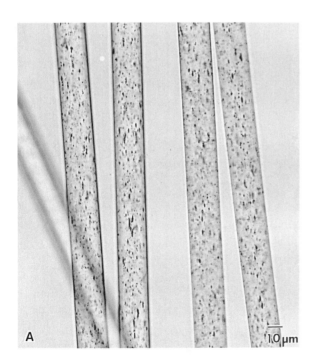

Fig. 5.1   Transmitted optical micrograph of a polyester textile fiber (A) shows cylindrical fibers containing dense pigment particles. A low magnification optical view (B) shows a fabric woven with Orlon fiber containing yarns.

micrographs showing longitudinal views of typical PET textile fibers (Fig. 5.1A) and a fabric composed of Orlon fibers (Fig. 5.1B). A range of fiber cross section sizes and shapes are shown in the optical micrographs in Fig. 5.2. A drawn polyethylene (PE) fiber seen in a cross section (Fig. 5.3) exhibits a fine spherulitic texture.

Birefringence, the difference between the refractive index parallel and perpendicular to the fiber axis, is an important quantitative measure of

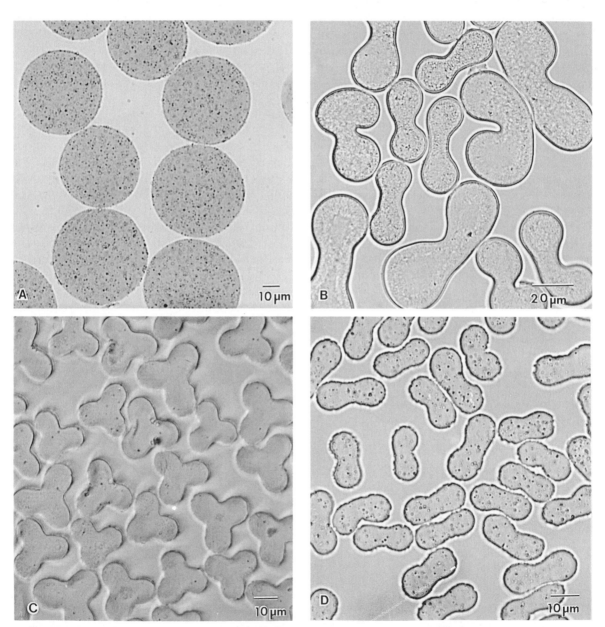

Fig. 5.2 Cross sectional views of representative fibers show the fiber shapes and dense pigment particles. The fibers are: (A) round polyamide, (B) irregularly shaped polyacrylonitrile, (C) trilobal shaped Orlon, and (D) dogbone shaped Orlon fiber sections.

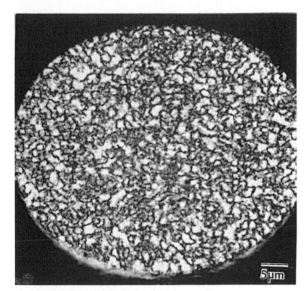

Fig. 5.3 A drawn polyethylene fiber cross section reveals a fine, spherulitic texture in polarized light.

molecular orientation [18]. Birefringence is measured by either the Becke line (immersion) method or a compensator method [2] (Sections 2.2.4 and 3.4.3). The Becke line method [19] measures the surface birefringence, whereas compensator methods measure the average birefringence of the fiber [20]. Combination of these methods provides a useful measure of the differential birefringence, a skin–core effect, if it is present.

When fibers are observed in the 45° position between crossed polarizers (polars), the change in thickness from the center to the fiber edge produces a series of polarization bands or fringes. An example of these fringes is shown in a PET fiber in Fig. 5.4 (color section). These fringes can be used to determine the retardation of the fiber and the birefringence equals retardation/thickness. If the fiber is round, its thickness is the same as its width. For low birefringence fibers, measurement of the retardation is straightforward, as few of these bands must be counted. However, for higher birefringence fibers there are many bands present which are difficult to count. Additionally, the zero order fringe must be identified for measurement of the birefringence.

It may be difficult to know which fringe is correct if the dispersion of the birefringence of the fiber is different to that of the compensator. A useful trick is to cut a wedge at the end of the fiber and count the number of fringes along the wedge, which is the number of full orders of path difference [21]. The additional partial order is measured with a compensator [22] (Section 3.4.3).

Birefringence provides a measure of the local orientation of a material, i.e. the mean orientation of monomer units. The relation between orientation and birefringence was known from early studies of polystyrene filaments which described both the theory and measurement [21, 23]. They showed that the orientation was greater at the surface than in the core. Mechanical properties, such as tensile strength and elongation at break, have been shown to increase with orientation. Other techniques, such as the measurement of shrinkage on heating, give information on the molecular orientation on a larger scale and provide a measure of the entanglement length. The combination of birefringence measurements and other techniques has been demonstrated, for example, in studies of thermal shrinkage [24] and of the effect of heat setting on the mechanical properties of PET fibers [25]. Birefringence measurements are effective in providing a structural parameter which may be used to relate process variables to mechanical properties.

5.1.2.2 *SEM of textile fibers*

The scanning electron microscope has proven to be a very useful instrument for the assessment of fiber morphology. The three dimensional images produced clearly show surface features, such as the presence of surface modifications, finish applications, wear and the nature and cause of fiber failure. The great depth of field, simple specimen preparation and high resolution have resulted in the SEM providing a major contribution to the study of textile fibers.

Textile fibers generally have a surface finish applied following spinning to aid handling of the fibers for production of yarns and fabrics and to provide special properties, such as flame retardancy. Fiber finishes have been observed by SEM of fiber surfaces [26] and in cross section by using x-ray techniques in the SEM [27]. Preparations are simple (Section 4.1.2) and are generally followed by the application of a conductive coating (Section 4.7.3). Unevenness and lack of uniformity of fiber coatings seen directly in the SEM have been correlated with spectroscopic and wettability studies [28].

Scanning electron micrographs taken in the normal mode do not always permit effective observation of features on fiber surfaces. Display modes such as deflection or Y modulation and imaging modes such as backscattered electron imaging (BEI) can provide clearer contrast, as shown in Figs 5.5 and 5.6. Heat aging of polymer fibers can cause cyclic trimers and oligomers to diffuse to the fiber surface. SEM (Fig. 5.5A) of heat aged octalobal fibers shows oligomers of crystalline appearance on the surface although deflection modulation provides a clearer view of the surface detail (Fig. 5.5B). Solid state BEI detectors can be set to emphasize either topographic contrast ('topo') or atomic number contrast ('compo') which suppresses the surface

detail in the image. Fiber surfaces are shown in both backscattered modes compared to normal SEM images in Fig. 5.6. A BEI 'compo' micrograph (Fig. 5.6A) shows some bright surface detail, which has higher mean atomic number than the fiber. This detail is not as obvious in either the 'topo' (Fig. 5.6B) mode or in the secondary electron image (Fig. 5.6C). This difference is more obvious at higher magnification in 'compo' (Fig. 5.6D) and SEI (Fig. 5.6E). The background texture in Fig. 5.6D is due to resolution of the metal coating.

5.1.2.3 *Fiber fractography*

Textile fiber fractography was initially developed at UMIST (University of Manchester Institute of Science and Technology), especially by Hearle. Fiber fractography and the classes of fracture were reviewed by Hearle and Simmens [15] and further defined later [29, 30]. These classes are shown in Table 5.1 with examples and appropriate references. The mechanism of fiber failure can be determined by fractography studies (Section 4.8.1) in the SEM. Typically, fibers broken during a standard physical test, such as

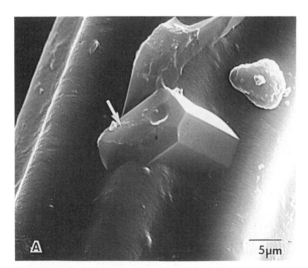 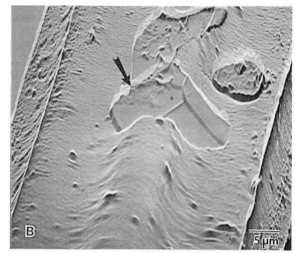

Fig. 5.5 Heat aging polyester fibers draws oligomers to the fiber surface. Crystalline oligomer particles are shown (arrow) in an SEM of an octalobal fiber (A). The curved surface of the fiber does not exhibit much detail other than the oligomers. The micrograph in (B) is of a similar region of this fiber in the Y modulation mode which accentuates the three dimensional surface material.

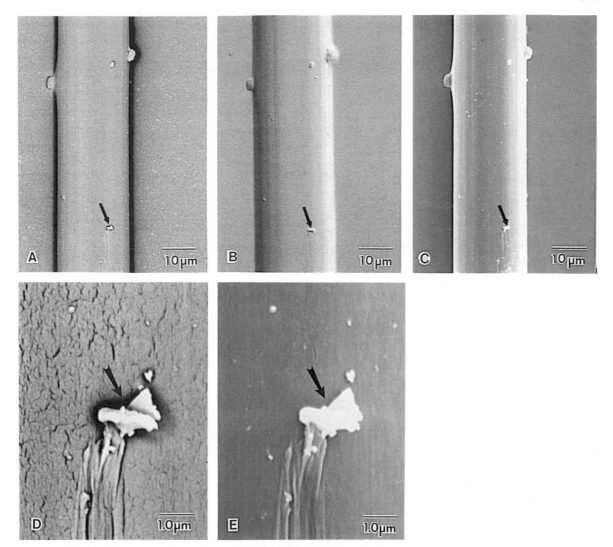

Fig. 5.6 Fiber surfaces are shown here taken in normal SEM and BEI modes. A BEI compositional image (A) reveals bright particles that are higher atomic number than the fiber itself. A BEI topographical image (B) shows that some particles are present on top of the fiber surface. The normal (SEI) image (C) does not exhibit as much detail. Higher magnification micrographs, taken in the BEI mode (D), and in SEI (E), show this difference in more detail. The cracked background (D) is the metal coating.

tensile testing in an Instron, are examined in the SEM for the nature of the failure mechanism and to identify the locus and cause of failure. Analytical microscopy (x-ray microanalysis) can be conducted in order to determine the chemical composition of any defects causing failure.

Aromatic polyamides (aramids) split longitud-

inally due to fibrillation, whereas nylon shows plastic deformation under the same conditions [37]. Gupta [38] showed skin–core effects in polyester fibers. Hearle and Wong [35] studied the fatigue properties of nylon 6,6 and PET. Fatigue is an important property as it is related to repeated loadings that are typical in general use.

Table 5.1 Fiber fracture

Fracture type	Forms of fracture	Polymer types	Ref.
Brittle fracture	Tensile failure; brittle failure	Elastomers, high modulus fibers, (e.g. aramids)	29
Ductile crack	Crack, draw, 'V' notch formation, catastrophic failure	Nylon, PET, acrylics	31
Axial fiber splitting	Split along length; tensile fracture, torsional fracture	Cotton, some acrylics	33
Fatigue splitting	Cracks along fiber initiate at surface and break with long thin tails	Nylon, PET, acrylics, aromatic polyamides	32, 34–36
Lateral failure	Failure normal to the fiber axis	Rayon, acrylics	15
Kink band	Compression inside of bend or by flexing at 45° to the axis; regions of reorientation due to shear forces	PET, nylon, aramids	15

Oudet and Bunsell [36] recently reported on the loading criteria for the fatigue failure of polyamide fibers, which, among other things, indicates continued interest in textile fiber fracture morphology studies. Examples of several of these fracture types will be discussed further.

Smook *et al.* [39] studied the fracture process of ultrahigh strength PE fibers by examination in the SEM. Fracture was shown to be initiated at surface kink bands leading to formation of a fibrillated fracture surface. Extensive study of fracture under various conditions showed a diameter dependence of the tensile strength which is consistent with the Griffith relation. Application of fractography to crosslinked fibers showed a change in fracture morphology from a fibrillar to a brittle mechanism.

Fracture mechanics considerations, summarized for polymers by Kausch [40, 41], permit

determination of the effect of defects on the fracture stress, or tensile strength, of bulk polymers and polymer fibers. It is important that such detailed study be performed on the original or *primary* fracture surfaces, which are the only surfaces which relate to the tensile stress. This is generally done by examining both failure surfaces to ensure that they are a matched pair. An example of matched, primary fracture surfaces was shown earlier (Fig. 4.34). In that case, a definite defect site was observed in the brittle fracture surface. Such assessment provides the structure–property information needed to modify the process and produce fibers with higher strengths.

Another example of typical fracture morphology, with a defect at the locus of failure near the fiber surface, is shown in Fig. 5.7. Once fracture is initiated, it slowly propagates radially away from the failure locus and continues into the adjacent polymer, providing a smooth surface, nearly perpendicular to the tensile stress, termed a 'mirror'. As the crack propagates away from the flaw site it accelerates causing crack branching. The region where the acceleration begins to occur is seen to have fine ridges, or 'mist', present whose size is dependent on the fiber microstructure. Crack propagation accelerates quickly through the fiber causing catastrophic failure. This 'fast fracture' morphology is generally characterized by ridges, or 'hackles', at an angle to the path of the original failure.

The fracture morphology in tensile fatigue is quite different as shown in the micrographs of a polyamide and a polyester fiber in Fig. 5.8 [42]. The fibrillar nature of a polyamide fiber results in a break with a tail, as seen in the broken ends (Fig. 5.8A and B). The typical fatigue failure has a crack initiated at the fiber surface with radial penetration into the fiber. The crack then runs along the fiber and finally the fiber fails once the cross section cannot support the applied load. In simple tensile failure, the two fracture surfaces are mirror images, whereas after fatigue testing, complementary fiber ends are quite different. Fatigue fracture of a polyester fiber appears somewhat different in morphology (Fig. 5.8C

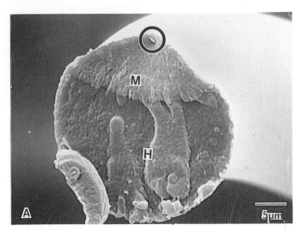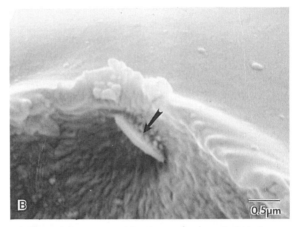

Fig. 5.7 SEM of fractured fibers reveals the flaws causing failure. A low magnification micrograph (A) shows the flat mirror (M) region, adjacent to the locus of failure, which is the region of slow crack growth. The fast crack growth region, or hackle (H), has large ridges. A higher magnification micrograph (B) reveals the flaw causing failure (arrow) in more detail.

and D). Although failure appears to be initiated at the fiber surface, the penetration is shallower. The crack extends along the near surface resulting in a very long tail on one fiber end. Tensile fatigue studies of synthetic polymer fibers have been discussed by Bunsell and coworkers [34, 36].

5.1.2.4 *SEM of fiber assemblages*

Yarns and fabrics are assemblages of fibers which have commercial application in the textile industry. The fabrics include those formed by weaving and also nonwovens. The geometry of the fabric, as well as the chemical composition of the polymer, influences mechanical properties and applications. The SEM is useful for evaluating: (1) construction, (2) coverage, (3) uniformity, (4) surface structure and (5) effects of wear.

Woven fabrics

The SEM was used to show yarns, composed of fibers twisted together, woven into a fabric (Fig. 5.9). Important features of the fabric, 'hand' and 'coverage', relate to the yarn geometry which can make a thick and comfortable fabric or a thin, uniform and open structure. Tilted side views

can be used to count the number and the length of protruding surface hairs which are known to affect the feel or hand of the fabric and its mechanical properties [43, 44]. The complex subject of fabric wear is particularly suited to SEM study, as the three dimensional structure and surface texture can be related to individual fiber failure [45–49]. In a review of the topic of wear in military clothing, Kirkwood [45] described SEM studies which showed the wear of cotton into long fibrils and nylon into shorter, thicker fibrils. The mechanism of attrition is known to be quite complicated, and it will not be discussed here, except to say that there are three elements involved in the mechanism of fabric attrition [50]: friction, surface cutting and fiber rupture. The effect of both testing and wear can be evaluated using the SEM. Nylon 6,6 among other fibers, has been tested for wear by measuring the depth of surface damage in the SEM following sliding the fabrics on rough hard surfaces [46]. Fiber debris, voids and pills were observed in blends of polyester and cotton [48] and viscose, acetate and nylon fabrics; then they were abraded and examined in the SEM [49]. The SEM also provides a useful tool for the evaluation of fibers and fabrics during the manufacturing process [51].

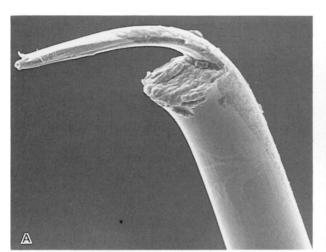

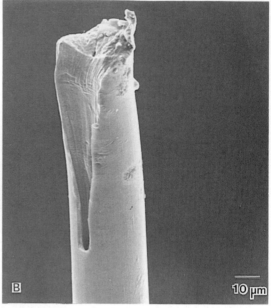

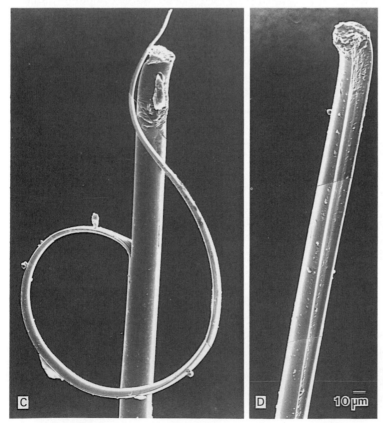

Fig. 5.8 Typical fatigue failure of a nylon 6,6 fiber is shown by SEM (A and B) of a fiber tested at 0–55% of nominal breaking load, with a lifetime of 1.3×10^5 cycles at 50 Hz. Fiber failure is initiated at the surface resulting in a tail on one side (A) and long furrow on the other side (B).

Fatigue failure of a polyester fiber, at 0–70% of nominal breaking load, with a lifetime of 3×10^5 cycles at 50 Hz, shows a somewhat different morphology. Fiber failure is initiated at the surface, but it continues along the surface resulting in a very long tail, on the one side (C), with splitting down that side as well, and a long shallow furrow on the other side (D). (From A. R. Bunsell, unpublished [42].)

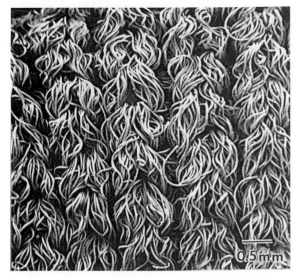

Fig. 5.9 SEM of a typical woven fabric shows twisted fibers in this fiber assemblage. The fibers overlap the space between the yarns providing coverage, or cover, an important fabric parameter.

Nonwoven fabrics

Nonwoven fabrics are another form of fiber assemblage, but they are less regular and uniform in structure than woven fabrics. A spray spun nonwoven is shown in side view (Fig. 5.10A) on an SEM stub, and a surface view (Fig. 5.10B) shows the fabric has a range of fiber diameters and shapes with many deformed, nonround fiber cross sections. The coverage, size and distribution of the open space and the individual fiber diameters can be important depending on the application. A more uniform, calendered nonwoven fabric is shown to have been spot calendered (Fig. 5.10C and D). The fabric has only been flattened out by the calendering process in local regions, whereas other regions are three dimensional. Many large, deformed fibers are observed of varying diameter.

5.1.2.5 *TEM of fibers*

Transmission electron microscopy techniques are very important for the elucidation of details of fiber microstructure. The types of detailed structures that are determined by TEM are:

(1) polymer structure;
(2) void size, shape and distribution;
(3) size, shape and distribution of fillers;
(4) local crystallinity;
(5) crystallite sizes.

Early TEM studies were by replica methods [18], as in a study of replicated and etched fiber surfaces [15]. Such studies are now conducted by SEM of external and bulk structures and by ultrathin sectioning for TEM. Microstructural studies generally require complementary optical and SEM study to understand the arrangement of the fine structural details within the macro-structure.

A fairly simple example defining the structure in an experimental fiber is described to show that, generally, just one microscopy technique does not provide the complete structural picture. Three different techniques/methods are shown in Fig. 5.11. SEM of the fractured fiber shows an overall view of the bulk structure and the fiber shape (Fig. 5.11A) and the internal porosity (Fig. 5.11B) although the size, shape and distribution of the voids is not fully defined. SEM of the outer fiber surface (Fig. 5.11C) shows that voids reach the fiber surface. TEM micrographs of ultrathin cross sections (Fig. 5.11D and E) clearly provide a description of the void sizes and their local distribution. A pore gradient is observed, with smaller voids at the surface and coarser voids within the fiber. The smaller voids are located within a micrometer sized band around the fiber periphery. Complementary microscopy has been shown to describe the experimental fiber microstructure. The void sizes and distribution are parameters that are both affected by process modifications and relate to the end use properties.

5.1.3 **Problem solving applications**

Characterization of fiber microstructure normally requires several microscopy techniques, as was

Fig. 5.10 A spray spun nonwoven fabric is shown in the SE images (A and B). A side view of the fabric is shown (A) on an SEM stub while a face view (B) shows there is a range of fiber diameters and shapes present with no specific pattern or arrangement. A calendered nonwoven surface is shown by SE images (C and D). The spots on the surface are regions of local melting that hold the fabric together (C). At higher magnifications (D) the fibers are seen to range from round to deformed shapes.

shown in the simple example in the previous section. An optical cross section of a fiber may have a dogbone shape (Fig. 5.2D), and yet this image does not reveal much about the internal fiber structure. On the other hand, a fracture surface of a fiber may reveal the presence of internal detail when viewed in the SEM (Fig. 5.11), and yet still not provide a complete picture

of the structure. Clearly, complementary microscopy techniques and non-microscopy techniques must be applied to solving structural problems. Specific problem solving examples are described here which are representative of the wide range of studies conducted and documented in the many journals that publish polymer research.

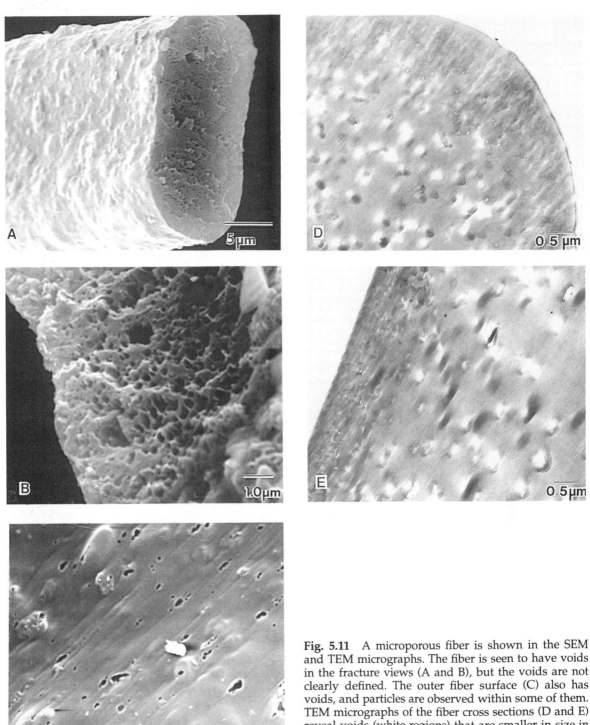

Fig. 5.11 A microporous fiber is shown in the SEM and TEM micrographs. The fiber is seen to have voids in the fracture views (A and B), but the voids are not clearly defined. The outer fiber surface (C) also has voids, and particles are observed within some of them. TEM micrographs of the fiber cross sections (D and E) reveal voids (white regions) that are smaller in size in an outer micrometer sized band than those in the central portion of the fiber.

5.1.3.1 *Characterization of textile fibers*

Delustrant and pigment particles are often used to provide modifications in the visual appearance of textile fibers and fabrics. For example, titanium dioxide particles are commonly added to polymers before fiber spinning to change the fiber luster. Additions to a polymer require monitoring the fiber formation process to assess the effect of the particles, for quality control of particle size and distribution and for failure analysis. Questions relate to the effect of particle size on mechanical properties and on the polymer structure. A wide range of practical problems can occur in any polymer process and these problems are easier to solve when the fiber microstructure is fully known. Poor surface texture, fiber breakage during spinning and nonuniformity in visual appearance are a few of the problems that may occur in any industrial plant producing textile fibers. Microscopic analysis would be helpful in most of these cases where comparison could be made between standard controls and problem fibers. Rather than dealing with any one of these specific problems, the example in the following section is a description of a typical, microstructural characterization of a polyester fiber containing titanium dioxide.

Direct optical observations

Direct observation of a fiber in the optical microscope or study of a fiber cross section provides information relating to the size and distribution of added particles. A longitudinal view of a PET fiber, in an OM micrograph (Fig. 5.1) reveals dense particles. The distribution of the particles can be seen, but their size and their relationship within the fiber are not clear at this magnification. Fiber cross sections (Fig. 5.2) show the particle distribution more clearly. Only a small, thin section about 5 μm thick is observed optically and a great number of such sections must be examined to define a statistical distribution.

Surface study

Electron microscopy techniques provide more resolvable detail than optical microscopy. SEM of a PET fiber surface (Fig. 5.12A) shows splits and delustrant particles. The particles were determined by x-ray microanalysis to contain titanium. A surface discontinuity is shown by SEM (Fig. 5.12B) and by high resolution secondary electron imaging (SEI) (Fig. 5.12C) taken in an AEM, a TEM with a scanning attachment. The high resolution SE image clearly shows particles protruding from holes in the fiber surface, although the image shows the limited depth of field in this technique, due in part to the short working distance. At higher magnifications (Fig. 5.12D and E), particles about 0.1 μm in diameter are seen adjacent to a fibrillar polymer texture. The microfibrils are about 50 nm wide and are oriented parallel to the fiber axis. Such structures are usually observed by TEM techniques [52]. These structures have only recently been resolved and identified by high resolution SEM [53].

Bulk study

Cross sections of particulate loaded fibers are required for accurate particle size determination and for studying the effect of the particles on the microstructure. Optical cross sections of PET fibers (Fig. 5.13A inset) have more particles present as the sections are significantly thicker (*c.* 5 μm) than TEM sections (Fig. 5.13) which only show a few aggregates of dense particles. Skin–core textures resulting from high speed spinning [54] may also be seen in some sections. Particles and the adjacent holes are observed in the section. An important question is whether the holes are caused by the process or are a result of the sectioning. Care must be taken in interpretation and thus follow up studies were conducted to determine the origin of the voids.

Further structural study can be conducted by simple peeling methods for SEM and by staining methods for TEM. The peeling of a segment of a fiber to reveal the internal structure, as first

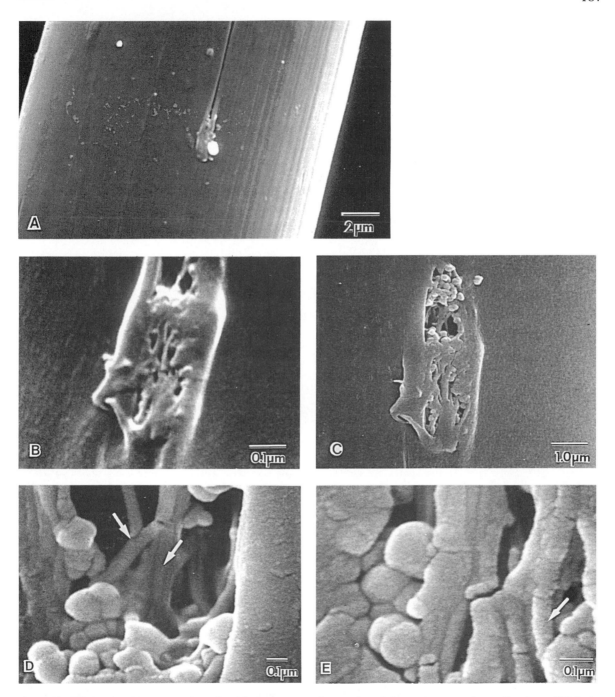

Fig. 5.12 Fibers containing titanium dioxide are commonly employed for various textile applications. SEM of a fiber surface, without finish (A), shows a surface split, likely caused by such a particle at or near the surface. A similar region at higher magnification is shown in both SEM (B) and high resolution SEI (C) micrographs. The defect region is not very clear in image (B) taken in a dedicated SEM, but the higher resolution image provides interesting detail (D and E). Particles, voids and microfibrils (arrows) are observed.

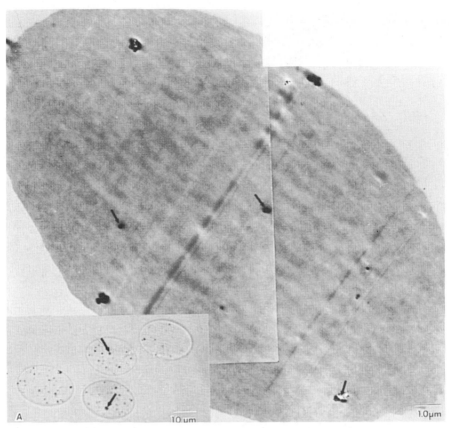

Fig. 5.13 TEM micrographs of polyester fiber cross sections reveal the size of the pigment particles. The optical inset (A) shows fibers with dense particles in greater amounts than seen by TEM due to the difference in section thickness. The dense particle aggregates are titanium dioxide particles. The fiber exhibits no major structural detail. The dense lines are knife marks produced during sectioning.

developed by Scott [18], has since been used to show the microfibrillar structure of nylon 11 and 12 [52] and of PE [55, 56]. The highly fibrillar structure which develops in PE fibers on drawing was correlated with the increasing crystalline orientation, as observed by x-ray diffraction, and with increased tensile strength and modulus. PET fibers peeled back to reveal their internal structure (Fig. 5.14) show the microfibrillar texture and the titanium dioxide particles in depressions within the fiber. As with the TEM sections, this morphology could be due to the deformation during peeling, although this study suggests that there are holes formed adjacent to the particles during the spinning process.

Staining–inclusion method

Another approach to the characterization of fiber microstructure is the isoprene inclusion method (Section 4.4.2.5). This was applied to the study of PET fibers [57] and to aramid fibers [58] for the purpose of showing their radial microporous and fibrillar texture. Any holes or voids are filled by inclusion of isoprene in the fiber. They are then stained by the reaction of osmium tetroxide with the included isoprene. Longitudinal sections of high speed spun PET are shown in the TEM micrographs in Fig. 5.15A and B of fibers before and after the reaction, respectively. Similar views at lower magnification were shown (Fig. 4.12) in

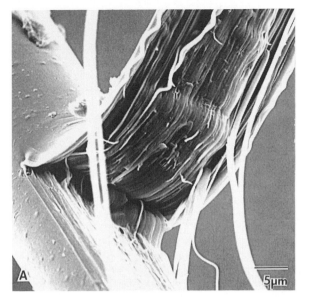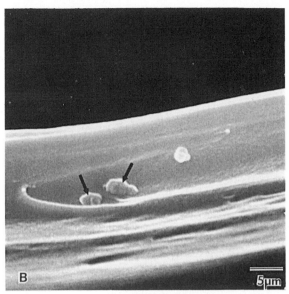

Fig. 5.14 PET fibers peeled back to reveal the internal textures are seen in these SEM micrographs. Important features include the microfibrillar texture and the particles of titanium dioxide. A low magnification view shows the peeled fiber (A) with a surface skin and loose fibrils. At higher magnification (B) particles are seen in furrows and voids are observed adjacent to the particles.

the section describing this method. The control fiber section (Fig. 5.15A) reveals holes or voids adjacent to the particles, as observed in SEM views of peeled fibers and fiber surfaces. After isoprene inclusion and staining, these regions (Fig. 5.15B) are electron dense and elongated parallel to the fiber axis. There is a fine pattern of elongated, dense regions, also aligned parallel with the fiber axis, which suggests there is an ordered arrangement of voids about 10 nm wide. Furthermore, voids are more prevalent near the outer fiber surface than within the fiber (Fig. 5.15C and D). The staining method has confirmed that voids are present in high speed spun PET fibers.

Summary

Characterization of the microstructure of high speed spun polyester fibers has been demonstrated using combined SEM of bulk peeled fibers and fiber surfaces, OM of thin sections and TEM of sections both stained and unstained.

The polyester fiber microstructure has been shown to be microfibrillar. It contains microvoids, elongated parallel to the fiber axis and has elongated voids adjacent to the particle aggregates, also aligned parallel to the fiber axis. The effect of the titanium dioxide delustrant particles is to produce voids adjacent to the particle aggregates which elongate on drawing. This morphology suggests a high degree of orientation with enhanced mechanical properties [56]. Study of PET fibers by x-ray scattering, infrared and birefringence [13] showed a microfibrillar structure, with microfibrils loosely held together in fibrillar units at least an order of magnitude larger in size. The microfibrillar structure shown here for PET fibers is similar to the structure shown for nylon stained with tin chloride [12].

5.1.3.2 *Metal loaded fibers*

Microscopy techniques can be used to evaluate the size and distribution of particles added to polymer fibers, such as metals that modify the

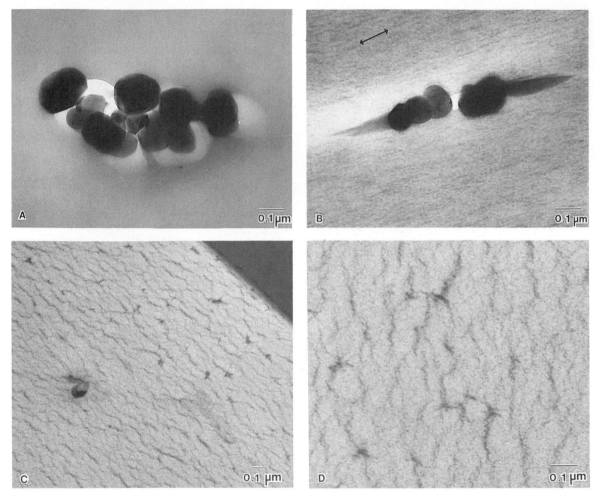

Fig. 5.15 TEM micrographs of longitudinally sectioned PET fibers taken before (A) and after (B) isoprene inclusion and staining reveal major differences. The untreated fiber exhibits no structural detail except for aggregated particles and holes within the aggregates. After treatment, dense regions of stained isoprene are observed adjacent to the particle aggregates confirming that these regions were holes in the original fiber. Finer dense elongated regions, aligned with the fiber axis (arrows), represent voids about 10 nm wide (B). There is a higher void density near the fiber periphery than in the core, as shown in cross section (C) and at higher magnification (D). A circumferential, skin–core arrangement of voids is observed.

physical, mechanical, or electrical properties. In general, ultrathin sections are examined in either STEM or TEM modes to reveal the particles within the polymer. Energy (EDS) and wavelength dispersive x-ray spectroscopy (WDS) methods are used to map for various elements in order to establish the relation between the particle morphology and chemical composition.

A specimen preparation method for x-ray analysis in the SEM is to use a trimmed block face, which remains after cutting thin sections, or to study a thick section.

As an example of such a study, a particle-loaded polymer fiber cross section is shown in TEM (Fig. 5.16A and B), BEI (Fig. 5.16C) and in an x-ray map (Fig. 5.16D). These images provide

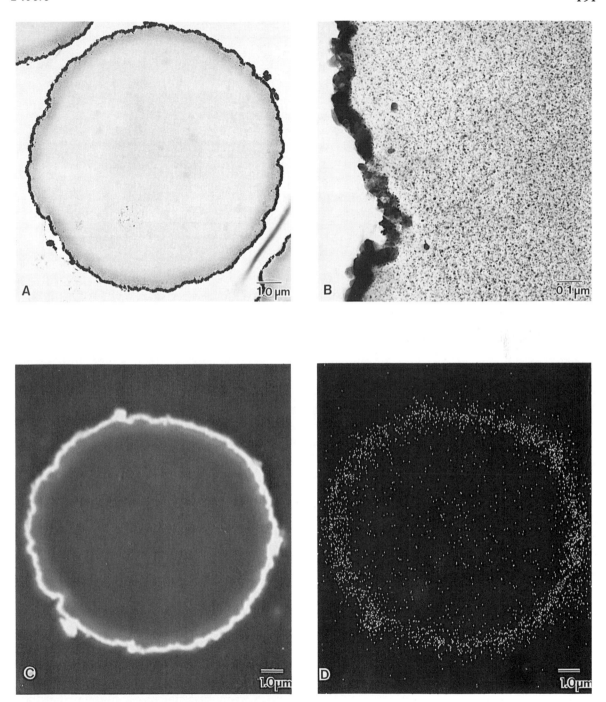

Fig. 5.16 A metal particle loaded fiber is shown by complementary microscopy techniques. Sharp detail is observed by TEM (A and B) which shows a dense band around the fiber periphery and fine dense particles adjacent to and within the band. The chemical composition of the particles is seen to differ from the polymer in the BEI compo image (C) and x-ray spectra and mapping (D) show the distribution of the metal.

a complementary assessment of the size, chemical nature and location of the particles. A dense band around the fiber periphery is seen by TEM (Fig. 5.16A and B) with dense particles (Fig. 5.16B) ranging from about 5 to 20 nm in diameter. The BEI image (Fig. 5.16C) shows that the dense outer layer contains higher atomic number material than the fiber. A WDS x-ray map (Fig. 5.16D) shows the elemental distribution. Thus, microscopy techniques provide a complete assessment of the particle size, chemical composition and location of particles added to polymer fibers to modify properties.

5.1.3.3 *Contamination*

Contamination studies are often required to understand process related problems and to investigate the cause of property deficiencies. In many cases the problem is not the polymer itself but the addition of some unknown material during handling or processing. Determination of the nature of such contaminants is a serious problem for the microscopist. An example of a contamination study relating to fibers will be described. Spinnerets used in fiber spinning can become clogged, stopping the process because of polymer plugging or contamination. The optical micrograph (Fig. 5.17A, color section) shows the surface of a spinneret with dark, rust colored material as well as white polymer. This material was scraped off the spinneret and dispersed on a carbon stub for EDS. The EDS spectra (Fig. 5.17B, color section) of the rust colored material showed that antimony (Sb), silicon (Si), titanium (Ti) and iron (Fe) were present. This showed inorganic contaminants to be the cause of clogging of the jet.

5.1.4 **Industrial fibers**

5.1.4.1 *Tire cords*

Major industrial applications of fibers are in the production of tire cords and belts. Tire cords are composed of yarns of polymer fiber, twisted together into cords, which are coated with adhesives. Wilfong and Zimmerman [59] discussed tire yarn property criteria and the polymer and fiber structural factors which control the properties. Polyester, rayon and nylon are common tire cord yarns. The tire cord composite is a complex system of multiple interfaces that must be well bonded to provide high strength. Characterization of tire cords by microscopy is required to understand the microstructure and apply that knowledge to solving problems. The problems encountered in tire cords relate to: (1) application of fiber finishes, (2) poor strength or strength reduction, (3) poor adhesion of fibers to adhesives, (4) fiber degradation, and (5) poor adhesion of the coated cord to the rubber.

Finishes must be applied uniformly to the fiber for protection, ease of handling and compatibility with the adhesive to limit strength reduction. The adhesive is applied to the cord to enhance cord integrity, protect the fibers from the rubber and to join the cord and rubber into a well integrated material. Essentially, then, there are several interfaces of interest: fiber–finish, fiber–adhesive and adhesive–rubber. The adhesives are generally resorcinol–formaldehyde–latex (RFL) dips, more fully described earlier [60]. An optical cross section (Fig. 5.18A, color section) provides an overview of an RFL coated tire cord. The outer surface of an RFL coated cord (Fig. 5.18B) and a surface close up is shown by SEM (Fig. 5.18C) and BEI (Fig. 5.18D). The normal SEM image does not provide much detail, except that a filmy coating is present, whereas the BEI image provides detail on the disposition of the RFL. Regions that are brighter are likely higher atomic number than the background. The adhesive coating appears to be 'puddled' into the interstices between the fibers more than on the outer fiber surfaces, which would be expected to result in incomplete adhesion to the rubber.

The combination of materials in a tire cord is a nightmare for the person performing the microtomy as each material has a different hardness and would be expected to pull apart during the cutting procedure. The ebonite specimen preparation procedure (Section 4.4.3) permits uniform hardening of the soft latex in the RFL and

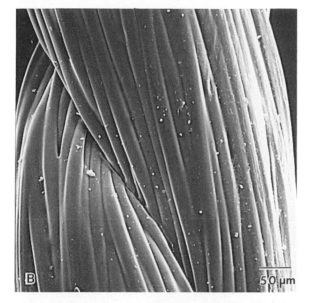

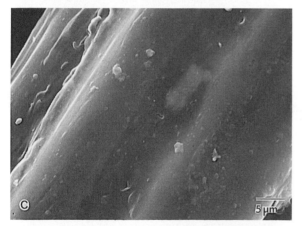

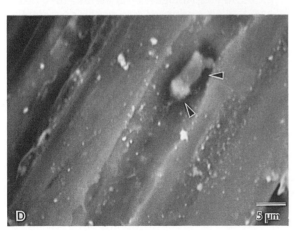

the rubber as was shown in a low magnification view of the structure in the TEM (Fig. 4.14). Detailed TEM images of cross sectioned tire cord specimens are shown in Fig. 5.19. An overview (Fig. 5.19A) shows that this cord has a two dip coating on the fiber surface with interfaces showing good integrity; thus there is good adhesion between the fiber and the RFL and between the RFL and the rubber. Specific morphological details of the two RFL layers are seen (Fig. 5.19B) that relate to the chemistry and behavior of the system. The fiber surface has a uniform RFL coating, containing dense specks which are depleted at the rubber interface (top) (Fig. 5.19C). A view of the rubber in a region with only a thin layer of adhesive (Fig. 5.19D) shows small dense specks and larger carbon black particles within the rubber phase. Carbon black is a common filler used to reinforce rubbers and enhance strength.

The ebonite diamond knife sectioning method provides excellent representation of the three interfaces for assessment of the morphology and correlation with mechanical properties. This method has been applied to optimize finish and adhesive application on tire cords and to better understand fiber–rubber adhesion, its influence on strength and strength loss. Tire cords exhibiting poor strength, strength loss and/or poor adhesion of the fiber to the adhesive have been observed, and fibers with good mechanical properties generally exhibit good adhesion overall.

Fig. 5.18 Tire cords are shown by several complementary microscopic methods. A cross section of an adhesive coated yarn is shown by optical microscopy (A) (color section). The surface of an RFL coated cord is shown by SEM (B). Comparative normal SEM (C) and BEI (D) images show the filmy adhesive coating. The BEI image shows that the surface has higher atomic number particles (arrows) and detailed surface structures that suggest that the adhesive is found in higher concentration in the interstitial regions, between fibers, than on the fiber surfaces in this experimental cord.

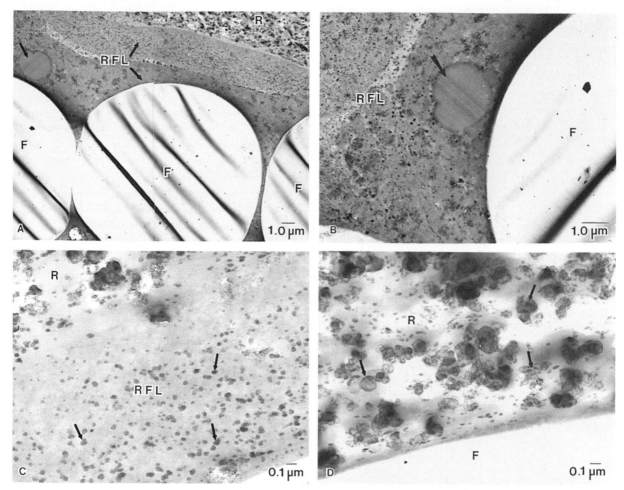

Fig. 5.19 Ebonite treated tire cords are shown in TEM micrographs of ultrathin cross sections. The montage (A) provides an overview of the double dip RFL coating the polyester fibers (F). The rubber (R) is seen coating the RFL. Note that the dark lines in the fibers are wrinkles due to the sectioning. At higher magnification (B) a phase separated, globular region is seen within the first RFL dip. Good adhesion is exhibited between the fiber and the RFL. A micrograph showing the fiber–RFL and the RFL–rubber interfaces (C) shows that dense specks are present but not near the rubber interface. A view of the rubber (D) shows that carbon black aggregates are present (arrows) as well as other dense particles. Adhesion is seen to be good at the RFL–rubber boundary.

5.1.4.2 *High modulus PE fibers*

Applications requiring superior mechanical properties combined with light weight have recently benefited from the use of polymers in fiber form. High performance organic fibers can be used to produce high tensile modulus, high strength and high energy absorption structures of much less weight than the equivalent parts made from structural metals or ceramics. The high modulus fibers are made from polyethylene (PE), aramids or thermotropic liquid crystalline polymers (TLCP). They are finding applications in cables, protective fabrics (bulletproof vests) and composites for automotive, marine and aerospace use.

Good reviews of these high performance polymer fibers include descriptions of their structure, morphology, production and properties [61–63]. All the fibers have a highly oriented and largely extended chain structure. This structure of closely packed rod-like molecules with weak intermolecular interaction produces a very anisotropic material that is weak in shear and compression. The tensile modulus along the fiber direction is largely controlled by the cross sectional area of the molecular chains, the chain stiffness (helical chains are soft, all-trans chains are stiff), and the perfection of chain orientation and extension. A high performance fiber may have the chain orientation distributed over a narrow range, so that the mean misorientation of molecular chains from the fiber axis is only one or two degrees. In diffraction, the fiber pattern then looks just like what would be obtained by rotating a single crystal about the chain axis. The mechanical properties of the chains are so anisotropic that even this small misorientation may reduce the axial modulus of the fiber by 30 or 50% from its ideal value. It may still be as much as one hundred times greater than the transverse modulus of the fiber. Composite concepts, such as the aggregate model of Ward [64] describe how the fiber mechanical properties are derived from the fiber structure. Fibers made of aramids or TLCPs are quantitatively modeled as an aggregation of molecules with a range of orientation, but this simple model does not work for PE.

The aramids and TLCP fibers are produced by melt or solution spinning in a liquid crystalline state. They are described in detail in Section 5.6.4. The rest of this section is concerned with high modulus fibers produced from conventional polymers. These may be made from a range of flexible chain linear polymers using a number of processing routes. For PE these include solid-state extrusion, die drawing, zone drawing and gel spinning [61, 65]. Commercial high modulus PE fibers, such as Dyneema (DSM) or Spectra (Allied), are produced from extremely high molecular weight polymer by forming a gel and spinning it. High molecular weight gives high strength along with the high modulus, and gel spinning gives the highest production rates – still very slow compared to those for conventional fibers. The polymer is first dissolved or highly swollen in solvent at elevated temperature, and the solution is wet spun into a quenching bath to form a gel filament. The gel can be dried by solvent extraction and then hot drawn to very high extension ratios. Alternatively, heating, drying by evaporation and drawing can be done simultaneously.

Fiber strength is affected by the presence of defects such as chain ends and their arrangement [54, 61, 62, 65]. Any flaw or cluster of defects that can act as a stress concentrator may reduce the strength. Therefore the highest strength fibers tend to have a close to homogeneous morphology, where the 'amorphous' material is very highly oriented. High modulus materials that do not have the highest strengths have a wider range of structural features, which may include lamellae [66]. Most high performance fibers are microfibrillar [61, 65], with morphology similar to the structure shown in Figs. 4.5 and 5.14. A natural model for the mechanical properties of the fiber is that of a fiber composite, with the more crystalline microfibrils the reinforcing, load bearing elements. The high alignment and extreme anisotropy of these fibers makes peeling or transverse splitting easy, so that internal surfaces are made accessible. These must be preferred fracture surfaces, planes of weakness, so they may not be completely representative of the whole fiber.

It is easy to analyze these internal surfaces by observing fibrillar features of some mean size, by SEM, TEM replica or SPM. If the preferred fracture planes are randomly distributed, there will be a wide range of feature sizes on the surface. A single microscopic technique will have some limit to resolution, and a limit to its field of view. Features approaching the size of the field of view must be rare in an image, so any single imaging technique can only show a limited range of sizes of object features. The most common will be a size above the resolution limit. Care must be taken to assess if there is a true hierarchy of structures or if this is due to sampling of a random continuum. Statistical image analysis of microfibril sizes from fracture surfaces has not been reported for flexible polymers, because it is very difficult to define the features unambiguously for numerical analysis. As so often in microscopy, a second technique or supporting information from a totally different method is required for certainty in interpretation.

TEM of stained fiber samples is such a second technique, but preparation of these samples is much more difficult. Schaper *et al.* [67] have made an extensive TEM study of surface grown high modulus PE fibers. They used various preparations including chlorosulfonation (Section 4.4.4) and ultramicrotomy. With computer image processing to improve feature visibility, they could clearly distinguish fibrils micrometers long and about 5–20 nm in width. They also saw defects, such as voids and kink bands similar to those seen in the LCP fibers (Section 5.6.4). The width of the fibril core was found not to vary with draw ratio, a result consistent with their model of formation of the extended chain fibrils. Calculations of tensile modulus based on the observed structure agreed well with the results of mechanical testing. This work, among many reviewed in ref. 67, supports the idea that there are microfibrils in this material that are the load bearing elements. A fibril width of 10–20 nm agrees with earlier work [68] and with small angle x-ray diffraction [69]. The earlier work used crystallographic DF contrast in the TEM on specially prepared thin samples of surface grown material. The crystal lengths in the fiber direction were shown to follow a most probable distribution; that is, a mean length could be determined, but there was no preferred length [68].

More recently, AFM [70–73] has been used to image the external and internal fracture surface of high modulus PE and other samples with high degrees of molecular orientation. Maganov *et al.* [70, 71] have studied cold extruded PE surfaces prepared with an ultramicrotome cutting along the extrusion direction. This controlled fracture gives the generally flat surface required for high resolution AFM. The rod-shaped samples were embedded in epoxy for the microtomy, and it is the cut surfaces, not the sections, that are used.

The image size ranged from 700×700 nm to the atomic scale. At the lower magnifications, microfibrillar structures were seen, with fibrils of 20–90 nm diameter and indefinite length. The authors claim to see the molecular chain structure at higher magnifications, and the images are quite impressive, apparently showing local details of the molecular arrangement. However, as described in more detail in Section 6.6, true atomic or molecular resolution of the undisturbed sample is exceptionally difficult to obtain in AFM. This is especially true when using soft materials like polymers. AFM instruments are still developing rapidly, with reduced noise levels allowing imaging at lower applied forces. Until very recently, imaging forces were always greater than 1 nN. This sounds like a very small force, but if a true resolution of 0.5 nm was obtained, the stress at the contact point would be > 4 GPa, resulting in gross distortion of the sample at the contact point.

Some experts believe that most of the AFM images at these contact forces that show atomic or molecular structures with periodic spacings smaller than 1 nm are due to multiple contacts between sample and probe tip. Multiple contacts between the sample and probe reduce the local load, and addition of multiple signals will still give the correct periodicity. Local detail will not be imaged correctly. If this is true, almost any local structure will appear by random interference of the multiple signals, so the effect of the operator selecting 'good' or 'interesting' images can be profound. A common method of reducing the distortion of the sample is to mount a thin layer on a rigid and smooth substrate. As was mentioned above, AFM requires a generally smooth surface, and if a regular fracture surface is used artifacts appear. Annis *et al.* [73] imaged fracture surfaces of extended chain crystals of PE grown at high pressure by both AFM and TEM replicas (Section 4.6.2). The images were very similar, with morphological details and features of the growing crystalline lamellae clearly shown. The AFM showed some sharp edges as rounded and could not image sharp protrusions. This is because the tip used to form the image is itself rounded, and the image is a convolution of the sample shape and the tip shape (Section 6.6).

STM is much more reliable in producing images with atomic resolution, but only for conducting samples. Bulk polymers must be coated with a thin layer of conducting metal. Even a very thin layer will obscure detail less than 1 nm in size, allowing a quantitative measurement of fibrillar or lamellar profiles, but hiding molecular structures [74].

5.2 FILMS AND MEMBRANES

5.2.1 Introduction

Films and membranes include a formidable array of materials that are widely used in a range of industrial applications. Films find application as coatings and packaging materials such as food wraps, and membranes are used for separations, controlled release, coating and packaging barriers and contact lenses. Structural studies of films fall into two major categories: model film studies and the study of commercial films. Model studies are generally conducted in university research laboratories where the thin, flat structure of a melt cast or drawn film provides an ideal specimen for study. Such thin films can be examined *directly* by TEM as well as by optical microscopy, SEM or SPM. Specimen preparation is minimal, and interpretation of the image is made easier for several reasons.

(1) Structures in very thin films are clearer because they do not overlap in the transmission image.
(2) Model cast films are made with no possibility of damage or deformation during specimen preparation by microtomy or fracture.
(3) Structures in model semicrystalline films may be larger and better defined than in commercial materials because of lower nucleation densities or lower quenching rates.

If the TEM is used, intepretation always requires care, even for this 'easier' case. Problems of radiation damage of the polymer [75] (Section 3.5) cannot be removed by using thin specimens.

Thin films may be divided into three classes, depending on their molecular orientation. Real films are biaxially oriented, with the molecular chain axis most likely to lie along the direction of film extrusion (the 'machine direction') and more likely to lie in the plane of the film than perpendicular to it. However, some films have little or no orientation, and their microstructure is similar to the bulk polymer (Section 5.3), with differences possibly due to the presence of the film surfaces. These allow rapid quenching and may increase nucleation of crystals. Other films are strongly stretched or drawn as part of the extrusion process, and molecular alignment along the machine direction dominates the properties. These films are fundamentally similar to fibers, although the geometry is different. The third class of films may be prepared by biaxial drawing, for example, to make the molecules lie in the plane of the film with no strong machine direction effect. These films have good properties in all directions in the sheet, and include for example, PET (Mylar) and polyimide (Kapton).

Model film studies are more appropriate and easier to relate to commercial films in cases where the orientation is low or primarily uniaxial, as biaxial orientation is difficult to mimic on a small scale. Films with large second phase particles also cannot be studied by this method as the true structure cannot be reproduced in a thin film. Nevertheless, many of the basic concepts used in describing microstructure and its development in films, particularly semicrystalline films, have been derived from model film studies.

Model studies have helped considerably in understanding blown PE films. The x-ray diffraction patterns from these films could not be interpreted unambiguously and their microstructure was unclear until Keller and Machin [76] produced the same structures in model drawn films. They described the structure and showed how it was formed during crystallization from an oriented melt. This helped in many ways as, for example, the effect of molecular weight distribution on film properties could be understood by its effect on the structure formation. The orientation in films is measured by optical birefringence techniques, and such studies have shown the effect of resin molecular weight in stretching of parisons in the orientation of bottles [77].

Membranes can be thought of as special types of films that provide specific end use characteristics. Membrane technology has replaced some conventional techniques for separation, concentration or purification [78]. Applications include desalination, dialysis, blood oxygenators, controlled release drug delivery systems and gas separation. Processing of polymer films and membranes is well known to affect the morphology, which in turn affects the physical and mechanical properties. As is true for all films, membrane separation properties are based on both the chemical composition and the structure resulting from the process. Membranes are produced in two major forms, as flat films and as porous hollow fibers, both of which will be discussed in this section.

A wide range of polymer chemical compositions is used in both films and membrane materials. A listing of commonly known polymers, including those found in films and membranes, is found in Appendix IV. The focus of this section is on a description of model film studies, industrial film applications and flat film and hollow fiber membranes, with examples of studies which show the morphology of the films and membranes.

5.2.2 Model studies

5.2.2.1 *Semicrystalline films*

Model studies of soluble semicrystalline polymers may be made quite simply by allowing a drop of a dilute solution of the polymer to fall onto a substrate, such as a glass microscope slide, and allowing the solvent to evaporate. The thickness of the resulting 'thin cast film' may be controlled by changing the solution concentration, and films about 20–100 nm thick are used for TEM. A hot solution or substrate may be required to control film thickness. If the film adheres to the glass too strongly to be removed

for mounting on a TEM grid without damage, an inert liquid or a water soluble material, such as NaCl, may be used as a substrate. For simple optical microscopy, the films are made thicker and not usually removed from their glass substrate.

Quiescently crystallized films may be used to determine the crystal morphology and, for example, its dependence on crystallization rate or polymer molecular weight. Drawing the films, while on a liquid or other deformable substrate, has been used to follow the spherulite to microfibrillar transition and so is clearly relevant to fiber drawing. Other industrial processes, such as fiber spinning, film extrusion and blowing and injection molding involve melting and deformation of the molten material and often crystallization under a high rate of extensional flow. Model films can be made under these conditions, and they can therefore be used to help understand the structure–property relations that result from this large class of industrial processes.

Early studies

Strained and unstrained films of natural rubber were examined by Andrews [79, 80] who showed that the spherulitic morphology of the unstrained films changes to a fibrillar morphology with crystalline units on the order of 6–25 nm wide. This is consistent with the work of Scott [18] who showed that strain-induced crystallinity results in fibrillar structures parallel to the strain direction. A further development of this idea by Keller and Machin [76] showed the morphology of melt extruded PE sheets to be an arrangement of lamellar crystals, normal to the stress direction, arranged in fibrillar units parallel to that stress. The concept of model film studies providing information related to commercially produced materials was described in this study.

Electron microscopy of drawn polypropylene (PP) films extended the range of these microfibrillar observations. Sakaoku and Peterlin [81] prepared thin films by evaporation on a water surface and then transferred the films onto Mylar. They then drew the Mylar films in a simple tensile deformation frame and the thin PP film was simultaneously deformed by the same amount. The PP films were etched by ion bombardment or with acid, for replication. Dark field imaging and electron diffraction showed microfibrils about 20 nm wide which had formed in micronecks in the original crystal lamellae. The stress field at the neck has a negative hydrostatic component (trying to increase the volume of the material), and in thin films longitudinal voids form with microfibrils bridging the gap, similar to the structures later found to exist in crazes (Section 4.8.4). The microfibrillar morphology of drawn materials, shown in the well known Peterlin model [6], has broad implications for the structures of both fibers and films. Tarin and Thomas [82] used gold decoration to show the deformation and transformation of a thin, spherulitic PE film, cast from decalin onto a hot mica sheet, into a microfibrillar morphology.

TEM examination of thin films and microtomed sections of iPS crystallized after deformation of the amorphous material [83] showed the formation of lamellar crystals growing perpendicular to the stress direction. They form as ordered stacks on long line nuclei, aligned parallel to the stress direction. SEM studies of PET films showed that a range of spherulitic morphologies could be induced by casting from trifluoroacetic acid. Differences in morphology resulting from different solvent evaporation rates were studied on cast and stretched films by polarizing optical microscopy, SEM and small angle light scattering [84]. Optical microscopy of model films has often been conducted using the hot stage and polarized light. This technique permits the determination of the growth habit and crystallization kinetics of polymer systems from thin films [85]. The disappearance of certain features as the temperature is raised can be correlated with melting peaks at the same temperature in DSC traces. This has been applied to oriented systems, particularly PE [86, 87]. There the fibrous nature of a high melting point component in materials grown from stirred solution or extruded at high pressure was confirmed and its concentration and distribution could be related to the prepara-

tion conditions. Thin PE samples which would be affected by adhering to the glass microscope slide or cover slip were embedded in a matrix of low melting point, e.g. low molecular weight PE.

Petermann and coworkers [88–94] conducted a series of model studies on the morphology and crystallization mechanisms of thin films produced by crystallization under a high rate of extensional flow. Two preparation methods were developed to produce oriented thin films (Section 4.1.3.6). An early method [88] involved drawing oriented fibers from molten film surfaces and annealing. Lamellar thickening occurred during annealing of PE [89]. A method still used to form high modulus thermoplastic films was developed by Petermann and Gohil [94] to produce highly oriented films by a longitudinal flow gradient which compares to the industrial use of high speed spinning processes. The method has been applied to the study of high density PE (HDPE) and isotactic PP. The structure of the oriented films is that of a stack of parallel lamellae on edge, which penetrate through the entire film. This makes the interpretation of transmission images particularly simple, as there is no overlapping of adjacent structural regions.

Recent studies

Thomas and coworkers investigated the structure of PE films by techniques that include TEM and STEM imaging. Yang and Thomas [95] studied the crystallization mechanisms and morphology of semicrystalline polymers by defocus imaging of oriented films [88, 93, 96] formed by the method described in the previous section. PE films were formed from solutions in xylene at temperatures between 122 and 130°C. Results showed that the melt–draw process does provide films containing highly oriented lamellar structure. Fig. 5.20 shows bright field defocus phase contrast electron micrographs of as drawn and annealed PE films [95]. The films consist of crystallites with the molecular axis aligned in the draw direction, as shown in the electron diffraction pattern inset. The bright field defocus phase contrast images consist of bright interlamellar

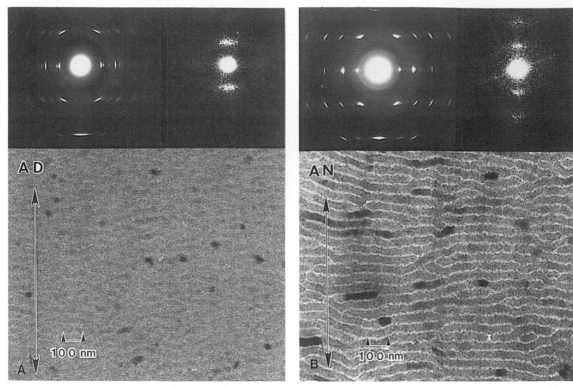

Fig. 5.20 Electron diffraction (top left) and defocus phase contrast micrographs of melt drawn polyethylene films. Optical diffraction patterns from the micrographs are at top right. As drawn film (A) is well oriented. Annealing (B) increases orientation and crystal size. Bright regions are interlamellar; the crystalline regions are gray or dark if they diffract. (From Yang and Thomas [95]; reproduced with permission.)

regions and gray crystalline lamellae. The as-drawn films (Fig. 5.20A) have shorter, less well oriented structures while annealing (Fig. 5.20B) causes an increase in crystallinity, orientation and lateral lamellar size. The long period determined by x-ray diffraction corresponds to the lamellar spacing observed in the electron microscope.

Petermann and Gohil [94] produced isotactic PP melts from *o*-xylene solutions at 130°C. Schultz *et al.* [97] and Schultz and Petermann [98] also used this method to study the oriented fibrous microstructure of annealed PP films crystallized from a highly extended melt. TEM was applied to the observation of the fibrillar to lamellar transformations. Finally, thin spherulitic films were formed [99] from a drop of a dilute PE solution in xylene placed on the surface of glycerol at 140°C. Following solvent evaporation,

the films were transferred to a Mettler FP-2 hot stage placed on an optical microscope, crystallized and quenched.

In the model studies discussed thus far, thin films were formed from solutions or melts, thus limiting preparation artifacts. Important model studies have also been conducted by methods more commonly employed to study the morphology of commercial materials. Two types of studies have been conducted on samples prepared by (1) production of microtomed and stained sections (Section 4.4) and (2) formation of replicas (Section 4.6) of acid etched (Section 4.5) materials. Both methods have been developed to the point where clear images of lamellar structures can be obtained. Bassett and co-workers have been responsible for developing the second method, and have reviewed its use for the study of crystallization and morphology in PE, other polyolefins and i-polystyrene [100–102]. For microtomy, polyolefins are stained first with chlorosulfonic acid [103], which makes them more rigid and easier to microtome [104–105]. Quantitative results for the crystal size distribution can be obtained [106–107] and compared to the results of SAXS or Raman spectroscopy [108]. These techniques show how the crystal morphology in melt cast films depends on the molecular weight, molecular weight distribution and the crystallization temperature.

5.2.2.2 *Amorphous films*

The discussion thus far has dealt with the morphology of semicrystalline polymers formed into films. Although it is not clear where a discussion of the morphology of amorphous or glassy polymer films should be found, especially in a text primarily on the *applications* of polymer microscopy, it is clear that this controversial topic must at least be summarized. The controversy centers around the interpretation of images of thin amorphous or glassy polymer films in the TEM. Some microscopists have taken the fine structure in such images to represent true structure in amorphous polymers, while others take them to be artifacts of preparation or imaging. Most might today take the second position, but there are papers [109, 110] that show that the issue is still open. Grubb [111] summarized the major arguments involved, making an important point that the cause for the disagreement might well be due to not distinguishing the different cases. In fact, the amorphous material discussed may be a crystallizable polymer, such as PET or polycarbonate, quenched to a glass and annealed below the glass transition temperature, or it may be an amorphous material which never shows crystalline order, such as atactic polystyrene or PMMA. Ordered regions in PET could possibly be explained by allowing that the TEM could detect incipient crystallization into nanometer size ordered regions, whereas bulk measurements,

such as x-ray diffraction linewidth, might show no change from the amorphous state. This however would not be a discovery of order in the amorphous state, and such an explanation could not hold for all atactic PS (aPS).

The earliest studies showing structures of some kind in amorphous films [112, 113] relied on observation of the film surface by replication, gold decoration and shadowing. In addition, Yeh and Geil [113] and later Yeh [114, 115] used transmission bright field images of very thin films of aPS, among other materials, and claimed to see regions of order 3–10 nm across, which were called 'nodules' and modeled as bundles of more or less parallel chains. Geil [116] summarized these studies, showing nodular structures by both surface and bulk preparation methods. The surface structures seen could not be misinterpretations of the image, but might be artifacts of preparation or real structures which are associated with the surface. In either case, again, the observation does not prove there is order in amorphous bulk material.

The fine structure seen in TEM of unstained films is much more difficult to interpret, and it has been suggested that it is merely random phase noise in the films made visible by a small defocus [111, 117, 118]. It is certainly true that modern high resolution microscopy, and for polymers this is high resolution, requires detailed descriptions of the microscope parameters, when such an image is obtained. An image from a model structure of some electron density fluctuation must be calculated, and only if it agrees with the experimental image will the model structure be taken seriously. This level of analysis does not exist for the amorphous polymer images. Polymers have the further problem that amorphous materials, such as unshadowed films, exhibit radiation damage [119] as do crystalline materials, and thus it is possible that structures exist, but that they may not be seen by microscopy [120]. Grubb [121] studied annealed isotactic polystyrene which contained small crystals and determined that radiation damage would make crystals smaller than 4 nm across undetectable by their diffraction.

Uhlmann [118] conducted electron microscopy studies of thin amorphous films and observed what he termed a typical 'pepper and salt' texture, characteristic of textures seen near the resolution limit in the electron microscope. For comparison Uhlmann and coworkers [118, 122] obtained small angle x-ray scattering (SAXS) data that are not consistent with a nodular texture in glassy polymers. The SAXS intensity measurements of glassy polymers such as PC, PMMA, PET, PVC and PS do not support such a domain structure. SAXS is a more suitable technique than TEM for detecting order, as a larger sample volume is statistically sampled.

Clearly, some authors do not interpret the textures observed for amorphous or glassy polymers as relating to any order and the SAXS data do not support the idea of an ordered structure in these materials. On the other hand, Geil and coworkers [109, 112, 113, 116] suggested that the nodular structures represent order in these amorphous or glassy polymers. The interpretation of the microstructures seen in amorphous glassy polymer films is clearly different in different laboratories. The issues are reviewed here, but no data supporting either view are fully described as this topic is beyond the scope of this book. This discussion is meant to draw attention to the issue of *interpretation* rather than drawing conclusions. Clearly, electron microscopy provides many useful observations; however, interpretation of the micrographs produced is nontrivial for structures near the resolution limit of the technique.

5.2.3 Industrial films

Industrial films of such chemical composition as polyethylene, polypropylene and polyester are manufactured for a wide range of applications. Accordingly, the morphology of these materials is studied to determine structure–property relations, to understand how to improve properties and also to control the quality of commercial products. Although model studies provide considerable detail relating to the structure, both before and after deformation of such films, model materials are generally thinner than commercial films, and thus the real product must also be evaluated. The types of preparation methods and instrumental techniques utilized closely parallel those described for polymer fibers. These techniques include: (1) measurement of birefringence, (2) measurement of crystallinity and orientation, (3) TEM of stained sections, (4) SEM of surfaces and bulk, and (5) SPM of surfaces.

Where the film is a coating, an added dimension to the study is the adhesion between the film and the substrate. Some industrial films have porous textures that are associated with the broad field of separation technology. These porous materials may be termed films or membranes and they will be discussed separately below.

5.2.3.1 *Optical microscopy*

The refractive index and birefringence of films can be measured in the optical microscope, which also allows details of texture in semicrystalline films to be resolved at the 0.2 μm level. Birefringence is the more common technique as it permits measurement of the molecular orientation. Refractive index can be used to help identify an unknown material, and a technique has been described to determine the density from the refractive index. Density relates to crystallinity; however, x-ray diffraction, heat of fusion measurements and direct density determination are all more common ways of obtaining measures of crystallinity. A biaxially oriented object has three refractive indices along its three axes (see Fig. 3.10). When the term 'the birefringence of a film' is used loosely, it normally refers to the difference between the refractive index in the machine direction and that in the transverse direction, as observed by viewing the easy way, perpendicular to the film plane. There will normally be a larger difference between the refractive indices in the machine direction and in the perpendicular direction, but for this to be seen directly a view in the film plane, along the transverse direction, would be required.

A major topic of interest relating to film

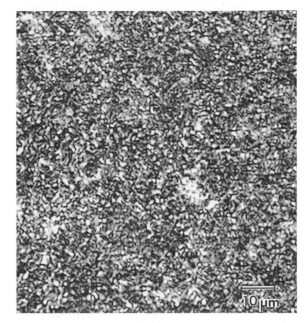

Fig. 5.21 A fine spherulitic texture is observed in this optical micrograph of a polyester film cross section taken in polarized light.

structure is the effect of crystallinity on the deformation mechanism. The optical properties of biaxially oriented films were studied in 1957 by Stein [123] who determined the full set of birefringences, by measuring the optical retardation as a function of the tilt of a PS film. Samuels [124] used complementary techniques of x-ray scattering, TEM of surface replicas and birefringence measurement in a study of the microstructure and deformation of isotactic polypropylene films. The familiar theme of deformation of spherulites to a fibrillar structure was again observed in this study. The application of refractive index measurements for anisotropic films has been described for the evaluation of film properties and processing variables. Samuels [125] described methods for determining the percent crystallinity, birefringence and refractive index distribution for commercial films. Bottle films have barrier characteristics related to the draw ratio used in processing. Paulos and Thomas [126] studied the effect of orientation on the structure and transport properties of a high

density blown polyethylene film. Birefringence and crystallinity measurements revealed that the decrease in transport properties, and thus enhanced barrier properties, was related more to the high level of orientation than to crystallinity. An example of the spherulitic texture observed by polarized light microscopy of a thin cross section of a polyester film is shown in Fig. 5.21. This texture is related to the crystallinity of the polymer film, and a range and distribution of spherulite sizes can be related to both process variables and applications.

5.2.3.2 *Electron microscopy*

The SEM is often quite useful for the observation of the surface structure of films. In order to evaluate the initial and deformed morphologies, Sherman [127] studied plastic deformation and tearing in high density PE blown films with varied molecular weight and melt index. High resolution SEM studies, by Tagawa and Ogura [128], have directly shown the lamellae in a blown PE film (Fig. 5.22) prepared by drawing in either the machine direction or at right angles to the machine direction. Most commercial films are

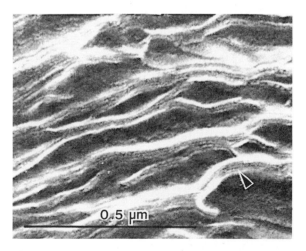

Fig. 5.22 Direct imaging of lamellae is shown in this enlarged SEM micrograph of a blown polyethylene film surface. The arrow shows a single lamella. (From Tagawa and Ogura [128]; reproduced with permission.)

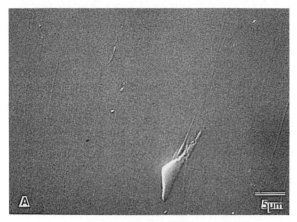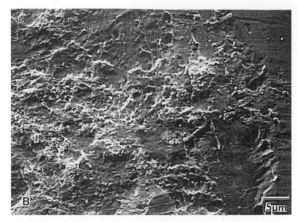

Fig. 5.23 The surfaces of cellulose acetate films containing fillers are shown in these SEM images where (A) is smooth and (B) is abraded in texture.

flat and smooth and the surfaces have little structure or topography present. SEM of cellulose acetate film surfaces reveals differences that relate to the fillers. Figure 5.23A shows a film surface that is very smooth in texture due to fine filler particles. After use, the film surface appears abraded (Fig. 5.23B) and nonuniform in texture. High tilt angles enhance the imaging of fine or shallow detail of film surfaces [129]. In addition, the nature of the fillers or contaminants on the surface may be determined by x-ray analysis.

Complementary microscopic techniques are useful in the elucidation of polymer film microstructures. Optical techniques provide information relating to the orientation and crystallinity, while SEM can be used for surface detail relevant to end uses. TEM techniques, similar to those used in model film studies and in fibers, are useful in describing the internal structures, especially of spherulites and their deformed counterparts, microfibrils. TEM studies of films and fibers continue to provide fundamental observations relating the structure to properties and applications.

5.2.3.3 *Scanning probe microscopies*

The surface textures of films examined by SEM are often uninformative, because small changes in height of the film surface do not give rise to significant image contrast. Scanning probe microscopy (SPM) is quite different. In these techniques the height of the film surface can be the primary output signal. The height resolution is extremely good, so that an accurate exaggerated relief map of the surface can be produced. The magnification perpendicular to the plane of the film can be ten or a hundred times the magnification in the plane of the film, and this brings out surface structure not otherwise visible. Atomic force microscopy can be applied to thin polymer films and to the surfaces of conventional films. Scanning tunneling microscopy requires that the film be extremely thin, be conducting or have a conducting coating applied to it. Many studies have been made of very thin or monolayer films of organic materials deposited on a flat substrate, which has commonly been HOPG. These films may be deposited directly, or made using Langmuir–Blodgett (LB) techniques in a trough. For example, Albrecht *et al.* [130] used LB methods to prepare extremely thin films of PMMA on graphite and characterized them by STM and AFM. These authors were interested in nanometer scale fabrication and information recording; they found that a voltage pulse applied to the STM tip caused a local modification of polymer fibrils.

In another early study, Marti *et al.* [131] prepared ordered ultrathin polymer films from monolayer LB films of the monomer that were polymerized by UV radiation. Samples of these films were transferred to collodion coated microscope grids for TEM. Microcrystalline domains of the same images showed parallel lines with spacing about 0.5 nm, taken to be rows of molecules. The films studied in the AFM need not be ordered or even solid. Mate *et al.* [132] have imaged liquid films of perfluorinated polyoxypropylene ether in the AFM. The thickness of films as thin as 2 nm could be measured, along with the force on the AFM due to wetting and the topography of the liquid/air interface.

Another example of the use of AFM on very thin films is the imaging of thin layers of poly(tetrafluoroethylene) (PTFE) that are deposited on glass simply by rubbing it with the solid polymer. Electron diffraction has shown that these films are highly ordered and very well oriented. AFM provides direct measurement of the thickness and continuity of these films [133]. Individual fibrils that were not attached to the substrate had different appearances when scanning in different directions. This is because the AFM tip pushes them around during scanning. Arrays of parallel rods with the intermolecular spacing of PTFE crystals are seen. The authors also claim to distinguish the helical structure of the individual molecules, and compare it to models derived from electron diffraction [133].

Two recent reviews provide further details of atomic force and scanning tunneling microscopy of organic surfaces [134] and thin films [135]. These scanning probe techniques must continue to be compared with more conventional methods in order to be able to interpret the image and to ensure that the sizes measured on structural details have not been modified, such as by the tip in AFM. Topics such as imaging of individual chemisorbed molecules, supported physisorbed molecular assemblies, biopolymers and bulk surfaces of polymers are shown imaged under vacuum, fluids and air and the local surface modifications imposed by the tip were described [134]. Hues *et al.* [135] reviewed many instrumental issues in regard to the imaging of thin films by AFM, describing principles of the

technique in some detail for the interested reader.

5.2.3.4 *Problem solving applications*

Tsukruk *et al.* [136, 137] conducted studies using SPM of organic and polymeric films, from self-assembled monolayers to composite molecular multilayers. Aspects of these films are described, including the surface morphology, surface defects and molecular scale ordering. Surface modification to the materials during scanning with the AFM tip are also considered. Topics such as measurements of the forces between surfaces, surface stability, wear, adhesion and elasticity are studied by AFM. Tsukruk [136] discussed molecular ordering, phase transformations in monolayer molecular films, fibrillar surface textures of polymers, such as PE, cellulose, aramids, polyimide fibers, latex dispersions and polymer blends. The stability and modification of polyglutamate LB bilayer films in the AFM were also discussed [137]. The possibilities for surface modification by the AFM tip were explored as holes were fabricated or written into the film surface. Bilayers were deposited on polished substrates cut from silicon wafers and AFM images were obtained. In this case the data on average thickness and macroscopic roughness was also shown by x-ray reflectivity measurements for comparison with AFM. The ability to use the AFM tip for lithography and also the potentially adverse effects of AFM imaging have been discussed [138]. Issues raised in a recent paper on biological molecules include some of the pitfalls of high resolution imaging in which authors are very 'optimistic' and overlook 'impurities, statistical significance and sensible physical mechanisms' for work with SPM [138]. The microscopist beware!

The value of low voltage SEM is shown, for example in Fig 5.24 of a PE biaxially blown film [139]. The film was cut, pulled until it necked and imaged uncoated at 800 eV. Fig 5.24A is an undeformed region showing lamellae. Fig. 5.24B is an image of the transition toward deformation.

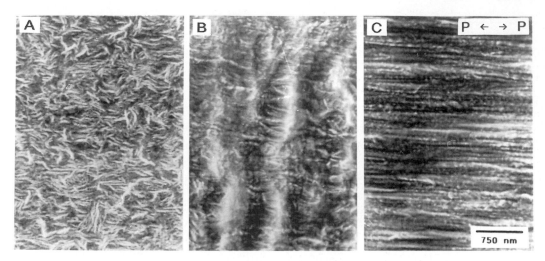

Fig. 5.24 Low voltage SEM of an uncoated PE biaxially blown film, pulled until it necked and imaged in an undeformed region showing lamellar deformation (B) and in a region showing necking (C). (Reproduced with permission, T. Reilly, unpublished [139]).

Fig. 5.24C is necked region with fibrils parallel to the applied load (P is direction of applied load).

Recently, application of STM and AFM imaging of lamellar structures in melt extruded polyethylene films was shown by Chen *et al.* with FESEM, birefringence and x-ray scattering measurements [141, 142]. Polyethylene and polypropylene extruded films with row lamellar structures have been utilized to produce microporous membranes, as will be described in more detail in the next section. It is important to understand the structure of such films as this is a precursor that controls, in part, the formation of the membrane structure. The direct visualization of the deformation processes in polyethylene has also been shown by high resolution electron microscopy of thin films [140]. Adams *et al.* [140] used a scanning transmission electron microscope to study high density polyethylene, formed by a melt drawing process and subsequently deformed at room temperature. That work shows the cavitation and formation of microfibers from the lamellae during deformation and the formation of a fibrillar morphology under higher deformation.

Birefringence measurements, using optical microscopy of the melt extruded films, shows that improved film orientation can be achieved by one of several methods, annealing, extruding at high speed or by the use of high molecular weight polymers. Imaging by the various scanning methods all clearly reveal the lamellar structures in the PE and PP films. X-ray scattering relates the increase of molecular alignment to changes in lamellar perfection and lamellar alignment during annealing. These techniques provide a means to establish structure–process–property relationships for the manufacture of microporous membranes [141, 142]. The materials investigated are film precursors of Celgard polyethylene flat sheet microporous membranes. The examples shown are all annealed and processed at high extrusion rate. Only the molecular weight is varied. Samples for FESEM were sputtered with 2 nm Pt and imaged at c. 5 kV. Figure 5.25A is an FESEM image showing fine lamellar structures in a lower molecular weight, annealed PE film precursor. The STM samples were mounted on silicon substrates and coated with 5 nm Pt using ion beam sputtering (see Section 4.7.3). The STM was a Nanoscope III (Digital Instruments) operated with a bias voltage of 100 mV and a tunneling current of 1 nA. Figure 5.25B, an STM image of a low molecular

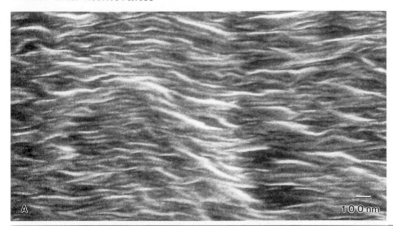

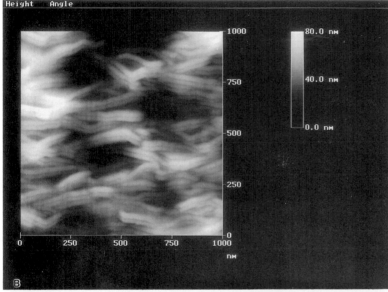

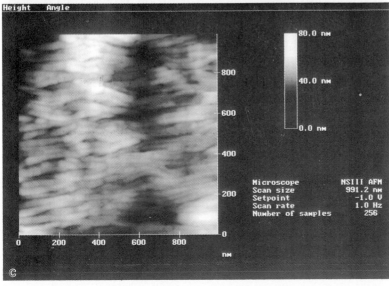

Fig. 5.25 Film precursors of polyethylene flat sheet microporous membranes, all produced using a high extrusion rate, and annealed, but with different molecular weights. An FESEM image taken at *c.* 5 kV (A) shows fine lamellar structures in a lower molecular weight, annealed PE film. The STM image (B), taken using a bias voltage of 100 mV and a tunneling current of 1 nA, of a low molecular weight, annealed PE film precursor reveals three dimensional lamellar structures similar overall to the textures in the FESEM image. An AFM contact image (C), taken using a long range scanner with a pyramidal Si_3N_4 tip, with the force monitored using a laser beam and a position sensitive photo-detector, of a higher molecular weight PE film, reveals similar detail as was shown by FESEM and STM. (From Chen *et al.* [141, 142]; reproduced with permission.)

weight, annealed PE film precursor illustrates the lamellar structures and spacings which are similar to those seen in the FESEM image. Finally, for AFM, imaging was performed on a Nanoscope III contact AFM (Digital Instruments) using a long range scanner with a pyramidal Si_3N_4 tip [141, 142]. In this instrument the force is monitored with a laser beam reflected from the cantilever supporting the tip and a position sensitive detector (Section 6.4). Figure 5.25C is an AFM image of a higher molecular weight sample, revealing similar detail to that shown by FESEM and STM. The study showed the effect of polymer molecular weight, film extrusion rate and annealing on lamellar textures was found in precursor films. The resulting microporous structures were shown to be a direct reflection of these precursor effects.

5.2.4 Flat film membranes

The technology of membrane separations is a new and growing field where the polymer membrane contributes unique separation properties based on its structure and, to some extent, on its chemical composition. Various manufacturing processes are used to create special structures in forms such as flat films and hollow fibers. Lonsdale [143] provides a review of the history and current status of separation media and their applications, and a text [144] provides a discussion of the materials science of synthetic membranes.

Conventional filters provide separation of particles in the range of 10 to 1000 μm. Microfilters, ultrafilters and reverse osmosis membranes provide separation on the scale from 1 μm down to several nanometers, as shown in Table 5.2. In order to provide such varied separation properties, the pore size, shape and distribution are significantly different in these membranes. Microfiltration involves the passage of water and dissolved materials while retaining micrometer sized suspended materials using homogeneous membranes. *Microporous membranes* can meet many separation demands when the pore size range is 0.05–1 μm, providing a range of applications [145]. *Ultrafilters* are surface permeable, passing water and salts while retaining macromolecular sized particles. The relation between the surface pores and flux for ultrafilters has been described [146]. *Reverse osmosis* (RO) filters pass water and retains both dissolved and particulate materials in the ionic size range. RO membranes are generally asymmetric, and anisotropic; that is, they have a gradient of pores ranging from a dense surface layer to a porous substructure, which provides mechanical strength, and they only work in one direction. *Composite membranes* are also asymmetric with a thin surface film sprayed or coated onto a porous substructure, supported by a synthetic fabric.

Electron microscopy has been applied to the determination of the structure of membranes for correlation with transport properties. The SEM provides both the best overall view and detailed,

Table 5.2 Membrane separation processes

Process	Filtration	Materials retained	Pore sizes
Conventional filtration	Coarse filters	Large particles	>2 (10–1000) μm
Microfiltration	Microporous membranes	Suspended matter	0.1–20 μm
Ultrafiltration	UF membrane	Macromolecules, colloids (passes salts)	0.01–0.5 μm
Reverse osmosis	Semipermeable membrane	Dissolved and suspended materials (ionic) saline	1–80 nm
Gas separation	Semipermeable membrane	Gases and vapors	0.2–1.5 nm

three dimensional images of these structures [147]. Optical microscopy can give a rapid overview without the possibility of a change in structure caused by an electron beam or a vacuum. Essentially, three structural types of membranes have been described: homogeneous, asymmetric and composite. In the homogeneous membrane, the pore structures are uniform throughout the cross section, whereas asymmetric membranes exhibit a pore gradient from a dense surface layer to macrovoids. Polycarbonate and polyacrylonitrile are examples of homogeneous membranes, while polybenzimidazole (PBI) and cellulose acetate (CA) are examples of asymmetric membranes. Composite membranes have a dense surface layer with a support structure composed of another polymer, often polysulfone.

Recently, FESEM and AFM have been used to image surface topography of membranes. Imaging with FESEM at low voltages does not require a metal coating, which might fill very fine pores if a thick layer is used. Low voltage is important to minimize beam damage of fine topography, again as the finer pores might be easily damaged or filled with carbon prior to the operator even being aware it has taken place. Comparison of AFM with SEM is most useful to ensure that interpretation of the AFM images is accurate. AFM does not require that the membrane be dry nor is the sample placed in a high vacuum system, although this is required for SEM. Environmental or high pressure SEMs permit hydrated materials to be imaged. However, this technique is not very useful at very high resolution, so its value depends on the level of detail that is of interest. Fritsche *et al.* [148, 149] have studied the structure of polyethersulfone ultrafiltration membranes using SEM and AFM and have compared the results. Although they concluded that size differences in the topography imaged were due to the metal coating and vacuum used in the SEM, they prepared the samples by different methods. The SEM samples were dried, frozen in liquid nitrogen and fractured, followed by Au coating and imaging at 80,000× using a 25 kV electron beam. The SEM

images clearly showed cracked metal coating, not mentioned by the authors, most likely due to a very thick Au layer. Of course the textures were significantly different in size compared to the textures observed in the AFM images of hydrated, uncoated samples. For such comparisons to be valid the specimen preparation must be similar and the conditions used should not produce artifacts.

5.2.4.1 *SEM – surface and bulk*

The SEM is used for the study of the surface and bulk structures of membranes. Membranes are prepared by attaching them to the specimen stub and applying a conductive surface coating. Bulk structures are observed for membranes fractured in air or liquid nitrogen, sectioned or critical point dried. Figures 5.26 and 5.27 are examples of the varied structures of typical membranes which can be imaged by this technique. An experimental HDPE microporous membrane has pores elongated in the draw direction (Fig. 5.26A) where the pores are less than 1 μm wide and have a range of lengths to about 2 μm. The pore volume is formed by stretching lamellae; the remaining unstretched lamellae are seen as flatter regions, perpendicular to the draw direction. Another example (Fig. 5.26B) is a membrane with a large, rounded and stretched porous network. A nucleopore membrane (Fig. 5.26C) has very discrete and rounded pores etched randomly into the film surface. This structure is similar to the polycarbonate nucleopore films [143] which have circular pores of constant cross section that run from the top to the bottom of the membrane. A cast membrane (Fig. 5.26D) has an open three dimensional network structure which appears formed by polymer in coated, particulate strings.

Polysulfone composite membranes provide a different chemical composition and structure to some of the examples shown. A polysulfone composite membrane is shown by SEM of cross sections (Fig. 5.27A and B) and of the top surface (Fig. 5.27C). A porous texture is seen (Fig. 5.27A) with larger macrovoids near the bottom surface. There is an open porous structure with a pore

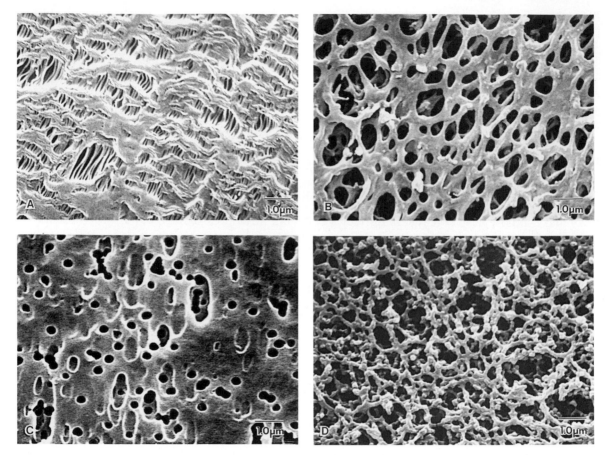

Fig. 5.26 SEM micrographs of several membrane surfaces reveal a range of pore structures that in turn result in a range of separation applications. An experimental, microporous, polyethylene membrane is shown (A) with elongated, stretched porous regions of various sizes, separated by fibrils, in the draw direction, and unstretched lamellae normal to the draw direction. This surface structure is quite different from three commercial membranes (B–D). One membrane (B) consists of a low density network of rounded pores, many of which are larger than 1 μm across. A nucleopore membrane (C) has more defined pore structure with rounded pores bored through from one side to the other. The morphology in (D) is an open network structure with the polymer in the form of strings of particles.

gradient, with smaller pores nearer the dense top surface (Fig. 5.27B). The asymmetric membrane has very fine surface pores, about 0.05–0.2 μm across (Fig. 5.27C) with an underlying open network composed of strings of polymer. The surfaces of composite membranes are generally dense, and SEM micrographs may not reveal any resolvable surface pores. Chemical etching of this dense surface layer is useful to observe the porous substructure (Fig. 5.27D).

5.2.4.2 *Reverse osmosis membranes*

Cellulose nitrate and cellulose acetate (CA) were among the first asymmetric, reverse osmosis membranes to be produced [150]. Plummer *et al.* [151] described 13 specimen preparation methods for the observation of CA membrane structures. They pointed out the lack of contrast in epoxy embedded sections and that one of the best stains, osmium tetroxide, reacts with the

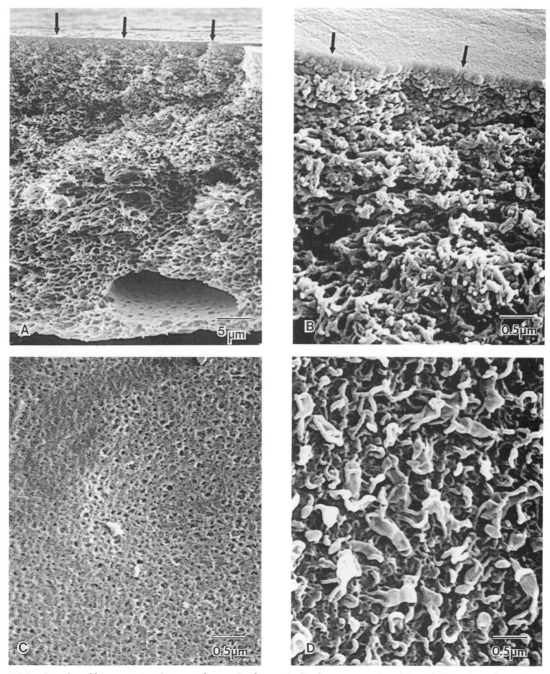

Fig. 5.27 A polysulfone composite membrane is shown in both cross section (A and B) and surface views (C). SEM images reveal large macrovoids on the bottom side of the membrane within a porous texture support layer. A dense surface layer (arrow) appears to be composed of granular particles of polymer with little pore volume. Some surface porosity is seen (C) but these pores are considerably smaller than those observed in the bulk of the membrane. Chemical etching of the top surface results in removal of the active surface (D), which gives another view of the bulk porous morphology.

polymer. Freeze fractured membranes were found by these authors to be of questionable value. In our experience, if care is taken, SEM study of fractured membranes can provide an informative view of the structure even though some structures collapse, and their sizes cannot be accurately determined. A method found acceptable was ultrathin sectioning of gelatin embedded wet membranes (TEM). The structure of CA membranes was shown by replication [152] and SEM [153].

Optical, scanning and transmission electron micrographs of a commercial cellulose acetate asymmetric membrane are shown in Fig. 5.28.

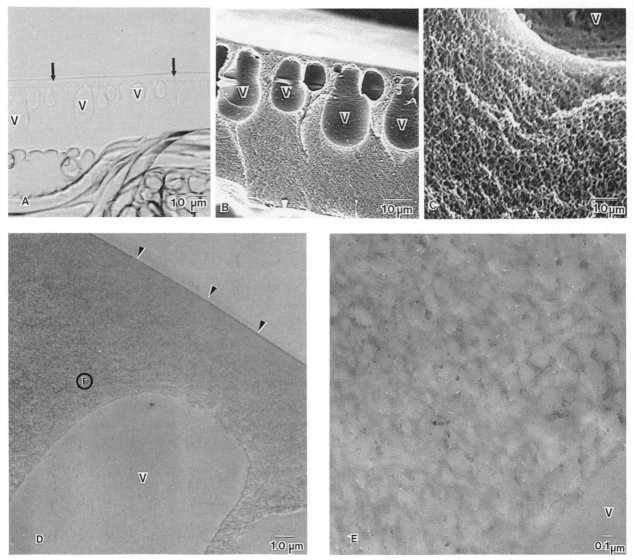

Fig. 5.28 Cellulose acetate membrane structures are shown by complementary techniques. The optical micrograph (A) shows an overview of the membrane, cast on a woven fabric support (bottom). A surface layer (arrows) is observed above large, rounded macrovoids (V). SEM cross sections reveal these macrovoids in more detail (B) and also show the nature of the fine pores (C). A TEM micrograph (D) of a section near the surface (arrows) reveals a dense layer, with a porous microstructure, shown more clearly at higher magnification (E).

Each view provides a different perspective on the membrane structure while, together, they give the complete structural model. Specimen preparation for OM and TEM cross sections was by microtomy of embedded membrane strips using a method developed to limit structural collapse (Section 4.3.4). An optical micrograph (Fig. 5.28A) shows the membrane cast on a woven support fabric. The active surface layer (top)

appears as a 'skin' several micrometers thick, with a support structure of rounded macrovoids. These micrographs are useful for assessing the macrovoid size and the membrane thickness and uniformity. SEM images (Fig. 5.28B and C) provide higher magnification views of a membrane cross section formed by fracturing in liquid nitrogen. The membrane is seen as a network of submicrometer sized pores (Fig. 5.28C). TEM

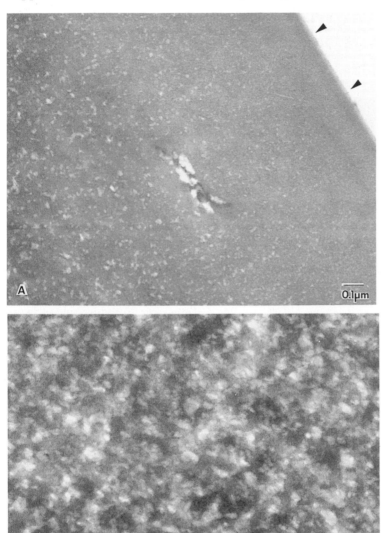

Fig. 5.29 The fine structure of a PBI asymmetric membrane is shown in TEM micrographs of cross sections. A dense surface layer (arrows) is observed in a micrograph (A) taken with the high brightness lanthanum hexaboride gun which shows no pores are resolved in the top 50 nm of the dense surface layer. Pores on the order of about 0.05 μm are clearly shown (B) within the membrane support structure.

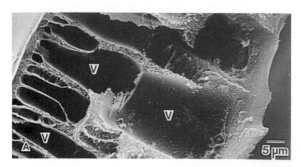
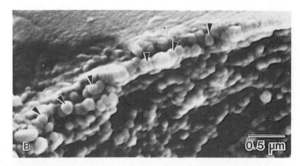

Fig. 5.30 High resolution SEM images of a critical point dried and fractured PBI membrane reveal the fine structure quite clearly. The overview micrograph (A) shows the macrovoids (V) and the porous walls within the membrane. The robustness of the macrovoids suggests that the method is useful for observation of the *in situ* structure. The dense surface layer is composed of spherical particles (arrows) that are deformed into a dense monolayer while the support structure below is formed by a more open network of these spherical particles (B).

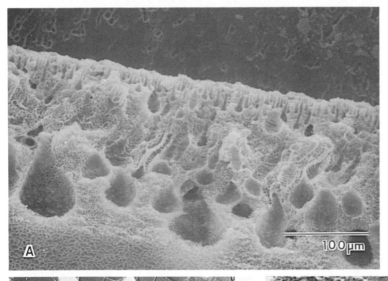

Fig. 5.31 An overview (A) of the asymmetric structure of a 'high nitrile resin' membrane reveals a porous substructure and a dense surface layer. TEM micrographs of sections of a surfactant treated and stained membrane reveal more details of the pores adjacent to the dense layer (B) and within the dense layer connecting to the outer surface (C). (From Vogele-Kliewer [159]; reproduced with permission.)

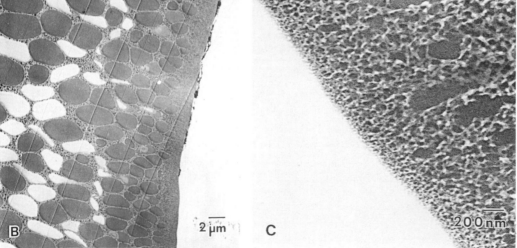

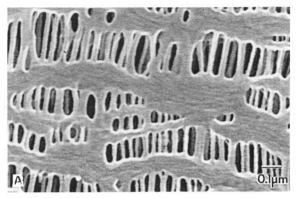 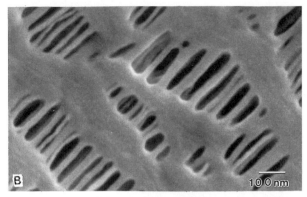

Fig. 5.32 Celgard 2400 microporous membrane surfaces are shown in SEM images taken using a LaB$_6$ gun (A), and at 5 kV accelerating voltage with a field emission gun (B). The fibrils, drawn in the machine direction define the pore boundaries. Rows of pores are seen across and into the membrane. (From M. Jamieson, unpublished [162].)

micrographs (Fig. 5.28D and E) show greater detail of the microstructure, although there is little contrast between the polymer and the epoxy embedding media. Smaller pores, not clearly resolved, are observed near the membrane surface (arrows) and larger pores are seen deeper within the asymmetric membrane (Fig. 5.28E). Resolution of the finest pores is limited due to the section thickness.

Asymmetric polybenzimidazole membranes have been developed for RO applications, in the form of hollow fibers [154] and flat film membranes [155] for water transport. By comparison with cellulose acetate, PBI has very attractive chemical, flammability and thermal properties. There are two problems encountered in attempting the preparation of such membranes for TEM: (1) deformation during drying and (2) lack of contrast. Often specific methods must be developed for each membrane type, although method development is quite time consuming.

A general method was developed [156] to limit drying deformation by directly embedding the wet membrane and removing water during resin infiltration (Section 4.3.4). TEM micrographs taken with a lanthanum hexaboride high brightness gun, for enhanced resolution, show pores less than 5 nm, but no pores are resolved in the top 50 nm surface layer (Fig. 5.29A) of a PBI membrane. Within the bulk membrane (Fig.

5.29B) there are much larger pores (about 50 nm). SEM images of a fractured, critical point dried membrane (Fig. 5.30) show robust, macrovoid structures. The top dense surface layer is clearly composed of a monolayer of densely packed, deformed particles, about 80 nm in diameter, packed so closely as to limit surface porosity. Less well packed particles form the more open bulk membrane texture. The structure shown confirms those hypothesized from earlier TEM replica studies of wet poly(amide–hydrazine) and dry polyimide asymmetric membranes [157, 158].

An experimental ultrafiltration membrane, identified as a high nitrile resin [159, 160], was prepared and examined by the technique of infiltration and post-staining of a surfactant [156]. An SEM image shows the structure (Fig. 5.31A) is porous with a thin, dense, surface layer. Surfactant filled and stained membrane sections are shown by TEM (Fig. 5.31B and C) which reveals the nature of the asymmetric pore structure with smaller pores near the surface active layer. The nature of the porous substructure within the dense layer connecting to the outer surface is also shown (Fig. 5.31C).

5.2.4.3 *Microporous membranes*

A method using staining and ultramicrotomy (Section 4.4.2) has been demonstrated that shows

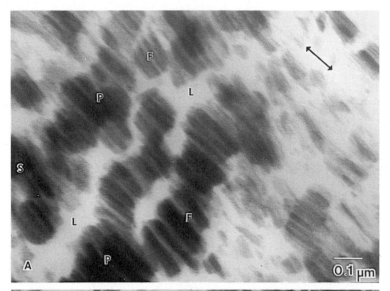

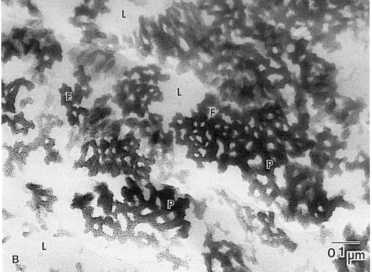

Fig. 5.33 TEM micrographs are shown of a surfactant treated and osmium tetroxide stained Celgard 2400 membrane cut along the three dimensions of the membrane: along the machine direction, across the machine direction and in the plane of the membrane along the face. A section cut along the machine direction (A) reveals fibrils (F) separating electron dense pores (P), filled with stained surfactant, arranged in rows elongated in the machine direction (arrow). Unstained lamellae (L) are white regions between these pores. The cross axial section (B) shows that these pores are arranged in networks that do not run straight across the film, but have a tortuous path.

the three dimensional structure of microporous membranes such as Celgard 2500 membrane (trademark, Hoechst Celanese Corp. [160]). Celgard is formed by film extrusion, annealing and stretching isotactic polypropylene. This produces an oriented crystalline structure with parallel arrays of pores [161]. The surface view of Celgard 2400 is shown by SEM (Fig. 5.32A) where rows of pores are aligned parallel to the machine direction. High resolution FESEM (Fig. 5.32B) reveals the drawn fibrils separating regions of undrawn

crystalline lamellae and defining the pore volume. The surface structure of Celgard 2400 has much smaller pores than Celgard 2500, but the overall three dimensional structures are similar.

TEM sections of Celgard 2400, prepared along the three axes, are shown in Fig. 5.33. Ultrathin sections cut along the axis, in the longitudinal direction (Fig. 5.33A), show the pores are oriented parallel to the machine direction. Dense regions are surfactant stained, nonporous regions are white, and gray regions result from the effect

Top Surface

TD   MD

Fig. 5.34 A three dimensional model of Celgard 2500 (trademark of Hoechst Celanese Corp.) is shown composed of sections cut along, across and in the plane of the machine direction. The surface is shown by an SEM micrograph. (From Sarada *et al.* [160]; reproduced with permission.)

of the section thickness [160]. The fibrils separating the pores do stain but not as much as the surfactant filled pores. The cross section (Fig. 5.33B) is composed of a network of pores, with little order, in agreement with the axial view. The TEM micrographs clearly show short parallel rows of pore channels, separated by unstretched lamellae and defined by the drawn fibrils. There is a random, tortuous, pore volume that provides unique microporous membrane applications. These micrographs could be combined to formulate a three dimensional model as was done earlier for Celgard 2500 (Fig. 5.34).

FESEM has provided increased information in the case of several experimental membranes [162] with very different pore structures. In the first case, the surface of a PTFE membrane exhibits a non-uniform series of rounded pores, as shown in Fig. 5.35A. At much higher magnification, FESEM images reveal the three dimensional nature of the pores as they extend into the bulk membrane (Fig. 5.35B). The surface of the membrane is seen to be wrinkled in texture but various attempts to view the image at lower magnifications, after high magnification imaging, did not reveal picture frame contrast that would

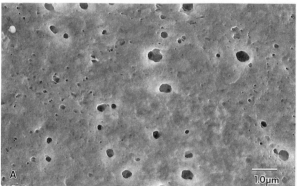
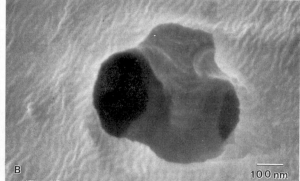

Fig. 5.35 The surface of a PTFE membrane imaged at 5 kV in an FESEM exhibits a non-uniform series of rounded pores, and a three dimensional nature as the pores extend into the bulk membrane. The surface of the membrane appears wrinkled in texture but imaging at lower magnifications, after high magnification imaging, did not reveal picture frame contrast that would suggest this texture is due to beam damage. (From M. Jamieson, unpublished [162].)

suggest that this texture is due to beam damage. In a final example [162] a two phase system was processed into a film which was fractured and imaged. Details of the texture shown in the FESEM images in Fig. 5.36, reveal very coarse, irregularly shaped pores and fine dispersed second phase particles.

5.2.5 Hollow fiber membranes

Hollow fiber membranes are treated in this section with the discussion of porous materials,

as they have more in common here than with fibers discussed earlier. The structures of hollow fiber membranes are somewhat analogous to those described for flat film or sheet membranes produced from similar polymers. Hollow fiber membranes [143, 163] have been produced from CA, PP, PBI, polysulfone, aromatic polyamides and polyacrylonitrile. Polysulfone hollow fibers were described [164] using an SEM study which showed a dense skin formed on the surface with a porous or spongy subsurface support structure. The fibers were prepared for SEM by breaking at

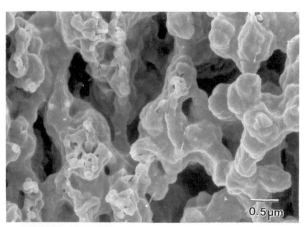
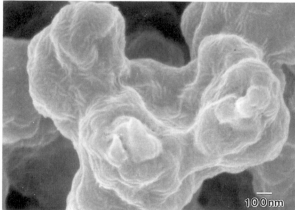

Fig. 5.36 FESEM images of a two phase polymer system, processed into a film which was fractured and imaged, reveals coarse, irregularly shaped pores and fine dispersed phase particles. (From M. Jamieson, unpublished [162].)

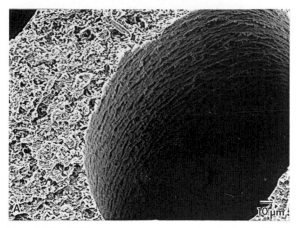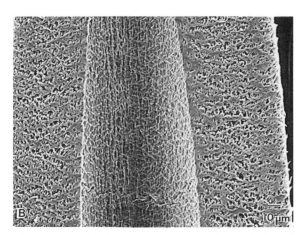

Fig. 5.37 Hollow microporous polyethylene fibers are shown in SEM images of cross sectional (A) and longitudinal (B) views which permit assessment of the structures.

liquid nitrogen temperature. The porosity of the hollow fibers was shown to be complex and asymmetric. Cabasso and Tamvakis [165] described composite hollow fiber membranes in some detail, showing the surface structure was composed of a polysulfone porous substrate coated with crosslinked polyethyleneimine or furan resin. A dense, semipermeable layer on top of the porous substructure is responsible for high salt rejection in this asymmetric RO membrane.

An example of a PE hollow fiber membrane will be described briefly. A hollow, microporous PE fiber is shown in the SEM micrographs in Fig. 5.37. The fibers were prepared by fracturing in liquid nitrogen to show the bulk cross sectional structure (Fig. 5.37A), and a cold razor blade was used to fracture the fiber for the longitudinal view (Fig. 5.37B). Combination of both views shows the dimensions and the porous structure.

5.3 ENGINEERING RESINS AND PLASTICS

5.3.1 Introduction

It is well known that the microstructure of engineering resins and plastics is determined by the manufacturing process. Thus resins and plastics produced by such processes as injection and compression molding and extrusion are

evaluated by microscopy techniques in order to determine their structure and to provide an understanding of structure–property relations. The range of polymers that are considered engineering resins and plastics is very large, and their applications are even broader. A listing of common commercial plastics and resins is provided in Appendix IV with trademarks and some applications. Information such as processing data, mechanical properties and other detailed discussions can be found in texts edited by Baijal [166], Margolis [167], Paul and Sperling [168] and Paul, Barlow and Keskkula [169].

5.3.1.1 *Resins and plastics*

Polymers in the category of engineering resins and plastics may be classified in several ways. They may be thermoplastic or thermoset. They may be crystalline or amorphous and they may be single phase or multiphase systems. This would allow for eight types of materials except that thermosets, because of their irregular crosslinked structure, are never crystalline. *Single phase* polymers do not have discernable second phase structures of different chemical composition. Thus homopolymers and random copolymers are single phase polymers, even if they are semicrystalline and so contain amorphous and crystalline regions. Blends of the few pairs of polymers that mix well (miscible blends) can also produce single phase material. The single phase polymers have a wide range of mechanical properties and morphologies, which depend on their specific characteristics, such as their flow behavior and their melting point, if they are crystalline. Examples of single phase thermoplastics are PE, iPP, PA and POM, which are all crystalline, and PS, PMMA, PVAC and PS–PPO blends which are all amorphous. Epoxies, unsaturated polyesters and phenolformaldehydes are examples of single phase materials which are amorphous thermosets.

Much information can be obtained by microscopy of crystalline thermoplastics, whereas microstructural study of single phase amorphous materials is not usually of much practical interest. This is why most microscopy studies of single phase polymers relate to crystalline materials, and amorphous polymers are mostly described in multiphase systems.

Multiphase or multicomponent polymers can clearly be more complex structurally than single phase materials, for there is the distribution of the various phases to describe as well as their internal structure. Most polymer blends, block and graft copolymers and interpenetrating networks are multiphase systems. A major commercial set of multiphase polymer systems are the toughened, high impact or impact modified polymers. These are combinations of polymers with dispersed *elastomer* (rubber) particles in a continuous matrix. Most commonly the matrix is a glassy amorphous thermoplastic, but it can also be crystalline or a thermoset. The impact modified materials may be blends, block or graft copolymers or even all of these at once.

As may be guessed from the names for these systems, the rubber particles are added to improve the mechanical properties of the matrix material, particularly to improve their low impact strength. The size of the rubber particles, their distribution, composition and compatability with the matrix all influence the mechanical properties of the final engineering resin. Typical multiphase polymers which include elastomers are:

(1) high impact polystyrene (HIPS);
(2) acrylonitrile–butadiene–styrene (ABS);
(3) poly(styrene–acrylonitrile) (SAN);
(4) acrylonitrile–chlorinated poly(ethylene–styrene) (ACS);
(5) poly(styrene–butadiene–styrene) (SBS);
(6) ethylene–propylene terpolymer (EPDM).

5.3.1.2 *Characterization*

A wide range of microscopy techniques are applied to the characterization of engineering resins and plastics. For example, crystalline polymers are viewed by polarized light microscopy to reveal the size and distribution of spherulites and the nature of the local orientation. Surface details, such as wear and abrasion are best viewed by SEM. For example, Vaziri *et al.* [170] conducted a detailed investigation of the wear of polymer materials. SEM of fractured and/or etched materials provides additional information on multicomponent structures. Combined OM and EM are applied to measure the size and distribution of the dispersed phases for correlation with mechanical properties such as impact strength. Particle size distributions are now routinely obtained with commercial image analyzers. Critical parameters, calibration and statistical sampling were considered in a study of latex particle size distribution (Section 4.9.2) using an automated image analyzing computer [171]. Dispersed phase particle size measurements of acrylonitrile blends were made directly from phase contrast optical micrographs and from TEM negatives [172] with a similar system.

Multiphase polymers are prepared for SEM by methods such as fractography, etching and extraction and for OM and TEM by thin sectioning methods. The dispersed phase in a multiphase polymer is often directly examined by SEM study of fractured surfaces. Polymers with large dispersed phases that adhere poorly to the matrix are the best candidates for direct analysis but are unfortunately the worst materials. Well adhered and small dispersed phases are often not visible directly in the SEM. Standard preparation methods and imaging techniques often do not provide sufficient contrast to describe the nature of multiphase polymers, and contrast enhancement methods are required. Chemical staining, chemical and electron beam etching [173], differential interference contrast [174, 175] and phase contrast TEM are all methods to increase image contrast. X-ray micro-analysis and backscattered electron imaging provide contrast based on the presence of high atomic number materials, such as chlorine in poly(vinyl chloride).

The method of differential radiation induced contrast depends on enhancement of contrast in multicomponent polymers where the components have different electron beam–polymer interactions [173]. Contrast has been observed in sections of styrene–acrylonitrile/poly(methyl methacrylate) (SAN/PMMA) polymers where the PMMA exhibits a high rate of mass loss compared to SAN, creating contrast between the phases. It is well known that electron irradiation results in chain scission and crosslinking, loss of mass and crystallinity [75]. Polystyrene, poly-acrylonitrile and SAN crosslink and thus are stable in the electron beam whereas polymers exhibiting chain scission, PMMA and poly(vinyl methyl ether), degrade in the beam. It is suggested that experiments be conducted on the homopolymers to determine the expected irradiation damage mechanism in the multi-component system [173].

Newer techniques also can aid imaging and interpretation of information for engineering resins, blends and plastics. FESEM at low voltages can replace conventional SEM, provid-ing similar information but with much more detail. Often such imaging reveals smaller dispersed phase sizes, interfacial regions and other details that preclude the need for the more laborious microtomy and staining required for conventional TEM. In the cases of 'molecular composites', of structures formed by spinodal decomposition, HREM can often image domains that were not seen previously and such materials may have been thought to be composed of a single phase. The scanning probe microscopies are also important to engineering resins as such imaging does not require a high vacuum system and finer surface detail may be imaged than by SEM, FESEM or TEM. Examples of the various imaging techniques will be provided in the sections that follow.

5.3.2 Extrudates and molded parts

This section will focus on examples of the structure of crystalline or crystallizable thermo-plastics formed by extrusion and molding pro-cesses, although there is clearly a wider range of extrudates and moldings that could be discussed, including RIM (reaction injection molding) [176] of polyurethane, nylon 6 and foams. Polymer melts solidified without deformation form struc-tural units, termed spherulites, composed of a central nucleus with a radiating array of lamel-lae. The size and nature of the spherulites is well known to be affected by the temperature of crystallization, as was shown for PET [177]. Standard production processes, such as extrusion and molding, however, often produce deforma-tion of the spherulitic structure, and the local polymer orientation is frozen in the final product. The crystallization of polymers is also affected by their composition and whether they are homo-polymers or multiphase polymers. A brief description of the relation of the extrusion and molding process to the engineering resin or plastic morphology follows. This discussion is meant to provide a basis for structure–property studies and is not an exhaustive description of polymer manufacturing processes.

5.3.2.1 *Process considerations*

Extrusion is a process which involves heating a polymer and forming a homogeneous melt which is then forced through a die by means of a rotating screw. This process causes deforma-tion, resulting in molecular orientation in the extrusion direction. The amount of orientation depends on the temperature and flow rate of the melt. Compression molding involves placing polymer powder or granules into a mold and softening it by heating. There is little orientation because the polymer flow is limited. In injection molding, the polymer melt is injected into a cooled or heated mold. The injection process causes deformation of the polymer and orientation in the flow direction. The flow pattern during mold filling [178] has a semicircular shaped advancing front, curving toward the mold wall, where the macromolecules orient parallel to the wall. Orientation induced by elongational and shear flow is found in the flow direction [179], especially near the surface. As with extrusion, the injection molding process causes formation of anisotropic structures. Anisotropic morphologies formed due to such flow fields are termed 'skin–core', multilayered, or banded to describe the variation in orientation in the final specimen.

The structures of injection molded semicrystalline polymers are quite heterogeneous resulting in substantial differences in mechanical and thermal properties at different points within a single molding. The interrelationships among processing, microstructure and properties of thermoplastics have been reviewed by Katti and Schultz [179]. The thickness of the various layers, especially the oriented surface skin, are affected by process variables, such as temperature and pressure. These authors described three temperatures that are well known to be important: the mold or wall temperature, the melt or barrel temperature and the freezing point. A cooler wall temperature will result in thicker skin and shear zones while a mold temperature near the melt temperature will lead to higher degrees of orientation. Melt and mold temperature are the most important processing parameters for both crystalline polypropylene and for amorphous polystyrene and HIPS. In general, higher melt temperature improves mechanical properties. Apparently, increases in pressure result in thicker surface layers [180]. Malguarnera and Manisali [181] reviewed the topic of weld line formation in injection molded thermoplastics. Weld lines occur when two or more polymer flow fronts unite and these regions can have

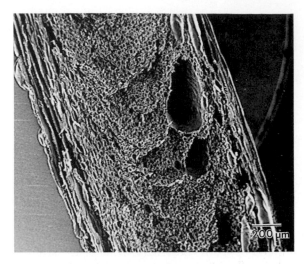

Fig. 5.38 SEM of a fractured, molded POM test bar, containing a high void level, shows a skin–core morphology. Elongation of the voids at the skin surface is due to high orientation whereas the more rounded voids and the semicircular flow front in the core results from less orientation in that region of the mold.

inferior mechanical properties. The effect of crystallinity [182] and the cooling rate from the melt [183], as expected, confirmed that differences in microstructure result in differences in tensile properties.

5.3.2.2 *Single phase polymers*

The morphology of molded articles depends on the chemical composition of the polymer, the process variables and the mold geometry. Standard molded tensile bars are discussed here for simplicity, but the principles are the same for any molding, although the nature of the specific flow field must be taken into account. The morphologies of injection molded tensile test bars of PE [184], POM [180, 185–187] and PP [188, 189] are similar and can be described as a complex, multilayered, skin–core structure. This structure is shown in the SEM micrograph of a molded POM test bar in Fig. 5.38. The molding shows the orientation of the polymer, emphasized here by the presence of voids which are highly oriented at the bar surface and less oriented within the core. Several intermediate layers are seen between the skin and the core. The relationships between process conditions, microstructure and mechanical properties of an injected molded thermoplastic have been reviewed [173, 191, 192]. Acid etching polished plaques and microtomy have been used to characterize the microstructure as a function of barrel and mold temperatures and injection pressure.

Bowman [192] conducted a systematic study of the structure–property relations of injection molded polyacetals (polyoxymethylene or POM) and observed correlations between process conditions, structures and mechanical properties. Barrel temperature effects were studied as they are known to influence both microstructure and mechanical properties [193]. Increased barrel temperature was shown to reduce the outer skin layer while increasing the extent of the equiaxed, unoriented core, resulting in a decreased tensile yield strength parallel to the injection direction.

5.3.2.3 *Multilayered structures*

An example of the multilayered structures common in polyacetals is shown in the polarized light micrographs (Fig. 5.39). They depict a uniformly nucleated crystalline structure formed due to mold filling and variations in the rate of cooling of the melt. The skin surface in the microtomed section (top in Fig. 5.39A) is birefringent, nonspherulitic and highly oriented. The molecular chains are oriented parallel to the injection direction. The central portion of the bar consists of a core (bottom Fig. 5.39A) with randomly oriented spherulites (Fig. 5.39C). It has no preferred molecular or lamellae orientation. There are usually one or more layers between the skin and core that are transitional shear zones with intermediate structure. In polyacetals this has been termed 'transcrystallinity' [194, 179]. Transcrystalline growth is controlled by the heat flow to the mold wall and is initiated by the cold mold wall. The melt at the wall cools rapidly, and dense spherulite nucleation takes place adjacent to the wall. Spherulites nucleated close to the wall, in the thermal gradient, have parabolic boundaries (Fig. 5.39B). Their continued growth inward can lead to a layer of columnar structure. The number and extent of the layers depend upon the specific processing conditions. For example, Bowman [192] identified five layers in acetal copolymers, but where the mold is thin there may be no core, and in a hot mold there may be no skin. Multilayered structures are also observed in

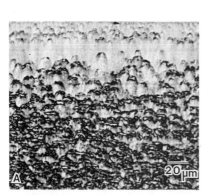

Fig. 5.39 Optical micrographs of polyacetal section show a spherulitic texture in polarized light. An overview of the outer mold region shows a birefringent skin (top) and an unoriented spherulitic core (A). Between the skin and core is a transition zone composed of spherulites with parabolic boundaries (B). Spherulites which are polygonal in shape due to impinging one another are seen in the core (C).

other molded plastics. PBT has been observed with four zones or layers: a nonspherulitic skin, regions with and without flow lines and a central core with many flow lines [195]. Rapid solidification of the polymer while filling the mold could explain the presence of the flow lines in this banded texture. Injection molded polypropylene was shown with four layers [196–198].

5.3.2.4 *Spherulitic textures*

Spherulites too small to measure by optical microscopy can be seen in the SEM, although untreated molded bar surfaces or fracture surfaces often do not reveal the spherulitic texture. Ion, plasma or chemical etching can reveal spherulites by differential etching of crystalline and amorphous material. A polarized light micrograph of an extruded polyacetal pellet section (Fig. 5.40A) has an oriented skin and a spherulitic core. Etching for 15 min in an oxygen plasma (162 °C) results in about 5% weight loss and a spherulite texture is seen (Fig. 5.40B). A 4 h treatment results in a coarser texture (Fig. 5.40C)

and only 10% of the original weight of the pellet remains. Smaller spherulites at the surface are revealed by the short etch and larger spherulites are uncovered by further etching.

Molded plastics generally have relatively smooth surfaces, as shown by SEM of a polyacetal (Fig. 5.41A). Therefore, etching is often quite useful in manufacturing processes to promote adhesion between the plastic and a surface coating, such as by electroplating [199]. Chemical etching for short times results in surface pit formation with pit shapes reflecting the microstructure (Fig. 5.41B and C). The effect of acid etching was to produce elongated pits in the direction of polymer flow into the mold. Longer etching times resulted in etching deeper into the unoriented core; thus, the bottoms of the pits are larger and more rounded than at the surfaces (Fig. 5.41C). Cross sectioned, etched and electroplated materials provide more information on the etch depth. Low magnification SEM (Fig. 5.41D) did not show the etched structure but the penetration of the electroplating metal is shown in an EDS map (Fig. 5.41E), and more detail is

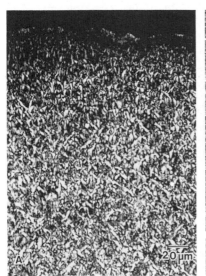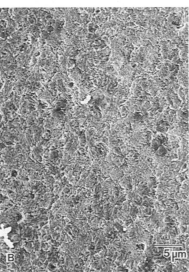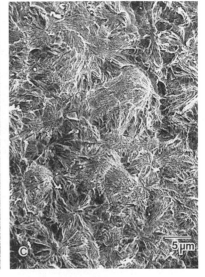

Fig. 5.40 Spherulites can be observed by both polarized light and in the SEM. Polarized light of a polyacetal section reveals a skin–core texture with fine, uniform spherulites in the core (A). Treatment in an oxygen plasma at 162 °C for 15 min uncovers the spherulite texture (B) and after a 4 h treatment larger spherulites are observed (C) within the molded specimen by SEM.

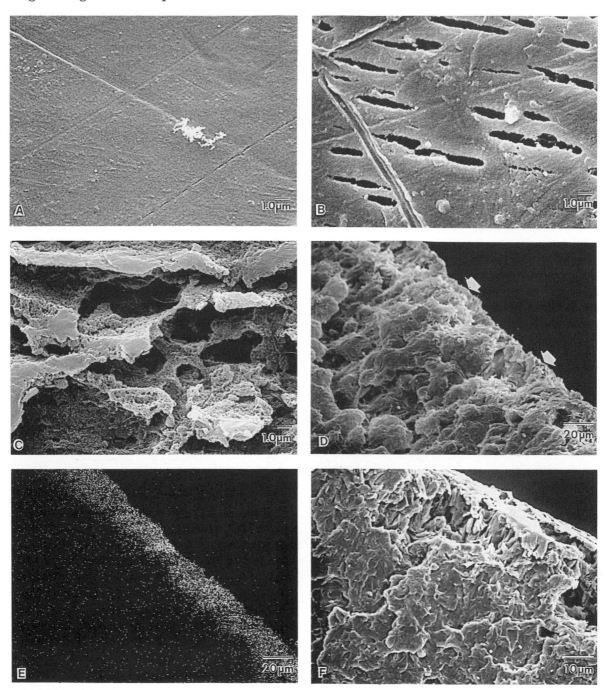

Fig. 5.41 SEM of a molded polyacetal surface shows a smooth texture (A) with little surface detail. Etching for short times results in elongated pits, oriented in the direction of polymer flow (B). Longer etching times result in surface pits deeper below the surface, due to etching larger spherulites in the core (C). Fractured cross sections of plated and etched surfaces do not show the structure near the surface (arrows) (D) except in EDS maps of the plating material (E) or at higher magnification (F).

seen at higher magnification (Fig. 5.41F). Such undercut structures are quite important for well adhered plated plastic parts.

Keith *et al.* [200] report that small concentrations of compatible polar polymers change the morphology of aliphatic polyesters. Crystalline polyesters such as poly(ϵ-caprolactone) have much larger spherulites when about 1% of PVC or poly(vinyl butyral) is added. These amorphous polar polymers act as anti-nucleating agents; normally these are polar low molecular weight compounds that preferentially adsorb on the nucleating impurities and have a low melting point. At the polymer crystallization temperature they keep a liquid surface on the particles that usually act as nuclei, suppressing their effect. This study, using transmitted light interference contrast microscopy to observe the banded spherulites, shows that miscible polymers can act in the same way.

5.3.2.5 *Skin–core structures*

In contrast to polyacetal, PE and PP, multilayered textures have not been observed for poly(butylene terephthalate) [201] or nylon. Nylon and poly(butylene terephthalate) moldings both exhibit a low crystallinity or amorphous surface layer, with little or no orientation, and a crystalline core which depends on mold conditions [192]. This amorphous skin is due to the rapid quenching of the polymer at the surface and the high glass transition temperature of these materials, not to the flow of the polymer. Thus, a simple skin–core texture rather than a multilayered texture has been observed for these polymers.

Polarized light micrographs show details of the spherulitic structure in molded nylon (Fig. 5.42A) which is similar to PBT [201]. The nonspherulitic skin, a transition zone and a spherulitic core region are observed. This is as expected, as the quench rate declines away from the surface, e.g. in the transition zone, and thus there is some nucleation of spherulites. A view of the transition zone (Fig. 5.42B) shows round spherulites in a fine textured matrix. The

equiaxed spherulites, especially adjacent to the skin, indicate there is little or no orientation. PBT skin thickness has been shown to increase with decreasing melt and mold temperatures, resulting in increased impact properties in bars with molded-in notches [201].

5.3.2.6 *Structure–property relations*

The structural heterogeneity resulting from injection molding plastics has been described as this process results in a higher degree of anisotropy compared to compression molding and extrusion, and similar methods are used to evaluate the microstructures formed by these processes. Voigt-Martin [105] and Bassett [100] showed structures similar to those produced by compression molding, and these were mentioned earlier (Section 5.2.2). The structure formed by the process affects the mechanical properties. The generally expected correlation of Young's modulus values increasing with orientation along the tensile axis has been observed [187, 189]. Izod impact strength values have been shown to be higher for specimens with increased skin and shear layer thickness, for post-notched PP [188, 189], and for increased skin thickness for molded-in notched PBT [201].

5.3.2.7 *Problem solving application*

Compact discs with digitally encoded music have taken a major share of the music recording business from more traditional analog media. In addition, CD-ROM and write once optical disc systems are also being marketed that permit inexpensive information storage at high density. A technical advantage of optical recording is that reading and writing with a focused laser beam has spot sizes less than 1 μm in size, thus there is a high density of information. Grooves or other features on the disc permit tracking of the data. The groove structure in polycarbonate (PC) substrates commonly used in the fabrication of optical discs has been studied [202] as have the pit structures in organic write once (WORM) optical data storage media [203], by SEM, TEM,

Fig. 5.42 Polarized light micrographs of a molded nylon cross section show a nonspherulitic skin (top in A) and rounded isolated spherulites in the transition zone. A classical Maltese cross extinction pattern is observed with black brushes showing the radial texture within the spherulites (B). (See also Fig. 1.3 in color section.)

FESEM and STM. In the WORM discs, organic thin films are spin coated onto the PC and then marked with diode lasers. Critical parameters in optical recording are the size, shape and depth of the features in which the information is coded. The symmetry of the groove geometry is important to the tracking as is the flatness of the land or groove bottom. Average sizes of the grooves are obtained by diffraction patterns of incident light but these are bulk average data and often there is a need to characterize the geometry in more detail, especially during development of new products.

A cross section of a PC disc would appear as a regular series of grooves and lands (raised, flat regions) with some periodicity. Depths can be on the order of 60 nm with periodicities around 1.5 μm and groove:land ratio of 1:3. The substrates [202] were injection molded commer-

cial discs with grooves created by the mold insert. Samples were cut and coated with thin gold and platinum films using ion beam sputtering (IBS) (Section 4.7.3) to form a conducting layer. Initially STM images were obtained using a 'pocket sized' STM [202]. Complementary images were obtained from single stage carbon replicas of the sample surfaces in the TEM. The PC was shadowed with Au–Pd at a shallow angle (*c.* 30°) and then a thin carbon layer was deposited in a vacuum evaporator. The disc was dissolved using methylene chloride and the replicas were placed on copper grids for TEM evaluation. In addition, substrates were thin sectioned about 100 nm thick perpendicular to the grooves for TEM study. SEM of the IBS coated disc showed general details and these were compared to FESEM images of uncoated discs.

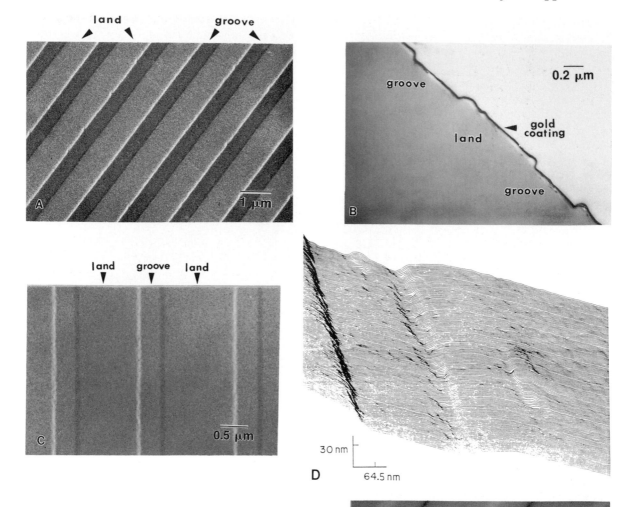

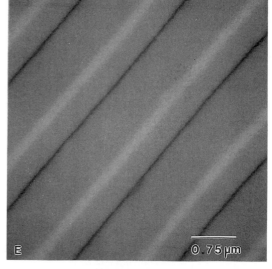

Fig. 5.43 The morphology of the grooves in a poly-carbonate substrate is shown in SEM, TEM, FESEM and STM images. An SEM micrograph (A) of a gold coated disc reveals the general groove and land morphology with *c.* 0.5 μm width grooves and 1.5 μm periodicity. TEM of ultrathin cross sections of gold coated discs (B) show a heavy dark line of the continuously coated surface; the thickness is in good agreement with the thin film thickness monitor. TEM of carbon replicas of the IBS gold coated disc (C) showed the grain structure and some gold particles about 10 nm in diameter. An early STM image (D) provides more detail regarding the depth of the groove than can be seen in the SEM image, with quantitation obtained by plotting a line scan of the STM image. FESEM at low voltage (1.2 kV) (E) of a similar optical disc with no metal coating does not readily show the depth of the grooves. (From Baro *et al.* [202]; reproduced with permission.

The morphology of the grooves in the substrate is shown in Figs 5.43A–E. A SEM micrograph in Fig. 5.43A [202] of a gold coated disc reveals the general groove and land morphology of the disc with *c.* 0.5 μm width grooves and 1.5 μm periodicity. Complementary TEM of ultrathin cross sections of gold coated discs are shown in Fig. 5.43B [202]). The heavy dark line shows the continuously coated surface; the thickness is in good agreement with the thin film thickness monitor. TEM of carbon replicas of the IBS gold coated disc showed the grain structure and some gold particles about 10 nm in diameter (Fig. 5.43C) [202]. Large particles (see arrow) are defects in the substrate surface. An early STM image shown in Fig. 5.43D [202] provides more detail regarding the depth of the groove that can be seen in the SEM image, with quantitation obtained by plotting a line scan of the STM image. Even FESEM at low voltage (1.2 kV), taken of a similar optical disc with no metal coating (Fig. 5.43E), does not readily show the depth of the grooves with the detail that is obtained in STM. The groove geometry observed by STM is also confirmed by the TEM images. The STM images can be obtained much more quickly with significantly less specimen preparation time than laborious microtomy techniques. In addition, imaging by STM is much easier and the images have improved considerably as this technology has matured since that first image was collected in 1987. An additional conclusion of this study [202] was that platinum coated samples produced much finer textures than the gold coatings, in agreement with prior expectations (Section 4.7.3).

The pits made in organic WORM media were also examined using STM and evaluated versus media performance [203]. Discs were prepared by spin coating an organic medium from an organic solvent onto polycarbonate substrates at varying spin speeds. Samples were laser marked at various energies and pulse times and then IBS coated. STM was used to determine the thickness of the organic layer and also the pit geometry. The thickness of a 'soft' organic layer is very difficult to determine by profilometry or even

TEM of microtomed sections, as there is no sharp step to aid measurement. As seen in Fig. 5.44, (color plate section) STM provides a direct measurement of the depth of a pit, that is the height difference between the media surface and the center of the pit [203]. Clearly, the STM image provides a means of accurate measurement of pit depths unavailable by SEM imaging, especially where the depths of structures are very shallow compared to the three dimensional geometry.

5.3.3 Multiphase polymers

The topic of multiphase polymers is vast with several books and review papers [166–170, 204–212] and hundreds of research papers describing the processes, morphologies, properties and applications of these important materials. The field of multiphase polymers has been driven by the realization that wholly new molecules are not always required for new applications and that blends can provide a rapid and economical means of development. As expected, processing of blends is critical to the microstructure and properties. Typical processing involves melt mixing using single and twin screw extruders and then the blends are extruded, injection molded, or blow molded by similar methods used for engineering resins. The morphology of these multiphase polymers depends on a range of parameters: the polymer molecules, interfacial tension, viscosity, shear mixing and phase separation kinetics. In the resulting blends, the size of the dispersed phase, the drop breakup and coalescence are all governed by the deformation process. A wide spectrum of properties can be obtained by the appropriate blending of polymers. White and Min [213], for example, have investigated the development of polymer blend morphology during processing and the role of interfacial tension and viscosity ratio as it affects phase morphology. Random copolymerization is used to modify the properties of a single phase, whereas graft and block copolymerization are used to modify the adhesion properties of the interface between the matrix and the dispersed phase. Improvement in mechanical properties often accompanies addition of one polymer to another; for instance, the tensile strength and modulus of PE are increased by the addition of PP [214]. There has also been considerable work in the area of interpenetrating polymer networks [169]. The features of major interest have been the two phase morphologies, including the size, shape and complexity of the 'phase within a phase' structure. Paul *et al.* [169] has shown that domains are formed by nucleation and growth mechanisms, and then they are modified by diffusion to lower energy structures.

Multiphase polymers are commonly toughened plastics which contain a soft, elastomeric or rubbery component in a hard glassy matrix or in a thermoplastic matrix. An example of the typical brittle fracture morphology of an unmodified thermoplastic is shown by SEM of nylon (Fig. 5.45A). Addition of an elastomeric phase modifies the brittle fracture behavior of the matrix, as shown in a fracture surface of a modified nylon (Fig. 5.45B). The modification depends on the composition and deformation mechanism of the material [204, 215], but normally it increases the fracture toughness and strength from that of the unmodified matrix resin. Impact strength, as measured for instance by an Izod impact testing apparatus, is affected by the dispersed phase

particle size, the rubber composition and adhesion to the matrix. Dispersed rubber particles 0.1–10 μm in size [216] and typically in the range of 0.1–2 μm [204] are generally best and good adhesion between the particles and the matrix is required for enhanced impact strength. Dispersed phase morphology can include wholly compatible phases: phases which are too compatible may not result in impact enhancement. However, there is no single particle size, chemical composition or elastomeric content that provides a formula for a successful polymeric system. The microstructure of multiphase polymers formed by extrusion and molding is discussed below.

5.3.3.1 *Rubber toughening*

Major factors that affect the size and distribution of dispersed phases in multiphase polymers are: elastomer content, compatibility, processing and viscosity. If the elastomer is present as the smaller volume fraction, it will most likely be the dispersed phase. As the volume fraction increases, the size of the dispersed phase can be larger and there is more likelihood of subinclusions of the matrix polymer in the elastomer [217, 218]. These subinclusions appear to enhance impact properties. The shape of the dispersed phase often changes with differences

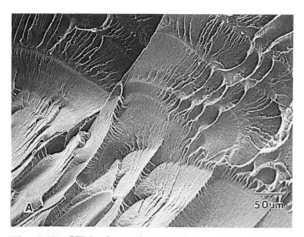
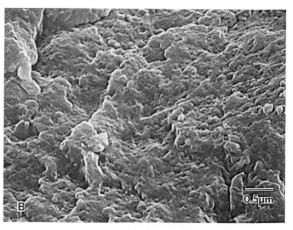

Fig. 5.45 SEM of a nylon fracture surface shows a brittle fracture that is consistent with poor mechanical properties of an unmodified polymer (A). A modified nylon has a significantly different morphology, although the modifier is not observed (B).

in composition [219, 220], mixing conditions, relative viscosities and with polymer orientation. For example, blends of 20% PE in PS form spherical domains, whereas at 50% PE the domains are elongated and significantly larger in size [221]. Dispersed phase particles tend to be elongated parallel to the flow axis, and they are elongated in shape with the short axis normal to the surface of a molded or extruded part while remaining spherical near the core. The particles tend to be oriented at a 45° angle in the shear region. Generally, the finer the particle size the better, although particles that are too small (much less than 0.2 μm) do not generally affect properties.

Processing plays a major role in the nature of the dispersed phase in multiphase polymers. Changes in the shear forces and the temperature provide different structures. In the case of PS modified with polyisoprene, TEM studies showed that smaller particles, broken down in size by melt shearing, resulted in lowered impact strength and increased tensile strength [222]. Particle dimensions have also been shown to be affected by the viscosity of the molten polymer and the concentration of the modifier. Heikens *et al.* [223] investigated copolymer modified PS and LDPE and ethylene–propylene copolymer with block and graft copolymers, and evaluated the mechanical properties and morphology, which showed that the graft and block copolymers increase the compatibility of the two phases.

Rubber toughening was discovered nearly sixty years ago, and the original theories to explain the phenomenon were proposed 25 years ago and have been reviewed [224]. Since that time, however, numerous deformation studies in the TEM have been conducted and reported by Kramer and his colleagues [218, 225, 226] and others [41] (Section 4.8.3). Control of the toughening process has depended on characterization of the structures resulting from the dispersion of the rubber particles, their grafting, crosslinking and copolymerization. Optical and electron microscopy allow measurement of the size and shape of the dispersed phase particles. Microscopy also permits observation of their internal

structures and the adhesion between the particles and the matrix. The particle size distribution is quite important as it is well known that there is an optimum 'window' of particle sizes for each type of matrix; 0.1 μm particles provide good toughening for poly(vinyl chloride) whereas the rubber particle size for PS is over 1 μm [227]. Particle size determination is simple if the polymer has discrete particles of the rubber dispersed in a matrix. However, the measurement is more difficult in cases where there is a core–shell or graft copolymer morphology or where there are subinclusions of the matrix resin present in the rubber phase. Subinclusions are commonly observed in ABS, HIPS and nylon blends and may be a result of the chemical process or the elastomeric content.

According to Paul *et al.* [169] the matrix polymers can be considered in two categories, brittle and ductile, each exhibiting specific requirements for reinforcing polymers. Brittle matrix polymers, PS, SAN, PMMA and epoxy, are all said [169] to have requirements for toughening as follows; interfacial adhesion, optimum particle sizes 0.1–0.3 μm depending on polymer matrix, chemical or physical cross linking of the rubber. The toughening mechanisms tend to be by crazing with some shear yielding and cavitation. These same authors [169] consider that ductile polymers, PC, polyamide, PP, PVC, PBT and PBO, have the following requirements: adhesion is not always required, optimum particle sizes are less than 0.5 μm, cross linking may be helpful and at times trace levels of rubber have large effects (e.g., in PVC). For more details on these statements the reader is directed to details in the reference [169]. Three main types of multicomponent polymer systems will be discussed: (1) combinations of two or more semicrystalline polymers, (2) impact modified thermoplastics and (3) impact modified thermosets.

5.3.3.2 *Optical characterization*

The microstructure of semicrystalline multicomponent polymers can often be determined by

Fig. 5.46 A polarized light micrograph of a polymer blend cross section, composed of two different poly-acetals, shows a nonuniform texture. A transcrystalline layer is seen adjacent to the skin and larger spherulites are seen in a matrix of finer textures.

polarized light microscopy of thin sections. A blend of two polyacetals, a homopolymer and a copolymer, is shown in the micrograph of a thin section (Fig. 5.46). The structure is rather interesting in that large spherulites of one phase are observed in finer spherulites of the other phase; a transcrystalline region is observed adjacent to the skin. However, semicrystalline multicomponent polymers can also appear very confusing in polarized light (Fig. 5.47A) as the spherulitic texture and the dispersed phase textures are superimposed and not distinguishable. Comparison in polarized light (Fig. 5.47A) and phase contrast (Fig. 5.47B) of a polyester–nylon copolymer shows that phase contrast clearly reveals the nature of the dispersed domains in this complex microstructure.

5.3.3.3 *SEM characterization*

Copolymers

Determination of dispersed phase morphology is most often conducted by SEM of fractured specimens. Fractures are prepared by manual

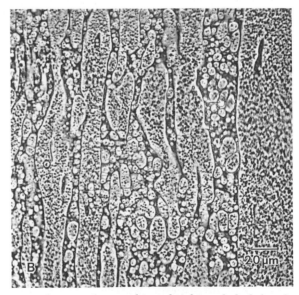

Fig. 5.47 A polarized light micrograph (A) of a polyester–nylon copolymer shows bright and dark bands obscuring the spherulitic texture. Phase contrast optical microscopy (B) reveals the dispersed phase texture of the copolymer which consists of multiple phases where the dispersed phase particles contain subinclusions of the matrix polymer.

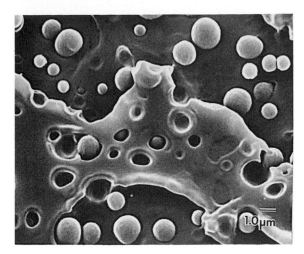

Fig. 5.48 Dispersed phase particles are observed in a SEM of a notched Izod fracture surface of a poly-acetal/polypropylene copolymer. Dispersed phase particles about 0.5–2 μm across and pullouts, holes where particles were pulled out during fracture, are observed.

methods, after immersion in liquid nitrogen, or by standard physical testing procedures. The microstructure of the homopolymers should be examined for comparison with the multiphase polymer. SEM of an Izod fracture surface of a POM/PP copolymer is shown in Fig. 5.48. The two phases are incompatible, i.e. they are present as two distinct phases. The dispersed phase particles range from less than 0.5 to 2 μm in diameter. The sample fracture path follows the particle–matrix interface and holes remain where particles have pulled out of the matrix, showing there is little adhesion between the phases.

The shape of dispersed phase particles is determined by the flow field and heat gradients that affect polymer orientation. For instance, the microstructure of copolymers of PE and PP is similar to the skin–core textures described for PE [228]. The orientation of the dispersed phase can affect the mechanical properties of the system. Spherical domains are more commonly formed in systems where phase separation occurs while the polymers are liquid. The SEM appears to reveal spherical dispersed phase particles (Fig. 5.48), although tilting can show they are actually elongated domains. As with typical fiber reinforcement (Section 5.4.2), the length of the dispersed phase protruding from the matrix is an indicator of the adhesion between that phase and the matrix.

Polymer blends have benefited from the use of low voltage field emission SEM as have other polymer imaging applications. From early work in the late 1980s showing the utility of improved contrast and reduced beam damage, even with metal coated samples [229, 230], advances have been made that permit imaging of uncoated specimens with excellent resolution [231, 232]. Recent imaging of polymer blends and copolymers (for example [233, 234]) have benefited from low voltage imaging (Section 6.3.1). Schwark *et al.* [233] imaged the surface morphology of styrene–butadiene block copolymers by low voltage high resolution SEM. Himelfarb and Labat [234] characterized polymer blends and block copolymers using both conventional and low voltage SEM and TEM of stained polymer blends. In this work they used preferential staining with ruthenium tetroxide and suggested that higher accelerating voltages (10–25 kV) are preferred for the measurement of particle size and shape. For high resolution images of surfaces topography, in this case 20 nm domains in hydrogenated styrene–butadiene–styrene block copolymers, the spatial resolution in low voltage imaging is comparable to conventional TEM.

An interesting comparison of low voltage imaging using a lanthanum hexaboride gun versus a field emission gun is shown in Fig. 5.49 for a three phase polymer blend [162, 230]. Figures 5.49A and B show similar fracture surfaces of the polymer imaged at 5 kV accelerating voltage with a lanthanum hexaboride gun (Fig. 5.49A) and FE gun (Fig. 5.49B). Even at this low magnification and with a thin platinum ion beam sputter coated surface, the sharpness of the FESEM image is clear as is the image detail at the interface of the large domains and the matrix.

Polyurethanes

Multiphase polymers containing polyurethane are common toughened polymers in which it has

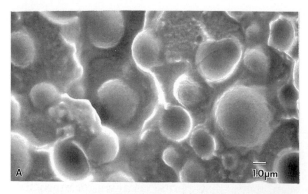

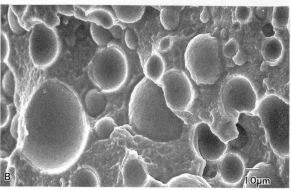

Fig. 5.49 Similar fracture surfaces of a polymer blend were imaged at 5 kV accelerating voltage with a lanthanum hexaboride gun (A) and with an FE gun (B) [162, 230]. (From M. Jamieson, unpublished [162].)

long been known that phase separation is important in determining structure–property relationships [235]. In some cases, as with a polyurethane modified polyester, a semi-interpenetrating network is formed, as is shown by TEM of osmium tetroxide stained sections [236]. Cryosectioning has been used to study the morphology produced by a relatively new polymer process, known as reaction injection molding (RIM), in polyester based polyurethanes [237]. The large temperature gradients in the mold were shown by Fridman *et al.* [237] to influence both molecular weight and morphology, and higher polymerization temperature resulted in better hard segment organization. A review of the structure of segmented polyurethanes has shown incompatibility to be a key factor in

determining morphology. Chen *et al.* [238] used both optical microscopy with a hot stage and video camera, and electron microscopy of cryosectioned specimens, to assess this morphology. Three crystal forms have been identified by complementary optical and TEM study of solution cast model films of polyurethane containing hard and soft segments [239].

Often polymers such as polyurethanes do not fracture at room temperature, or the fracture is ductile, and both impact testing and specimen preparation must be conducted below room temperature. Additionally, etchants may be required in order to bring out the dispersed phase morphology. Demma *et al.* [240] studied the morphology and properties of blends of polyester-based thermoplastic polyurethanes with ABS, PS, SAN copolymer and with an ASA terpolymer (acrylonitrile–styrene–acrylic ester terpolymer). Specimens did not break when Izod impact testing was performed at room temperature, and thus lower temperature testing was required. For SEM study specimens were immersed in liquid nitrogen for 10 min and then fractured. Fractured ABS specimens did not reveal any particles. Chemical etching of the ABS blends was used in an attempt to show the size distribution of the phases. Etching with methyl ethyl ketone (4 h) at room temperature dissolved the SAN copolymers in ABS [241]. THF vapor treatment (1 h) was said to be a solvent for the thermoplastic polyurethane [242]. Micrographs of fractured and etched (methyl ethyl ketone) polyurethane blends revealed holes where the SAN copolymer was located. However, a major problem with etching preparation methods, as noted by these authors [240], is that often the etchant has adverse effects on the matrix as well as the dispersed phase.

SEM micrographs of a polyacetal/polyurethane multiphase polymer are shown in Fig. 5.50. The outer surface and a fractured internal surface of this extrudate were chemically etched in order to determine the nature of the dispersed phase. The surface (Fig. 5.50A) shows a complex structure due to etching. The fracture surface after solvent extraction (Fig. 5.50B) is complex, as

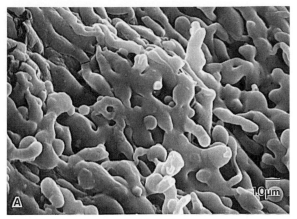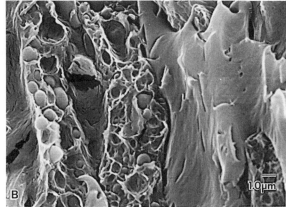

Fig. 5.50 SEM images of liquid nitrogen fractured polyacetal–polyurethane blend show a complex, network morphology (A) made more complicated by chemical etching (B). SEM of the etched fracture surface (B) suggests that the etchant has affected both the dispersed phase and the matrix.

it appears that both the matrix and the dispersed phase particles have been affected by the etchant. When using etching other methods should also be used to confirm the nature of the microstructure.

Toughened thermoset resins

Toughening of polymers with rubber has seen greatest application in thermoplastic resins. However, the technology has also been extended to thermosetting resins, such as epoxies [243, 244]. The topic of rubber modified thermoset resins has been described and reviewed comprehensively [245]. Epoxy resins toughened with rubber particles have enhanced properties like toughened thermoplastics, although the theory of such toughening is not as well understood. Bucknall [204] and Kunz-Douglass *et al.* [246] and others have discussed toughening models; however, these are beyond the scope of the present discussion. In any model of toughness, a major parameter is the size distribution of the dispersed rubber particles. Microscopy provides this information in the same way as described for thermoplastic polymers.

Thermoset epoxy resins were toughened by small elastomeric inclusions of a carboxyl terminated butadiene–acrylonitrile (CTBN) random copolymer by Visconti and Marchessault [243], who showed the variation in size as a function of CTBN content by TEM and light scattering. A major study of rubber modified epoxy resins has been reported by Manzione *et al.* [247, 248], who showed a range of morphologies which result in a range of mechanical properties, even for a single polymer. An amine cured rubber modified epoxy was characterized by STEM imaging and quantitative methods were developed to determine the volume fraction of dispersed phase [249] as this is known to be critical to enhanced toughness. Sayre *et al.* [249] stained the samples in THF containing osmium tetroxide.

Examples of rubber toughened epoxy resins are shown in Figs 5.51 and 5.52. The fracture morphology in Fig. 5.51 is typical of glassy or brittle fracture with spherical holes due to poor adhesion of the rubber particles. A fractured multiphase rubber toughened epoxy resin which has a more complex microstructure is shown in Fig. 5.52. The fracture morphology is brittle and glassy with failure occurring across the well adhered dispersed phase particles. Subinclusions of the resin are observed within the dispersed phase particles, likely due to the high concentration of the rubber phase. Voids, large and small smooth holes, are observed within the matrix and also within the dispersed phase particles.

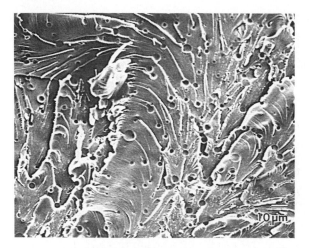

Fig. 5.51 An SEM image of a fractured rubber toughened epoxy resin exhibits brittle fracture. Holes from the dispersed phase particles show the rubber is incompatible with the matrix resin and there is poor adhesion resulting in rubber particles being pulled out during fracture.

Biodegradable polymers

In recent years there has been an increasing interest in the development of degradable polymers [250–254]. Much of this interest has been driven by increasing volumes and visibility of plastic waste from commodity and packaging products. Methods to reduce the volume of waste include reduction of use, increased recycling of polymers and the development of biodegradable materials. Biodegradable materials have another quite separate field of use, in biomedical products, such as degradable sutures that do not need to be removed. Other possibilities include screws and pins to repair broken bones that degrade as the bone heals, and devices for the controlled release of drugs into tissue [251]. The definition of biodegradable is in dispute, but we can distinguish between fully degradable materials and materials which quickly break down into more stable small fragments. The latter would remove the visibility of plastic waste, but would not be suitable for medical devices described here.

The fully degradable materials that are currently in production are polyesters that are sensitive to hydrolysis. This is the reverse of the condensation polymerization reaction, and these materials will depolymerize in the presence of water all the way back to oligomers which can be digested by microbes. Biopol (ICI) is a material of this type, a copolyester of poly 3-hydroxybutyrate (PHB) and poly 3-hydroxyvalerate (PHV), that is produced by bacterial fermentation which will slowly degrade and is compostible [252]. The bacterial polyesters, because of their biological preparation, are much purer than the average commercial polymer. The homopolymers are very highly crystalline and have very large spherulites, and even samples for optical microscope may contain no effective nuclei [253]. The commercial product is a copolymer because it has reduced crystallinity and spherulite size, and therefore better mechanical properties. Other polyesters used for biomedical applications are poly(lactic acid), poly(glycolic acid), and poly(p-dioxanone). These generally degrade faster; degradable sutures for example, must be packed individually in vacuum sealed foil, and used immediately on opening.

From the point of view of the microscopist, these materials can be treated like any other semi-crystalline polymer, with the added problems that they must be kept in a dessicator, and like other polyesters, they can be affected by transesterification. So far microscopy has not contributed much to study of the degradation process in these materials. The water diffuses into the amorphous regions and causes chain scission there. The result is that the molecular weight and the strength are reduced significantly before there is much change in the morphology. Standard techniques to assess crystallinity, such as x-ray diffraction and birefringence, are used to evaluate materials after exposure to water.

The materials that degrade into small stable particles are blends; often starches are blended with non-degradable thermoplastics such as PE. The same considerations noted in this section on multiphase polymers hold true for these blends. For instance, the interfacial energy and rheology play a role in the size of the dispersed phases. These blends can often be processed by standard melt forming polymer techniques, by compounding and extrusion into films and fibers and injection or blow molded to form polymer

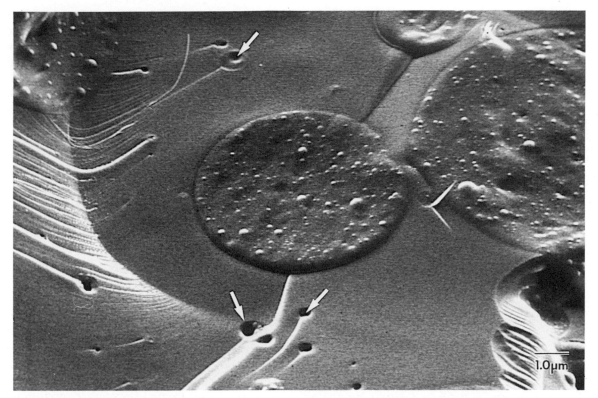

Fig. 5.52 SEM of a rubber toughened epoxy resin shows that brittle fracture occurs through both the matrix and the dispersed phases. Voids (arrows) are observed within the dispersed phase and also within the matrix. Small subinclusions are seen within the dispersed phases.

products. Accordingly, process and material variables that affect multiphase polymers affect these degradable blends. Naturally, blends can also be designed that are fully biodegradable, for example by blending poly(hydroxybutyrate) based materials with natural poly(saccharides) and synthetic (polycaprolactone) polymers, as described by Yasin and Tighe [254]. This review provides information on the strategies for the design of such systems, to meet user requirements for a long shelf life and rapid degradation.

Scanning electron microscopy is the primary technique used to evaluate the blends as formed, as well as sections examined by OM or TEM. The change in the structure as a result of the hydration or degradation process is of major interest so examination of the material both before and following a hydration or degradation process is appropriate. Dynamic imaging techniques such as environmental or high pressure

SEM (HPSEM) to image wet specimens during water induced swelling and degradation would seem to be the preferred technique for such studies. Video taping images of dynamic experiments has value for understanding the failure of biodegradable systems and understanding the mechanisms of degradation. However, HPSEMs are used to image in a wet environment at the expense of resolution.

An example of the effect of degradation on a starch-based polymer blend film imaged in an HPSEM is shown in Fig. 5.53 [255]. The biocomponent films were prepared by a proprietary method under development for accelerating biodegradation and then they were slightly oxidized using a plasma etcher and exposed to thermophilic compost for 60 days. Upon completion of the compost cycle the samples were rinsed with isopropanol to remove loose debris and dried at room temperature. The HPSEM images

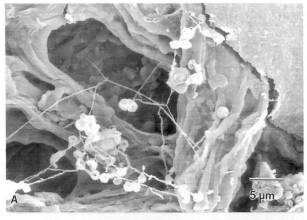

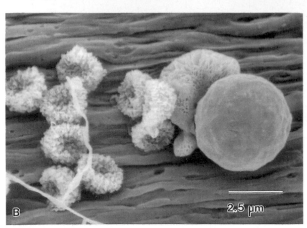

Fig. 5.53 The HPSEM images were taken at 10 kV of compost surfaces that were evaporated with chromium. An HPSEM image (A) of the film has a porous surface due to the removal of one of the blend components during the composting stage. Micro-organisms and white rot fungus remain on the moderately hydrophilic surface. A porous extruded film surface (B) in more detail has pores due to composting; robust micro-organisms and yeast colonies adhere to the hydrophilic film surface. (From G. Loomis, M. Izbicki, C. Kliewer, D. R. Sawyer, unpublished [255].)

were all taken at 10 kV of surfaces that were evaporated with chromium. Figure 5.53A is an HPSEM image of the film which has a porous surface due to the removal of one of the blend components during the composting stage. Micro-organisms and white rot fungus remain on the

moderately hydrophilic surface. Figure 5.53B shows the porous extruded film surface in more detail, again with pores due to the removal of one of the blend components during composting; robust micro-organisms and yeast colonies adhere to the hydrophilic film surface. Although the HPSEM is not a technique for high resolution imaging, it does allow imaging wet materials and dynamic experiments.

5.3.3.4 *TEM characterization*

There are multiphase polymers where OM and SEM techniques cannot fully describe the microstructure due to a combination of small particle size (less than 0.5 μm) and good adhesion between the dispersed phase and the matrix. Additionally, broad particle size distributions are often encountered, and in these cases a combination of techniques is required to describe the microstructure. TEM requires ultrathin specimens, about 50–500 nm or less in thickness, which are prepared by film casting or ultrathin sectioning. Films formed by casting or dipping methods provide a much easier specimen preparation method than ultrathin sectioning of bulk plastics. However, a major question in such studies is always whether the microstructure is the same as in bulk polymers of industrial interest. Specific stains are often required to provide contrast between the dispersed phase and the matrix polymer.

Cast thin films

Widmaier and Meyer [256] studied the structure of an ABA polystyrene–isoprene block copolymer as a function of temperature by osmium tetroxide staining thin cast films. Hsiue and Yang [257] studied the morphology and properties of α-methylstyrene–butadiene diblock copolymer films cast from several solvents. Films at a 0.1% concentration were cast on water and stained with 2% osmium tetroxide solution for 1 h. The microstructure was shown to differ for films cast from different solvents as there is a polymer–

solvent interaction. Reich and Cohen [289] studied the phase separation of polymer blends in thin films and compared the behavior to that of the bulk material, as it is well known that phase transformations in thin, nonpolymeric, solid films differ from those in the bulk material [259]. Polystyrene–poly(vinylmethyl ether) blends were shown [259] to be affected by the substrate used in dip coating when the film thickness was less than 1 μm. Handlin *et al.* [260] studied the morphology of four ionomers and discovered that solvent casting produced artifacts but no information about ionic domains, whereas microtomed sections of sulfonated EPDM showed phase separated regions. Bates *et al.* [261] studied the microphase structure of solvent cast diblock copolymers and copolymer–homopolymer blends containing spherical microdomains. Some artifacts in the size of the dispersed phases were caused by solvent casting. In conclusion, solvent casting of thin films for TEM of multiphase structures is simpler than sectioning, but the potential artifacts suggest caution must be exercised in interpretation. For evaluation of bulk industrial materials, sectioning appears to be the method of choice for determining microstructure.

Microtomed and stained sections

Many of the studies of multiphase polymers are conducted on unsaturated rubbers which are adequately stained by osmium tetroxide, which reveals the nature of the dispersed phase domains. Polymers with activated aromatic groups have been selectively stained by reaction with mercuric trifluoroacetate (Section 4.4.8). Hobbs [262] has successfully used this technique to provide contrast in blends of poly(2,6-dimethyl-1,4-phenylene oxide) and Kraton G (SBS block copolymer). Although this stain is effective in enhancing contrast, a drawback of the method is that the material is not hardened or fixed by the stain.

Early morphological studies to determine the nature of multiphase polymers and blends were reviewed by Folkes and Keller [263]. Many studies were of extruded block copolymers of materials such as SBS where the dispersed phase, an unsaturated rubber stained with osmium tetroxide, was observed in the form of spheres, cylinders or lamellae [264]. These were referred to earlier (Section 4.4). An excellent example of such studies is shown in Fig. 5.54. A TEM micrograph of a thin section of a poly(styrene–butadiene) diblock copolymer, stained with osmium tetroxide [265], depicts the (100) projection of a body centered cubic lattice.

A TEM study of poly(vinyl chloride)/chlorinated polyethylene (PVC/CPE), for assessment of the dispersed phase particle morphology and correlation with impact properties [266], is a good example of current applications of such morphological methods. Microtomed sections of the blend were stained by a two stage osmium tetroxide method to reveal the CPE phase [267]. As the CPE concentration increased, the discrete two phase morphology changed to a continuous CPE network resulting in a transition from brittle to ductile impact fracture (shown by SEM) and increased impact strength. The behavior of these blends was also studied [268] by deformation under an optical microscope and by TEM study of stained sections. Another example of the use of complementary techniques was combined hot stage OM, TEM and quantitative analysis of particle size for the determination of the nature of phase separation dynamics and morphology, especially the late stage mechanism of phase separation of polymer blends [269]. Discrete particles were shown to form early in the process but they may change by aggregation and coalescence.

A recent series of experiments by Hobbs and coworkers [270, 271], on toughened blends of poly(butylene terephthalate) and bisphenol A (BPA) polycarbonate (PC) toughened with a core/shell impact modifier, have benefited from imaging by TEM and SEM. Selective staining with ruthenium and osmium tetroxide and etching with diethylene triamine were used to assess the distribution of the blend components and to investigate the effects of thermal history on morphology. In this work, the impact modifier

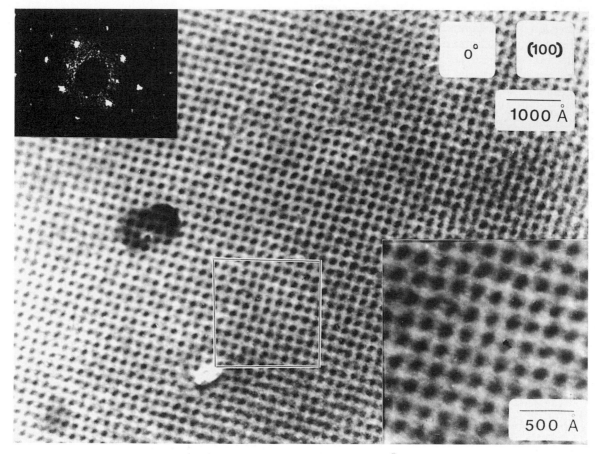

Fig. 5.54 A transmission electron micrograph of an osmium tetroxide stained thin section of a poly(styrene–butadiene) diblock copolymer (16.1 wt% polybutadiene shows the (100)) projection of a body centred cubic lattice. (From Kinning *et al.* [265]; reproduced with permission.)

was differentiated from the PBT and PC by reaction with OsO_4 which forms chemical complexes with double bonds. The PC was imaged preferentially by its greater ability to absorb RuO_4. In these blends the PBT was observed as the continuous phase with the impact modifier observed isolated in islands of the PC with an interpenetrating network formed above 40% PC. The SEM studies of specimens etched with diethylene triamine showed information about the melt miscibility and phase separation of the resins. The overall morphological study provides evidence which supports the strong interfacial region and aids understanding of the properties of this blend system. The toughening mechanism

of these blends was also evaluated by combination of notched impact testing and morphological evaluation [270, 271]. Differences in behavior were discussed in terms of microscopic failure processes. Finally, Dekkers *et al.* [272] also studied the effect of morphology on the properties of blends of PBT, PC and poly(phenylene ether) using electron microscopy. The dispersed particles of rubber-modified PPE were observed encapsulated by thin envelopes of PC and embedded in a PBT matrix.

Toughening of a semicrystalline polymer by a phase segregated block copolymer introduces several levels of complexity, as shown in an example of a nylon 66 toughened with a Kraton

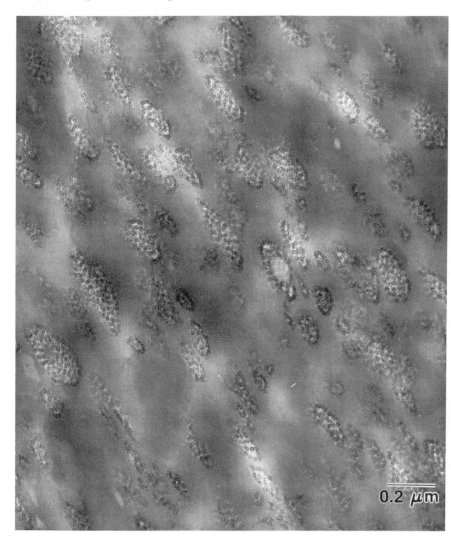

Fig. 5.55 TEM image of a thin section of nylon 6,6 with a thermoplastic elastomer, stained with RuO$_4$ and PTA. (From Wood [273]; reproduced with permission.)

G SEBS triblock polymer in Figure 5.55 [273]. Double staining was used to reveal the microstructure of both the nylon matrix with PTA and the styrene rich regions of the triblock polymer with RuO$_4$. There is selective absorption of PTA by the amorphous nylon phase so it becomes dense and the individual crystalline lamellae are seen as meandering white regions in the matrix. Regions of Kraton up to 200 nm in size are seen to consist of styrene rich (black) cylinders in a matrix [273].

Examples

Applications showing the range of microscopy techniques and specimen preparation methods used on commercial impact polymers will be described. Changes in polymer morphology are expected upon addition of an elastomer; for instance, such addition is expected to cause a decrease in the spherulite size as the elastomer domains can act as nucleating sites [274]. This has been observed for many polymers including modified nylon [275]. Characterization of an EPDM impact modified nylon 6,6 has been reported [276] by the use of osmium tetroxide staining (one week) followed by TEM imaging to show the 'core–shell' microstructure.

Phase contrast optical microscopy (Fig. 5.56A), TEM of a stained cryosection (Fig. 5.56B and C) and STEM imaging of unstained ultrathin sections (Fig. 5.56D) all show elliptically shaped, dispersed phase particles in a matrix. The contrast in STEM is a result of a difference in radiation damage between the two polymers;

nylon likely crosslinks in the beam, whereas the elastomer phase exhibits mass loss due to chain scission [173]. Advantages of STEM are that there is no need for any stain or etchant and the image can be rapidly obtained and processed. A disadvantage of STEM imaging is that the specimen is quickly changed and then destroyed by the electron beam unless great care is taken to limit the electron dose. The particle sizes and distribution are similar in the STEM images of room temperature sectioned nylon and the TEM sections of cryosectioned and stained polymer.

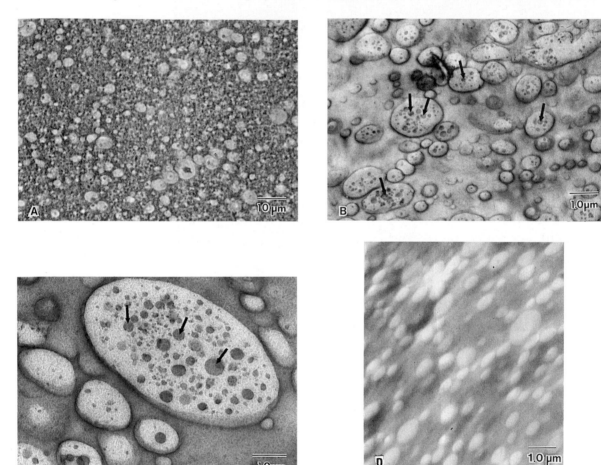

Fig. 5.56 A phase contrast optical micrograph (A) of an impact modified nylon shows the fine dispersion of modifier in the matrix. TEM micrographs of a cryosection, stained with ruthenium tetroxide (B and C) show more detail and finely dispersed subinclusions (arrows) within the elastomeric phase. STEM (D) of an unstained cryosection shows less dense regions in a darker background due to mass loss of the rubber phase during exposure to the electron beam, resulting in contrast enhancement.

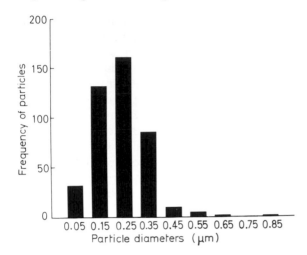

Fig. 5.57 A histogram showing the distribution of particles less than 1 μm was obtained by quantitative analysis of the TEM micrographs.

The TEM image permits observation of the smaller sub-included nylon phase in the elastomer, due to the effect of staining. A particle size distribution (Fig. 5.57) obtained by image analysis of the TEM micrographs shows that a broad range of particle sizes are present.

5.3.3.5 *AFM and complementary characterization*

Atomic force microscopy has now been used for the evaluation of multiphase polymer materials [233, 278–281]. Collin *et al.* [278] imaged specimen surfaces of copolymers prepared using a Mettler hot stage. The free surface images of samples prepared at varied annealing times were obtained using a Digital Nanoscope II AFM in air in the constant force mode. This imaging provided a better understanding of the formation of islands in the copolymer that was similar to previous optical microscopic observations, but much more detailed and quantitative. Dikland *et al.* [279] used AFM to investigate the dispersion of low molecular weight compounds in ethylene–propylene copolymer rubbers. Light microscopy could not resolve the details and the high vacuum in the electron microscopes caused problems with volatility of the low molecular weight compounds.

Schwark *et al.* [233], Annis *et al.* [280] and Vezie *et al.* [281] have recently provided the first detailed, complementary studies of diblock copolymers using conventional TEM, low voltage high resolution HRSEM and AFM. Examples taken from these studies are shown in Fig. 5.58 (supplied by D. L. Vezie). The samples in these figures are polystyrene–polybutadiene diblock copolymers, both of molecular weight 40,000 and designated SB 40/40. The bulk samples were cast from dilute solution in toluene over a period of seven days and then annealed at 115°C for seven days. At this composition, the block copolymer phase separates into a lamellar structure with a 60 nm repeat with a corrugated surface structure. Figure 5.58A is a bright field TEM image [233] of samples stained with OsO$_4$ for 24 h, embedded in epoxy and cryo-ultramicrotomed perpendicular to the surface to give a cross section showing the epoxy block copolymer interface. The image was taken at 200 kV at slight underfocus with the beam parallel to the lamellae. The dark regions are polybutadiene and a thin layer of polybutadiene covers the entire surface as it has the lower surface energy. Figure 5.58B and C are low voltage FESEM images [233] taken of the bulk samples after staining with OsO$_4$ for 24 h to enhance contrast and sputter coating with Au–Pd for 15 s. Images were taken at 1.0 kV in an FE immersion lens SEM, in this context a high resolution SEM (HRSEM), with the beam nearly perpendicular to the free surface (Fig. 5.58B) and with a tilt of 40° about the axis shown (Fig. 5.58C), resulting in a decrease in the lamellar contrast. Figure 5.58D is an image of an unstained sample taken at 1.0 kV [281], under the same conditions as above, showing contrast is present but is weaker without the stain. Figure 5.58E is an AFM image [280] of the free surface of an unstained sample, imaged using a silicon nitride pyramidal tip in repulsive mode with medium range forces on a Digital Instruments, Inc., Nanoscope II. The height profile of the corrugated surface structure is shown in Fig. 5.58F [280]. The AFM is being used at comparatively low resolution here, so there are none of the difficulties of interpretation that bedevil the molecular scale AFM images, and the surface profile can be accepted as accurate. The complementary nature of this outstanding study makes the results clearer and easier to understand.

5.3.4 Failure analysis

A major objective of microstructural analyses on multiphase and other polymers is failure analysis, i.e. determination of the mode or cause of failure. Failure analysis generally involves characterization of a material which has failed, either in service, or in a physical test. Controls are not always available, and timing is often critical. In some cases, the types of analysis required may well be similar to those described above: phase contrast, SEM and TEM. Other microstructural techniques that are valuable in solving such materials problems are chemical contrast imaging and elemental x-ray mapping.

Plastic pipes made from poly(vinyl chloride) (PVC) and high density PE are used in many applications, including pipes for water and gas transport where brittle failure obviously limits their use. A method for inducing brittle failure for testing was developed as a plane strain fracture toughness test [282, 283]. The test involved notching the specimen and using a razor blade induced fracture to replace fatigue cycling tests. The fracture toughness measured was a function of the resistance to brittle failure in a sharp crack region. Microscopy of the fracture zone was used to characterize the nature of the fracture surface.

A toughened nylon with low impact strength is seen in the SEM image in Fig. 5.59. The fracture surface is brittle in nature, and a classical flaw–mirror–mist–hackle fracture pattern is observed. At higher magnification, the fracture is shown to have propagated away from a defect site and has formed fracture ridges where the fracture accelerated (Fig. 5.59B). An enlarged view of the defect site (Fig. 5.59C) shows it to be a round contaminant fiber, which itself exhibits brittle failure.

X-ray microanalysis and BEI imaging provide information regarding the elemental composition of dispersed phase particles and also of the contaminants reponsible for failure. These techniques are useful for the study of multiphase polymers, if one of the phases contains an element not contained by the other. Price *et al.* [284] used SEM/EDS to study the interface between two polymers, one of which was PVC, to measure the local composition during interdiffusion. Failure of elastomer compounds used in tank track pads, natural rubber, SBR, EPDM and 50/50 SBR/BR was studied as a function of elastomer cure and degradation in service [285]. Andrade *et al.* [286] and Hobbs and Watkins [287] reviewed the application of BE imaging to study microtomed multiphase polymers. Flat specimens generally do not provide topography necessary for informative SEM imaging. Andrade *et al.* [286] used BEI of osmium tetroxide

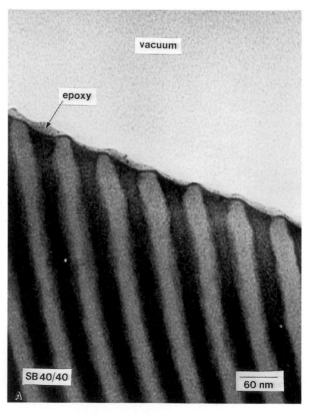

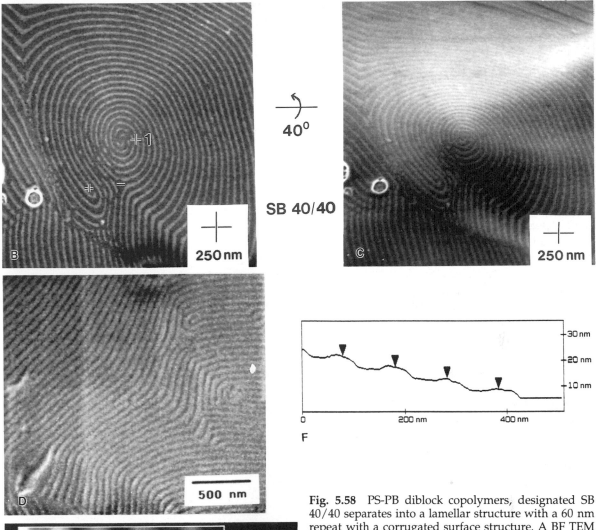

Fig. 5.58 PS-PB diblock copolymers, designated SB 40/40 separates into a lamellar structure with a 60 nm repeat with a corrugated surface structure. A BF TEM image (A) [233] of samples stained with OsO$_4$ shows the epoxy block copolymer interface. The dark regions are PB and a thin layer of PS covers the entire surface as it has the lower surface energy. Low voltage FESEM images at 1.0 kV in an FE immersion lens SEM were taken of the bulk samples after staining with OsO$_4$, with the beam nearly perpendicular to the free surface (B) and with a tilt of 40° about the axis shown (C), resulting in a decrease in the lamellar contrast [233]. An image (D) of an unstained sample taken at 1.0 kV [281], under the same conditions as above, shows weaker contrast. An AFM image of the free surface of an unstained sample is shown in (E) and a height profile of the corrugated surface is shown in (F) [280]. (From Vezie *et al.* [233, 281]; reproduced with permission.)

stained PS/PB to provide images which showed the dispersed and stained polybutadiene as a result of atomic number contrast. Hobbs and Watkins [287] used the inherent differences in atomic number of multiphase polymers for imaging by SEM and BEI. They stained polymers of low inherent chemical contrast with either osmium tetroxide or 15% bromine in methanol (30 s).

The combination of normal SEM, BEI and elemental mapping provides identification of the chemical composition and distribution of elements within polymers that is quite useful for failure analysis. For example, SEM (Fig. 5.60A) shows a fractured polymer surface containing large dispersed phase particles, about 10 μm across, and submicrometer sized particles as well as large holes. The BEI image (Fig. 5.60B) shows that both the large and the smaller particles are brighter than the background polymer, and thus they are composed of elements of higher atomic number than the polymer. Elemental mapping shows that the small particles contain antimony (Fig. 5.60C), whereas the larger particles contain chlorine (Fig. 5.60D). The less distinct map (Fig. 5.60C) is due to small particles being present below the surface. These are not observed in SEM but *are* sampled in the x-ray detection volume. Care must be taken in elemental mapping and x-ray analysis, in general, to allow for the fact that the detection volume for x-rays is always greater than that for scattered electrons and the actual detection volume is a function of both the atomic number of the matrix and the accelerating voltage of the electrons.

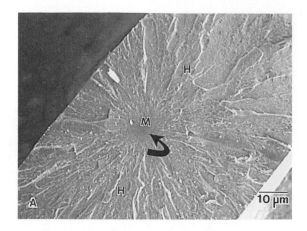

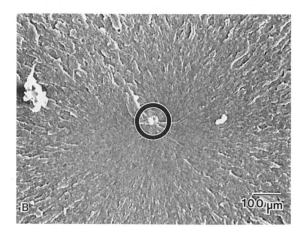

Fig. 5.59 SEM study of brittle failure in an impact modified nylon molded article reveals classical fracture morphology. The locus of failure (arrow) is seen (A) with surrounding mirror (M), mist and ridged hackle (H) regions propagating out into the bar. A higher magnification view of the flaw and mirror is shown (B). The flaw (C) appears to be a round fiber, likely a contaminant.

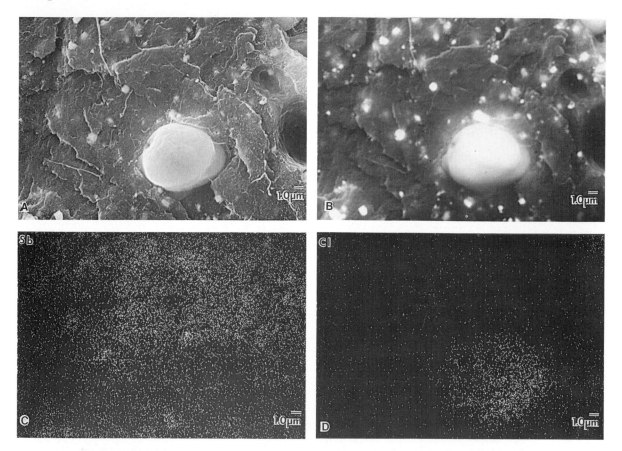

Fig. 5.60 SEM of a multiphase polymer (A) shows large dispersed phase particles and submicrometer sized particles and holes. BEI (B) shows that the dispersed phase particles have higher atomic number than the matrix. Elemental mapping shows that the small particles contain antimony (C) whereas the larger particles contain chlorine (D). Once the specific size/shape of the particles are identified by mapping, BEI imaging can be used to study the specimen surface.

5.4 COMPOSITES

5.4.1 Introduction

Polymers containing rigid fillers are termed composites. Composites generally contain short or continuous fibers or nonfibrous particles, usually minerals. Additives such as carbon black are also included in this discussion. Thermoplastic homopolymers, multiphase polymers and thermosets may be used as the matrix for composite materials. Rigid fillers are added to composites either to *fill* the polymer and perhaps modify some physical property, or to *reinforce* the matrix, that is to bear some of the load. Fiber composites are used in aircraft, small boat hulls and automotive bodies because of their light weight and high strength. Applications of carbon fiber reinforced composites are in tennis and racket ball rackets, fishing poles and in the medical/orthopaedic field for joint replacement [288], as well as in aircraft. Today, there is a trend to a greater proportion of the aircraft being composed of polymer matrix composites [289]. Continuous graphite (carbon), glass and Kevlar fibers in epoxy resins have been the major composite materials applied to such components.

Incorporation of short glass or carbon fibers into many engineering thermoplastics improves their strength, stiffness and heat deflection. The strength and fatigue crack resistance depend on the volume fraction of fiber in the composite, the mechanical properties of the fiber and polymer and particularly the adhesion between the two components. Nonfibrous fillers find application in low cost composites where they may enhance physical properties or simply replace the more expensive polymer with minimal loss of mechanical integrity.

Cellulosic materials have been used as fillers in polymer composites (bakelite) and as cellulose derivatives (collodion) since the very beginnings of the plastics industry. More recent applications for cellulose fillers include the use of wood flour with isotactic PP for interior automotive panels, wood polymer composites for construction material, improved modulus composites based on cellulosic fiber with thermosets and thermoplastics for automotive, construction materials, furniture and road making among the other applications [290]. As with all fillers, a major reason for using these natural fibers is to reduce the cost of the product, although they are of high modulus, and may increase the stiffness if their aspect ratio is kept high enough during processing. Recent reviews describe current trends in the use of cellulosics in general [290], prospects for the use of wood cellulose as reinforcements in polymer composites [291], composites that contain natural polymers, including cellulose derivatives [292] and a discussion of wood consumption compared to plastics [293].

The processing of composites is quite complex as the resin, the fibers and, most importantly, the interfacial bond must all be controlled to provide the required properties. The lack of fiber to resin bonding can result in a decrease in the stress transfer from the resin to the fiber and limited reinforcement. It is beyond the scope of this book to describe or even outline the important topics in the processing stage and the reader is directed to the references in this section and specifically to several texts [294–297]. The relevant microscopy techniques and examples of their application to morphological study are described here and elsewhere [298–300].

5.4.2 Composite characterization

Optical microscopy and SEM are the techniques usually employed to evaluate the structure of composite materials. The hardness of composites generally precludes microtomy, so polishing is used to provide thin sections for transmitted light microscopy or thick sections for reflected light study. Since about 1967 [301] fractured composites have been studied by SEM techniques, primarily due to the great depth of focus of this technique. Specimens are fractured by normal physical testing procedures or in use, and SEM is used to evaluate such features as the size, distribution and adhesion of the filler fibers or particles. The length of protruding fibers and the adhesion to the matrix play a major role in the strength and toughness enhancement. Quantitative microscopy techniques can be used to determine the length distribution of the fibers in a composite or the length of fibers protruding from a fracture surface. These are important parameters which relate to mechanical properties. Fiber length distribution information is important, as there is a *critical fiber length* required for reinforcement [302–303]. After injection molding, there is a distribution of lengths with many fibers less than this critical length, resulting in poor composite properties. Earlier studies by Sawyer [304] described a method of removal of fibers from composites, without damage, followed by quantitative analysis of the fiber length distribution. Such methods are used to optimize the glass fiber length during processing.

Oron [305] described dynamic evaluation of fracture mechanisms in composite materials by direct observation in the SEM. In this systematic study tensile loading of notched carbon fiber epoxy composites was conducted in the SEM. The three stages of fracture first described by Tetelman [306] are: (1) the tip of the notch is blunted and microcracks form there, (2) micro-

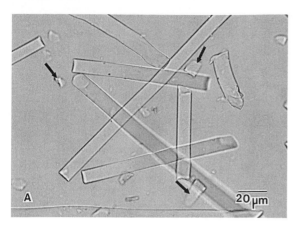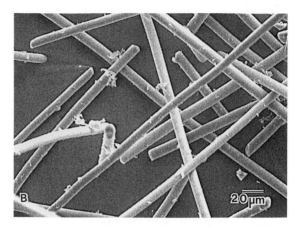

Fig. 5.61 An optical micrograph (A) of glass fibers removed from a filled engineering resin by a low temperature ashing method shows that they are isotropic, clear cylinders which tend to break into bits or shards (arrows). A low magnification SEM (B) shows the surface texture and shape of the fibers.

cracks grow with fiber debonding and (3) failure occurs with cracks widening followed by catastrophic failure. The effect of short glass fibers and particulate fillers on fatigue cracking in polyamides was evaluated in the SEM [307]. Improvement in fatigue crack propagation was found with increased glass fiber loading and reinforcement.

Conductive reinforcements in thermoplastics are a new application of composites [289]. High temperature resistant thermoplastics, such as PPS, PEI, PPO and liquid crystal polyesters, have been combined with conductive fibers, generally chopped graphite and nickel coated graphite. Fiber distribution, orientation and fracture morphology were determined by optical and SEM techniques. Poor bonding in the nickel coated graphite composite, as observed in SEM tensile fractures, resulted in lowered tensile and impact properties.

Fiber and particulate fillers can be obtained prior to compounding, or they can be removed from the composite by high or low temperature ashing, plasma ashing, microwave ashing or etching. High temperature ashing in a muffle furnace can change the phase of the filler or cause embrittlement and breakage. Low temperature ashing with an RF plasma is used to remove the organic matter without such phase changes or

embrittlement [304]. It is important to know the morphology of the fillers before studying their composites. This aids in understanding the complex morphologies. Examples of fiber and particulate fillers, removed or free from resin, are shown in Figs 5.61 and 5.62. A transmitted light micrograph (Fig. 5.61A) shows the general nature of short glass fibers. Glass fibers removed from a filled resin by low temperature ashing are shown by SEM in Fig. 5.61B. Flat mica flakes (Fig. 5.62A) are quite different from irregularly shaped talc particles (Fig. 5.62B) and finely divided clay particles (Fig. 5.62C). Polarized light microscopy of such minerals also aids their identification, if the filler in the composite is an unknown. In that case, the 'Particle Atlas' by McCrone and Delly [308] is an invaluable source of representative optical micrographs of common minerals.

Composites may also be examined by transmission EM methods, such as by bright field, dark field and electron diffraction of ultrathin sections. Sections of carbon fiber composites are quite difficult to obtain, but the technique is possible and has been described by Oberlin [309, 310]. Important information that can be obtained relates to the fiber–resin interface which is known to be critical to composite properties and is often adversely affected by environmental conditions.

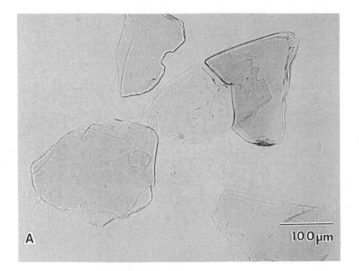

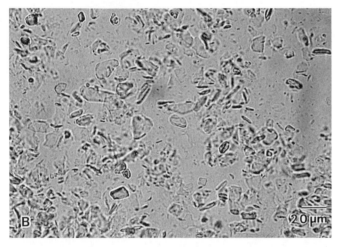

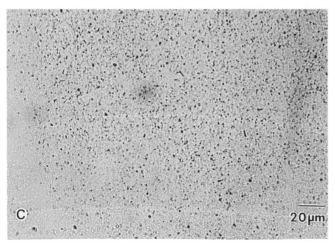

Fig. 5.62 Optical micrographs show the size and shape of several minerals used as fillers for composites: (A) mica flakes appear platy in shape with irregular boundaries, (B) talc particles have a much finer, platy texture and the particles exhibit a range of shapes from nearly fibrous to platy and (C) clay particles are very fine grained with no characteristic shape.

5.4.3 Fiber composites

The strength of a short fiber reinforced composite depends upon the fiber length and orientation, the volume fraction of fibers, the matrix and its mechanical properties and the interfacial bond between the fiber and the matrix resin. A text by Folkes [311] provides an excellent review of this complex topic. The major topics of this discussion are the microscopy techniques and preparation methods. There are four imaging techniques and methods that are typically applied to assessment of composite microstructure: (1) reflected light microscopy of polished surfaces, (2) transmitted light microscopy of thin sections, (3) contact microradiography (x-ray imaging) and (4) SEM analysis of both polished and fractured specimens. Applications of microscopy to the study of glass, carbon or graphite fiber composites will be explored, especially relating to the effects of processing, fiber length and interfacial bonding.

5.4.3.1 *OM characterization*

Reflected light microscopy is a technique which provides an overview of composite microstructures. Specimens are polished (Section 4.2) and examined in reflected light, generally at magnifications about 30–500×. Glass fibers appear as light colored round or elliptical sections. The orientation and loading of glass fibers in various regions of a composite can be shown by such studies. It is worth noting that the ratio of the major and minor axes of the ellipse produced when the cylindrical fibers are intersected during cutting and polishing can be used to determine the angle of the fibers (their orientation) within the matrix. Image analyzing computer systems can rapidly measure the orientation of many fibers within the matrix in this way, as long as the fibers are exactly cylindrical. For most fibers, a thin section viewed in transmission is required, and this is difficult with a normal optical microscope. A very thin section is difficult to prepare and includes long sections of few fibers; on the other hand a thicker section is not all in focus at one time, and may contain overlapping

fibers. However, only a clear image can give good results, as the computer system must be able to distinguish fiber boundaries and discard overlaps. The laser confocal scanning microscope (LCSM), can overcome these problems. The instrument is described in Section 6.2.1; see also ref. [312]. It has the ability to form 'optical sections', that is, sharp images of a particular plane within a composite, as long as the matrix material is transparent. Observation of one such sectional image gives information on the orientation of fibers within that plane, but the associated computer systems can put together the information in many such images, formed as the sample is scanned along the optic axis of the instrument. The result is a full determination of fiber orientation, for a depth of up to 20 or 30 μm into the composite [313, 314].

Transmitted light microscopy permits assessment of the polymer matrix as well as the fiber orientation. Polished thin sections can be used for bright field, polarized light and phase contrast microscopy, as shown for a glass fiber reinforced polyamide (Fig. 5.63). In bright field (Fig. 5.63A), in a section cut perpendicular to the flow axis of the test bar, the fibers are round, showing that they are oriented along the bar axis. The regions between the fibers contain the fine textured polymer. In polarized light (Fig. 5.63B), the polymer is birefringent and composed of fine spherulites while the fibers are isotropic, as expected. Phase contrast imaging (Fig. 5.63C) shows particles (white) are present that are different in refractive index compared with the polymer. Overall, combination of the three optical microscopies shows that the reinforced polyamide is composed of glass fibers, oriented with the flow axis of the mold, and that dispersed, second phase particles are also present.

Waddon *et al.* [315] have extensively studied the crystal texture of PEEK, PEK and PPS, including melt grown spherulites using polarized light microscopy. They formed thin films by heating the polymer and the fibers on microscope slides and pressing them or smearing between the slides and coverslips. Polymers were chosen

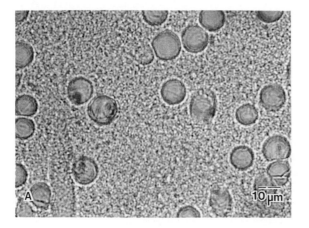

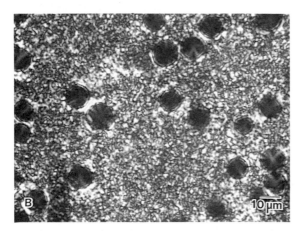

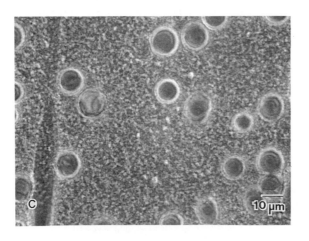

with low density of crystal nuclei to form larger spherulites and they were also crystallized isothermally at low supercoolings and rapidly cooled. Images taken in polarized light showed that transcrystalline textures were induced by the fibers and this texture varied depending on the fiber loading and thus their distance apart. Transcrystallinity is known to be caused by preferential nucleation on the surface of a substrate. Clearly, the crystallization of the polymer in composites will not only affect the crystalline morphology but also the composite properties.

5.4.3.2 *Contact microradiography*

Contact microradiography is a method that permits assessment of the length and orientation of glass fibers in composites, although it cannot be used for aramid or graphite fibers. Sections are cut 50–150 μm thick with a low speed diamond cutting saw and exposed to an x-ray beam providing an image on an underlying photographic plate [311]. The resulting photograph has more contrast than the optical micrograph. The method was reviewed by Folkes [311] and more fully described by Darlington *et al.* [316]. The effect of molding conditions on fiber length and distribution has been studied by contact microradiography. Flow geometry and injection speed have been shown to affect fiber orientation and skin thickness of the molding. Bright *et al.* [317] used contact microradiography of glass filled PP to show the effect of the flow field on the fibers. They explained the flow field according to the 'fountain' model [178] which shows that skin orientation is due to extensional flow while shear

Fig. 5.63 Transmitted light micrographs of a glass fiber reinforced polyamide polished thin section shown by three optical techniques. Imaging in bright field (A) reveals clear, round fibers aligned perpendicular to the section plane and a mottled textured matrix. Polarized light (B) shows the glass fibers are isotropic (black) whereas the polymer is birefringent and composed of finely textured spherulites. Phase contrast (C) shows that there are small, white regions of different refractive index than the matrix.

flow is observed in the core. This extensional flow causes a high degree of molecular orientation along the flow axis in fiber filled composites in much the same way as was described earlier for unfilled polymers (Section 5.3). A high injection rate causes an increase in skin thickness, and fibers in the core are aligned transverse to the flow axis. Low injection rates result in a thicker core with fibers parallel to the flow axis. Mold temperature also affects the nature of the skin and core, and thus fiber orientations as colder mold temperatures give thicker skin layers. Fiber alignment resulting from three flow conditions was shown using contact microradiography by Crowson *et al.* [318]. Convergent flow causes high fiber alignment while diverging flow causes fibers to align perpendicular to the flow direction. Shear flow leads to alignment nearly parallel to the flow direction. Overall, the process conditions affect polymer orientation and filler fibers are generally oriented with the polymers. The interested reader is directed to the references for details relating to the x-ray method.

5.4.3.3 *SEM of glass fiber composites*

Scanning electron microscopy is the most widely used imaging technique for the study of both short and continuous fiber composites. Fracture surface micrographs will be described here along with well known general principles relating to their interpretation. The nature of the adhesion between the matrix and the resin and information relating structure to mechanical properties can be obtained by SEM assessment of the composite fracture surface. Voloshin and Arcan [319] showed by SEM inspection that debonding of the fiber and matrix in unidirectional glass fiber–epoxy composites is the cause of shear failure under bidimensional stress. Wu *et al.* [320] studied the impact behavior of short fiber–liquid crystalline polymer composites by SEM evaluation of instrumented impact tested and modular falling weight tested materials. They observed a high degree of anisotropy in both the morphology and mechanical properties of the unfilled polymer. Adding short fibers reduced this

anisotropy. The SEM observations also suggested failure mechanisms in these composites.

SEM images of composite surfaces of two different specimens resulting from notched Izod impact testing of glass fiber filled thermoplastics are shown in Fig. 5.64A–F. Fibers are aligned parallel to the surface in the skin, and they are not aligned in the core (Fig. 5.64A). Fibers are shown which exhibit poor fiber wetting, as shown by the clumps of fibers with no resin rather than single fibers distributed within the matrix (Fig. 5.64B). A typical 'hackle' morphology is observed in the fractured resin (Fig. 5.64C), between fibers and larger hackle or ridged patterns within the matrix. Short fibers with resin coating (Fig. 5.64D) suggest there is good bonding whereas fibers are seen debonded from the fracture surface (Fig. 5.64E). The surface of an individual glass fiber shows a thin layer of resin bonded to the fiber surface (Fig. 5.64F). Adhesion is controlled by a compatible surface finish or size put on the fibers to enhance bonding. Classical brittle fracture of the glass fibers is seen (Fig. 5.65A and B), and the bonding at the fiber surface is shown in detail (Fig. 5.65C) in a glass fiber reinforced resin.

The fiber lengths of fillers or reinforcements are dramatically diminished during injection molding, and many studies have been conducted on both minimizing that breakage and determining the critical fiber length required for reinforcement [302]. Most of these studies have been conducted on glass fibers, although the principles are similar for all reinforcements. There is also a great range of variability of orientation and fiber length in composites. However, critical applications require precise predictions of mechanical behavior which has been addressed [311, 321]. A model of microstructure–property relations has been developed to predict mechanical behavior. Quantitative microscopy was applied to determine the location and orientation of the fibers in a polished specimen and to measure fiber pullout lengths in SEM fracture surfaces. The critical fiber length was assumed to be about four times the mean observed pullout length. It is important to know the length of the fibers in the composite

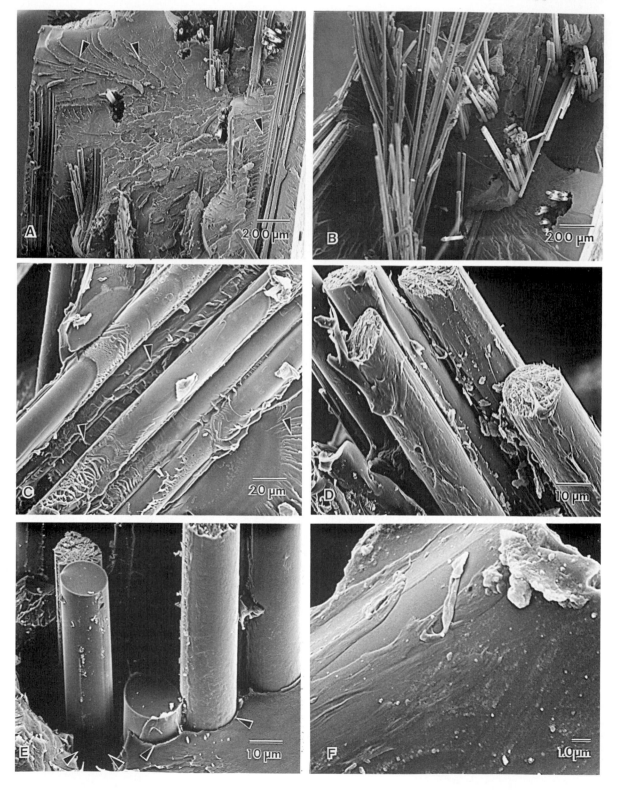

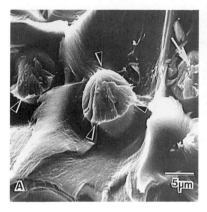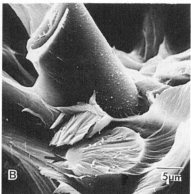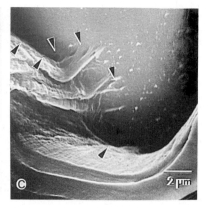

Fig. 5.65 SEM of a glass fiber reinforced nylon shows excellent compatibility between the fiber surfaces and the matrix. Short, nearly lateral failure across the matrix and fiber is observed. Some glass fibers exhibit classical brittle fracture (arrow) (B). Adhesion of the matrix resin is seen (arrows) at the fiber surfaces (C).

compared to the critical fiber length for that system, in order to know if the fibers are providing reinforcement.

5.4.3.4 *Carbon/graphite fiber composites*

The microscopy techniques described for the evaluation of glass fiber composites are widely used to determine the microstructure of carbon and graphite fiber composites. Microscopy of crack propagation in carbon fiber reinforced composites is also very important in understanding mechanical properties. Test specimens and actual composite products are often evaluated to determine the distribution of the fibers in the resin, typically epoxy, and the degree of resin wetting of the fibers. Voids in the composite can be the locus of failure, and their identification and cause are quite important to mechanical property evaluation.

An example of reflected light microscopy of a polished specimen is shown in Fig. 5.66 which elucidates the complex engineering involved in fabrication of a tennis racket. The micrographs show the nature of the continuous fabric, or prepreg, used for fabrication with layers of fiber bundles arranged at angles to one another (Fig. 5.66A). Black regions are voids where there is no reflection of light. The polished fiber cross sections in the micrographs appear white due to their high reflectivity. Gray regions are devoid of fibers and are termed resin rich. Single panels of unidirectionally oriented yarn bundles, or tows, are uniform in both distribution and resin wetting (Fig. 5.66B). Higher magnification of fibers which are nearly parallel (Fig. 5.66C) and perpendicular (Fig. 5.66D) to the polished surface is shown here.

Specific surface features identified by SEM examination of fractured composites have been more widely documented in the case of carbon fiber than for glass fiber composites. Fatigue damage in graphite–epoxy composites was investigated by Whitcomb [322] where delamination and cracking were found to result during fatigue tests. This is an important result as such

Fig. 5.64 SEM of Izod impact fractured, glass fiber filled thermoplastic test specimens show nonuniform distribution of fibers in the two different specimens (A, C, D and B, D, F). The fibers (A) appear aligned parallel to the skin and the matrix exhibits brittle failure as hackle marks (arrows) are seen. The fibers (B) protruding appear long and poorly wetted with the resin. Hackle or ridged patterns (arrows) are observed (C). Resin is also seen on the fiber surfaces in some regions (D and F) whereas cleaner fiber surfaces and less well bonded regions are also observed (E).

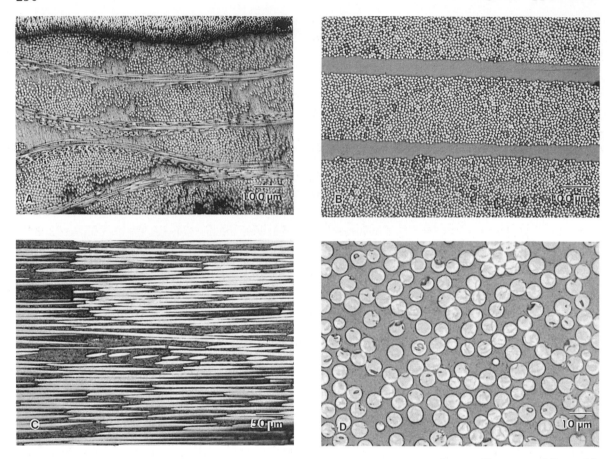

Fig. 5.66 Reflected light micrograph of polished carbon fiber composites are shown: fibers are white, voids (holes) are black, and the resin is gray in color due to surface reflectivity. A section of a graphite fiber reinforced composite (A) shows a complex arrangement of fibers with uniform wetting and packing. Single ply carbon fiber panels (B) embedded in epoxy and polished are uniformly packed. Higher magnification micrographs show detail is possible by this technique for fibers oriented parallel (C) and perpendicular (D) to the polished surface.

testing relates to long term behavior of the composite. Richards-Frandsen and Naerheim [323] described graphite/epoxy composites where three point bending induced delamination fatigue failure. Fracture surfaces were examined in the SEM in order to characterize the structure. Several common features were noted: matrix cleavage, hackle formation (in the epoxy) and wear. Defects causing crack propagation were caused by resin rich inclusions and voids. Tensile fracture in high modulus graphite fiber composites, subjected to off axis tensile loading, showed several fracture features [324], such as matrix

lacerations or hackle, fiber pullout, resin free fibers, matrix cleavage and matrix debris. Fracture modes were identified by Sinclair and Chamis [324] who showed smooth surfaces with matrix lacerations in intralaminar shear, and smooth surfaces with matrix cleavage in transverse tensile fracture. Although fractography studies of brittle fibers can be conducted to determine the locus and cause of failure by classical fracture mechanics, such a simple determination cannot be made in the case of a composite [325]. Accordingly, single fiber testing of graphite fibers in thermosetting resins has been conducted in an optical microscope in order to determine the nature of the stress concentration at the fiber–matrix interface [326].

The polished surface of a specimen with carbon fibers and a resin contains several voids and some regions show poor fiber wetting, whereas adjacent regions exhibit good wetting and fiber distribution (Fig. 5.67A and B). Fibers which exhibit poor resin wetting are shown by SEM of carbon fiber epoxy fracture surfaces (Fig. 5.68). The carbon fiber fabric has long fibers protruding from the surface (Fig. 5.68A). Fracture along the bundles shows hackle formation in the resin matrix (Fig. 5.68B). Higher magnification micrographs (Fig. 5.68C and D) show fibers protruding from the surface with little resin on the striated surfaces, and resin is seen pulled away from the fiber surfaces. Overall, this particular composite exhibits poor wetting of the fibers with resin, and thus fracture occurs at the fiber–matrix interface.

5.4.3.5 *Hybrid composites*

Hybrid composites are formed from a mixture of fibers and a bonding matrix [327], and these include carbon–glass and carbon–aramid fibers. Hybrid composites have economic, physical and mechanical property advantages that are exploited for various applications. For instance, composites containing expensive carbon fibers, for added strength and stiffness, may include glass fibers for increased toughness. Kalnin [328] evaluated glass–graphite fiber/epoxy resin composites. Nylon 6 polymerized around unidirectionally aligned carbon and glass fibers showed reinforcement properties [329]. Kirk *et al.* [330] measured the fiber debonding length in a model hybrid carbon and glass fiber composite and evaluated the fracture energy as a function of the carbon fiber/glass fiber ratio. They determined that the rule of mixtures prediction overestimates the fracture energy in the hybrid composites. Hardaker and Richardson [327] have described the theoretical and empirical developments in this area. Characterization of hybrid composites is conducted using the same methods described above.

5.4.4 Particle filled composites

Minerals and other particles are used as fillers in plastic moldings either to enhance mechanical properties or to reduce shrinkage and flammability. The minerals and other particles used as commercial fillers are: mica, clay, talc, silica, wollastonite, glass beads, carbon black and calcium carbonate. Theberge [331] summarized

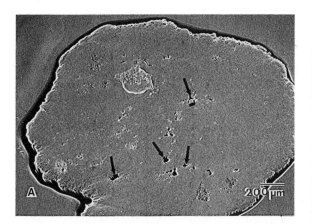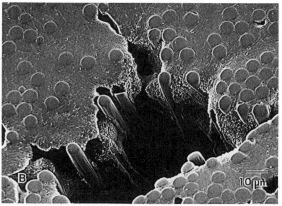

Fig. 5.67 SEM of a polished carbon fiber composite specimen reveals details relating to fiber uniformity and packing. Wetting of the fibers with resin is seen clearly. The overview (A) shows several voids (arrows) in the specimen which are seen to be regions where the fibers are poorly wetted with resin (B).

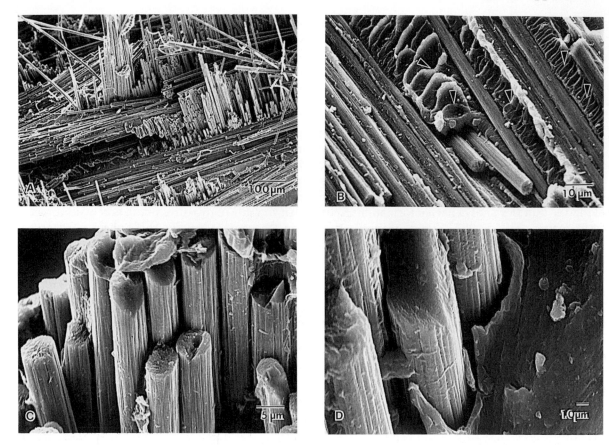

Fig. 5.68 Fracture surfaces of a carbon fiber composite are shown in the SEM images. An overview (A) shows the fibers are in the form of a fabric, with yarns aligned at 90° to one another. Matrix hackle, or ridges (arrows), are seen adjacent to the fibers in one orientation (B). Poor fiber wetting is shown as clean, striated fiber surfaces (C) pulled away from the matrix (D, arrows).

product advances in thermoplastic composites. Small mica flakes are known to boost mechanical properties in composites, but color changes are caused by their brown color. Mica was used to reinforce polypropylene with optimum performance obtained for particles in the range of 80–280 μm diameter [332]. Factors found important in this study were concentration, aspect ratio, surface treatment, processing, mean size and size distribution of the mica flakes. Garton *et al.* [333] discussed modification of the interface in mica reinforced PP. Adhesion of the filler is quite important to improving properties, and therefore fillers are usually surface treated with coupling agents to enhance the interfacial bonding and

wetting [331]. Mineral fibers have also been used in the reinforcement of polypropylene and nylon 6 [334]. In some cases the properties are similar to those of glass fiber composites. An important factor is the extent of fiber length loss during processing. Mineral fillers which are not fibers, with an aspect ratio (length/width) less than 10:1, reduce warpage of thermoplastics but they also result in a loss of tensile strength and other mechanical properties. Miller *et al.* [335] studied the role of the interface on properties in model composites using SEM to examine fracture surfaces of glass-sphere filled polyethylene. The SEM provided a qualitative view of the bonding process and the nature of the region of modified matrix surrounding the glass spheres. High degrees of chemical modification to the glass surfaces resulted in the formation of an interfacial layer and a change in failure mechanism from adhesive to cohesive failure.

Mica is often used as a filler or a reinforcement as this mineral can provide enhanced properties. SEM fracture surfaces of a mica composite and a glass fiber–mica composite are shown in Fig. 5.69. The mica composite has a fibrillar fracture surface texture, due to the thermotropic copolyester matrix, with the platy mica flakes aligned with the polymer and split along these plates during fracture (Fig. 5.69A). A glass fiber–mica composite fracture surface (Fig. 5.69B) shows the mica has fractured and the particles and the glass fibers appear well adhered to the matrix. These factors suggest good adhesion and reinforcement, leading to enhanced mechanical properties. Mica flakes can be dozens of micrometers in diameter, and thus they lend themselves to optical observation. Sections of a mica filled polyester are shown in optical bright field micrographs in Fig. 5.70A and B. The size and shape of the mica flakes after processing may be determined from such images. Measurement of the thickness of the thin, platy flakes is much more difficult, as they tend to align in any matrix, and they are much thinner than they are wide. TEM of an ultrathin section (Fig. 5.70C) shows

dense mica cross sections which have broken during diamond knife sectioning.

Some particulate fillers are known to cause changes to the polymer matrix, for example calcium carbonate affects the nucleation of some thermoplastics. The addition of any particulate filler has the potential of causing nucleation, although specific nucleating agents are also added to polymers to affect crystallization. Chacko *et al.* [336] have observed no large scale order, i.e. there are no spherulites present upon addition of calcium carbonate to PE. An optical micrograph of POM appears spherulitic (Fig. 5.71A), whereas addition of calcium carbonate appears to result in a loss of spherulitic texture (Fig. 5.71B). Addition of fillers can also have deleterious effects, e.g. the formation of a nonuniform or matte surface finish on a molded part, as shown in the SEM micrograph (Fig. 5.71C). Such particles in craters could be caused by a problem with the mold surface, but EDS analysis provided confirmation of the presence of calcium, as shown in the EDS map (Fig. 5.71D). Apparently, the calcium carbonate particles at the surface of the mold were not fully wetted with the polymer under the specific molding conditions used, and therefore a nonuniform matte surface finish resulted. Thus, processing polymers with particulate or fibrous materials does not preclude changes in the polymer morphology during the

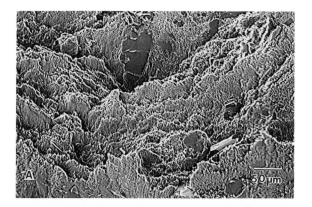 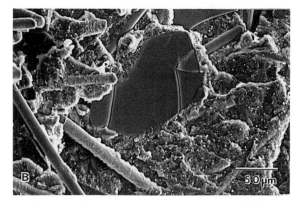

Fig. 5.69 SEM of a mica filled plastic bottle (A) and a mica–glass fiber composite (B) both show the platy shape of mica. Although the mica fracture surfaces do not appear resin coated there is good adhesion of these particles with the matrix. The mica is aligned with the oriented polymer (A).

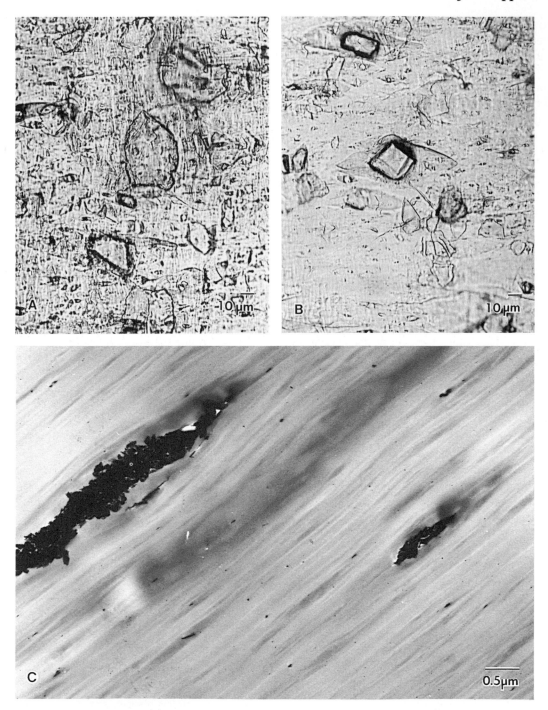

Fig. 5.70 Sections of a mica filled thermoplastic are shown in the optical (A and B) and TEM (C) micrographs. The platy mineral filler particles are aligned with the polymer. The particulate texture of the mica (black) in the TEM cross section reflects the effects of diamond knife fracture of individual mica flakes.

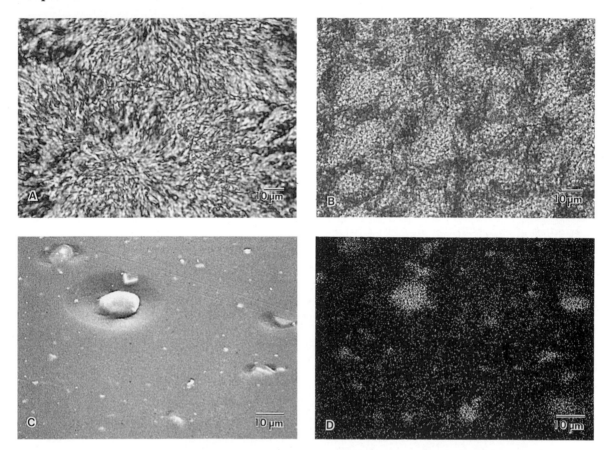

Fig. 5.71 Semicrystalline thermoplastics, such as a polyacetal, shown in polarized light (A) do not appear spherulitic after the addition of calcium carbonate (B). The surface of such a filled, molded part can exhibit a matte finish due to poor wetting of the particles with the polymer. SEM observation of the molded surface (C) shows pock marks which are particles in cavities not filled with polymer. EDS mapping (D) shows the particles contain calcium.

process, especially if the filler is added either prior to polymerization or before crystallization.

5.4.5 Carbon black filled rubber

The microscopy of rubbers generally involves the characterization of the particle size distribution of additives like carbon black. The topic of microscopy of rubbers is found in many sections of this chapter. Much of the discussion of multiphase polymers (Section 5.3.3) involves rubber domains and particles that must be characterized to understand the performance of

the polymer. This characterization is conducted by optical microscopy, SEM and TEM, and will not be further discussed here. Often the rubber particles are composites or blends themselves containing additives and fillers required to modify properties. Carbon black is a common particulate filler added to stiffen rubbers, for conductivity, and also simply to provide black coloration to a molded part. The behavior of the rubber is very dependent upon the size distribution of the carbon black and its adhesion to the rubber matrix. Optical microscopy and TEM are employed to study the structure of filled rubbers, and these have been reviewed extensively by Kruse [337] and Hobbs [338]. Preparation methods are similar to those discussed for rubber in tire cords, the ebonite method (Section 4.4.3), ultrathin cryosectioning (Section 4.3.5) and staining methods (Section 4.4). SEM analysis is more difficult than might be assumed, as often the filler is well adhered to the rubber and thus is not visible. The fracture morphology of elastomers has been studied by SEM, e.g. the flaw controlled fatigue fracture of styrene–butadiene rubber (SBR) and EPDM rubber by Eldred [339]. De and coworkers [340–342] are well known for using SEM in their fracture studies of rubbers, such as SBR and EPDM.

Carbon black particles are generally aggregates that appear as fine dense particles when observed in sections viewed in an optical microscope (Fig. 5.72A). However, processing problems can occur which result in significantly larger particles (Fig. 5.72B) which can be the locus of failure in the molded part. Particle size distribution is also critical for conducting polymers. It can be difficult to determine, as typically the large particles would have poor adhesion to the matrix and a fracture surface might well contain only a hole where the particle was located before failure. If such failures occur, optical assessment can provide leads to the problem, if larger than usual aggregates of

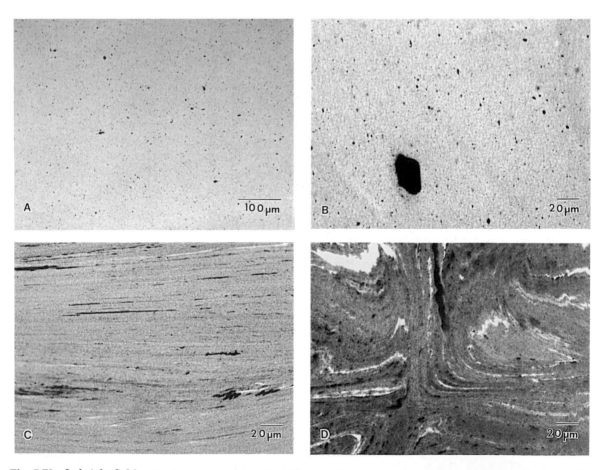

Fig. 5.72 In bright field optical micrographs, carbon black particles appear as dense particles (A). Such particles can aggregate (B) and provide a locus of failure. Streaks in molded parts, due to poor mixing of the carbon black, are shown in sections of the molded specimens (C and D).

carbon black are observed. At times, black molded parts are seen which are streaked, i.e. the black color appears nonuniform, and different gray levels may be observed (Fig. 5.72C). A more subtle surface variation, matte and gloss surfaces on the same part, may also be caused by poor or improper mixing of the carbon black during processing (Fig. 5.72D). Poor distribution can be a problem in any fiber filled or particle composite and is likely to cause variations in both physical and mechanical properties as a result. Optical microscopy shows the carbon black distribution and processing variants and

optimization can be monitored by optical techniques for both quality control and problem solving.

It is well known that carbon black particles are aggregates and the TEM can be used to image the individual particles in ultrathin sections (Fig. 5.73A). The dense particles are aggregates of individual carbon black particles. This micrograph is a good control for study of a multiphase polymer containing carbon black. A TEM micrograph of a black, polyurethane filled polyacetal is shown in Fig. 5.73B. Interestingly, the carbon black particles have enhanced the image contrast

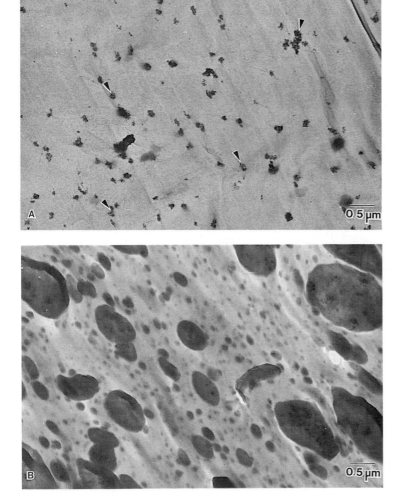

Fig. 5.73 Carbon black particles (arrows) are shown in a sectioned carbon black filled polymer (A) to be aggregates of smaller particles less than 0.1 μm in diameter. Interestingly, a black, multiphase polymer, shown in a TEM micrograph of an ultrathin section (B), has carbon black particles within the dispersed phase domains.

as they are located within the dispersed poly-urethane, and thus the dispersed phase can be observed without staining.

5.5 EMULSIONS AND ADHESIVES

5.5.1 Introduction

Emulsions, dispersions of one liquid in another liquid, find broad application in the fields of food, medicine, pesticides, paints and coatings, paper, printing and cosmetics. Emulsions include a broad range of liquids which consist of a stable, continuous liquid phase in which a second discontinuous immiscible liquid phase is present [343, 344]. In a broad sense, emulsions can be further classified as macroemulsions, latexes, colloids and microemulsions. The two types of emulsion, based on the size of the dispersed particles, are *macroemulsions*, where particles range from 0.2 to 50 μm, and *microemulsions*, with particles from 10 to 200 nm. Particle size also determines the optical clarity of the emul-sion. Macroemulsions tend to be milky white, with particles less than 1 μm ranging from blue–white to semitransparent, and microemulsions, with particles less than 50 nm are generally transparent. A *latex* is a special case of a water emulsion of rubber or polymer particles which is applied extensively in the paints and coatings areas. *Colloids* are defined as any particle, liquid or solid, which has some linear dimension between 1 nm and 1 μm and which when dissolved in a liquid will form a suspension. Thus, either 'emulsion' or 'colloid' can represent this entire group of liquids, and in fact, emulsions can be considered as colloidal suspensions.

Microemulsions [345, 346] are special types of emulsions that form spontaneously and have very small particles. Microemulsions are opti-cally clear, thermodynamically stable dispersions of two immiscible liquids obtained by the use of carefully adjusted surface active molecules (sur-factants). Both liquids in a microemulsion will be present in regions of the same order of magni-tude, with the 'dispersed' phase on the order of 10–100 nm. Aggregates of surface active mole-cules, or *micelles*, form into colloidal sized clusters in such a way that hydrophilic groups are directed toward the water. These definitions [347] are general in nature, but they suffice for the present purpose and the interested reader is directed to texts on this topic for a more rigorous discussion [343, 346, 348, 349].

A paper by Ninham [350] clearly shows that the understanding of emulsions is not trivial. That work on the theoretical forces due to liquid structures at interfaces is well beyond the scope of this book, but it stands as an example of the state of this rather complex field which must deal with the dynamics of liquid–liquid and solid–liquid interfaces.

The major structural unit of interest in emul-sions, microemulsions, colloids and latexes is the 'particle'. It is well known that the particle shape, size and distribution of a latex controls the properties and end use applications. Many latexes are manufactured with a controlled and sometimes monodisperse distribution of particle sizes. Polymer liquids, in the form of emulsions and adhesives, are wet and sticky, and therefore specimen preparation for electron microscopy is very difficult. As a result of the importance of the determination of particle size distribution, micro-scopy techniques have focused on specimen preparations which do not alter this distribution or which alter it as little as possible. Methods have included special cryotechniques (Section 4.9), staining-fixation methods (Section 4.4), microtomy (Section 4.3) and some simple meth-ods (Section 4.1) such as dropping a solution onto a specimen holder. This section is meant to provide a brief survey of the types of microscopy applications which have been found useful in the evaluation of emulsions and latexes.

5.5.2 Latexes

The structure and morphology of multiphase polymers have been discussed (Section 5.3) and the particle size and distribution have been shown to be quite important for both mechanical properties and applications. In many polymers (e.g. ABS), the particle size distribution is determined during the rubber manufacturing process. The surfactant concentration during emulsion polymerization controls the size distribution of the rubber latex, and subsequent grafting increases the size further. Particle sizes can be controlled to yield a range of sizes or a monodisperse latex. Particles that are larger than 1 μm in diameter are difficult to produce because they tend to coagulate. Crosslinking of the butadiene can occur during the process, but this can be controlled somewhat by the addition of inhibitors. Styrene and acrylonitrile monomers are added to the polybutadiene latex in the second stage of the emulsion polymerization process and new particles can form or polymer can be deposited on the surface of the polybutadiene where grafting can take place. The end result of such processes is a range of particle compositions, sizes and morphologies. Core–shell morphologies are commonly encountered in ABS latex where the polybutadiene is the core and the shell is styrene–acrylonitrile copolymer (SAN). Subinclusions of SAN can also form within the rubber phase. Subinclusions have also been shown to form in other polymers, such as in nylon containing polymer blends with rubber (Section 5.3).

Molau and Keskkula [351] were among the first to study the mechanism of particle formation in rubber containing polymers. They showed that phase separation occurs between the rubber and a vinyl polymer during the polymerization of solutions of rubber in vinyl monomers which is followed by formation of an oil-in-oil emulsion. During phase inversion of the emulsion, rubber solution droplets are formed which change into solid rubber particles in the final polymer. Structural investigations by phase contrast optical microscopy, shown in this chapter (Section 5.3), reveal dispersed particle size and distribution. Ugelstad and Mork [352] reported on new diffusion methods for the preparation of emulsions and polymer dispersions where the size and distribution of the latex particles were monitored by very simple optical, SEM and TEM methods. A microemulsion polymerization has been reported for the first time [353] with

spherical latex particles produced about 20–40 nm in diameter.

5.5.2.1 *OM, SEM and TEM characterization*

Optical and SEM techniques have been used to image micelles and latex particles by some straightforward preparation methods. Particles in suspension or air dried may be imaged directly by optical microscopy when the particles are well over 1 μm in diameter. Rather large poly(vinyl acetate) particles (Fig. 5.74A) and poly(vinylidene chloride) particles (Fig. 5.74B) are shown in optical micrographs. Film forming latexes cannot generally be examined by simple air drying as they tend to form a film. One example of a technique to prepare emulsion particles has been described (Section 4.9) by Katoh [354]. Poly(ethyl acrylate), with a softening temperature below room temperature, was freeze dried and the particles were coated with carbon and gold while at low temperature. Examination in the SEM showed the particles to be spherical. Direct observation of monodisperse latexes has been shown by a special optical technique [355] developed for understanding the stability behavior of monodisperse systems. A metallurgical microscope with a pinhole plate of aluminium foil positioned at the field aperture iris of the illumination tube was used to increase the resolving power. The concentrated sample was placed into a Pyrex glass tube (15 mm diameter) with a thin glass window. After equilibration, the cell was examined using an oil immersion objective. The image was displayed on a TV monitor and video recorded for measurement of interparticle distances. Kachar *et al.* [356] developed a technique of video enhanced differential interference contrast to study aggregates and interactions of colloids by real time experiments with a high resolution TV camera connected to an optical microscope equipped for Nomarski differential contrast.

The formation of micelles, or colloidal particles, by block copolymers in organic solvents has been described and reviewed by Price [357]. The molecular weight of polystyrene was estimated

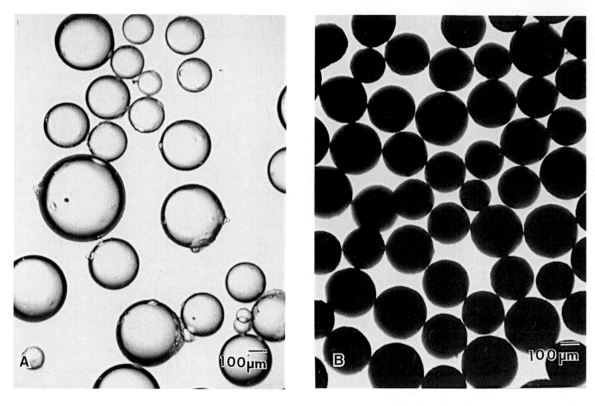

Fig. 5.74 Optical micrographs show the size and shape of poly(vinyl acetate) (A) and poly(vinylidene chloride) (B) beads.

from specimens prepared by spraying and evaporation for TEM. Freeze etching a drop of solution rapidly frozen with liquid nitrogen [358] was described (Section 4.9.4) where the solvent was allowed to evaporate and a replica produced of the fracture surface. Another method [359] was to allow a drop of an osmium tetroxide stained micellar solution to spread and evaporate onto a carbon film for TEM. Price *et al.* [360] investigated micelles from a polystyrene–poly(ethylene/propylene) block copolymer in a lubricating oil. The specimens were prepared for TEM either by casting a film on water and picking up the section on a carbon coated grid, followed by shadowing, or by painting the solution onto a freshly cleaved mica surface and coating the surface with carbon/platinum. Copolymers developed by an emulsion polymer-

ization process, from monomers with differences in hydrophilicity, were shown to be heterogeneous [361]. Lee and Ishikawa [362] prepared ultrathin, osmium tetroxide stained cross sections for TEM examination of 'inverted', core–shell latexes. Overall, there are a few specimen preparation methods that have been found successful in characterizing particle morphology.

TEM has proven to be the most effective technique for the characterization of the particle size distribution in emulsions. A dilute solution cast on a carbon film and metal shadowed shows an agglomerate of latex particles in a commonly encountered drying pattern where the shadowing method shows the particles are flat and obviously not separate or discrete (Fig. 5.75A). The area in Fig. 5.75B would also be difficult to use for particle size measurement, whereas in

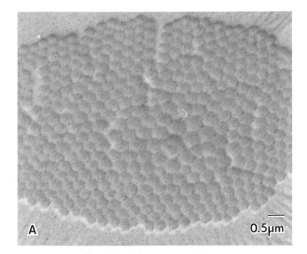

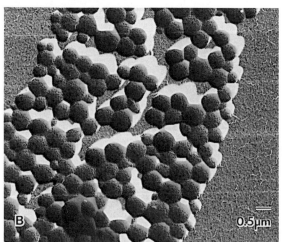

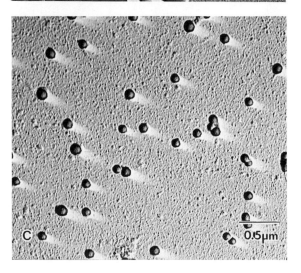

Fig. 5.75C the emulsion particles are well separated and the shadows show they are three dimensional in shape due to freeze drying.

A combination of staining, to enhance contrast, and a cold stage in the microscope has been applied to the study of latexes by TEM in order to limit such flattening and aggregation. Shaffer *et al.* [363] developed a phosphotungstic acid (PTA) staining method (Section 4.4.5) where the latex was added to a 2% PTA stain and then dropped onto a TEM grid for imaging. A cold stage was used in the microscope to limit any change in the particles during examination. The effect of the cold stage and PTA staining was shown earlier (Fig. 4.16A and B). TEM micrographs (Fig. 5.76) show the structure of a latex prepared by this method [364]. The two micrographs are printed to show the details of the structure; the less dense print (Fig. 5.76A) shows that a definite core–shell morphology is present whereas the denser print (Fig. 5.76B) emphasizes the network arrangement of the particles.

Shaffer *et al.* [365] have continued to modify staining techniques for TEM of latex particles. Recent work on structured latex particles prepared by seeded emulsion polymerization focused on the effects of changes in polymerization variables, such as batch versus semicontinuous, core–shell ratio, shell thickness and shell composition. In this system the core was poly(*n*-butyl acrylate) and the shell was poly(benzyl methacrylate–styrene). A few drops of the latex was combined with a few drops of a 2% uranyl acetate solution which serves as a negative stain. A drop of that mixture was deposited on a stainless steel formvar-coated grid. After drying it was stained in ruthenium tetroxide vapor to differentiate the rubbery core, which is not

Fig. 5.75 TEM micrographs of several emulsion particle samples show a range of aggregation. An air dried droplet (A) resulted in agglomerated flat particles. More three dimensional particles would still be difficult to measure as they are touching (B). The emulsion particles (C) are well dispersed and shadowing with chromium clearly shows that they are discrete spheres after freeze drying.

stained, from the dark shell which stains due to its ring structure. A method was more fully described and a figure shown (Fig. 4.18) of latex particles prepared by staining with OsO_4, RuO_4, and phosphotungstic acid to reveal full details of the core–shell particles.

Replicas of microemulsions have shown micelles to be spherical particles. Biais *et al.* [366] applied the method of freeze replication to show the micellar morphology by TEM. These authors used Freon 22 to fast freeze the microemulsion by methods already described (Section 4.9.4). Spherical structures, about 10–25 nm, have been revealed for oil–water or water–oil microemulsions stabilized by graft copolymers. Rameau *et al.* [367] used the same freeze fracture replica method as Biais and coworkers to image a copolymer of poly(vinyl-2-pyridinium) and polystyrene. A lengthy discussion of microemulsions [368] provides the key types of experimental results from such studies: phase diagrams, structures, thermodynamic considerations and a discussion of interfacial tension.

The film formation of latexes is well known. This process has been followed by both SEM [369] and TEM [370] as it is important in understanding how latex polymers form films coatings applications. Films of varied thickness, of 60/40 poly(styrene–butadiene), were aged for various times and the aging subsequently stopped by bromination [369]. They were then examined in the SEM [371]. The 'further gradual coalescence' process [372, 373] was shown to result in a change of particle shape as a function of aging [371]. A detailed TEM study of the effect of surfactants on the film formation of poly(vinyl acetate) (PVAC) and poly(vinyl acetate–butyl acrylate) (PVAC/BA) latexes has been reported when the rate of film coalescence was monitored by film replication [370]. TEM micrographs are shown as an example of both the replication method and the effect of aging on film formation. The surface of a film of PVAC/BA latex is shown after 8 h (Fig. 5.77A and B) and after 15 days (Fig. 5.77C); both were prepared with the same surfactant. Clearly, the particulate nature of the film is still obvious after 8 h but the film texture

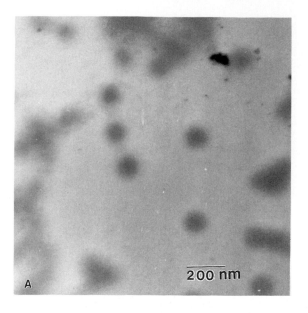

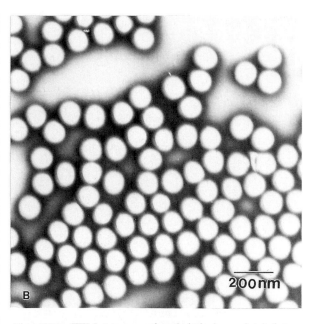

Fig. 5.76 TEM images of poly(ethyl acrylate) latex particles taken using a cold stage (A) has indistinct particle edges, whereas the same latex negatively stained with 2% aqueous PTA has clearly imaged edges (B). (From O. L. Shaffer, unpublished [364].)

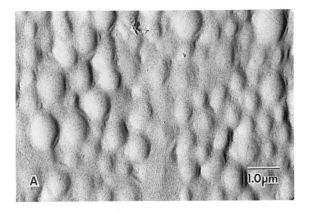

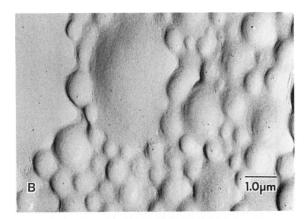

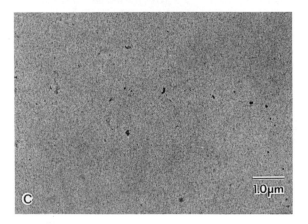

Fig. 5.77 Vinyl acetate latex film coalescence is shown by TEM of platinum–palladium–carbon replicas from aged latex films cast on glass. TEM micrographs show the particulate nature of a film aged for 8 h (A and B) compared to the flat film observed after 15 days aging (C).

has changed with time, and after 15 days there were no obvious surface details remaining.

5.5.2.2 *Particle size measurements*

Latex particle size measurements are generally conducted by either light scattering or by electron microscopy. Rowell *et al.* [374] reported rigorous measurements of polystyrene latex particles by both techniques, and the average values agreed within one per cent. The preparation for TEM involved the drying of a drop of the diluted latex onto a replica of a diffraction grating, shadowing it with carbon and taking direct measurement from enlarged micrographs. Micelles of block copolymers of polystyrene and poly(dimethyl siloxane) in *n*-alkanes were shown by a similar preparation when the diluted solution was dropped onto a carbon coated grid for TEM [375].

Image analyzing computers are now routinely applied to the measurement of structures observed in microscope images. Quantitative microscopy involves several steps which all must be considered as part of the analysis. The sample must be prepared and the actual microscopy performed. Key factors are to ensure that the specimen does not change dimension during preparation or during microscopy, that a representative sample is prepared and analyzed and that the calibration of the microscope is accurate. Gratings are used to calibrate the microscope, and standard polystyrene latex is used to have a check on the change in particle size as a function of the method, the vacuum and other microscope conditions. Analysis directly from TEM negatives avoids an additional step of printing micrographs. The actual measurement of particle dimensions with an automatic image analyzer is trivial compared to these earlier steps, although due consideration must be given to statistical experimentation, sampling and presentation of the distribution data. An example of the determination of particle size distribution which is generally applicable is the study of a latex prepared by freeze drying (Section 4.9.2) [171]. Figure 5.78 shows TEM micrographs of air and

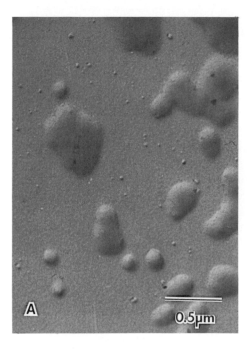

Fig. 5.78 A typical film forming latex is shown in the air dried preparation (A) where discrete particles are not obvious compared to a freeze dried and shadowed preparation (B).

freeze dried latex specimens. The flat film forming nature of the air dried latex (Fig. 5.78A) is in sharp contrast to the three dimensional particles and shadows seen after freeze drying (Fig. 5.78B).

5.5.2.3 *AFM characterization*

The specimen preparation required for atomic force microscopy of polymers is minimal. The ability to examine a wet specimen eliminates artifact formation due to drying and the effect of vacuum. In the case of latex samples, the measurement of the particle diameter and distribution is also enhanced by the ability to resolve fine details and to easily make digital measurements. Issues with AFM imaging relate to instrumental parameters, especially whether the AFM is in the contact or non-contact mode (see Chapter 6). Two images are shown in Fig. 5.79 of polystyrene and poly(ethyl methacrylate) (PS/PEMA) latex, imaged in the non-contact mode with a Park Autoprobe AFM. Figure 5.79A

shows an uncleaned latex to have some surfactant remaining on the surface of the particles and interfering with packing whereas Fig. 5.79B shows a cleaned surface with little extraneous detail and excellent packing.

Another example of AFM imaging compared to FESEM imaging [376], is shown in Figs. 5.80 and 5.81 of latex particles which are used for modifying epoxy to improve toughness [377]. Figure 5.80 is an FESEM image [376] of a whitened region from a three point bend fracture surface of epoxy modified with carboxyl-terminated butyl nitrile rubber (CTBN) particles. AFM of the same specimen shows the particle size correlates well with the FESEM image. The AFM image in Fig. 5.81A [376] provides much more detail at the particle–epoxy interface which can be compared with the fracture toughness measurements. Figure 5.81B–C is a line scan analysis of an interesting feature which shows the epoxy forming a bridge between the rubber particles. The height of the feature is about 530 nm.

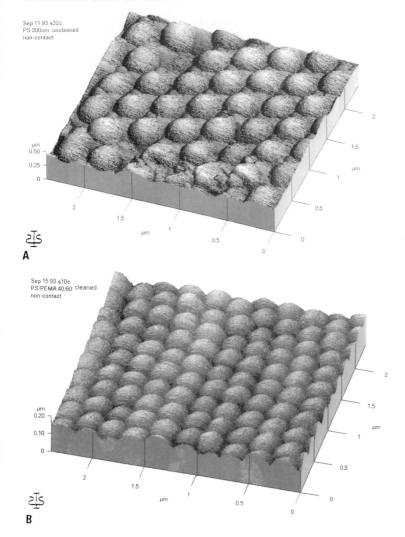

Fig. 5.79 AFM in the non-contact mode of (A) an uncleaned PS/PEMA latex and (B) a cleaned PS/PEMA latex. (From O. L. Shaffer, unpublished [376].)

5.5.3 Wettability

The spreading of a liquid onto a substrate relates to such applications as the coating of liquids on paper and the spreading of binders and finishes on fibers. Such wettability studies are affected by the roughness of the solid surface and the manner in which the liquid spreads on the surface. Polymers may provide the solid substrate, the spreading liquid, or both. Mason and coworkers [378–381] have described interesting theoretical and experimental results relating the effect of surface roughness to wetting. These authors have explored the concept of wettability theoretically [379], and they have described the equilibrium contact angle which a liquid surface makes with a solid it contacts as a measure of that wettability [380]. The SEM was used to demonstrate and confirm these theoretical projections [378, 381]. In the early study [378], molten drops of PE and PMMA were allowed to spread and solidify on a paper substrate prior to standard preparation for SEM. Later, poly(-phenyl ether) (PPE) vacuum pump oil (Santovac-5-Monsanto Chemicals), which is not volatile, was used for dynamic studies. The PPE was fed

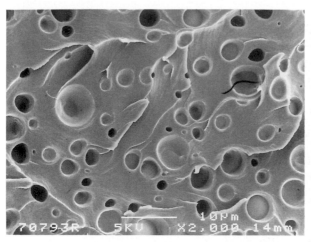

Fig. 5.80 An FESEM (5 kV) image of a whitened region from a three point bend fracture surface of epoxy modified with carboxyl-terminated butyl nitrile rubber (CTBN) particles. (From O. L. Shaffer, unpublished [376].)

through a hole in the sample stub, from outside the specimen chamber, and the wetting experiment was recorded on video tape [381]. Mori *et al.* [382] used freshly cleaved mica surfaces as steps 60 nm in height to inhibit wetting. Surface roughness has a major effect on the local contact angle between the liquid and the substrate of interest.

5.5.4 Adhesives and adhesion

All adhesives are polymers, and they are used in many ways, for example in composites, automotive tire cords, plywood, tapes and labels. A particularly demanding application is the cementing of metal joints in military aircraft with polymers such as epoxy resins. The interfaces in such materials must be characterized to determine the strength of the adhesive bond and the relation of such properties as peel strength with morphology. The morphology of the adhesive fracture surfaces is generally investigated in the SEM. The topic of adhesion science and adhesion and absorption of polymers has been described in volumes edited by Lee [383–385].

Smith and Kaelble [386] conducted a study to

determine the adhesive failure of a metal–polymer system. A multidisciplinary study was used to describe the aluminium alloy and the epoxy adhesive. Ellipsometry was used to estimate oxide film thickness and optical properties. Optical microscopy, SEM and TEM established the topography, and wettability parameters were calculated from contact angle measurements and bond strengths. Morphological assessments of the polymer adhesive–metal joint are made by SEM of the fracture surface or by production of transverse sections. Hahn and Kotting [387] prepared transverse sections by machining followed by ion etching with argon to remove the smeared structure of the phenolic and epoxy resins. The SEM showed a variation in adhesive morphology as a function of location with respect to the metal part which further depended upon the adhesive type, cure, condition and nature of the metal surface.

Brewis and Briggs [388] showed pretreatments to be important in adhesion. Surface pretreatments for polyolefins include chlorine, ultraviolet radiation, dichromate/sulfuric acid, hot chlorinated solvents and corona discharge. These authors reviewed the nature of the changes and mechanisms associated with pretreatments, including references to SEM and TEM studies of the resulting structures. Surface modification of PE by radiation induced grafting resulted in improved wet peel strength, and optical and SEM techniques showed a change in bond failure in a wet environment [389]. A surface graft with good adhesive bonding to epoxy adhesives was produced by vapor phase grafting of methyl acrylate onto PE. The SEM was also used to observe the surfaces of molded, roughened and etched polyolefins, PVC and ABS both before and after metallization [390]. The study showed that the metal bonds to the plastic mainly by mechanical anchoring.

Much study of adhesion has focused on developing improved adhesives and composites for aerospace applications [391, 392]. Surface analysis techniques such as SEM and XPS are commonly employed to analyze fracture surfaces, and contact angle measurements have also

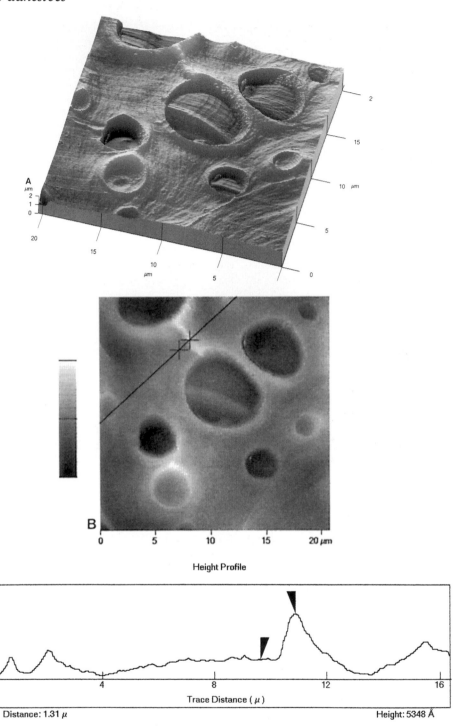

Fig. 5.81 AFM of the same specimen as in Fig. 5.80 shows the particle size correlates well with the FESEM image. The AFM image (A) [376] provides much more detail at the particle–epoxy interface which can be compared with the fracture toughness measurements. A line scan analysis (B–C) of an interesting feature shows the epoxy forming a bridge between the rubber particles. The height of the feature is about 530 nm. (From O. L. Shaffer, unpublished [376].)

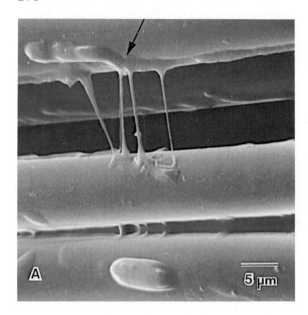
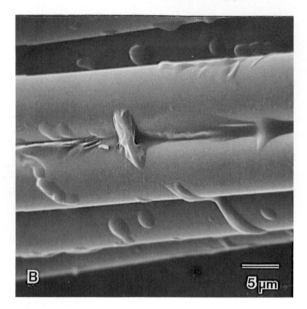

Fig. 5.82 SEM images show the morphology of surface finish coated glass fibers. The finish is seen in several morphologies; as a thin film coating on the surfaces, as lumps of material on the surfaces and connecting fibers, and as etchings or fibrils spanning across the fibers.

supported such studies. Adhesives and adhesion are quite important in two other major areas of application: as adhesives bonding rubber in automotive tires and as sticky tapes and labels. Two papers, by two separate groups working simultaneously, have described the structure of the resorcinol–formaldehyde–latex (RFL) adhesives used for bonding rubber to tire cords. Rahrig [393] suggested that the RFL adhesives are two phase systems with an interpenetrating network morphology based on his dynamic mechanical and thermal analysis studies. Meantime, Sawyer *et al.* [394] reported on a method and its application to actual automotive tires, whereby the fine structure of the RFL and, more importantly, the interfacial morphology, were clearly shown. Micrographs were described earlier (Figs 4.14 and 5.19); these reveal the complex RFL morphology by an ebonite staining (Section 4.4.3) microtomy method.

Two more examples of the application of microscopy techniques to adhesive structural studies are fiber 'finishes' and adhesive labels. Glass fibers are usually pretreated with a polymer coating which protects the fibers during

handling and which may be designed to maximize adhesion in composites. The SEM was used extensively during the early 1970s, first to evaluate and then to control finish application [395] on glass fibers. SEM of glass fiber surfaces (Fig. 5.82) show coatings ranging from thin and uniform to 'lumpy'; these coatings hold fibers together by ductile 'strings'.

Sticky tapes and labels are obvious uses of adhesives that cover a wide range of everyday applications. A relatively new type of adhesive has apparently been used in the manufacture of Post-it (trademark, 3M) products, small slips of paper with a reusable adhesive strip at one edge. SEM images show the adhesive partly coating the paper fibers (Fig. 5.83) [396]. Rounded domains of about 5–50 μm form a contact on the applied surface (Fig. 5.83A). Regions between these rounded domains have a fine particulate structure which does not adhere, permitting the label to be removed easily. In order to image the adhesive nature of these materials, two adhesive strips were attached and then peeled back and examined in the SEM. (The adhesive strip and a piece of paper peel away from each other too

Fig. 1.3 A thin section of bulk crystallized nylon, in polarized light, reveals a bright, birefringent and spherulitic texture. At high magnification a classic Maltese cross pattern is seen, with black crossed arms aligned in the position of the crossed polarizers. (A) was isothermally crystallized, with large spherulites. (B) has been quenched during crystallization, giving large spherulites surrounded by smaller ones

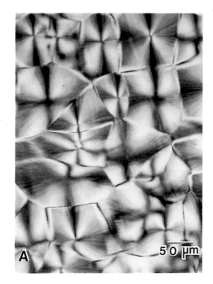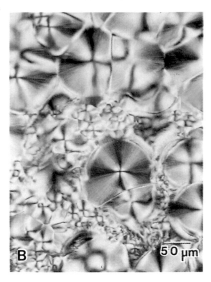

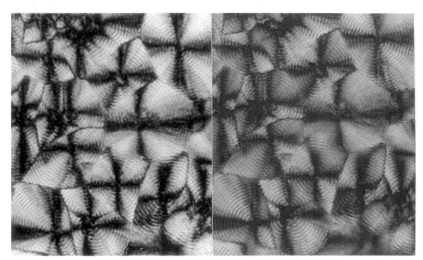

Fig. 3.13 Two images of a thin melt-cast film of high density polyethylene. The region is $200 \times 250\mu m$. The left-hand image was taken under crossed polars. For the right-hand image, a first order red plate was also used. The spherulites in this material are quite symmetric and regular, and have circumferential bands caused by a regular twist of the crystal orientation as it grows. The blue and yellow colors produced by the first order red plate show that the spherulites are negatively birefringent (see text).

Fig. 3.14 Two images of a thin melt-cast film of polycaprolactone. As in Fig. 3.13, the region is $200 \times 250\mu m$, the left-hand image was taken under crossed polars and for the right-hand image, a first order red plate was also used. The spherulites in this material are much less regular, and some show colors under crossed polars. The first order red plate image can still show that the spherulites are negatively birefringent.

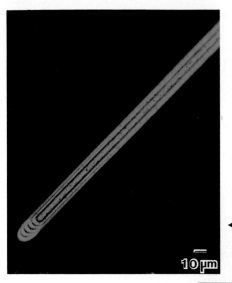

Fig. 5.4 A polyester fiber in polarized light, aligned at 45° to the crossed polarizers. The dark bands are fringes that reflect the high order birefringence. In the orthogonal position the same fiber would exhibit

Fig. 5.17 Combination of optical microscopy and EDS analysis permits the identification of contaminants plugging a spinneret. Rust colored material present on the spinneret (A) was scraped off to give the EDS spectra (B). The plugging material contains silicon, phosphorus, antimony, titanium, chromium and iron. The copper is due to the specimen holder.

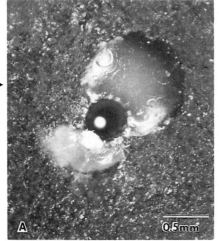

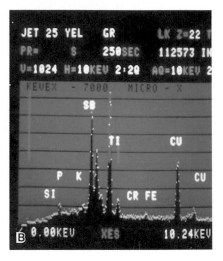

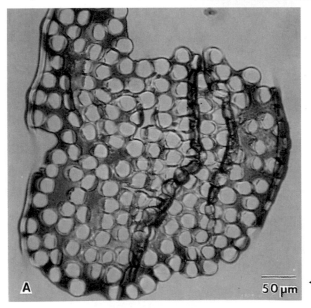

Fig. 5.18A Tire cords: a cross section of an adhesive coated yarn is shown by optical microscopy.

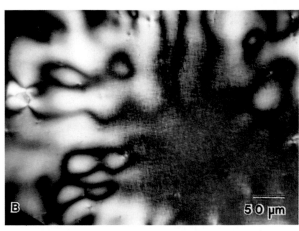

Fig. 5.44 STM provides a direct measurement of the ► depth of a pit. This is the height difference between the media surface and the center of the pit, as seen in this image, especially where the depths of structures are very shallow compared to the three dimensional geometry. The total height variation in the image is 77 nm divided into 15 different colors; the depth of the pit is *ca*. 41 nm. (From H. A. Goldberg *et al.*, unpublished [203].)

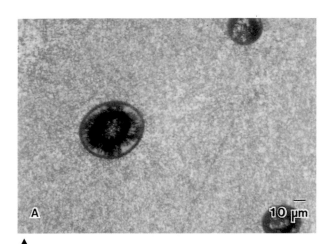

▲
Fig. 5.85 Thick and thin regions of a thermotropic melt structure in polarized light. In a thick region (A) the fine structure is not too clear but the onset of decomposition is shown by the round bubbles. A thinner region (B) shows thread-like detail and a nematic texture with four brushes.

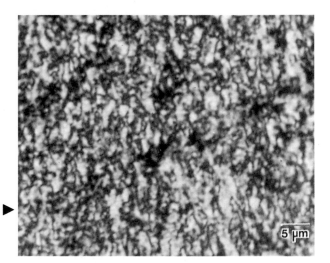

Fig. 5.86 A polarized light micrograph of a section cut ► from a molded article reveals a complex, fine nematic texture with no obvious orientation. The color in the image enhances the detail.

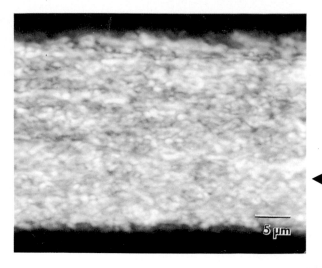

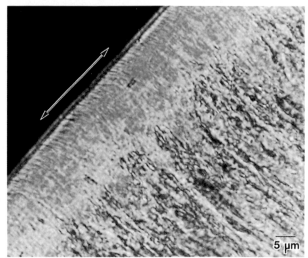

Fig. 5.87 Incomplete extinction in uniaxially oriented TLCP fibers, ribbons and films gives a 'salt and pepper' texture which is seen as individual domains less than 0.5μm across. The similar polarization colors in polarized light suggest the domains are within the same order, and thus have similar birefringence.

Fig. 5.89 Polished thin sections of a low orientation extrudate observed in polarized light. An unoriented free fall strand shows banding normal to the strand axis (arrow) and away from the slightly oriented skin.

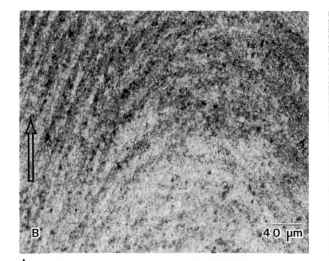

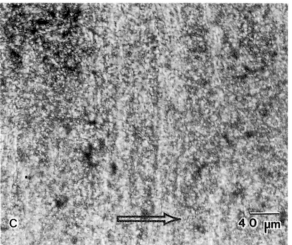

Fig. 5.91 Lateral, curved flow patterns are seen in a polarized light micrograph overview of a thin section (B). The flow layers are nearly normal to the flow direction (arrow) in the center of the bar (C).

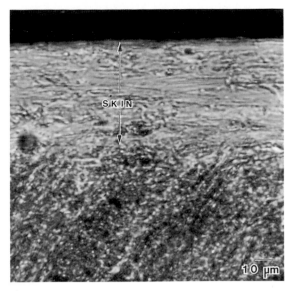

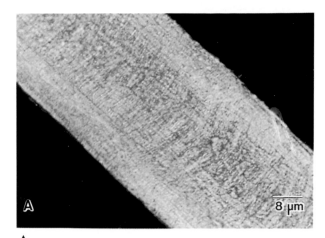

Fig. 5.95A Free fall TLCP strand is shown in polished section in circularly polarized light.

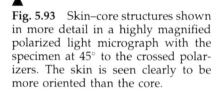

Fig. 5.93 Skin–core structures shown in more detail in a highly magnified polarized light micrograph with the specimen at 45° to the crossed polarizers. The skin is seen clearly to be more oriented than the core.

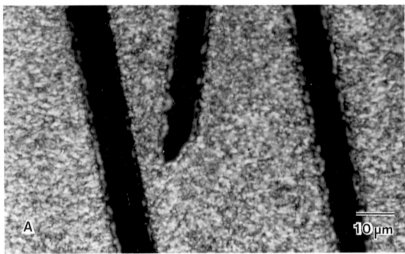

Fig. 5.99A A glass fiber reinforced LCP composite is shown to have interesting morphology. A polished thin section is shown in polarized light to exhibit a fine domain texture with some orientation of the polymer on the glass surfaces.

Fig. 5.104 Optical micrograph of an ultrathin longitudinal section of an extruded experimental PBZT film (on a TEM grid) in polarized light with the fiber at 45° to the crossed polarizers. A definite skin–core texture is observed as the skin appears bright yellow and the core appears blue and less oriented.

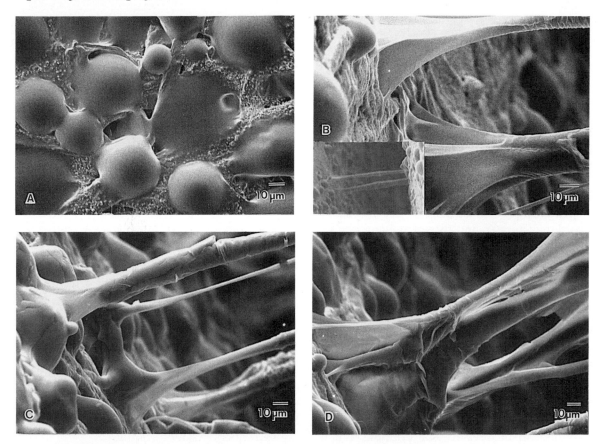

Fig. 5.83 The morphology of an adhesive layer on a Post-it (3M trademark) product is shown in these SEM micrographs. The surface of the adhesive coating appears to be composed of oblate spheroidal shaped particles (A). A thinner, particulate filled coating appears to cover the paper fibers between these adhesive particles. Two strips of adhesive were attached to one another and partially pulled apart, as shown in the insert (B). Strings of adhesive are seen to connect the two strips (B–D). (From D. R. Sawyer, unpublished [396].)

easily to permit imaging of the adhesive mode.) Figure 5.83B–D shows this adhesive action, with an overview of the specimen inset in B. Thus, the adhesive has a low contact area compared to the total surface area, permitting easy removal. The SEM clearly shows the morphology of such adhesives.

5.6 LIQUID CRYSTALLINE POLYMERS

5.6.1 Introduction

The development of high performance polymers, such as high modulus fibers and super-tough polymer blends, has accelerated in recent years

as a direct result of increased knowledge of process–structure–property relationships. Highly oriented materials have been produced by modification of conventional polymers [65, 397] and by the design of rod-like, liquid crystalline polymers. A liquid crystalline polymer (LCP) is one which forms a partially ordered state on heating (thermotropic LC) or in solution (lyotropic LC). Liquid crystalline states in polymers are generally classified in the same way as LC states in small molecules [398, 399]. The degree of molecular order in liquid crystals is intermediate between the three dimensional order in solid crystals and the disorder of an isotropic liquid. A liquid crystal can be nematic, cholesteric or smectic due to their degree of molecular order. Nematic crystals are ordered in one dimension; the long axes of the molecules are parallel and the local direction of alignment is called the 'director'. Cholesteric crystals have the director ordered in a spiral fashion, and colors appear if the twist of the period of the spiral is the wavelength of light. Smectic crystals have their molecules parallel and arranged in layers.

Heating a thermotropic liquid crystal results in decreasing the molecular order. The general pattern is as follows below, but not all possible phases may appear, and there are many types of smectic crystals. In addition, the LC phase may appear upon cooling rather than upon heating.

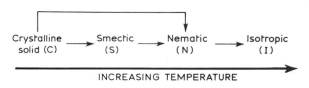

In many cases unique optical textures are observed for the various orientations and structures of the three classes of liquid crystals. Thin films of nematic crystals, for example, can be identified by the pattern of dark threads (isogyres) which can appear in the optical microscope in transmission with crossed polarizers. Hot stage polarizing optical microscopy is often used to identify the phases and the transition temperatures. In some cases, the optical texture is not uniquely identifiable and x-ray diffraction and thermal analysis by DSC are used to complement the microscopy.

Liquid crystalline polymers have been discussed in many texts and review papers [65, 400–413] during the last decade, in which the synthesis, processing, morphology, orientation and structure–property relations are described. The major applications of these materials have been as high modulus fibers and films, with unique properties due to the formation of ordered lyotropic solutions or thermotropic melts which transform easily into highly oriented, extended chain structures in the solid state.

Thermotropic polymers are melt processable and thick extrudates and molded parts are formed with high strength characteristics as in fiber reinforced thermoplastics.

There are thousands of liquid crystalline polymers that can be considered in three commercially important classes:

(1) aromatic polyamides;
(2) 'rigid rod' polymers;
(3) aromatic copolyesters.

High modulus fibers from lyotropic aromatic polyamides, poly(p-phenylene terephthalamide) (PPTA), were first commercialized under the Kevlar trademark by DuPont [414]. The aromatic polyamides, or *aramids*, are produced by a dry jet–wet spinning process where the nematic structure in solution is responsible for the high modulus fiber performance [415–419]. Another class of lyotropic fibers, also produced by dry jet–wet spinning, are the rigid rod polymers developed as part of the U.S. Air Force Ordered Polymers Program [420–424]. The most common of these ordered polymers, poly(p-phenylene benzobisthiazole) (PBZT), is difficult to process, but it exhibits the highest tensile properties of all the LCP fibers produced to date.

Thermotropic aromatic copolyester fibers are produced by melt spinning since the combination of an aromatic backbone and flexible segments results in an LCP which can be melt processed. Copolyesters, such as p-hydroxybenzoic acid (PHBA) modified poly(ethylene terephthalate) (PET), 60/40 PHBA/PET (X7G) polymers, have been produced by the Tennessee Eastman Company [425–427]. Other melt processable nematic thermotropic LCPs (TLCP), based on combinations of 2,6-naphthalene dicarboxylic acid (NDA), 2,6-dihydroxynaphthalene (DHN) and 6-hydroxy-2-naphthoic acid (HNA), and referred to as Vectra LCP resins, have been commercialized (trademark by Hoechst Celanese Corporation [406]).

Mobile LCPs may be mounted between a glass slide and a glass cover slip for polarized light microscopy. It is important to realize that the glass surfaces can have a strong effect on the orientation and structure of the LCP. Polishing the glass can align the director along the polishing grooves and rigorous cleaning encourages the director to be normal to the glass surfaces. This orientation is called 'homeotropic' because the specimen, viewed along the director, may appear dark, as if it were isotropic (e.g. ref. 428). On cooling thermotropic LCPs, the structure of the liquid crystalline state can be 'frozen in' and studied in solid samples. Specimens for OM may then be prepared by microtomy and polishing. Even if the material becomes more ordered, that is semicrystalline, the structure is closely related to that in the LC state. SEM of solidified LCP fibers follows standard preparation which includes analysis of the surfaces, peelbacks (Section 4.3.1) and fractures (Section 4.8). TEM requires thin specimens such as those produced during sonication, dispersion and disintegration (Section 4.1.3) or ultrathin sectioning (Section 4.3.4).

Optical microscopy, SEM and TEM studies of LCPs have been reported [429–430]. Recently, imaging techniques with higher spatial resolution have been applied to determine the size, shape and organization of microfibrillar structures [431–435]. FESEM at low voltages, and scanning tunneling microscopy (STM), are capable of imaging regions from 1 nm to many micrometers on the same specimen. FESEM imaging follows standard SEM preparation methods, although if a metal coating is used it must be thin and very fine grained or the metal grain structure will be resolved in the image (Section 4.7). In the case of STM images the same fine grain metal coating can be used. Ion beam microsputtering platinum at <5 nm thickness worked well for samples placed on HOPG surfaces. For AFM imaging no special coating is needed as non-conductive specimens may be imaged. Sections, ultrasonicated materials, peeled fibers or films can be imaged by FESEM, STM or AFM techniques. The experimental observations obtained using these novel characterization techniques have been correlated and compared to results of observations from more traditional imaging methods, such as polarized light microscopy (PLM) and TEM, to aid in the interpretation of structures in images from these newer techniques. This section will include an extended evaluation of LCPs in order to provide an example of the application of both the older and new techniques.

5.6.2 Optical textures

Characterization of the optical textures of LCPs is key in the identification of the specific phases present and in understanding the structure and its relation to the solid state properties. Dynamic hot stage microscopy experiments with video tape recording provide images of the texture changes associated with phase changes, as a function of temperature and time. The majority of the optical textures reported in the literature are of either melt or melt quenched structures, although the optical textures present in solid fibers, moldings and extrudates have been described [429, 430]. Examples from studies reported in the literature will be reviewed and optical structures shown.

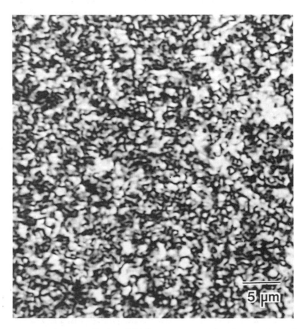

Fig. 5.84 A polarized light micrograph of a sectioned nematic TLCP reveals a schlieren texture.

Anisotropic nematic, thread-like textures have been observed for both thermotropic [429, 430, 435] and lyotropic [436] LCPs. It is interesting that rheo-optical studies [437] of a slow cooled thermotropic LCP and a high density non-LCP polyethylene both exhibit similar textures in polarized light [437]. These are shown in a nematic LCP (Fig. 5.84) and for comparison in polyethylene (Fig. 5.3). The textures appear similar except for their size scales.

Mackley *et al.* [438] observed nematic threads, or isogyres, in the X7G mesomorphic polymer. Viney *et al.* [439] also studied X7G in polarized light and observed changes as a function of temperature when heating thick sections between glass slides. Thick and thin regions form within the melt, with the thin regions frozen to one surface of the glass upon cooling. The appearance of these regions is quite different, but the differences are due to the superposition of structures. In crossed polarizers a thick region (Fig. 5.85A, color section) has a thread-like texture with little detail compared to a thin region (Fig. 5.85B, color section). Observation of the thin film reveals a definite nematic texture with line singularities, seen as points where generally two or four dark brushes meet; in this specimen four brushes are observed. It is important to recognize the variation in textures with specimen thickness as sections of LCP articles are thick and quite complex, as shown in a TLCP molding (Fig. 5.86, color section), compared to 1 μm thick sections produced by microtomy or in a thin melt cast film (Fig. 5.85B, color section). Sections produced by a range of methods have similar domain textures dependent only on thickness.

Optical studies of uniaxially aligned TLCP fibers, films and ribbons observed in the orthogonal position in polarized light exhibit a 'salt and pepper' texture and incomplete extinction. Close examination of this structure shows a fine *domain* texture. The individual domains, about 0.5 μm in diameter, are regions of local order which can be seen in micrographs of thin sections (Fig. 5.87, color section). There is a slight color variation between domains which suggests they

are distinct units with similar birefringence, and they may be the result of a serpentine molecular trajectory [430]. The fiber viewed at 45° to the polarization direction appears highly oriented, as expected.

5.6.2.1 *Banded structures*

Banding has been observed in both lyotropic and thermotropic polymers examined by optical and electron imaging techniques [430, 431, 439–446]. Incomplete extinction has been observed for some of the aramids where 'bands', normal to the fiber axis, are observed in polarized light [440] (Fig. 5.88). It is now well known that the aramids exhibit axial banding having periodicities of about 500 nm, which is observed by dark field TEM [447]. Simmens and Hearle [440] have proposed that the optical observations and the pleated sheet model of Dobb *et al.* [447] are compatible and that the optical bands are the bends or folds between the pleats which might well exhibit the local density differences observed by DF TEM.

A lateral banded texture was also observed in the solid state core of poorly oriented extrudates

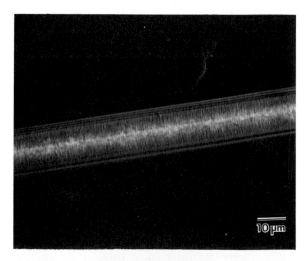

Fig. 5.88 Incomplete extinction is also observed for PPTA fibers where bands arranged laterally across the fiber are thought to be a function of the pleated sheet structure.

[430] and for many model studies in which the aromatic copolyesters were sheared for sample preparation [439, 441–442, 444]. Viney *et al.* [439] observed bands in thin melt formed copolyester films after rapid cooling and suggested they were due to a periodic variation in orientation. Recently, Viney and Putnam [445] and Putnam and Viney [446] characterized formation of bands by shearing LCPs between parallel glass plates on a light microscope stage. They measured the rate of evolution of the banded microstructure. In this study, in hydroxypropyl cellulose (HPC), banded textures were observed to form after shearing stopped [445, 446]. More studies of this type are required to fully understand the genesis of the banded textures.

An example of the banded structures observed in oriented TLCPs is shown in a polished thin section of an extruded rod with low average orientation (Fig. 5.89, color section). Banding is exhibited in regions away from the outer surface or skin with band widths of about 500 nm. Banding has also been seen in aramid fibrils in polarized light [447] and microbanding has been observed in TEM bright field images of spun aromatic copolyester fibrils [430], as shown in Fig. 5.90. This microbanding, on a scale less than 10 nm, has not been observed for heat treated Vectran LCP fibers which had increased orientation.

This periodicity of about 500 nm has been reported for banded textures observed in both the thermotropic copolyesters and the aramids [407, 430, 439–445]. The lateral banded textures exhibited by some of the thermotropes and the pleated sheet textures exhibited by some of the aramids are observed only in materials which have a poorer degree of molecular orientation. The more highly oriented thermotropic and lyotropic fibers do not exhibit these textures. For instance, heat treated Vectran fibers do not exhibit any lateral banding. Likewise, Kevlar 149 exhibits a higher tensile modulus than Kevlar 49 [448], nearly 80–90% of theoretical predicted values. This is consistent with increased crystallinity and crystallite size, and without a pleated sheet structure. The relationship of the optical

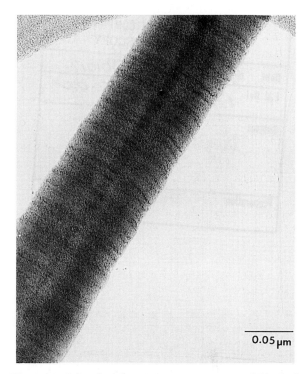

0.05 μm

Fig. 5.90 Microbanding is seen in a sonicated fibril of a spun TLCP fiber in a bright field TEM micrograph. Fibrils sonicated in ethanol are dropped onto holey carbon coated grids and carbon coated. Heat treated, high modulus fibers do not exhibit such banding.

textures, the domains, to the fibrillar textures is also important. The polarized light micrograph of the tape in Fig. 5.87 (color section), shows that the domains appear to be aligned along the fiber axis. The elongated, worm-like domains consist of fibrils which meander in and out of the plane of the section, parallel to the director or fiber axis [435].

The interpretation of Donald and Windle [441, 442, 444] and Sawyer *et al.* [430, 435] is that the bands are associated with a serpentine path of the molecules along the shear direction, consistent with the fine domain structure in the highly oriented TLCP fibers. The analogy to the meander reported for poly(*p*-phenylene benzobisthiazole) (PBZT) [442] and the sharp path caused by the pleated sheet structures of some aramid fibers [447] are all consistent. It is likely

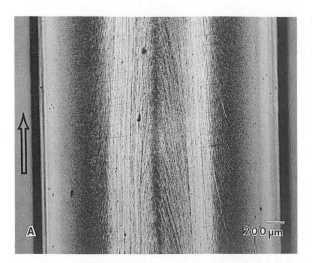

Fig. 5.91 A reflected light micrograph (A) shows the layered structures in a molded bar aligned parallel to the flow direction (arrow). Variation in density and color reflect variation in orientation from layer to layer. Lateral, curved flow patterns are seen in a polarized light micrograph overview of a thin section (B) (color section). The flow layers are nearly normal to the flow direction (arrow) in the center of the bar (C) (color section).

that the diffuse, dark zones outlining the domains as shown by TEM are domain walls [442]. Anwer *et al.* [449] suggested that these boundaries are analogous to the walls observed for small molecule liquid crystals [398, 399]. Finally, it is clear from TEM images of ultrathin sections of fibers and the STM images of peeled fibers, both shown in the next section, that the fibrils and microfibrils are not perfectly aligned within the sections. 'Y' shaped junctions are observed which provide strong evidence that there is a meander of the microfibrils, although on average the microfibrils are very well aligned in the extrusion direction.

5.6.3 Solid state structures

5.6.3.1 *Molded parts and extrudates*

Thermotropic LCPs are melt processed by injection molding and extrusion to form highly oriented rods, strands and molded articles. It is

well known [425, 450, 451] that extrudates have high molecular orientation which develops as a result of the effect of the flow field on the easily oriented extended chain molecules. Molded articles composed of thermotropic LCPs have properties that are better than short fiber reinforced composites, and they have been termed 'self-reinforcing' [452, 453]. Highly anisotropic physical properties are explained by the highly anisotropic structures: layers, normal to the flow direction; bands, parallel to the flow; and a skin–core structure. The skin–core structure observed for LCPs is similar to the structure of typical thermoplastics (Section 5.3) with orientation a maximum at the surface (skin), due to elongational flow, and at a minimum in the core, due to shear flow [178, 451, 453, 454, 455]. The layered structures observed in moldings are apparent by eye due to the color variations associated with local orientation differences between the layers. Microscopy permits assessment of structure–property correlations as the highly anisotropic structures are process dependent and relate to mechanical properties.

Thermotropic LCP molded bars exhibit a layered structure as shown by reflected light (Fig. 5.91A) of a cut and polished bar. Thin sections of a molded bar show fine, nematic domains with superimposed flow lines (Fig. 5.91B, color section), especially near the center of the bar (Fig. 5.91C, color section). Skin–core morphologies are obvious in injected molded bars and extrudates with domains aligned in the flow direction. Complementary SEM assessment of fractured injection molded bars provides an overall view of the layered structure (Fig. 5.92A), the surface skin (Fig. 5.92B) and the internal fibrillar structures of the inner skin (Fig. 5.92C) and core (Fig. 5.92D).

Complex skin–core and banded textures are observed in extrudates, where the orientation is a function of the draw ratio and the final diameter, with higher orientation in finer strands. The orientation and incomplete extinction in the skin are shown in a section taken with the flow axis at 45° from this position (Fig. 5.93, color section). Nematic domains are seen in the core, and to

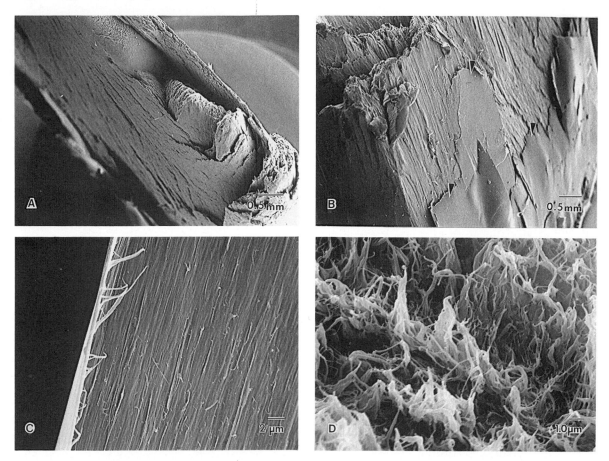

Fig. 5.92 The layered structure of molded bars is shown by SEM of fractured LCP specimens. An overview (A) shows the concentric layers are thin while the outer, coherent skin (arrows) is layered or sheet-like in structure (B). Fibrillar structures are observed in the skin (C) and core (D).

some extent in the skin, although the extinction (black) regions in the skin (Fig. 5.93) reflect higher orientation.

The common woody or fibrillar fracture of extrudates is clearly seen in the SEM micrograph in Fig. 5.94. This fracture morphology is controlled, to some extent, by the orientation of the strand. A micrograph of a polished thin section of a slightly oriented TLCP rod photographed in circularly polarized light reveals a nematic structure, banding and evidence of skin orientation (Fig. 5.95A, color section). SEM fracture surfaces reveal a more coherent, fibrillar structure in the slightly oriented rod (Fig. 5.95C) compared to the unoriented rod (Fig. 5.95B).

Etching experiments were also conducted in order to elucidate the fine structure of the extrudates. Plasma etching of a polished extrudate was monitored by high resolution SEM. An overview (Fig. 5.96A) shows the flow lines of the polymer in a region midway between the skin and core of the strand, and at higher magnification, grain-like domains consist of parallel structures (Fig. 5.96B and C) not parallel with adjoining domains. A detailed view (Fig. 5.96D) shows the fine fibrillar texture of these local regions. The etching experiments appear to reveal domain structures, which have a high degree of nematic order aligned by the flow process.

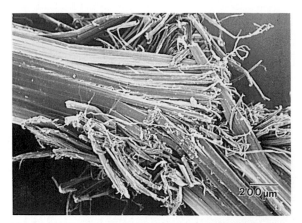

Fig. 5.94 Fractured extrudates appear woody in texture when viewed in the SEM, and coarse and fine fibrils are observed.

Recent work on imaging magnetically aligned thermotropic polyesters, using low voltage FES-EM and AFM, has added information about the fine structure of LCPs [281, 456–458]. The sample shown in Fig. 5.97 is a random semiflexible terpolyester, heated to the nematic phase and aligned in a 13.2 Tesla magnetic field at the Francis Bitter National Magnet Lab at MIT [456]. Samples were quenched to freeze in the order and then prepared according to the lamellar decoration technique developed by Thomas and

Wood [457]. in which the sample is annealed, above the LCP glass transition but below the melting point. It then crystallizes in the form of lamellae, which grow perpendicular to the local chain axis and thus 'decorate' the molecular director field. Thomas and Wood [457] used this method for the TEM study of the molecular director pattern in flow oriented thin films. When nematic LCPs are aligned in a magnetic field they form defects known as *inversion walls*, in which the molecules on each side of the wall are rotated 180° with respect to one another. The study of such walls using low voltage HRSEM, has shown them to be three-dimensional, as also confirmed by AFM. Figure 5.97A is a low magnification HRSEM image of several inversion walls, which appear as dark lines as they are actually valleys [281]. Figure 5.97B (upper micrograph) is a higher magnification HRSEM image [281, 458]. Figure 5.97B (lower micrograph) is an AFM image taken on a Park Scientific AFM in the repulsive mode with a silicon nitride pyramidal tip with medium range forces [281]. The images in Fig. 5.97B are taken at the same magnification for comparison [281].

Structural models describing LCP extrudates and moldings [429, 430, 453, 455, 459, 460] have been derived from microscopy techniques. Thapar and Bevis [453, 455] showed a schematic of

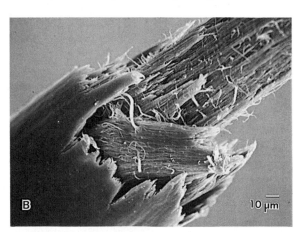

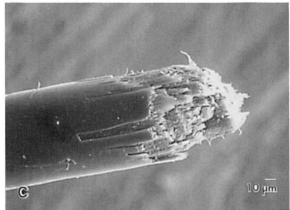

Fig. 5.95 Free fall TLCP strands are shown in polished sections in circularly polarized light (A) (color section) and also by SEM of fractures (B, C). The less oriented strand appears more uniform in domain texture and also exhibits a coarser woody fracture (B). Some orientation is observed in the more highly oriented strand (A) and the fracture morphology is more uniform (C).

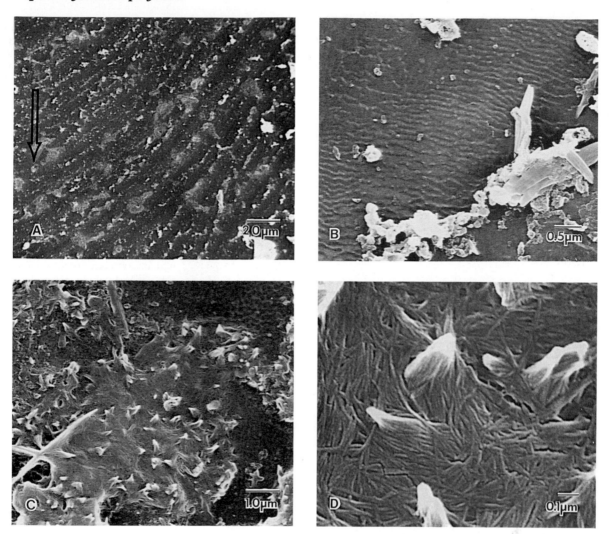

Fig. 5.96 Plasma etching with argon reveals the fine internal structure of an extrudate in these SEM images. The overview (A) shows the flow lines of the polymer. Grain-like domains are observed in a region midway between the skin and core (B) with internal fine structural detail (C) which suggests local orientation. Microfibrillar structures are observed at higher magnification (D).

the skin–core and layered structures of injection molded thermotropic LCPs by polishing and etching methods for SEM. The micromorphology of thermotropic copolyester injection molded bars, 1–20 mm thick, was shown by Thapar and Bevis in an excellent study [455]. The samples were cut, polished and etched with concentrated sulfuric acid for about 20 min, washed and cleaned ultrasonically in acetone. This work

clearly showed the process dependence of the multiple layers which exhibited continuous changes in topography from the sample edge to its center. The structure could be related to the observed increase in modulus with decreasing molded thickness. Baer and co-workers [459, 460] developed a similar model for an LCP extrudate by SEM fracture studies. Sawyer and Jaffe [429–431, 434, 435] have developed a

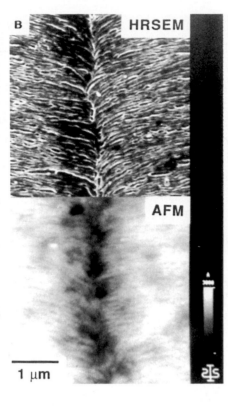

15 µm

1 µm

Fig. 5.97 A random, semi-flexible terpolyester, aligned in a magnetic field at the Francis Bitter National Magnet Lab at MIT [456] and decorated [457] is shown to form inversion walls, in which the molecules on each side of the wall are rotated 180° with respect to one another. A low magnification, low voltage HRSEM image (A) is shown of several inversion walls, which appear as dark lines as they are actually valleys [281]. The upper micrograph in (B) is a higher magnification HRSEM image [458, 281] and the lower micrograph in (B) is an AFM image taken at the same magnification [281]. AFM showed the wall is a valley *c.* 200 nm deep. (From Vezie *et al.* [458, 281]; reproduced with permission.)

general structural model after applying a wide range of microscopy techniques to the study of LCP extrudates and moldings as well as to highly oriented fibrous materials. This model (Fig. 5.98) shows skin–core, layered and banded macrostructures and fine, hierarchical fibrillar microstructures which range from macrofibrils, on the order of 5 µm across, to microfibrils, on the order of 0.05 µm (50 nm) wide and about 5 nm thick [430].

5.6.3.2 *Composites*

Thermotropic LCPs in the form of engineering resins are sold for molding into parts, such as interconnect devices. This requires high flow into the mold and high thermal stability to withstand post molding processes and end uses at elevated temperatures. The resins are sold as composites in extruded pellet form, compounded with fillers, such as glass fibers, carbon fibers, minerals and pigments. The product literature contains exten-

sive information regarding formulations and properties – mechanical, physical, thermal and electrical – that are well beyond the scope of this book. Recent papers on this topic [461–463] and several texts and chapters have been referenced which discuss the orientation developed upon processing [408–413]. Kenig *et al.* [462] provide an example of work that has been done to follow orientation development mechanisms for comparison with properties in glass fiber filled LCP composites. That discussion is similar to the one presented here for the solid state structures without fillers, i.e. that the flow and process variables affect both the structure and properties of the product [429, 430, 453, 455, 460]. A recent rather thorough review on both filled and unfilled thermotropic LCPs, including studies of Vectra moldings, by Plummer *et al.* [463], shows results consistent with the overall picture presented in the literature.

A variety of techniques were used to reveal the complex structures. Samples were molded under

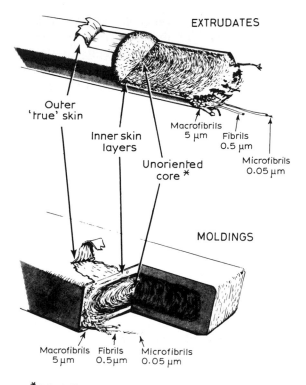

EXTRUDATES

Outer
'true' skin

Inner skin
layers

Unoriented
core *

Macrofibrils
5 μm

Fibrils
0.5 μm

Microfibrils
0.05 μm

MOLDINGS

Macrofibrils   Fibrils   Microfibrils
5 μm   0.5μm   0.05 μm

*Orientation depends upon:
  - polymer
  - draw ratio
  - diameter
  - process

Fig. 5.98 A structural model is shown which provides a schematic of the macrostructures in moldings and extrudates – structures such as layers, bands and skin–core textures. Process changes appear to affect the macrostructures while the nature of the fine structures appear similar overall. (From Sawyer and Jaffe [430]; reproduced with permission.)

various injection conditions and SEMs of polished surfaces etched in concentrated sulfuric acid for 10 min were compared to liquid nitrogen fractured samples. Thin sections were also prepared for optical microscopy with the final polish being carried out using 1 μm diamond paste. In addition, removal of successive layers from moldings was followed by measurement of the modulus profile and the orientation distribution was also measured by WAXS. The unfilled Vectra moldings were shown to have a highly oriented outer skin layer and a banded texture within the

inner skin and core layers, as has been noted [430, 431, 435] and is due to shear induced orientation. Moldings with fillers, such as wollastonite, were also evaluated [463] and fillers were seen to affect the various layers and bands. Care must be taken in the interpretation of any specific molding. The layers, their orientation and the sizes of the fine structures have all been known to vary with the polymer, with conditions of injection molding and even with the distance from the gate. There is a wide range of topics of interest for composites of polymers and inorganic fillers for reinforcement. This includes adhesion, cohesion, interphase formation at the interface, the length or shape of the filler and the arrangement of the fillers as determined by the flow pattern (Section 5.4).

In LCP composites the fibers can reduce local orientation of the polymer matrix. A polarized light micrograph of a polished thin section of a glass fiber reinforced Vectra molding shows fibers (black isotropic) in a nematic matrix (Fig. 5.99A, color section). There is some indication of orientation of the polymer on the glass fiber surfaces which is confirmed by SEM examination of a fracture surface (Fig. 5.99B–D). Additionally, there are composites which are reinforced either with LCP fibers in thermoplastic or thermoset matrices. For instance, aramid fibers are used in composites with epoxies for aerospace applications which require high tensile strength and modulus. Vectran fibers have been used in composites with a wide range of thermoplastics, as described, for example, by Brady and Porter [464]. They compounded Vectran as spun and heat treated fibers with polycarbonate and used SEM of fractured moldings to show the interface for correlation with physical properties.

5.6.3.3 *Blends with LCPs*

There have been numerous studies of blends of LCPs with other LCPs and with thermoplastics during the last decade, generally to reduce cost and to modify the physical properties. SEM of fracture surfaces has been the morphology technique used to image the blend morphology

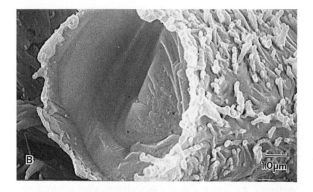

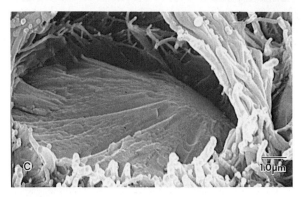

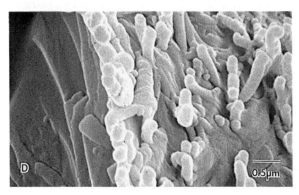

Fig. 5.99 A glass fiber reinforced LCP composite is shown to have interesting morphology. A polished thin section is shown in polarized light (A) (color section) to exhibit a fine domain texture with some orientation of the polymer on the glass surfaces. SEM fracture views (B–D) show the tenacious adhesion of the LCP to the fibers. Fibrillar structures are oriented parallel to the fiber surface and submicrometer sized domains are observed (D).

and to reveal the nature of the dispersed phase, orientation and anisotropy in much the same way as was described in Section 5.3. In the special case of LCPs there are two issues worth noting. First, from a point of view of the microscopy, it is difficult to microtome cross sections of LCPs and this limits routine observation to an assessment of fracture surfaces. Secondly, from a processing point of view the LCPs readily deform into elongated fibrils when they are dispersed in another polymer. The self reinforcing nature of the LCPs in neat form is enhanced by the elongational flow field in blends with other polymers to yield these fibers within a matrix.

Processing affects the morphology of the blends and a wide range of morphologies are observed which depend on the ratio of LCP to thermoplastic, process conditions, temperature, shear rate, etc.

Many authors have written on this topic, and several examples are referred to here [410, 465, 466–471]. In these various studies, the relation between microstructure developed during injection molding and compositions were studied. Characterizations included microscopy, rheology and physical properties. The blend morphologies change dramatically in the core and skin region of the moldings with a tendency to elongated fibers in the skin layers and more globular domains in the core of the moldings. The lower the level of LCP in the blend the more likely that the more fibrillar morphology is present. Blends in which the LCP is the major phase tend to exhibit globular domains of the thermoplastic within the more fibrillar LCP.

There has been work on the development of molecular composites. For example, Chuah *et al.* [469] have shown PBZT and nylon 6,6, to undergo thermally induced phase separation when heated to the nylon melting temperature. Backscattered electron images in an SEM were used to show the phase behavior of the composites [469] as the heavy sulfur atoms provide contrast for PBZT-rich domains. Akhtar *et al.* [470] have done extensive work on blends of two thermotropic LCPs processed by melt mixing and evaluated by thermal, rheological, mechanical and morphology studies, the latter by SEM. Pracella *et al.* [471] studied blends of PPS with a commercial LCP, Vectra B resin, by blending in a Brabender mixer. The biphasic morphology was studied by SEM which showed good contact between matrix and dispersed particles in blends with 5–20 weight % LCP. Effects on PPS crystallization rate and crystal size were also reported [497].

A TEM thin section of an unstained LCP with 5% polycarbonate is shown as an example in Fig. 5.100, as prepared by Wood [273]. Ultramicrotomy of the blend shows the typical banded texture of the LCP with the dark submicrometer isotropic polycarbonate domains uniformly dispersed in the ordered matrix. It is worth noting that although there is extensive research going on in university laboratories today there are few major commercial products which are blends with LCPs. In most cases this is due to the mismatch in thermal and rheological properties which results in materials which have physical properties that resemble the thermoplastics rather than the liquid crystalline polymers.

5.6.4 High modulus fibers

5.6.4.1 *Aromatic polyamides*

Aromatic polyamide fibers are produced by spinning liquid crystalline polymer solutions of PPTA–sulfuric acid dopes into a water coagulation bath [414], resulting in the formation of a crystalline fiber with a surface skin. Variations in the structure produced by annealing at elevated temperature are known to increase the fiber modulus due to a more perfect alignment of the molecules [472]. The chemistry and physics of the aromatic polyamide fibers have been reviewed [419].

The structure of the aramid fibers has been studied by Dobb and Johnson [447, 473–477] and summarized by many others [406, 415, 417–419, 478]. Dobb *et al.* [473] first showed a lattice image for fibrillar fragments produced by sonica-

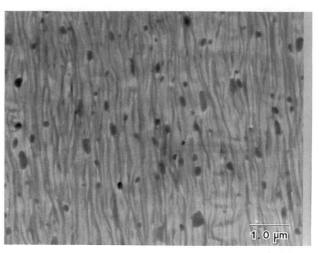

Fig. 5.100 A TEM thin section of an unstained LCP with 5% PC is shown prepared by ultramicrotomy. (From Wood [273]; reproduced with permission.)

tion. Sonicated fibrils suspended over a holey carbon coated grid are shown with an electron diffraction pattern (Fig. 5.101A). A lattice image of an aramid formed from a fibrillar fragment [475] is shown in Fig. 5.101B. Bright field (Fig. 5.102A) and dark field (Fig. 5.102B) TEM micrographs of an ultrathin section show the fibrillar and banded textures in an aromatic polyamide fiber. Dobb *et al.* [475] described the structure of several aramids by TEM and correlated increased modulus and preferred orientation. They observed the general 'hierarchies' of structure that are known to relate to mechanical properties. Measurement of crystallite sizes from bright and dark field images agreed with crystalline sizes calculated from x-ray diffraction data.

A schematic of the aramid structure has been shown by Dobb *et al.* [447] and Jacquemart and Hagege [479]. Ultrathin sectioning and sonication were used in the former study while in the latter fibers were peeled for SEM, bright field and dark field TEM techniques. Dobb *et al.* [447] showed the fibers consist of radially arranged sheets regularly pleated along the long axes, with each sheet composed of extended molecular chains connected laterally by hydrogen bonds. Small angle x-ray scattering results showed a variation

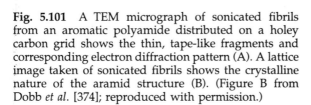

Fig. 5.101 A TEM micrograph of sonicated fibrils from an aromatic polyamide distributed on a holey carbon grid shows the thin, tape-like fragments and corresponding electron diffraction pattern (A). A lattice image taken of sonicated fibrils shows the crystalline nature of the aramid structure (B). (Figure B from Dobb *et al.* [374]; reproduced with permission.)

in electron density which according to TEM is due to microvoids [476]. The voids, elongated along the fiber axis, were observed by the silver sulfide insertion method which enhanced the contrast between the silver filled voids and the fiber (Section 4.4.7) and showed the pleated sheet structure [58].

Aromatic polyamide fibers fail in tension by axial splitting, resulting in fine fibrils formed along their length. The fibrils can be very long or they may be woody or sheet-like in morphology depending on the process history. Cracks propagate parallel to the fiber axis due to rupture of

hydrogen bonds. The mode of deformation and failure have been studied [480, 481]. Pruneda *et al.* [482] suggested that tensile failure involves shear induced microvoid growth along the fiber axis leading to crack propagation, splitting and failure. The fracture mechanism in fatigue reflects poor compressive properties [480, 481, 483] as failure is transverse to the fiber axis. Dobb *et al.* [477] studied the compression behavior of aramid fibers by SEM and TEM and proposed a mechanism for the formation of kink bands consistent with a loss of tensile strength after compression. SEM of composite fracture surfaces

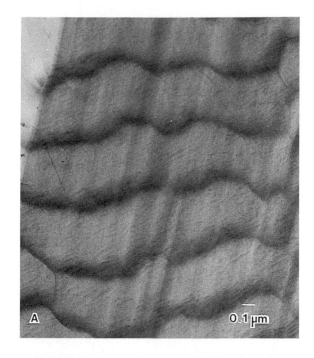

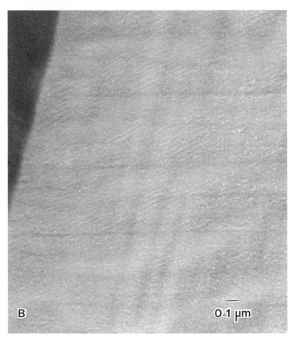

Fig. 5.102 TEM micrographs of ultrathin longitudinal sections of an aromatic polyamide fiber show the banded structures in bright field (A) and the crystallites in dark field (B).

shows transverse failure and fiber splitting (Fig. 5.103A). Kink bands and poor adhesion are observed between the fiber and the matrix (Fig. 5.103B and C).

The aramid fibers have been commercially available for nearly twenty years, and the excellent work in university laboratories, such as by Dobb *et al.* at Leeds has continued [484, 485]. New types of aramid fibers have been studied to compare the tensile behavior and the structure, by electron microscopy and x-ray diffraction studies [484]. SEM images, taken of fractured fiber surfaces, recovered after breaking in glycerol and cleaning in hot water, showed a range of internal defects in the fibers and further information was gained from ultrathin sections of fibers first impregnated with silver sulphide (Section 4.4.7) [486]. This work reinforced the view that the tensile properties of aramid and other LCP fibers are due to a wide number of structural features. Clearly, high molecular orientation results in high modulus, but additionally these authors suggest that low skin content and large crystallite size also increase tensile modulus in the aramids. Defects are well known to adversely affect tensile strength. Thus the substantial, internal helical cracks in the high modulus variant Kevlar 149 were shown to result in decreased tensile strength [484–486]. In another study [487], the ultraviolet resistance of these fibers with large cracks was said to decrease dramatically. This was perhaps due to enhanced penetration of the UV light to the already weakened fiber. In Kevlar 981 a thicker skin was said to favor high strength, but this fiber also has a smaller diameter and one wonders if there is a diameter effect on strength as well [484, 485]. Interestingly, this fiber also has a large number of needle-shaped microvoids aligned parallel to the fiber axis and arranged in a band or skin around the fiber periphery. The filling of the microvoids with silver sulfide and subsequent testing showed a loss in tensile properties and more transverse failure but there was improvement in compressive resistance [484]. This improvement is interesting as it suggests that compressive strength may be affected by a

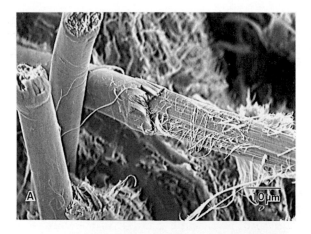

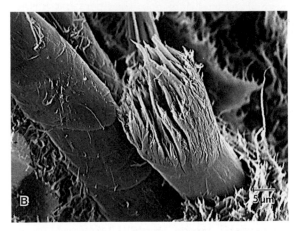

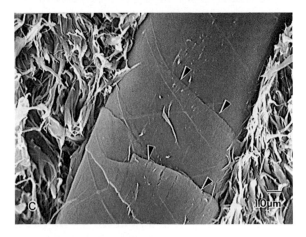

Fig. 5.103 SEM of aramid containing composites show a combination of axial splitting and fatigue type fracture mechanism. Poor adhesion is observed as are kink bands (arrows) and local axial splitting.

'glue' holding the microfibrils together whereas there has been little advantage observed for molecular 'glue' methods, such as cross linking and branching. Finally, the internal structure of aramid fibers is being studied by AFM methods (for instance [434, 435, 488, 489]), which should provide insights into the nanostructures present.

5.6.4.2 *Rigid rod polymers*

Poly(*p*-phenylene benzobisthiazole) is one of a group of polymers with rod-like molecules, spun into high strength fibers as part of the Air Force Ordered Polymers Research Program [420]. Allen *et al.* [490] described the development of high mechanical properties from anisotropic solutions of the polymer. The structure–property relations of PBZT fibers have been reviewed [423]. The rigid rod molecules group to form fibrils, observed in peeled fibers in the SEM [491]. Optical microscopy has been important in defining the size and distribution of the macrovoids present in the fibers [491].

Odell *et al.* [424] studied the role of heat treatment, which is known to enhance mechanical properties, using PBZT (model) films. The extruded films have a rather poorly developed crystallinity. On heat treatment the films change color from light straw to blue. Dark field EM studies show that the increase in modulus with heat treatment is likely due to the increase in crystal size and perfection. Electron diffraction shows nearly perfect alignment along the fiber direction. However, the fibers exhibit only two dimensional order [421, 424, 492, 493], even though higher modulus can be obtained than from PPTA [422].

Ultrathin sections of PBZT on TEM grids were used for an OM study. Optical micrographs are shown of an extruded film section taken in polarized light (Fig. 5.104, color section). Selected area electron diffraction patterns were taken of the sections in the TEM in the core (Fig. 5.105A) and the skin (Fig. 5.105B). The diffraction results show a variation in orientation; the core is much less oriented than the skin. This is consistent with the results of Minter *et al.* [492] who used successive dark field micrographs of spun and heat treated PBZT films. The speckle observed in dark field micrographs of ribbon fragments from those films, shown in Fig. 5.106, reveal the crystallites [492]. Shimamura, Minter and Thomas [493] produced lattice images (Fig. 5.107) of fibers fragmented by detachment replication which show evidence for local three dimensional order. Lattice imaging is becoming important in microstructural studies, (see Section 6.4.1) although only a few polymers, including PPTA [473, 474] (Fig. 5.101B) have been imaged by this technique.

Structural investigations have been carried out by a number of authors, including Allen *et al.* [494] who used wide angle x-ray diffraction, mechanical testing and SEM imaging of fractured as spun and heat treated PBZT fibers. Tension applied during heat treatment at elevated temperatures resulted in increased crystallite size perpendicular to the fiber axis with increases in tensile strength. The aramids have a lower modulus because the chains are not straight due to the large bend in the CO−NH angle. PBZT and PBO (also called PBZO) are strictly straight, so they are stiffer in a similar aggregation of molecules [495]. PBO and ABPBO fibers have also been examined by x-ray scattering, SEM and TEM, as part of the U.S. Air Force Ordered Polymers program [496]. Heat treatment

of the as spun fibers produces transversely broadened crystallites which are unlike uniaxially elongated crystallites in heat treated PBZT or aramids. PBO exhibits greater three dimensional crystallinity and long range order than PBZT but less than PPTA. SEM of fractured PBO fibers show there are bundles of fibrils present in which microfibrils, *c.* 0.2–2 μm in diameter, are observed. TEM dark field images of samples produced by detachment replication using collodion reveal the size and arrangement of the crystallites. Larger crystallites are observed for heat treated fibers. The conclusion of these authors [494] is that PBO fibers are composed of fibrillar bundles of about 5 μm in diameter and of round or ribbon-like fibrils about 0.2 μm in size. Details of the morphology of PBZT and PBO have also been published by Martin and Thomas [497] who used a combination of scattering techniques, electron microscopy and HREM. The diameter of the bundles and fibrils is said to decrease as a result of heat treatment although the crystallites grow in size during this process. Molecules and crystallites are shown to be highly oriented and a skin–core effect is observed.

5.6.4.3 *Aromatic copolyesters*

Thermotropic aromatic copolyesters have a major advantage over the lyotropes as the thermo-

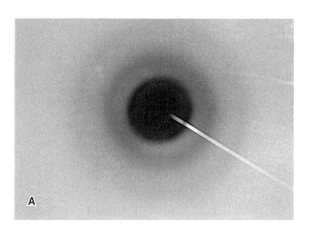 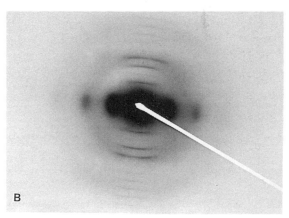

Fig. 5.105 Selected area diffraction patterns of the sample in Fig. 5.104 (color section) show the core (A) is not oriented compared to the skin (B) in a PBZT thin section.

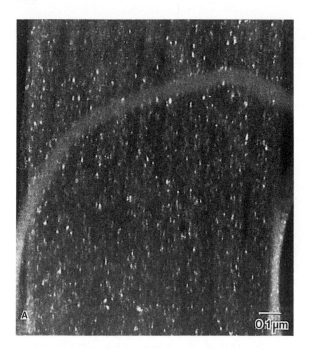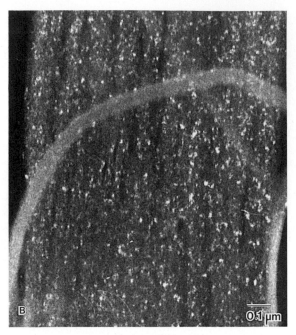

Fig. 5.106 Successive equatorial dark field electron micrographs of fragments from heat treated PBZT film show bright spots which are the crystallites. (From Minter *et al.* [492]; reproduced with permission.)

tropes can be melt processed. Two of the types of aromatic copolyesters that can be processed into fibers are the PHBA modified polyesters, such as X7G, and Vectra LCPs. The modified polyesters [425] are quite well known. Molding temperature affects the orientation and the mechanical properties, and the copolyesters have been shown to be biphasic by SEM [498–500], optical and TEM [501–503] methods. The biphasic structure of X7G has been reported [503] for extruded fibers by optical and EM imaging and microdiffraction. Incomplete extinction in polarized light shows that there are domains with a range of orientations. TEM micrographs of ultrathin longitudinal sections reveal a dense dispersed phase elongated along the fiber axis (Fig. 5.108). Microdiffraction, from regions 20–100 nm across, show the dispersed phase domains (inset 1) and the oriented fiber matrix (inset 2).

A wide range of microscopy techniques has been applied to the characterization of oriented TLCP extrudates [430], including etching, sonication, ultramicrotomy, fracture and peeling. Ex-

amples of the types of structures observed for thermotropic copolyesters are shown in Figs. 5.108–5.110. Shear banding, typically at about 45° to the fiber axis, is common in all types of LCP fibers, as shown in polarized light in Fig. 5.109. A fibrillar texture is known to exist in all nematic LCPs and has been observed by several techniques. Sonicated fibrils examined in the TEM appear sheet-like and are wider than they are thick. Ultrathin sections examined in the TEM show fibrillar textures, as seen in a micrograph of a longitudinal section in Fig. 5.110.

A general structure model developed for highly oriented liquid crystalline fibers [429, 430] is shown schematically in Fig. 5.111. The model was initially defined for the developmental Vectran LCP fibers, but it has been extended by study of the aromatic polyamides and the major features appear to be general in nature. The model extends the structure hierarchy proposed by Dobb, Johnson and Saville [475] for the aramids. Three fibrillar elements have been noted: microfibrils < 50 nm in size;

fibrils on the order of 500 nm in size; and macrofibrils, about 5 μm (5000 nm) across. Two of these structures have been seen the most clearly by a range of techniques. The microfibrils are the smallest structural units that have been observed and they often appear to be clustered into bundles, or fibrils, in the 500–1000 nm size range. The model describes the structure of the fibers and the highly oriented portions of the TLCPs in moldings and extrudates. In these materials, process history and temperature affects macrostructures, such as skin–core, bands and layering (Fig. 5.98). The model is only an aid to the understanding of the relationships between process, structure and properties in LCPs. The new, higher resolution imaging devices have recently been used to describe the microfibrillar textures in more detail and have permitted us to begin to understand the role of the microfibrils to the mechanical properties, especially the most limiting property of the highly oriented LCPs,

compressive strength. The next section is an example of a structure–property study using a combination of new and established imaging techniques.

5.6.5 Structure–property relations in LCPs

5.6.5.1 *Microfibrillar structures*

In the last decade, major technological developments have occurred in the production of polymer fibers with high mechanical strength and stiffness. In concert with these efforts, studies have been directed toward a better understanding of the relationship among chemical composition, physical structure and mechanical properties. One goal is to develop predictive structure–property models to develop marketable technologies. The discussion that follows includes examples of the types of microscopy

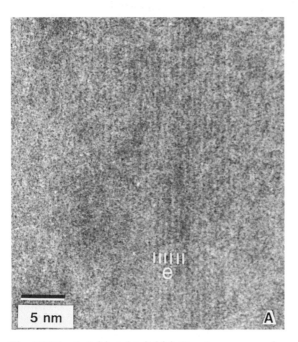 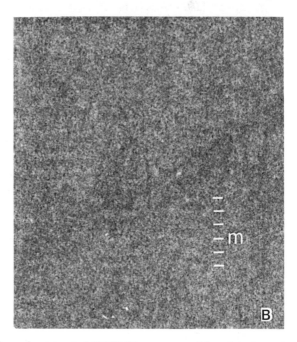

Fig. 5.107 Axial bright field lattice images are shown for a heat treated PBZT fiber prepared by detachment replication. Note that the lattice fringes are most easily visible when the figure is viewed obliquely down the parallel lines drawn on the figure. The polymer is translationally disordered and does not form a true three dimensional crystal. This is why the fringes marked 'm' are wavy. (From Shimamura *et al.* [493]; reproduced with permission.)

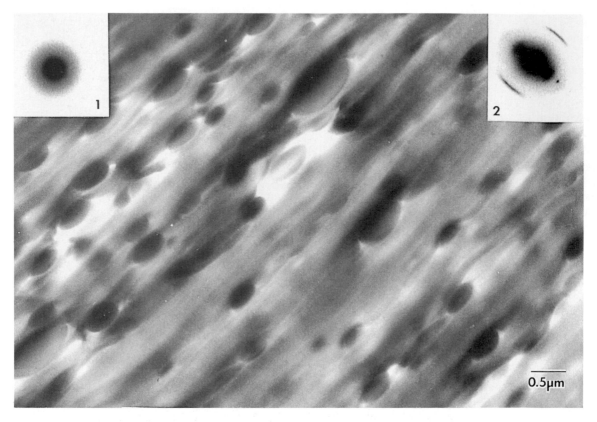

Fig. 5.108 An example of the biphasic structure of X7G formed by specific thermal history is shown by TEM. An ultrathin, longitudinal section of the fiber is shown to have dense elongated dispersed phase particles present in a fibrillar matrix. Microdiffraction of the dispersed phase shows it is amorphous (inset 1) and likely PET, and a pattern for the matrix (inset 2) appears oriented and is likely the PHBA.

techniques which can be helpful for such studies in general. Of course, LCPs cannot be addressed simply by microscopy, and the interested reader should study the extensive literature on this topic for a more complete exposition.

One general concept that has received much attention is the notion that the microfibril is the fundamental building block in polymers made from flexible linear molecules. A decade ago Sawyer and George [504] proposed a basic microfibrillar building block for both natural and synthetic materials. The microfibrils have been known to exist in the natural materials since the 1950s [505, 506] when they were imaged in the earliest TEMs. This minimum stable sized structure appears to be the building block of

polymers and potentially this is the unit that can 'aggregate' and account for mechanical properties [507, 508]. One factor is common to materials as dissimilar as cellulose, polyesters, lyotropic aramids and thermotropic liquid crystal polymers is that the molecular chain is rather long compared to its width and thickness. Thus it is possible that the microfibril is simply a replication of the molecular chain. This replication, and the straightness of the chain is, of course, dependent upon the specific chemical composition.

A fibrillar texture has been shown to exist for many LCPs including the aromatic copolyesters, the aramids and the 'rigid rod' polymers. A common manifestation of the highly oriented fibers is poor compressive properties, demonstrated by the kink bands which can be seen in the SEM images, for instance in Fig. 5.103. Kink bands have been studied by Dobb *et al.* [477] and many others [480–483] and a mechanism for their formation, consistent with tensile loss, was proposed. However, only conjecture has been made as to the size scale that initiates the cause of compressive failure until recently. Martin [497, 509, 510] addressed this issue by use of HREM and Sawyer *et al.* [433–435] imaged textures on the microfibril scale using combined TEM, FESEM and STM imaging techniques.

Martin used HREM to image the deformation and disorder in extended chain polymer fibers, such as PBZT and PBO [509] showing that deformation occurs by strain localization into kink bands. The ultrastructure of the fibers as a function of process conditions showed structural

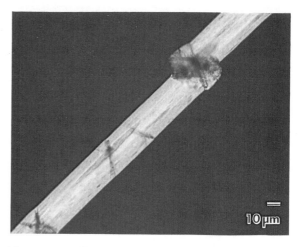

Fig. 5.109 All LCP fibers have a tendency to form shear bands at an angle to the fiber surface. Shear banding is shown for a thermotropic copolyester in polarized light. Stressed fibers tend to fracture at such shear bands.

Fig. 5.110 The *in situ* structure of a spun Vectran fiber is shown in a TEM micrograph of an ultrathin, longitudinal section. The orientation is shown by the microdiffraction pattern inset.

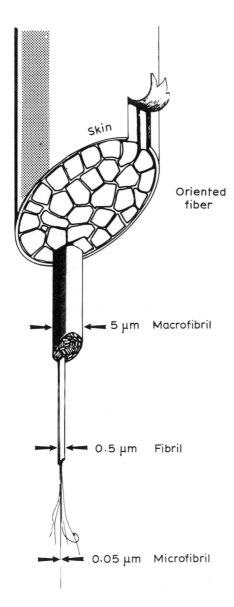

Skin

Oriented
fiber

5 μm   Macrofibril

0.5 μm   Fibril

0.05 μm   Microfibril

Fig. 5.111 The structures observed in uniaxially oriented LCP fibers, ribbons and films can be summarized by this structural model. The model defines the nature of the fibrillar textures into three categories based upon size: macrofibrils, fibrils and microfibrils. In each case the sizes of structures have been determined from complementary microscopy techniques. (From Sawyer and Jaffe [430]; reproduced with permission.)

features such as grain boundaries, dislocations and the molecular details of the deformation processes. HREM images revealed the crystallite size, shape, orientation and internal perfection. The nature of the disorder within a kink band was imaged, modeled and compared to diffraction data. Kinks were observed by Martin and Thomas [510] during TEM investigation of PBZT and PBO fiber fragments. Figure 5.112A [510] shows an equatorial DF image of a kink in PBO heat treated at 600°C. Bright areas in the image are crystallites that are scattering electrons into the objective aperture. Rotation of the sample showed that the material within the kink bands was misoriented but still crystalline. Figure 5.112B [510] is a low dose HREM lattice image. The (200) lattice planes of spacing 0.55 nm are imaged of a kink in a PBO heat treated at 665°C. The detailed nature of the deformation within the kink band is revealed, including the non-uniformity of strain and local, high angle bending or buckling of the stiff polymer chains. Low dose HREM of the kink interior revealed high angle changes in chain orientation, indicative of covalent bond bending or breaking [509–510].

Recent work of Sawyer *et al.* [433–435] used high resolution FESEM and STM with other complementary techniques to describe the microfibrillar and kink structures. Fibers and films were peeled back, ion beam sputter coated with platinum and examined by FESEM and STM. FESEM images were acquired from a highly oriented Vectran fiber which was peeled to reveal kinked regions in the bulk of the fiber (Fig. 5.113A). Heat treated Vectran fibers appeared more ordered (Fig. 5.113B) and this is also the case for a Kevlar fiber (Fig. 5.113C) which exhibits a fibrillar texture. A peeled-back Vectran fiber clearly reveals kinked regions in a specimen observed by FESEM (Fig. 5.113D). Peeled-back Vectran fibers were also imaged in the STM in Fig. 5.114 of a region 500 nm in the x and y axes with a maximum z axis range of 50 nm. The oriented microfibrillar texture is clear in the STM image, but the local alignment is seen to vary with two types of local disorder evident. First, a kink band is seen in the image and there are clear

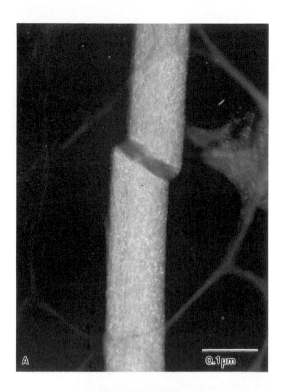

Fig. 5.112 An equatorial dark field image of PBO fiber heat treated at 600 °C (A) shows well defined and localized boundaries between the kink and the deformed fiber. A low dose, HREM 0.55 nm (200) micrograph (B) of a kink in PBO heat treated at 665 °C show (200) lattice fringes of spacing 0.55 nm. The kink band clearly contains very localized high angle bending or buckling of the stiff polymer chains. (From Martin and Thomas [510]; reproduced with permission.)

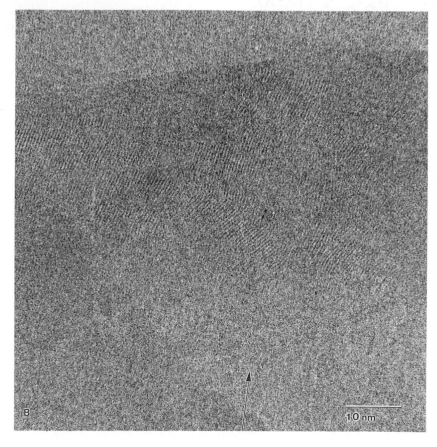

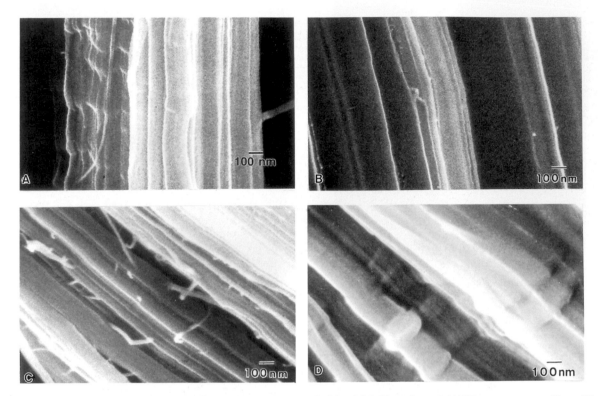

Fig. 5.113 FESEM images show the internal texture in peeled-back highly oriented (A) Vectran as spun fiber, (B) Vectran heat treated fiber and (C) Kevlar fiber which also exhibits a fibrillar texture. Kink bands are also observed in a Vectran fiber (D). (From Sawyer *et al.* [435]; reproduced with permission.)

discontinuities across individual microfibrils. The damaged microfibrils are within the kink band and suggest bond breaking has occurred. The second type of disorder observed is the 'Y' shaped arrangement of the microfibrils, also observed by TEM (Fig. 5.110).

The sonication procedure described earlier was used to prepare samples of Vectran and aramid fibers and highly oriented tapes for TEM and STM imaging. The sonicated fibrils were sha-dowed with metal (Fig. 5.115A) and fibrils are observed which are long and are seen to fibrillate into finer microfibrils. No clear interfibril tie fibrils were observed by high resolution imaging. Twisting and a cotton-like flat, or tape-like fibril structure is observed. The aramid fiber is shown to fibrillate into units less than 10 nm wide (Fig. 5.115B) and smaller fibrils can be seen within the larger ones. Although the origin of the fibrils

remains a question these images appear to show microfibrils within larger fibrils.

STM images of sonicated Vectran fibers (Fig. 5.116) shows two individual microfibrils which were measured to be about 10 nm wide and about 3 nm in the thickness or z-height direction. This image was the first to show us the power of STM imaging of structures at one level above the atomic scale and at perhaps the resolution level which is the most relevant to mechanical proper-ties. The shape is tape-like, about 3 nm by 10 nm, not round. More revealing images of a Vectran LCP fiber (Fig. 5.117) and Kevlar (Fig. 5.118) are shown in the image series in which smaller and smaller regions were consecutively scanned. Interestingly a periodic texture of about 50 nm is observed across a group of microfibrils, arranged normal to the microfibril axis; this texture appears similar in size and spacing to

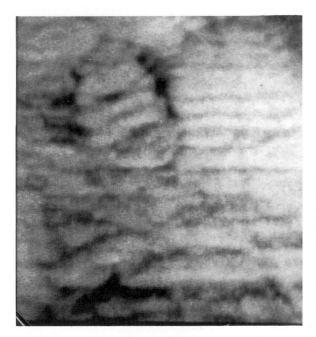

appear to consist of oriented fibers which are further composed of smaller microfibrils. The presence of domain walls is consistent with the worm-like arrangement of the fibrils in and out of the plane of the section. This meander is further illustrated in STM images. Thus the organization of the thermotropic LCPs appears to include microfibrils arranged within fibrils.

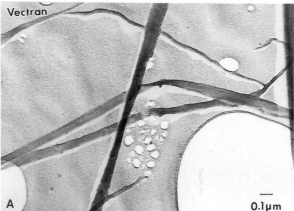

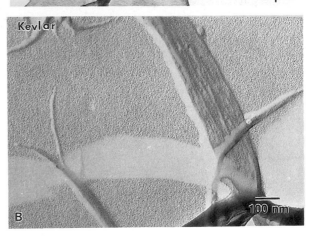

Fig. 5.114 Peeled-back Vectran fibers are shown in a top down view in the STM of a region 500 nm in the x and y axes and a maximum z axis range of 50 nm. The fibrillar texture and local disorder is observed. A kink band is observed as are clear discontinuities across individual microfibrils and damaged microfibrils within the kink band. (From Sawyer *et al.* [435]; reproduced with permission.)

the 'nonperiodic layer (NPL) crystals' observed by Donald and Windle [511]. The STM images of Kevlar fibers (Fig. 5.118) most clearly reveal the nature of the microfibrils. Figure 5.118A shows a bundle of uniform microfibrils about 10 nm wide within a larger aggregated structure. Microfibrils are also shown in Fig. 5.118B and measurements of the thickness of a few of the smallest microfibrils were made directly in the STM. The smallest microfibrils are *c.* 3–5 nm wide and about 1 nm thick with a tape-like shape.

Polarized light microscopy has shown micrometer-sized domains in the thermotropic LCPs aligned along the fiber axis. The meander of the domains is consistent with their polarization colors (Fig. 5.87, color section). The domains

Fig. 5.115 TEM micrographs of sonicated fibrils, shadowed with metal in order to provide information about the three dimensional structure of (A) Vectran and (B) Kevlar fibers, illustrate the tape-like structure. A range of fibrils are observed which are long and are seen to fibrillate into finer fibrils. Twisting and a cotton-like flat, twisted or tape-like fibril structure is observed. The aramid fiber (B) is shown to fibrillate into units less than 10 nm wide. (From Sawyer *et al.* [435]; reproduced with permission.)

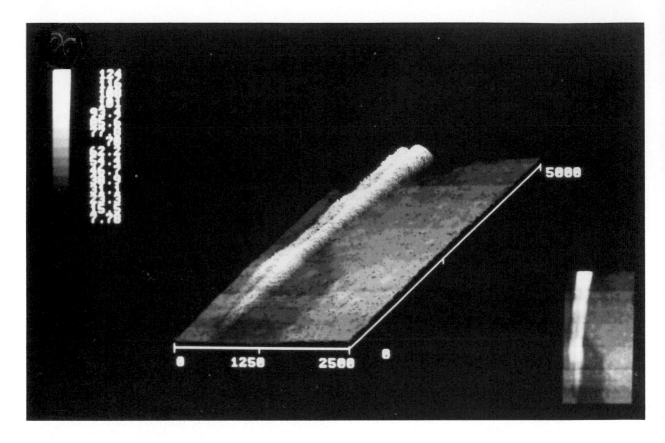

Fig. 5.116 STM image of sonicated LCP fibrils in an image 250 nm × 500 nm. Two microfibrils are observed about 10 nm wide and about 3 nm thick. (From Sawyer *et al.* [435]; reproduced with permission.)

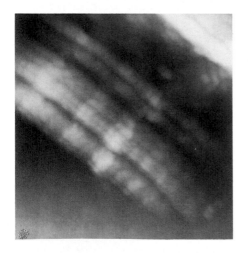

The shape of the fibrils and microfibrils appears to be flat or tape-like and some twisting is also observed. The images suggest that the smallest microfibrils are < 30 nm wide and < 5 nm thick for the lyotropic and thermotropic fibers, measured using the same techniques.

5.6.5.2 *LCP structure model*

A modified model, incorporating the information gained from the newer techniques is shown in

Fig. 5.117 A STM image of a Vectran heat treated fiber reveals microfibrils within a larger unit; area scanned is 500 nm across. A 50 nm periodicity is observed normal to the microfibril axis. (From Sawyer *et al.* [435]; reproduced with permission.)

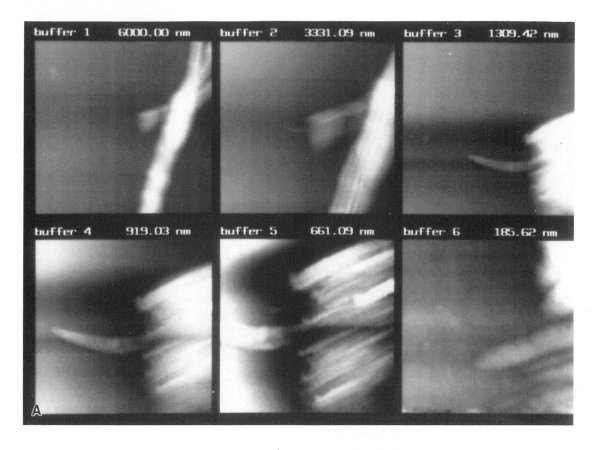

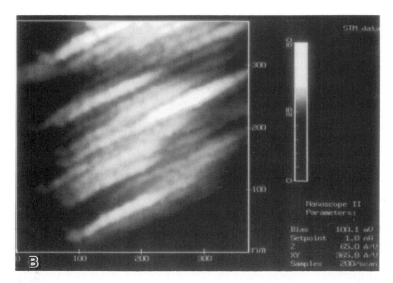

Fig. 5.118 STM images of Kevlar most clearly reveal the nature of the fibrillar hierarchy at the macromolecular level. A bundle of uniform microfibrils about 10 nm wide is observed (A) within a larger aggregated fibrillar structure unit in images taken at successively higher magnifications. Finer fibrils are also clearly seen (B). (From Sawyer *et al.* [435]; reproduced with permission.)

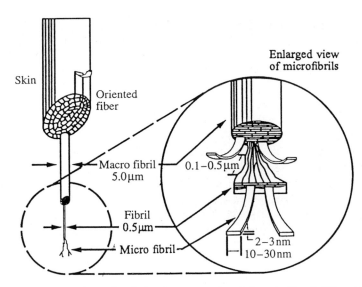

Fig. 5.119 A new structure model is shown [435] that has more detail than in Fig. 5.111 published earlier [430]. The model suggests there is a hierarchy, at least on the scale from 500 nm to the 1 nm size scale, specific to the liquid crystalline polymers. The key element shown is the microfibril, the same microstructural unit basic to melt spun and drawn flexible polymers. (From Sawyer *et al.* [435]; reproduced with permission.)

Fig. 5.119. The model suggests there are structures on the scale from 500 nm to < 50 nm, specific to the liquid crystalline polymers. The key element is the microfibril, the same microstructural unit basic to melt spun and drawn flexible polymers. The orientation of the microfibrils is along the fiber or elongational axis and this results in extremely high tensile modulus values for these materials. However, on a local scale it is clear that the microfibrils meander along the path of the director and are not literally rigid rods.

It is interesting that the very high orientation values calculated from x-ray scattering do not differ significantly for fibers with quite different tensile modulus. Ward [512] has stated that x-ray diffraction alone does not provide the complete answer for understanding the tensile modulus properties of the LCPs. In the case of the aramids, the x-ray results, combined with the presence or absence of the pleated sheet structure, are consistent with the tensile modulus. Chain rigidity and strong intermolecular cohesive forces in the solid state, owing to hydrogen bonding [513] are responsible for the properties.

It is interesting to note that two polymorphs have been observed for the aramids [514, 515] with the pseudo-orthorhombic structure exhibiting two chains per unit cell, in agreement with Ward [512, 516] and Windle [517] who suggest the same two chain structure for the thermotropes.

The final mechanical parameter of interest is compressive strength which is very poor for the highly oriented LCPs. Low compressive strength is consistent with high orientation. As mentioned earlier, the failure mode in compression is via kink band formation which is an intrinsic feature of highly oriented fibers. Adams and Eby [518] reported very high tensile moduli and strengths for PBO fibers, and much lower compressive strength, with failure occurring by buckling in local kink bands. Tensile recoil measurements for compressive strength were made by Allen [519]. The recent work of Martin [509–510] and Lee [520] confirms the critical nature of these kink band defects and their relation to mechanical properties. The images show the onset of failure at the microfibrillar scale.

Efforts to visualize microfibrils have been successful [435, 509, 521–523]. Cohen and Thomas

[522] showed the transformation of a monodomain nematic solution into solid state fibrillar textures. TEM of epoxy impregnated PBZT fibers and films showed an interconnected network of oriented microfibrils with a typical width of 10 nm [522]. Buckling under compression resulted in buckling of individual microfibrils. Martin [521] provided evidence of intermolecular twist defects in extended chain fibers, such as PBO, by use of molecular modeling. The macromolecules in as spun and annealed aramid fibers were recently imaged by AFM [523], showing the periodicities to be consistent with x-ray diffraction and computer simulation results.

The LCPs appear to follow the aggregate model [507, 508, 516, 524, 525, 528]. The good tensile mechanical properties of the aramids are thought to be associated with chains in crystallographic registry, i.e. the persistence length of the molecular chain is greater than the average axial distance between the crystallites. The best mechanical properties are associated with aramids which do not exhibit the axially periodic bands or defects. The thermotropic copolyesters, such as Vectran LCP fibers, with the best mechanical properties also do not exhibit a lateral banded texture, suggesting that better molecular chain registry is associated with improved properties. However, in the case of the Vectran fibers no true three dimensional crystallinity has been observed although some level of crystallinity has been calculated from x-ray diffraction (for instance, [504]) and Donald and Windle [511, 526–528] have observed the poorly formed crystals which they named 'non-periodic layer' (NPL) crystals. TEM studies with electron diffraction show these platelet shaped structures result in equatorial reflections and thus are associated with intermolecular ordering with the thin axis parallel to the molecular chain axis [527]. Finally, flow was shown to induce textures that are oriented and appear to be similar to these poorly formed crystals by use of etching and SEM imaging [528].

Many questions remain that have not been resolved that relate to the mechanisms of formation of microfibrils and the role they play in mechanical properties [529]. These are beyond the theme of this chapter which was to provide an example of structure studies conducted by a range of new and conventional microscopy techniques.

REFERENCES

1. J. C. Guthrie, *J. Text. Inst.* **47** (1956) 248.
2. J. L. Stoves, *Fibre Microscopy* (Van Nostrand, Princeton, 1958).
3. R. D. Van Veld, G. Morris and H. R. Billica, *J. Appl. Polym. Sci.* **12** (1968) 2709.
4. J. W. S. Hearle and R. H. Peters, *Fiber Structure* (The Textile Institute, Manchester, 1963) p. 209.
5. W. O. Statton, *J. Polym. Sci.* **(C)20** (1967) 117.
6. A. Peterlin, *Polym. Sci. Symp.* **32** (1971) 297.
7. A. Peterlin, *J. Appl. Phys.* **48** (1977) 4099.
8. P. Barham and A. Keller, *J. Polym. Sci. Polym. Lett. Edn.* **13** (1975) 197.
9. D. C. Prevorsek, R. H. Butler, Y. D. Kwon, G. E. R. Lamb and R. K. Sharma, *Text. Res. J.* **47** (1977) 107.
10. D. C. Prevorsek, Y. D. Kwon and R. K. Sharma, *J. Mater. Sci.* **12** (1977) 2310.
11. D. C. Prevorsek, P. J. Harget, R. K. Sharma and A. C. Reimschuessel, *J. Macromol. Sci., Phys.* **B8** (1973) 127.
12. A. C. Reimschuessel and D. C. Prevorsek, *J. Polym. Sci., Polym. Phys. Edn.* **14** (1976) 485.
13. D. C. Prevorsek, G. A. Tirpak, P. J. Harget and A. C. Reimschuessel, *J. Macromol. Sci., Phys.* **B9** (1974) 733.
14. J. W. S. Hearle, J. T. Sparrow and P. M. Cross, *The Use of the Scanning Electron Microscope* (Pergamon Press, Oxford, 1972).
15. J. W. Hearle and S. C. Simmens, *Polymer* **14** (1973) 273.
16. H. R. Billica and R. D. Van Veld, *Surface Characteristics of Fibers and Textiles*, edited by M. J. Schick (Marcel Dekker, New York, 1975).
17. McAllister Isaacs, *Textile World Manmade Fiber Chart*, Textile World (1984).
18. R. G. Scott, Symposium on Microscopy, ASTM Special Publ. 257 (1959) p. 121.
19. R. C. Faust, *Proc. Phys. Soc.* **68** (1955) 1081.
20. J. M. Preston and K. I. Narasimhan, *J. Text. Inst.* **40** (1949) T327.
21. E. F. Gurnee, *J. Appl. Phys.* **25** (1954) 1232.

22. M. A. Sieminski, *The Microscope* **23** (1975) 35.

23. R. D. Andrews, *J. Appl. Phys.* **25** (1954) 1223.

24. G. M. Bhatt and J. P. Bell, *J. Polym. Sci., Polym. Phys. Edn.* **14** (1976) 575.

25. V. B. Gupta and S. Kumar, *J. Appl. Polym. Sci.* **26** (1981) 1865.

26. M. K. Tomioka and S. H. Zeronian, *Text. Res. J.* **44** (1974) 1.

27. W. R. Goynes, J. H. Carra and J. D. Timpa, in *Proc. 35th Annu. Mtg EMSA* (Claitors, Baton Rouge, 1977) p. 168.

28. G. Gillberg and D. Kemp, *J. Appl. Polym. Sci.* **26** (1981) 2023.

29. J. W. S. Hearle, B. Lomas and A. R. Bunsell, *Appl. Polym. Symp.* **23** (1974) 147.

30. J. W. S. Hearle, in *Proc. Textile Inst., Inst. Textile de France Conf., Paris* (1975) p. 60.

31. J. W. S. Hearle and P. M. Cross, *J. Mater. Sci.* **5** (1970) 507.

32. A. R. Bunsell, J. W. S. Hearle and R. D. Hunter, *J. Sci. Instrum.* **4** (1971) 860.

33. J. W. S. Hearle and J. T. Sparrow, *Text. Res. J.* **41** (1971) 736.

34. A. R. Bunsell and J. W. S. Hearle, *J. Mater. Sci.* **6** (1971) 1303.

35. J. W. S. Hearle and B. S. Wong, *J. Text. Inst.* **68** (1977) 89.

36. Ch. Oudet and A. R. Bunsell, *J. Mater. Sci. Lett.* **3** (1984) 295.

37. R. C. Laible, F. Figucia and B. H. Kirkwood, *Appl. Polym. Symp.* **23** (1974) 181.

38. V. B. Gupta, *J. Appl. Polym. Sci.* **20** (1976) 2005.

39. J. Smook, W. Hamersma and A. J. Pennings, *J. Mater. Sci.* **19** (1984) 1359.

40. H. H. Kausch, *Polymer Fracture* (Springer Verlag, Berlin, 1978).

41. H. H. Kausch, Ed., *Crazing in Polymers*, Adv. Polym. Sci. Ser. 52/3 (Springer Verlag, Berlin, 1983).

42. A. R. Bunsell, unpublished.

43. L. Fourt and N. R. S. Hollies, *Clothing: Comfort and Function* (Marcel Dekker, New York, 1970).

44. H. N. Yoon, L. C. Sawyer and A. Buckley, *Text. Res. J.* **54** (1984) 357.

45. B. H. Kirkwood, *Text. Res. J.* **44** (1974) 545.

46. J. H. Warren and N. S. Eiss, Jr, *Trans. ASME* **100** (1978) 92.

47. G. S. Egerton and K. M. Fisher, *Text. Res. J.* **40** (1970) 780.

48. C. L. Warfield and J. F. Stone, *Text. Res. J.* **49** (1979) 250.

49. M. L. El Gaier and G. E. Cusick, *J. Text. Inst.* **12** (1975) 426.

50. S. Backer, *Text. Res. J.* **21** (1951) 453.

51. R. S. Chouhan, *Colourage* (1980) 12.

52. W. G. Perkins and R. S. Porter, *J. Mater. Sci.* **16** (1981) 1458.

53. L. C. Sawyer, in *Proc. 39th Annu. Mtg. EMSA* (Claitors, Baton Rouge, 1981) p. 334.

54. A. Ziabicki and A. H. Kawai, Eds, *High-Speed Fiber Spinning: Science and Engineering Aspects* (Wiley–Interscience, New York, 1985).

55. W. L. Wu, V. F. Holland and W. B. Black, *J. Mater. Sci. Lett.* **14** (1979) 252.

56. G. M. Sze, J. Spruiell and J. L. White, *J. Appl. Polym. Sci.* **20** (1976) 1823.

57. B. Catorie, R. Hagege and D. Meimoun, *Bull. Inst. Text. Fr.* **23** (1969) 521.

58. R. Hagege, M. Jarrin and M. Sotton, *J. Microsc.* **115** (1979) 65.

59. R. E. Wilfong and J. Zimmerman, *J. Appl. Polym. Sci.* **17** (1973) 2039.

60. G. Gillberg, L. C. Sawyer and A. Promislow, *J. Appl. Polym. Sci.* **28** (1983) 3723.

61. M. Jaffe, High Modulus Polymers, in *Encyclopedia of Polymer Science and Engineering*, edited by H. F. Mark, N. M. Bikales, C. G. Overberger and G. Menges, Vol. 6, 2nd Edn (John Wiley, New York, 1986) p. 699.

62. E. Baer and A. Moet, Eds, *High-Performance Polymers* (Hanser, New York, 1991).

63. K. Hood, *Trends in Polymer Science* **1** (1993) 129.

64. I. M. Ward, *Structure and Properties of Oriented Polymers* (John Wiley, New York, 1975).

65. A. Zachariades and R. S. Porter, Eds, *The Strength and Stiffness of Polymers* (Marcel Dekker, New York, 1983).

66. J. Odell, D. T. Grubb and A. Keller, *Polymer* **19** (1978) 312.

67. A. Schaper, D. Zenke, E. Schulz, R. Hirte and M. Taege, *Phys. Stat. Sol.* **116** (1989) 179.

68. D. T. Grubb and M. J. Hill, *J. Cryst. Growth* **48** (1980) 407.

69. D. T. Grubb and K. Prasad, *Macromolecules* **25** (1992) 4575.

70. S. N. Magonov, K. Qvarnstrom, V. Elings and H. J. Cantow, *Polym. Bull. (Berlin)* **25** (1991) 689.

71. S. N. Magonov, K. Qvarnstrom, D. M. Huong, V. Elings and H. J. Cantow, *Polym. Mater. Sci. Eng.* (1991) 163.

72. L. M. Eng, H. Fuchs, K. D. Jandt and J. Petermann, *Helv. Phys. Acta* **65** (1992) 870.

73. B. K. Annis, J. R. Reffner and B. Wunderlich, *J. Poly. Sci.: Part B: Poly. Phys.* **31** (1993) 93.

74. D. H. Reneker, *et al.*, Observation of oligomers, polymers and their crystals by scanning tunneling and atomic force microscopy, in *Polymer Characterization*, edited by M. Dosiere (Kluwer Academic Press, London, 1993).

75. D. T. Grubb, *J. Mater. Sci.* **9** (1974) 1715.

76. A. Keller and M. J. Machin, *J. Macromol. Sci. Phys.* **B1** (1967) 41.

77. C. Bonnebat, G. Roullet and A. J. De Vries, *Polym. Eng. Sci.* **21** (1981) 189.

78. J. McDermott, *Industrial Membranes* (Noyes Data Corp., Park Ridge, NY, 1972).

79. E. H. Andrews, *Proc. R. Soc.* **A270** (1962) 232.

80. E. H. Andrews, *Proc. R. Soc.* **A277** (1964) 562.

81. K. Sakaoku and A. Peterlin, *J. Polym. Sci.* **9** (1971) 895.

82. P. M. Tarin and E. L. Thomas, *Polym. Eng. Sci.* **18** (1978) 472.

83. J. Dlugosz, D. T. Grubb, A. Keller and M. B. Rhodes, *J. Mater. Sci.* **7** (1972) 142.

84. C. M. Chu and G. L. Wilkes, *J. Macromol. Sci., Phys.* **B10**(2) (1974) 231.

85. J. D. Hoffman, L. J. Frolen, G. S. Ross and J. I. Lauritzen, *J. Res. Nat. Bur. Stand.* **75A** (1975) 671.

86. A. J. Pennings, *J. Polym. Sci. Symp.* **59** (1977) 55.

87. D. T. Grubb, J. A. Odell and A. Keller, *J. Mater. Sci.* **10** (1975) 1510.

88. J. Petermann and H. Gleiter, *Phil. Mag.* **28** (1973) 1279.

89. J. Petermann, M. J. Miles and H. Gleiter, *J. Macromol. Sci., Phys.* **B12** (1976) 393.

90. M. J. Miles, J. Petermann and H. Gleiter, *J. Macromol. Sci., Phys.* **B12** (1976) 523.

91. M. J. Miles, J. Petermann and H. Gleiter, *Prog. Colloid Polym. Sci.* **62** (1977) 6.

92. R. M. Gohil and J. Petermann, *J. Polym. Sci, Polym. Phys. Edn.* **17** (1979) 525.

93. M. J. Miles and J. Petermann, *J. Macromol. Sci. Phys.* **B16** (1979) 1.

94. J. Petermann and R. M. Gohil, *J. Mater. Sci.* **14** (1979) 2260.

95. D. C. Yang and E. L. Thomas, *J. Mater. Sci.* **19** (1984) 2098.

96. E. L. Thomas, in *Structure of Crystalline Polymers*, edited by I. H. Hall (Elsevier–Applied Science, London, 1984) p. 79.

97. J. M. Schultz, J. S. Lin, R. W. Hendricks, J. Petermann and R. M. Gohil, *J. Polym. Sci., Polym. Phys. Edn.* **19** (1981) 609.

98. J. M. Schultz and J. Petermann, *Colloid Polym. Eng.* **262** (1984) 294.

99. V. P. Chacko, W. W. Adams and E. L. Thomas, *J. Mater. Sci.* **18** (1983) 1999.

100. D. C. Bassett, *Principles of Polymer Morphology* (Cambridge University Press, Cambridge, 1981) p. 102–14.

101. D. C. Bassett, *CRC Crit. Revs., Solid State Mater. Sci.* **12** (1984) 97.

102. V. Vittoria and D. C. Bassett, *Coll. Polym. Sci.* **267** (1989) 661.

103. G. Kanig, *J. Crystal Growth* **18** (1980) 303.

104. D. T. Grubb and A. Keller, *J. Polym. Sci., Polym. Phys. Edn.* **18** (1980) 207.

105. I. G. Voigt-Martin, *J. Polym. Sci., Polym. Phys. Edn.* **18** (1980) 1513.

106. I. G. Voigt-Martin, *Adv. Polymer Sci.* **67** (1985) 195.

107. I. G. Voigt-Martin and L. Mandelkern, *J. Polym. Sci., Part B: Polym. Phys.* **27** (1989) 967.

108. I. G. Voigt-Martin, G. M. Stack, A. J. Peacock and L. Mandelkern, *J. Polym. Sci., Part B: Polym. Phys.* **27** (1989) 957.

109. T. Takahama and P. H. Geil, *J. Polym. Sci., Polym. Phys. Edn.* **21** (1983) 1247.

110. S. Lee, H. Miyaji and P. H. Geil, *J. Macromol. Sci., Phys.* **B22** (1983) 489.

111. D. T. Grubb, in *Developments in Crystalline Polymers – 1*, edited by D. C. Bassett (Applied Science, London, 1982), p. 26–7.

112. P. H. Geil, *J. Polym. Sci.* **20** (1967) 109.

113. G. S. Yeh and P. H. Geil, *J. Macromol. Sci., Phys* **B1** (1967) 235.

114. G. S. Yeh, *Crit. Rev. Macromol. Chem.* **1** (1972) 173.

115. G. S. Yeh, *J. Macromol. Sci., Phys.* **B6** (1972) 451.

116. P. H. Geil, *J. Macromol. Sci., Phys.* **B12** (1976) 173.

117. E. L. Thomas and E. J. Roche, *Polymer* **20** (1979) 1413.

118. D. R. Uhlmann, *Disc. Faraday Soc.* **68** (1979) 87 and 120.

119. D. Vesely, A. Low and M. Bevis, in *Developments in Electron Microscopy and Analysis*, edited by J. Venables (Academic Press, New York, 1976) p. 333.

120. M. Mayer, J. B. Vander Sande and D. R. Uhlmann, *J. Polym. Sci., Polym. Phys. Edn.* **16** (1978) 2005.

121. D. T. Grubb, *Disc. Faraday Soc.* **68** (1979) 125.

122. D. R. Uhlmann, A. L. Renninger, G. Kritchevsky and J. B. Vander Sande, *J. Macromol. Sci., Phys.* **B12** (1976) 153.

123. R. S. Stein, *J. Polym. Sci.* **24** (1957) 383.

124. R. S. Samuels, *J. Polym. Sci.* **3** (1965) 1741.

125. R. J. Samuels, *J. Appl. Polym. Sci.* **26** (1981) 1383.

126. J. P. Paulos and E. L. Thomas, *J. Appl. Polym. Sci.* **25** (1980) 15.

127. E. S. Sherman, *Polym. Eng. Sci.* **24** (1984) 895.

128. T. Tagawa and K. Ogura, *J. Polym. Sci., Polym. Phys. Edn.* **18** (1980) 971.

129. A. Garton, D. J. Carlsson, R. F. Stepaniak, P. Z. Sturgeon and D. M. Wiles, *J. Mater. Sci. Lett.* **14** (1979) 2775.

130. T. R. Albrecht, M. M. Dovek, C. A. Lang, P. Grutter, C. F. Quate, S. W. J. Kuan, C. W. Frank and R. F. W. Pease, *J. Applied Phys.* **64** (1988) 1178.

131. O. Marti, H. O. Ribi, B. Drake, T. R. Albrecht, C. F. Quate and P. K. Hansma, *Science* **239** (1988) 50.

132. C. M. Mate, M. R. Lorenz and V. J. Novotny, *J. Chem. Phys.* **90**(12) (1989) 7550.

133. P. Dietz, P. K. Hansma, K. J. Ihn, F. Motamedi and P. Smith, *J. Mater. Sci.* **28** (1993) 1372.

134. H. Fuchs, *J. Molec. Structure* **292** (1993) 29.

135. S. M. Hues, R. J. Colton, E. Meyer and H. J. Guntherodt, *Mater. Res. Soc. Bull.* Jan. 1993, p. 41.

136. V. V. Tsukruk and D. H. Reneker, submitted to *Polymer.*

137. V. V. Tsukruk, M. D. Foster, D. H. Reneker, A. Schmidt, H. Wu and W. Knoll, *Macromolecules* **27** (1994) 1274.

138. R. J. Wilson, K. E. Johnson, D. D. Chambliss and B. Melior, *Langmuir* **9** (1993) 3478.

139. T. Reilly, unpublished.

140. W. W. Adams, D. Yang and E. L. Thomas, *J. Mater. Sci.* **21** (1986) 2239.

141. R. T. Chen, M. G. Jamieson and R. Callahan, *Proc. 50th Ann. Mtg. EMSA* (San Francisco Press, San Francisco, 1992) p. 1142.

142. R. T. Chen, C. K. Saw, M. G. Jamieson and T. R. Aversa, *J. Appl. Poly. Sci.* **53** (1994) 471.

143. H. K. Lonsdale, *J. Membrane Sci.* **10** (1982) 81.

144. D. R. Lloyd, Ed., *Membrane Materials for Liquid Separations,* ACS Symp. Ser. 269 (American Chemical Society, Washington, DC, 1985).

145. P. Meares, Ed., *Membrane Separation Processes* (Elsevier Scientific, Amsterdam, 1976).

146. A. G. Fane, C. J. D. Fell and A. G. Waters, *J. Membrane Sci.* **9** (1981) 245.

147. W. Pusch, *J. Membrane Sci.* **10** (1982) 325.

148. A. K. Fritsche, A. R. Arevalo, A. F. Connolly, M. D. Moore, V. B. Elings and C. M. Wu, *J. Appl. Poly. Sci.* **45** (1992) 1945.

149. A. K. Fritsche, A. R. Arevalo, M. D. Moore, C. J. Weber, V. B. Elings, K. Kjoller and C. M. Wu, *J. Appl. Poly. Sci.* **46** (1992) 167.

150. S. Loeb and S. Sourirajan, *ACS Adv. Chem. Ser. 38* (American Chemical Society, Washington, DC, 1962) p. 117.

151. H. K. Plummer, C. D. Melvin and E. Eichen, *Desalination* **15** (1974) 93.

152. M. Katoh and S. Suzuli, in *Synthetic Membranes, Vol. 1, Desalination* ACS Symp. Ser. 153. (American Chemical Society, Washington, DC, 1981) p. 247.

153. K. N. Kapadia and V. P. Pandya, *Desalination* **34**(3) (1980) 199.

154. F. S. Model and L. A. Lee, *Am. Chem. Soc. Div. Org. Coatings Plast. Chem., Prepr.* **32** (1972) 384.

155. F. S. Model, H. J. Davis and J. E. Poist, in *Reverse Osmosis and Synthetic Membranes,* edited by S. Sourirajan (National Research Council Canada, 1977).

156. L. C. Sawyer and R. S. Jones, *J. Membrane Sci.* **20** (1984) 147.

157. M. Panar, H. H. Hoehn and R. R. Hebert, *Macromolecules* **6** (1973) 777.

158. C. W. Alegrane, D. G. Pye, H. H. Hoehn and M. Panar, *J. Appl. Polym. Sci.* **19** (1975) 1475.

159. C. Vogele-Kliewer, in *Proc. 43rd Annu. Mtg. EMSA* (San Francisco Press, San Francisco, 1985) p. 86.

160. T. Sarada, L. C. Sawyer and M. Ostler, *J. Membrane Sci.* **15** (1983) 97.

161. H. S. Bierenbaum, R. B. Isaacson, M. L. Druin and S. G. Plovan, *Ind. Eng. Chem., Prod. Res. Dev.* **13** (1974) 2.

162. M. Jamieson, unpublished.

163. H. K. Lonsdale and H. E. Podall, *Reverse Osmosis Membrane Research* (Plenum Press, New York, 1972).

164. I. Cabasso, K. Q. Robert, E. Klein and J. K. Smith, *J. Appl. Polym. Sci.* **21** (1977) 1883.

165. I. Cabasso and A. P. Tamvakis, *J. Appl. Polym. Sci.* **23** (1979) 1509.

166. M. D. Daijal, Ed., *Plastics Polymer Science and Technology* (Wiley–Interscience, New York, 1982).

167. J. M. Margolis, Ed., *Engineering Thermoplastics* (Marcel Dekker, New York, 1985).

168. D. R. Paul and L. H. Sperling, Eds, *Multicomponent Polymer Materials,* Adv. Chem. Ser. No. 211 (American Chemical Society, Washington, DC, 1986).

169. D. R. Paul, J. W. Barlow and H. Keskkula, in *Encyclopedia of Polymer Science*, H. F. Mark, N. M. Bikales, C. G. Overberger and G. Menges, Eds, 2nd Edn, Vol. 12 (Wiley, New York, 1988) p. 399.

170. M. Vaziri, R. T. Spurr and F. H. Stott, *Wear* **122** (1988) 329, 148a.

171. L. C. Sawyer, B. Strassle and D. J. Palatini, in *Proc. 37th Annu. Mtg. EMSA* (Claitors, Baton Rouge, 1979) p. 620.

172. E. L. Lawton, T. Murayama, V. F. Holland and D. C. Felty, *J. Appl. Polym. Sci.* **25** (1980) 187.

173. E. L. Thomas and Y. Talmon, *Polymer* **19** (1978) 225.

174. P. Prentice and J. G. Williams, *Plast. Rubber Process. App.* **2** (1982) 27.

175. P. Prentice, *Polymer* **23** (1982) 1189.

176. J. Kresta, Ed., *Reaction Injection Molding Polymer Chemistry and Engineering*, ACS Symp. Ser. 270 (American Chemical Society, Washington, DC, 1985).

177. N. C. Watkins and D. Hansen, *Text. Res. J.* **68** (1968) 388.

178. Z. Tadmor, *J. Appl. Polym. Sci.* **18** (1974) 1753.

179. S. S. Katti and J. M. Schultz, *Polym. Eng. Sci.* **22** (1982) 1001.

180. E. S. Clark, *SPEJ* **23** (1967) 46.

181. S. C. Malguarnera and A. Manisali, *Polym. Eng. Sci.* **21** (1981) 586.

182. D. G. M. Wright, R. Dunk, D. Bouvart and M. Autran, *Polymer* **29** (1988) 793.

183. W. Y. Chiang and M. S. Lo, *J. Appl. Poly. Sci.* **34** (1987) 1997.

184. V. Tan and M. R. Kamal, *J. Appl. Polym. Sci.* **22** (1978) 2341.

185. E. S. Clark and C. A. Garber, *Int. J. Polym. Mater.* **1** (1971) 31.

186. E. S. Clark, *Appl. Polym. Symp.* **20** (1973) 325.

187. E. S. Clark, *Appl. Polym. Symp.* **24** (1974) 45.

188. M. R. Kantz, *Int. J. Polym. Mater.* **3** (1974) 245.

189. M. R. Kantz, H. D. Newman and F. H. Stigale, *J. Appl. Polym. Sci.* **16** (1972) 1249.

190. J. Bowman, N. Harris and M. Bevis, *J. Mater. Sci.* **10** (1975) 63.

191. J. Bowman and M. Bevis, *Plast. Rubber Mater. Appl.* (1976) 177.

192. J. Bowman, *J. Mater. Sci.* **16** (1981) 1151.

193. D. E. Scherpereel, *Plast. Eng.* (Dec. 1976) 46.

194. D. R. Fitchmun and S. Newman, *J. Polym. Sci. Polym. Lett. Edn.* **7** (1969) 301.

195. J. E. Callear and J. B. Shortall, *J. Mater. Sci.* **12** (1977) 141.

196. D. R. Fitchmun and Z. Mencik, *J. Polym. Sci., Polym. Phys. Edn.* **11** (1974) 951.

197. Z. Mencik and D. R. Fitchmun, *J. Polym. Sci., Polym. Phys. Edn.* **11** (1974) 973.

198. G. Menges, A. Wubken and B. Horn, *Colloid Polym. Sci.* **254** (1976) 267.

199. C. B. Bucknall, *Toughened Plastics* (Applied Science, London, 1977) p. 333.

200. H. D. Keith, F. J. Padden, Jr. and T. P. Russell, *Macromolecules* **22** (1989) 666.

201. S. Y. Hobbs and C. F. Pratt, *J. Appl. Polym. Sci.* **19** (1975) 1701.

202. A. M. Baro, L. Vazquez, A. Bartolame, J. Gomez, N. Garcia, H. A. Goldberg, L. C. Sawyer, R. T. Chen, R. S. Kohn and R. Reifenberger, *J. Mater. Sci.* **24** (1989) 1739.

203. H. A. Goldberg, F. J. Onorato, K. Chiang, M. Jamieson, Y. Z. Li, R. Piner and R. Reifenberger, *Proc. of SPIE*, Vol. 1078 (Soc. of Photo Optical Instrumentation Engineers, Washington, January 1989) p. 170.

204. C. B. Bucknall, *Toughened Plastics* (Applied Science, London, 1977).

205. D. R. Paul and S. Newman, Eds, *Polymers Blends*, Vols. I and II (Academic Press, New York, 1978).

206. J. A. Manson and L. H. Sperling, *Polymer Blends and Composites* (Plenum, New York, 1976).

207. L. H. Sperling, Ed., *Recent Advances in Polymer Blends, Grafts and Blocks* (Plenum, New York, 1974).

208. N. A. J. Platzer, Ed., *Multicomponent Polymer Systems* ACS Adv. Chem. Ser. 99 (American Chemical Society, Washington, DC, 1971).

209. N. A. J. Platzer, Ed., *Copolymers, Polyblends, and Composites*, ACS Adv. Chem. Ser. 142 (American Chemical Society, Washington, DC, 1975).

210. D. Klempner and K. C. Frisch, Eds, *Polymer Alloys: Blends, Blocks, Grafts and Interpenetrating Networks* (Plenum, New York, 1977).

211. S. L. Cooper and G. M. Estes, Eds, *Multiphase Polymers*, ACS Adv. Chem. Ser. 176 (American Chemical Society, Washington, DC, 1979).

212. E. V. Thompson, in *Polymer Alloys 2: Blends, Blocks, Grafts, Interpenetrating Networks*, edited by D. Klempner and K. C. Frisch (Plenum, New York, 1977); *Polym. Sci. Technol.* **11** (1980) 1.

213. J. L. White and K. Min, *Makromol. Chem., Macromol. Symp.* **16** (1988) 19.

214. A. P. Plochocki, *Polimeri* **10** (1965) 23.

215. W. J. Coumans, D. Heikens and S. D. Sjoerdsma, *Polymer* **21** (1980) 103.

216. H. Keskkula and W. J. Frazer, *J. Polym. Sci.* **7** (1968) 1.

217. C. B. Bucknall, *Toughened Plastics* (Applied Science, London, 1977) p. 9.

218. A. M. Donald and E. J. Kramer, *J. Mater. Sci.* **17** (1982) 2351.

219. F. J. McGarry, *Proc. R. Soc. Lond.* **A319** (1970) 59.

220. C. B. Bucknall and T. Yoshii, *Third Int. Conf. on Deformation Yield and Fracture of Polymers* (March 29–April 1, 1976) 13.

221. A. Aref-azar, J. N. Hay, B. J. Marsden and N. Walker, *J. Polym. Sci., Polym. Phys. Edn.* **18** (1980) 637.

222. H. Keskkula and S. G. Turley, *Polymer* **19** (1978) 797.

223. D. Heikens, N. Hoen, W. Barentsen, P. Piet and H. Laden, *J. Polym. Sci., Polym. Symp.* **62** (1978) 309.

224. C. B. Bucknall, *Toughened Plastics* (Applied Science, London, 1977) Chap. 7.

225. B. D. Lauterwasser and E. J. Kramer, *Phil. Mag.* **39A** (1979) 469.

226. A. M. Donald, T. Chan and E. J. Kramer, *J. Mater. Sci.* **16** (1981) 669.

227. C. B. Bucknall, *Toughened Plastics* (Applied Science, London, 1977) pp. 9, 185.

228. M. J. Henke, C. E. Smith and R. F. Abbott, *Polym. Eng. Sci.* **15** (1975) 79.

229. S. J. Krause, W. W. Adams, S. Kumar, T. Reilly and T. Suzuki, *EMSA Proc.* **45** (1987) 466.

230. L. C. Sawyer and M. Jamieson, *EMSA Proc.* **47** (1989) 334.

231. J. Pawley and D. C. Joy, *EMSA Proc.* **50** (1992) 1278.

232. S. J. Krause and W. W. Adams, *MSA Proc.* **51** (1993) 866.

233. D. W. Schwark, D. L. Vezie, J. R. Reffner, E. L. Thomas and B. K. Annis, *J. Mater. Sci. Lett.* **11** (1992) 352.

234. P. B. Himelfarb and K. B. Labat, *Scanning* **12** (1990) 148.

235. S. L. Cooper and A. V. Tobolsky, *J. Appl. Polym. Sci.* **10** (1966) 1837.

236. D. J. Hourston and Y. Zia, *Polymer* **20** (1979) 1497.

237. I. D. Fridman, E. L. Thomas, L. J. Lee and C. W. Macosko, *Polymer* **21** (1980) 393.

238. C. H. Y. Chen, R. M. Briber, E. L. Thomas, M. Xu and W. J. MacKnight, *Polymer* **24** (1983) 1333.

239. R. M. Briber and E. L. Thomas, *J. Macromol. Sci., Phys.* **B22**(4) (1983) 509.

240. G. Demma, E. Martuscelli, A. Zanetti and M. Zorzetto, *J. Mater. Sci.* **18** (1983) 89.

241. B. Chavvel and J. C. Daniel, in *Copolymers, Polyblends and Composites*, ACS Adv. Chem. Ser. 142, edited by A. P. Z. Platzer (American Chemical Society, Washington, DC, 1975) p. 159.

242. J. A. Koutsky, N. V. Hien and S. L. Cooper, *Polym. Lett.* **8** (1970) 353.

243. S. Visconti and R. H. Marchessault, *Macromolecules* **7** (1974) 913.

244. C. B. Bucknall, *Toughened Plastics* (Applied Science, London, 1977) pp. 31, 82.

245. C. K. Riew and J. K. Gillham, Eds, *Rubber-Modified Thermoset Resins*, ACS Adv. Chem. Ser. 208 (American Chemical Society, Washington, DC, 1984).

246. S. Kunz-Douglass, P. W. R. Beaumont and M. F. Ashby, *J. Mater. Sci.* **15** (1980) 1109.

247. L. T. Manzione, J. K. Gillham and C. A. McPherson, *J. Appl. Polym. Sci.* **26** (1981) 889.

248. L. T. Manzione, J. K. Gillham and C. A. McPherson, *J. Appl. Polym. Sci.* **26** (1981) 907.

249. J. A. Sayre, R. A. Assink and R. R. Lagasse, *Polymer* **22** (1981) 87.

250. R. D. Gilbert, V. Stannett, C. G. Pitt and A. Schindler, in *Developments in Polymer Degradation*, N. Grassie, Ed. (Applied Science Publishers, London, 1982) p. 259.

251. M. Chasin and R. Langer, *Biodegradable Polymers as Drug Delivery Systems* (Marcel Dekker, Inc., New York, 1990).

252. E. R. Howells, *Chem and Ind.* (1982) 502.

253. P. J. Barham, *J. Mater. Sci.* **19** (1984) 3826.

254. M. Yasin and B. J. Tighe, *Plastics, Rubber and Composites Processing and Applications* **19** (1993) 15.

255. G. Loomis, M. Izbicki, C. Kliewer and D. R. Sawyer, unpublished.

256. J. M. Widmaier and G. C. Meyer, *J. Polym. Sci., Polym. Phys. Edn.* **18** (1980) 2217.

257. G. Hsuie and C. Yang, *J. Appl. Polym. Sci.* **25** (1980) 1715.

258. S. Reich and Yu. Cohen, *J. Polym. Sci., Polym. Phys. Edn.* **19** (1981) 1255.

259. S. Mader, *Thin Solid Films* **35** (1976) 195.

260. D. L. Handlin, W. J. MacKnight and E. L. Thomas, *Macromolecules* **14** (1981) 795.

261. F. S. Bates, C. V. Berney and R. E. Cohen, *Macromolecules* **16** (1983) 1101.

262. S. Y. Hobbs, *J. Macromol. Sci., Rev. Macromol. Chem.* **C19** (1980) 221.

263. M. J. Folkes and A. Keller, in *Physics of Glassy Polymers*, edited by R. N. Haward (Applied Science, London, 1973).

264. J. Dlugosz, M. J. Folkes and A. Keller, *J. Polym. Sci., Polym. Phys. Edn.* **11** (1973) 929.

265. E. L. Thomas, D. J. Kinning, D. B. Alward and C. S. Henkee, *Macromolecules* **20** (1987) 2934.

266. A. Siegmann and A. Hiltner, *Polym. Eng. Sci.* **24** (1984) 869.

267. D. Fleischer, E. Fischer and J. Brandrup, *J. Macromol. Sci., Phys.* **B14** (1977) 17.

268. A. Siegmann, L. K. English, E. Baer and A. Hiltner, *Polym. Eng. Sci.* **24** (1984) 877.

269. B. Z. Jang, D. R. Uhlmann and J. B. Vander Sande, *Rubber Chem. Technol.* **57** (1983) 291.

270. S. Y. Hobbs, M. E. J. Dekkers and V. H. Watkins, *J. Mater. Sci.* **23** (1988) 1219.

271. M. E. J. Dekkers, S. Y. Hobbs and V. H. Watkins, *J. Mater. Sci.* **23** (1988) 1225.

272. M. E. J. Dekkers, S. Y. Hobbs and V. H. Watkins, *Polymer* **32** (1991) 2150.

273. B. A. Wood, in *Advances in Polymer Blends and Alloys Technology*, Vol. 3 (Technomic Publishing, Lancaster, 1992) p. 24.

274. E. Martuscelli, C. Silvestre and G. Abate, *Polymer* **23** (1982) 229.

275. R. W. Hertzberg and J. A. Manson, *Fatigue of Engineering Plastics* (Academic Press, New York, 1980).

276. M. T. Hahn, R. W. Hertzberg and J. A. Manson, *J. Mater. Sci.* **18** (1983) 3551.

277. C. B. Bucknall, *Toughened Plastics* (Applied Science, London, 1977) p. 117.

278. B. Collin, D. Chatenay, G. Coulon, D. Ausserre and Y. Gallot, *Macromolecules* **25** (1992) 1621.

279. H. G. Dikland, S. S. Sheiko, L. van der Does, M. Moller and A. Bantjes, *Polymer* **34** (1993) 1773.

280. B. K. Annis, D. W. Schwark, J. R. Reffner, E. L. Thomas and B. Wunderlich, *Makromol. Chem.* **193** (1992) 2589.

281. D. L. Vezie, W. W. Adams and E. L. Thomas, *Polymer* **36** (1995) 1761.

282. J. F. Mandell, A. Y. Darwish and F. J. McGarry, *Polym. Eng. Sci.* **22** (1982) 826.

283. J. F. Mandell, D. R. Roberts and F. J. McGarry, *Polym. Eng. Sci.* **23** (1983) 404.

284. F. P. Price, P. T. Gilmore, E. L. Thomas and R. L. Laurence, *J. Polym. Sci., Polym. Symp.* **63** (1978) 33.

285. D. W. Dwight and J. E. McGrath, Final Report for Dept. Army Contract DAAK30-78-C-0098, US Army Tank, Automotive Research and Development Command, Warren, MI Dec. 1979.

286. J. D. Andrade, D. L. Coleman and D. E. Gregonis, *Makromol. Chem. Rapid Commun.* **1** (1980) 101.

287. S. Y. Hobbs and V. H. Watkins, *J. Polym. Sci., Polym. Phys. Edn.* **20** (1982) 651.

288. L. N. Gilbertson, in *Proc. of SEM Conf.*, Vol. 1. (1977) p. 109.

289. L. J. Buckley, I. Shaffer and R. E. Trabocco, *SAMPE Q.* 16 (July 1984) 1.

290. D. Maldas and B. V. Kotka, *TRIP* **1** (1993) 174.

291. P. Zadorecki and A. J. Michell, *Polym. Compos.* **10** (1989) 69.

292. B. Westerlind, M. Rigdahl and A. Larson, in *Composite Systems from Natural and Synthetic Polymers*, edited by L. Salmen, A. de Ruvo, J. C. Seferis and E. B. Stark (Elsevier Science Publishers, 1986) p. 83.

293. D. D. Stokke, *Materials Research Soc. Symp. Proc.* **266** (1992) 47.

294. L. J. Broutman and R. H. Krock, Eds, *Modern Composite Materials* (Addison Wesley, Reading, MA, 1967).

295. D. Hull, *An Introduction to Composite Materials* (Cambridge University Press, New York, 1981).

296. J. C. Seferis and L. Nicolais, Eds, *The Role of the Polymeric Matrix in the Processing and Structural Properties of Composite Materials* (Plenum, New York, 1983).

297. A. R. Bunsell, C. Bathias, A. Martrenchar, D. Menkes and G. Verchery, Eds, *Advances in Composite Materials*, Proc. Third Int. Conf. on Composite Materials, Paris, 26–29 August 1980 (Pergamon, Oxford, 1980).

298. K. Friedrich, *Colloid. Polym. Sci.* **259** (1981) 808.

299. J. F. Mandell, D. D. Huang and F. J. McGarry, *Polym. Composites* **2** (1981) 137.

300. J. F. Mandell, F. J. McGarry, D. D. Huang and C. G. Li, *Polym. Composites* **4** (1983) 32.

301. J. D. Fairing, *J. Composite Mater.* **1** (1967) 208.

302. H. L. Cox, *Br. J. Appl. Phys.* **3** (1952) 72.

303. M. J. Folkes, *Short Fibre Reinforced Thermoplastics* (Research Studies Press, New York, 1982) p. 19.

304. L. C. Sawyer, *Polym. Eng. Sci.* **19** (1979) 377.

305. M. Oron, *Proc. of the SEM, IITRI 6* (1973) 94.

306. A. S. Tetelman, *Composite Materials: Testing and Design*, ASTM STP 460 (1969) p. 473.

307. R. W. Lang, J. A. Manson and R. W. Hertzberg, *Polym. Eng. Sci.* **22** (1982) 982.

308. W. C. McCrone and J. G. Delly, *The Particle Atlas,*

2nd Edn (Ann Arbor Science, Ann Arbor, 1973).

309. A. Oberlin, *Carbon* **17** (1979) 7.

310. M. Guigon, J. Ayache, A. Oberlin and M. Oberlin, in *Advances in Composite Materials*, edited by A. R. Bunsell, C. Bathias, A. Martrenchar, D. Mankes and G. Verchery (Pergamon Press, Oxford, 1980) p. 223.

311. M. J. Folkes, *Short Fibre Reinforced Thermoplastics* (Research Studies Press, New York, 1982).

312. T. Wilson, *Confocal Microscopy* (Academic Press, London, 1992).

313. A. Clarke, N. Davidson and G. Archenhold, *J. Microscopy* **171** (1993) 69.

314. J. L. Thomason and A. Knoester, *J. Mater. Sci. Lett.* **9** (1990) 258.

315. A. J. Waddon, M. J. Hill, A. Keller and D. J. Blundell, *J. Mater. Sci.* **22** (1987) 1773.

316. M. W. Darlington, P. L. McGinley and G. R. Smith, *J. Mater. Sci. Lett.* **11** (1976) 877.

317. P. F. Bright, R. J. Crowson and M. J. Folkes, *J. Mater. Sci.* **13** (1973) 2497.

318. R. J. Crowson, M. J. Folkes and P. F. Bright, *Polym. Eng. Sci.* **20** (1980) 925.

319. A. Voloshin and L. Arcan, *J. Composite Mater.* **13** (1979) 240.

320. J. S. Wu, K. Friedrich and M. Grosso, *Composites* **20** (1989) 223.

321. F. J. Guild and B. Ralph, *J. Mater. Sci.* **14** (1979) 2555.

322. J. D. Whitcomb, *ASTM Symp. on the Fatigue of Fibrous Composite Materials, San Francisco, CA* (May 22–23, 1979).

323. R. Richards-Frandsen and Y. N. Naerheim, *J. Composite Mater.* **17** (1983) 105.

324. J. H. Sinclair and C. C. Chamis, in *Proc. 34th Annu. SPI 22-A* (1979) p. 1.

325. R. A. Kline and E. H. Chang, *J. Composite Mater.* **14** (1980) 315.

326. K. Mizutani and T. Iwatsu, *J. Appl. Polym. Sci.* **25** (1980) 2649.

327. K. M. Hardaker and M. O. W. Richardson, *Polym. Plast. Technol. Eng.* **15** (1980) 169.

328. I. L. Kalnin, *Composite Materials Testing Design*, ASTM STP 497 (1972) p. 563.

329. T. J. Bessell and J. B. Shortall, *J. Mater. Sci.* **12** (1977) 365.

330. J. N. Kirk, M. Munro and P. W. R. Beaumont, *J. Mater. Sci.* **13** (1978) 2197.

331. J. E. Theberge, *Polym. Plast. Technol. Eng.* **16** (1981) 41.

332. J. P. Trotignon, B. Sanschagrin, M. Piperaud and J. Verdu, *Polym. Composites* **3** (1982) 230.

333. A. Garton, S. W. Kim and D. M. Wiles, *J. Polym. Sci., Polym. Lett. Edn.* **20** (1982) 273.

334. J. Kubat and H. E. Stromvall, *Plast. Rubber: Process.* (June 1980) 45.

335. J. D. Miller, H. Ishida and F. H. J. Maurer, *J. Mater. Sci.* **24** (1989) 2555.

336. V. P. Chacko, F. E. Karasz, R. J. Farris and E. L. Thomas, *J. Polym. Sci., Polym. Phys. Edn.* **20** (1982) 2177.

337. J. Kruse, *Rubber Chem. Technol.* **46** (1973) 1.

338. S. Y. Hobbs, in *Plastics Polymer Science and Technology*, edited by M. D. Bayal (Wiley–Interscience, New York, 1982) p. 239.

339. R. J. Eldred, *J. Polym. Sci., Polym. Lett. Edn.* **10** (1972) 391.

340. A. K. Bhowmick, G. B. Nando, S. Basu and S. K. De, *Rubber Chem. Technol.* **53** (1980) 327.

341. S. K. Setua and S. K. De, *J. Mater. Sci.* **18** (1983) 847.

342. P. K. Pal and S. K. De, *J. Appl. Polym. Sci.* **28** (1983) 3333.

343. K. J. Lissant, Ed., *Emulsions and Emulsion Technology* (Marcel Dekker, New York, 1974).

344. M. J. Rosen, *Surfactants and Interfacial Phenomena* (Wiley–Interscience, New York, 1978).

345. L. M. Prince, Ed., *Microemulsions: Theory and Practice* (Academic Press, New York, 1977).

346. I. D. Robb, Ed., *Microemulsions* (Plenum, New York, 1982).

347. G. Gillberg, personal communication.

348. S. Friberg, Ed., *Food Emulsion* (Marcel Dekker, New York, 1976).

349. M. El-Aasser, *Advances in Emulsion Polymerization and Latex Technology* (Lehigh University, Bethlehem, PA, 1979).

350. B. W. Ninham, *J. Phys. Chem.* **84** (1980) 1423.

351. G. E. Molau and H. Keskkula, *Appl. Polym. Symp.* **7** (1968) 35.

352. J. Ugelstad and P. C. Mork, *Adv. Colloid Interface Sci.* **13** (1980) 101.

353. S. S. Atik and J. K. Thomas, *J. Am. Chem. Soc.* **104**(22) (1982) 5868.

354. M. Katoh, *J. Electron Microsc.* **28** (1979) 197.

355. K. Furusawa and N. Tomotsu, *J. Colloid Interface Sci.* **93** (1983) 504.

356. B. Kachar, D. F. Evans and B. W. Ninham, *J. Colloid Interface Sci.* **100** (1984) 287.

357. C. Price, in *Developments in Block Copolymers – 1*, edited by I. Goodman (Applied Science, London, 1982) p. 39.

358. C. Price and D. Woods, *Eur. Polym. J.* **9** (1973) 827.

359. C. Booth, V. T. De Naylor, C. Price, N. S. Rajab and R. B. Stubbersfield, *J. Chem. Soc., Faraday Trans.* **74** (1978) 2352.

360. C. Price, A. L. Hudd, R. B. Stubbersfield and B. Wright, *Polymer* **21** (1980) 9.

361. D. Distler and G. Kanig, *Colloid Polym. Sci.* **256** (1978) 793.

362. D. I. Lee and T. Ishikawa, *J. Polym. Sci., Polym. Chem. Edn.* **21** (1983) 147.

363. O. L. Shaffer, M. S. El-Aasser and J. W. Vanderhoff, in *Proc. 41st Annu. Mtg. EMSA* (San Francisco Press, San Francisco, 1983) p. 30.

364. O. L. Shaffer, unpublished.

365. I. Segall, O. L. Shaffer, V. L. Dimonie and M. S. El-Aasser, *Proc. 51st MSA* (San Francisco Press, San Francisco, 1993) p. 882.

366. J. Biais, M. Mercier, P. Lalanne, B. Clin, A. M. Bellocq and B. Lemanceau, *C. R. Acad. Sci. Paris* **C285** (1977) 213.

367. A. Rameau, P. Marie, F. Tripier and Y. Gallot, *C. R. Acad. Sci. Paris* **C286** (1978) 277.

368. A. M. Bellocq, J. Biais, P. Botherol, B. Clin, G. Fourche, P. Lalanne, B. Lemaire, B. Lemanceau and D. Roux, *Adv. Colloid Interface Sci.* **20** (1984) 167.

369. M. S. El-Aasser and A. A. Robertson, *J. Paint Technol.* **17** (1975) 50.

370. B. R. Vijayendran, T. Bone and L. C. Sawyer, *J. Dispersion Sci. Technol.* **3**(1) (1982) 81.

371. M. S. El-Aasser and A. A. Robertson, *Kolloid Z. Z. Polym.* **252** (1973) 241.

372. E. B. Bradford and J. W. Vanderhoff, *J. Macromol. Chem.* **1** (1966) 335.

373. E. B. Bradford and J. W. Vanderhoff, *J. Macromol. Sci., Phys.* **B6** (1972) 671.

374. R. L. Rowell, R. S. Farinato, J. W. Parsons, J. R. Ford, K. H. Langley, J. R. Stone, T. R. Marshall, C. S. Parmenter, M. Seaver and E. B. Bradford, *J. Colloid Interface Sci.* **69** (1979) 590.

375. J. V. Dawkins and G. Taylor, *Macromol. Chem.* **180** (1979) 1737.

376. O. L. Shaffer, unpublished.

377. J. Qian, R. Pearson, M. S. El-Aasser and V. Dimonie, *SAMPE* **25** (1993) 40.

378. J. F. Oliver and S. G. Mason, in *The Fundamental Properties of Paper Related to its Uses*, Trans. Symp. Mtg. 1973, Vol. 2, edited by F. M. Bolam (Ernest Benn, London, 1976) p. 428.

379. C. Huh, M. Inoue and S. G. Mason, *Can, J. Chem. Eng.* **53** (1975) 367.

380. C. Huh and S. G. Mason, *J. Colloid Interface Sci.* **60** (1977) 11.

381. J. F. Oliver and S. G. Mason, *J. Colloid Interface Sci.* **60** (1977) 480.

382. Y. H. Mori, T. G. M. Van de Ven and S. G. Mason, *Colloids Surf.* **4**(1) (1982) 1.

383. L. H. Lee, Ed., *Recent Advances in Adhesion* (Gordon and Breach, New York, 1973).

384. L. H. Lee, Ed., *Adhesion Science and Technology* (Plenum, New York, 1975).

385. L. H. Lee, Ed., *Adhesion and Adsorption of Polymers* (Plenum, New York, 1980).

386. T. Smith and D. H. Kaelble, in *Treatise on Adhesion and Adhesives*, Vol. 5, edited by R. L. Patrick (Marcel Dekker, New York, 1981).

387. O. Hahn and G. Kotting, *Kunstatstoffe* **74**(4) (1984) 238.

388. D. M. Brewis and D. Briggs, *Polymer* **22** (1981) 7.

389. S. Yamakawa and F. Yamamoto, *J. Appl. Polym. Sci.* **25** (1980) 25.

390. M. Kadreva, *Physicochemical Aspects of Polymer Surfaces*, Vol. 2, edited by K. L. Mittal (Plenum, New York, 1983) p. 125.

391. J. P. Wightman, *SAMPE Q.* **13** (1981) 1.

392. B. Beck, *An Investigation of Adhesive/Adherend and Fiber/Matrix Interactions, Part B, SEM/ESCA Analysis of Fracture Surfaces*, NASA Report NAG1-127, January 1983.

393. D. B. Rahrig, *J. Adhesion* **16** (1984) 179.

394. G. Gillberg, L. C. Sawyer and A. L. Promislow, *J. Appl. Polym. Sci* **28** (1983) 3723.

395. L. C. Sawyer, unpublished.

396. D. R. Sawyer, unpublished.

397. A. Ciferri and I. M. Ward, Eds, *Ultra-high Modulus Polymers* (Applied Science, London, 1979).

398. P. G. deGennes, *The Physics of Liquid Crystals* (Oxford University Press, Oxford, 1979).

399. D. Demus and L. Richter, *Texture of Liquid Crystals* (Verlag Chemie, New York, 1978).

400. P. J. Flory, Molecular theories of liquid crystals, in *Polymer Liquid Crystals*, edited by A. Ciferri, W. R. Krigbaum and R. B. Meyer (Academic Press, New York, 1982) Chap. 4.

401. A. Ciferri, W. R. Krigbaum and R. B. Meyer, Eds, *Polymer Liquid Crystals* (Academic Press, New York, 1982).

402. J. F. Johnson and R. S. Porter, Eds, *Liquid Crystals and Ordered Fluids* (Plenum, New York, 1970).

403. G. W. Gray and P. A. Winsor, *Liquid Crystals and*

Plastics, Vols 1 and 2 (Horwood, Chichester, 1974).

404. J. H. Wendorff, Scattering in liquid crystalline polymer systems, in *Liquid Crystalline Order in Polymers*, edited by A. Blumstein (Academic Press, New York, 1978), p. 41.

405. J. L. White and J. F. Fellers, in *Fiber Structure and Properties*, edited by J. L. White; Appl. Polym. Symp. 33 (1978) 137.

406. G. Calundann and M. Jaffe, in *Proc. Robert A. Welch Conf. on Chemical Research, XXVI, Synthetic Polymers* (Houston, Texas, Nov. 15–17, 1982) p. 247.

407. M. G. Dobb and J. E. McIntyre, *Properties and Applications of Liquid-Crystalline Main-Chain Polymers*, Adv. in Polym. Sci. Ser. 60/61 (Springer Verlag, Berlin, 1984).

408. G. Calundann, M. Jaffe, R. S. Jones and H. N. Yoon, in *Fibre Reinforcements for Composite Materials*, edited by A. R. Bunsell (Elsevier, Amsterdam, 1988) ch. 5.

409. A. E. Zachariades and R. S. Porter, Eds, *High Modulus Polymer* (Marcel Dekker, New York, 1988).

410. R. A. Weiss and C. K. Ober, Eds, *Liquid Crystalline Polymers*, ACS Symposium Series 435 (American Chemical Society, Washington, DC, 1990).

411. C. L. Jackson and M. T. Shaw, *Int. Mater. Rev.* **36** (1991) 165.

412. W. A. MacDonald, in *Liquid Crystalline Polymers* edited by A. A. Collyer (Elsevier, London, 1992) p. 407.

413. A. M. Donald and A. H. Windle, *Liquid Crystalline Polymers* (Cambridge University Press, Cambridge, 1992).

414. For example, S. L. Kwolek, U.S. Patent 3,600,350 (1971).

415. M. Panar, P. Avakian, R. C. Blume, K. H. Gardner, T. D. Gierke and H. H. Yang, *J. Polym. Sci., Polym. Phys. Edn.* **21** (1983) 1955.

416. M. G. Northolt, *Polymer* **21** (1980) 1199.

417. J. R. Schaefgen, T. I. Bair, J. W. Ballou, S. L. Kwolek, P. W. Morgan, M. Panar and J. Zimmerman, in *Ultra-high Modulus Polymers*, edited by A. Ciferri and I. M. Ward (Applied Science, London, 1979) p. 173.

418. J. R. Schaefgen, in *The Strength and Stiffness of Polymers*, edited by A. Zachariades and R. S. Porter, (Marcel Dekker, New York, 1984) p. 327.

419. M. Jaffe and R. S. Jones, in *High Technology Fibers, Part A, Handbook of Fiber Science and Technology,*

Vol. III, edited by M. Lewin and J. Preston (Marcel Dekker, New York, 1985) p. 349.

420. T. E. Helminiak, *Am. Chem. Soc., Div. Org. Coatings Plast. Chem. Prepr.* **40** (1979) 475.

421. E. J. Roche, T. Takahashi and E. L. Thomas, in *Fiber Diffraction Methods*, edited by A. D. French and K. H. Gardner, ACS Symp. Ser. 141 (American Chemical Society, Washington DC, 1980) p. 303.

422. E. J. Roche, R. S. Stein and E. L. Thomas, *J. Polym. Sci., Polym. Phys. Edn.* **18** (1980) 1145.

423. S. R. Allen, A. G. Fillippov, R. J. Farris and E. L. Thomas, in *The Strength and Stiffness of Polymers*, edited by A. Zachariades and R. S. Porter (Marcel Dekker, New York, 1983) p. 357.

424. J. A. Odell, A. Keller, E. D. T. Atkins and M. J. Miles, *J. Mater. Sci.* **16** (1981) 3309.

425. W. J. Jackson, Jr and H. F. Kuhfuss, *J. Polym. Sci., Polym. Chem. Edn.* **14** (1976) 2043.

426. W. C. Wooten, Jr, F. E. McFarlane, T. F. Gray, Jr and W. J. Jackson, Jr in *Ultra-high Modulus Polymers*, edited by A. Ciferri and I. M. Ward (Applied Science, London, 1979) p. 227.

427. J. Economy and W. Volksen, in *The Strength and Stiffness of Polymers*, edited by A. Zachariades and R. S. Porter (Marcel Dekker, New York, 1983) p. 293.

428. C. Noel, C. Fridrich, F. Laupretre, J. Billard, L. Bosio and C. Strazielle, *Polymer* **25**(2) (1984) 263.

429. L. C. Sawyer and M. Jaffe, *Proc. ACS Div. Polym. Mater. Sci. Eng.* **53** (1985) 485.

430. L. C. Sawyer and M. Jaffe, *J. Mater. Sci.* **21** (1986) 1897.

431. L. C. Sawyer and M. Jaffe, in *High Performance Polymers*, edited by E. Baer and A. Moet (Carl Hanser Verlag, Germany, 1991) p. 56.

432. I. H. Musselman and P. E. Russell, *Microbeam Analysis—1989*, P. E. Russell, Ed. (San Francisco Press, San Francisco, CA) p. 535.

433. I. H. Musselman, P.E. Russell, R. T. Chen, M. G. Jamieson and L. C. Sawyer, in *Proceedings of the XIIth International Congress for Electron Microscopy, Seattle*, August 1990, edited by W. Bailey (San Francisco Press, San Francisco, 1990) p.866.

434. L. C. Sawyer, R. T. Chen, M. Jamieson, I. H. Musselman and P. E. Russell, *J. Mater. Sci. Lett.* (1992) 69.

435. L. C. Sawyer, R. T. Chen, M. Jamieson, I. H. Musselman and P. E. Russell, *J. Mater. Sci.* **28** (1993) 225.

436. H. Aoki, Y. Onogi, J. L. White and J. F. Fellers,

Polym. Eng. Sci. **20**(3) (1980) 221.

437. T. Asada, in *Polymer Liquid Crystals,* edited by A. Ciferri, W. R. Krigbaum and R. B. Meyer (Academic Press, New York, 1982) p. 247.

438. M. R. Mackley, F. Pinaud and G. Siekmann, *Polymer* **22** (1981) 437.

439. C. Viney, A. M. Donald and A. H. Windle, *J. Mater. Sci.* **18** (1983) 1136.

440. S. C. Simmens and J. W. S. Hearle, *J. Polym. Sci., Polym. Phys. Edn.* **18** (1980) 871.

441. A. M. Donald and A. H. Windle, *Colloid Polym. Sci.* **261**(10) (1983) 793.

442. A. M. Donald and A. H. Windle, *J. Mater. Sci.* **18** (1983) 1143.

443. A. Zachariades, P. Navard and J. A. Logan, *Mol. Cryst. Liq. Cryst.* **110** (1984) 93.

444. A. M. Donald and A. H. Windle, *J. Mater. Sci.* **19** (1984) 2085.

445. C. Viney and W. S. Putnam, *Proc. 52nd. MSA* (San Francisco Press, San Francisco, 1993) 864.

446. W. S. Putnam and C. Viney, *Molecular Crystals and Liquid Crystals* **199** (1991) 189.

447. M. G. Dobb, D. J. Johnson and B. P. Saville, *J. Polym. Sci., Polym. Phys. Edn.* **15** (1977) 2201.

448. S. J. Krause, D. L. Vezie and W. W. Adams, *Polymer Communications* **30** (1989) 10.

449. A. Anwer, R. J. Spontak and A. H. Windle, *J. Mater. Sci. Lett.* **9** (1990) 935.

450. Y. Ide and Z. Ophir, *Polym. Eng. Sci.* **23**(5) (1983) 261.

451. S. Garg and S. Kenig, *Proc. ACS Div. Polym. Mater. Sci. Eng.* **52** (1985) 90.

452. Z. Ophir and Y. Ide, *Polym. Eng. Sci.* **23**(14) (1983) 792.

453. H. Thapar and M. Bevis. *J. Mater. Sci. Lett.* **2** (1983) 733.

454. Z. Tadmor and C. G. Gogos, *Principles of Polymer Processing* (Wiley–Interscience, New York, 1979).

455. H. Thapar and M. J. Bevis, *Plastics and Rubber Processing and Applications* **12** (1989) 39.

456. S. D. Hudson, D. L. Vezie and E. L. Thomas, *Makromol. Chem., Rapid Commun.* **11** (1990) 657.

457. E. L. Thomas and B. A. Wood, *Faraday Discuss. Chem. Soc.* **79** (1985) 229.

458. W. W. Adams, D. L. Vezie and E. L. Thomas, *Proc. of 50th Annu. EMSA* (San Francisco Press, San Francisco, 1992) p. 266.

459. E. Baer, A. Hiltner, T. Weng, L. C. Sawyer and M. Jaffe, *Proc. ACS Div. Polym. Mater. Sci. Eng.* **52** (1985) 88.

460. T. Weng, A. Hiltner and E. Baer, *J. Mater. Sci.* **21** (1986) 744.

461. P. D. Frayer, *Polymer Composites* **8** (1987) 379.

462. S. Kenig, B. Trattner and H. Anderman, *Polymer Composites* **9** (1988) 20.

463. C. J. G. Plummer, B. Zulle, A. Demarmels and H. H. Kausch, *J. Appl. Poly. Sci.* **48** (1993) 751.

464. R. L. Brady and R. S. Porter, *J. Thermoplastic Composite Mater.* **3** (1990) 252.

465. S. Garg and S. Kenig, in *Strength and Stiffness of Polymers,* edited by R. S. Porter and A. E. Zachariades (Marcel Dekker, New York, 1988).

466. R. A. Weiss, W. Hue and L. Nicholais, *Polym. Eng. Sci.* **27** (1987) 684.

467. D. Beery, S. Kenig and A. Siegmann, *Poly. Eng. Sci.* **31** (1991) 459.

468. D. Dutta, R. A. Weiss and K. Kristal, *Polymer Composites* **13** (1992) 394.

469. H. H. Chuah, T. Kyu and T. E. Helminiak, *Polymer* **28** (1987) 2130.

470. S. Akhtar and A. I. Isayev, *Polymer Eng. Sci.* **33** (1993) 32.

471. M. Pracella, P. Magagnini and L. L. Minkova, *Polym. Networks Blends* **2** (1992) 225.

472. H. Blades, US Patent 3767756 (1973).

473. M. G. Dobb, A. M. Hendeleh, D. J. Johnson and B. P. Saville, *Nature* **253** (1975) 189.

474. M. G. Dobb, D. J. Johnson and B. P. Saville, *Phil. Trans. R. Soc. (London)* **A294 (1980) 483.**

475. M. G. Dobb, D. J. Johnson and B. P. Saville, *J. Polym. Sci., Polym. Symp.* **58** (1977) 237.

476. M. G. Dobb, D. J. Johnson, A. Majeed and B. P. Saville, *Polymer* **20** (1979) 1284.

477. M. G. Dobb, D. J. Johnson and B. P. Saville, *Polymer* **22** (1981) 960.

478. P. Aviakian, R. C. Blume. T. D. Gierke, H. H. Yang and M. Panar, *Am. Chem. Soc., Div. Polym. Chem., Polym. Prepr.* **21** (1980) 8.

479. J. Jacquemart and R. Hagege, *J. Microsc. Spectrosc. Electron.* **3** (1978) 427.

480. A. R. Bunsell, *J. Mater. Sci.* **10** (1975) 1300.

481. M. M. Lafite and A. R. Bunsell, *J. Mater. Sci.* **17** (1982) 2391.

482. C. O. Pruneda, R. J. Morgan and F. M. Kong, *29th Nat. SAMPE Symp.*, April 3–5, 1984, p. 1213.

483. J. H. Greenwood and P. G. Rose, *J. Mater. Sci.* **9** (1974) 1804.

484. M. G. Dobb and R. M. Robson, *J. Mater. Sci.* **25** (1990) 459.

485. M. G. Dobb, C. R. Park and R. M. Robson, *J. Mater. Sci.* **27** (1992) 3876.

486. M. Sotton and A. M. Vialard, *Textile Res. J.* (1981)

842.

487. M. G. Dobb, R. M. Robson and A. H. Roberts, *J. Mater. Sci.* **28** (1993) 785.

488. D. Snetivy, G. J. Vancso and G. C. Rutledge, *Macromolecules* **25** (1992) 7037.

489. S. F. Y. Li, A. J. McGhie and S. L. Tang, *Polymer* **34** (1993) 4573.

490. S. R. Allen, A. G. Fillippov, R. J. Farris, E. L. Thomas, C. P. Wong, G. C. Berry and E. C. Chenevey, *Macromolecules* **14** (1981) 1135.

491. S. R. Allen, A. G. Fillippov, R. J. Farris and E. L. Thomas, *J. Appl. Polym. Sci.* **26** (1981) 291.

492. J. R. Minter, K. Shimamura and E. L. Thomas, *J. Mater. Sci.* **16** (1981) 3303.

493. K. Shimamura, J. R. Minter and E. L. Thomas, *J. Mater. Sci. Lett.* **2** (1983) 54.

494. S. R. Allen, R. J. Farris and E. L. Thomas, *J. Mater. Sci.* **20** (1985) 2727, 4583.

495. D. T. Grubb, in *Materials Science and Technology*, Vol. 12, edited by E. L. Thomas (VCH Publishers, Weinheim, 1993).

496. S. J. Krause, T. B. Haddock, D. L. Vezie, P. G. Lenhert, W. Hwang, G. E. Price, T. E. Helminiak, J. F. O'Brien and W. W. Adams, *Polymer* **29** (1988) 1354.

497. D. C. Martin and E. L. Thomas, in W. W. Adams, R. Eby and D. McLemore, Eds, *Mater. Sci. and Eng. of Rigid Rod Polymers*, Mat. Res. Soc., Symp. Proc. 134 (1989).

498. B. P. Griffin and M. K. Cox, *Br. Polym. J.* **147** (December, 1980).

499. F. N. Cogswell, *Br. Polym. J.* **170** (December, 1980).

500. A. E. Zachariades, J. Economy and J. A. Logan, *J. Appl. Polym. Sci.* **27** (1982) 2009.

501. D. G. Biard and G. L. Wilkes, *Polym. Eng. Sci.* **23**(11) (1983) 632.

502. A. Zachariades and J. A. Logan, *Polym. Eng. Sci.* **23**(15) (1983) 797.

503. L. C. Sawyer, *J. Polym. Sci., Polym. Phys. Edn.* **22** (1984) 347.

504. L. H. Sawyer and W. George, in *Cellulose and Other Natural Polymer Systems: Biogenesis, Structure, and Degradation*, edited by R. Malcolm Brown, Jr. (Plenum, New York, 1982) p. 429.

505. E. Baer and A. Moet, Eds, *High Performance Polymers* (Hanser, New York, 1991).

506. A. Frey-Wyssling, *Science* **119** (1954) 80.

507. M. J. Troughten, A. P. Unwin, G. R. Davies and I. M. Ward, *Polymer* **29** (1988) 1389.

508. D. I. Green, A. P. Unwin, G. R. Davies and I. M. Ward, *Polymer* **31** (1990) 579.

509. D. C. Martin, Direct imaging of deformation and disorder in extended chain polymer fibers, PhD. dissertation (1991).

510. D. C. Martin and E. L. Thomas, *J. Mater. Sci.* **26** (1991) 5171.

511. A. H. Donald and A. H. Windle, *J. Mater. Sci. Lett.* **4** (1985) 58.

512. I. M. Ward, personal discussions.

513. S. Kwolek, W. Memger and J. E. Van Tromp, *Inter. Symp. on Polymers for Adv. Techn.*, edited by M. Lewin, Intern. Union of Pure and Applied Chemistry (1987) p. 421.

514. M. G. Northolt, *Eur. Polymer J.* **10** (1974) 799.

515. K. Haraguchi, T. Kajiymama and M. Takayanagi, *J. Appl. Polymer Sci.* **23** (1979) 915.

516. R. A. Allen and I. M. Ward, *Polymer* **32** (1991) 203.

517. A. H. Windle, personal discussions.

518. W. W. Adams and R. K. Eby, *Mat. Res. Soc. Bulletin* **XII**(8) (1987) 22.

519. S. R. Allen, *J. Mater. Sci.* **22** (1988) 853.

520. C. Y. C. Lee, *Poly. Eng. Sci.* **33** (1993) 907.

521. D. C. Martin, *Macromol.* **25** (1992) 5171.

522. Y. Cohen and E. L. Thomas, *Macromol.* **21** (1988) 433.

523. D. Snetivy, G. J. Vancso and G. C. Rutledge, *Macromol.* **25** (1992) 7037.

524. G. R. Davies and I. M. Ward, Structure and properties of oriented thermotropic liquid crystalline polymers, in *High Modulus Polymers*, edited by A. E. Zachariades and R. S. Porter (Marcel Dekker Inc., New York, 1988).

525. D. I. Green, G. R. Davies, I. M. Ward, M. H. Alhaj-Mohammed and S. Absul Jawad, *Polymers for Advanced Technologies* **1** (1990) 41.

526. T. J. Lemmon, S. Hanna and A. H. Windle, *Polymer Commun.* **30** (1989) 2.

527. R. J. Spontak and A. H. Windle, *J. Mater. Sci.* **25** (1990) 2727.

528. S. E. Bedford and A. H. Windle, *Polymer* **31** (1990) 616.

529. H. Y. Yoon, *Colloid and Poly. Sci.* **268** (1990) 230.

CHAPTER SIX

New techniques in polymer microscopy

6.1 INTRODUCTION

A wide range of techniques in microscopy has either appeared within the past few years, or has only recently been applied to the study of polymer systems. These new forms of microscopy are distinct from the continuing evolution of the microscope. This is currently rapid in the direction of ease of use, computer control and increased use of digital image storage and processing, even for optical microscopy [1, 2]. Some examples of these new forms of microscopy as they have been applied to polymers are included in the previous two chapters. In the case of novel types of scanning electron microscopy and high resolution transmission electron microscopy, the principles of microscope operation and image formation are described in Chapters 2 and 3, respectively.

In this chapter the new methods that are being applied to polymer systems are described in more detail, with particular attention paid to traps for the unwary user. Some techniques are still in the process of development, but clearly hold out promise of application to polymer science and engineering. Each section begins with some general references to texts and reviews on the technique being described. The reader who becomes anything more than a casual user of the new techniques will need more information than can be provided here.

6.2 OPTICAL MICROSCOPY

6.2.1 Confocal scanning microscopy

The confocal scanning microscope (sometimes LCSM, for laser *confocal scanning microscope*) has many forms, but all are optical microscopes with some form of scanning added to the regular optics. A normal optical microscope produces poor images when the sample surface is rough or the signal comes from a range of depths in a transparent sample. An LCSM does not have this limitation and has a slightly better lateral resolution than the regular optical microscope. The basic principles of confocal instruments, and their practical design are described by Wilson [3]. A later text gives more applications [4], and a handbook apparently directed at biologists is extremely useful [5]. Both basic principles and new developments are covered in a very recent book with the ambiguous title *Multidimensional Microscopy* [6].

In a confocal microscope a small aperture is placed in the plane where rays coming from a particular plane in the object come to a point (see Fig. 6.1, the plane is called the *confocal* plane). A light detector is placed behind the aperture. The aperture selects one point (x, y) in the plane, and also cuts out most of the light coming from other planes in the specimen. If the illumination is focused onto the selected point in the object, then information comes from the point (x, y, z) only. Scanning the illumination and the confocal aperture together over x and y builds up a scanned image of the selected plane.

Mechanically registering two scanning systems would be difficult, so commercial LCSMs are reflection microscopes, where (as shown in Fig. 6.1) the beam passes through the same scanner twice. The beam is scanned on the specimen, then de-scanned onto a fixed confocal pinhole. Confocal microscopes may have a transmission mode, but currently this will not be confocal. It may use a separate regular illumination system, or the scanned illumination. In the latter case the

transmitted light detector has a large area and no confocal aperture in front of it. Two schemes for transmission confocal microscopy have been suggested. In one the specimen is scanned in x and y through a stationary beam [7, Chapter 5 of ref. 6]. In the other the transmitted beam is collected, passed back around the specimen and into the reflected light detector with another beam splitter. A polarization scheme keeps the reflected and transmitted signals separate.

A confocal microscope can be made which uses regular illumination and has a rapidly rotating disc containing many holes in the confocal plane. These holes act as confocal apertures and mechanically scan over the object, but much light is lost, and the output is integrated by the eye or by film. It is not in a suitable form for digital processing. The laser light source naturally produces a bright illuminated spot, and is scanned in the normal square raster by mechanical scanning of mirrors, or by acoustic-optical crystal deflectors. Alternatively, the illumination may be formed into a line, which only needs a single scanning device. A slit is then the confocal aperture [8]. A slit is not so efficient at rejecting light from different heights in the specimen, so a line scanning instrument will not have the vertical resolution that a spot scanning instrument can have [9]. An instrument with a single scanning device and a linear detector (such as a diode array) to record a line of data at once is simpler and can collect data faster.

So far the description of the instrument implies that the image formed is a slice or section of the specimen at a particular height. (The LCSM has been described as an 'optical microtome'). If the specimen has a rough reflective surface, such an image would contain a bright contour line where the surface intersects the selected plane, with the rest of the field dark. To get more information the specimen is scanned in the z direction – not continuously, but in steps between image scans. Storing all these images gives a three-dimensional image of the specimen that can be processed to show and measure three-dimensional features. This is an extremely powerful tool requiring considerable image processing

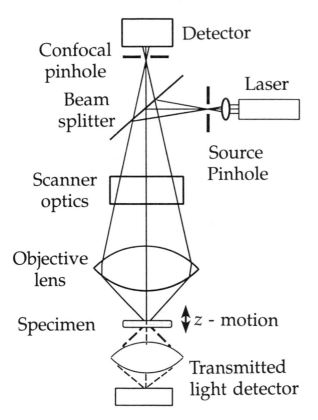

Fig. 6.1 Schematic of a laser confocal scanning microscope. The laser illumination spot is scanned across the sample. Reflected light is de-scanned and passed through the beam splitter and the confocal aperture to the detector. The detector of transmitted light may be present as an accessory, but it gives a normal, not a confocal, optical microscope image.

power [10]. It is mostly applied to transparent specimens, as an opaque reflective surface can be treated in a simpler way. A low density of features in the transparent specimen is helpful in simple reflection mode, so that the focusing of the illuminating probe is not disrupted.

In the simpler modes of operation for a single surface the specimen is still step-scanned in the *z*-direction. But as each image is collected, the intensity values are either added or set to the maximum obtained during the *z*-scan. The results of these operations are similar, a single image that is in focus at every point (if the *z*-scan has a sufficient range). At each point the bright contour line that appears when the surface is at the focus level dominates the image. Normally the same (x, y) point in each *z* image is compared or summed as it is collected. Using adjacent points, $(x - z, y)$ for example, gives a view from a different direction, and a stereo pair of images can be generated.

For metrology, an image is formed by setting the value of each pixel to the value of *z* at which the image was brightest. This will produce a topographical map of the most reflective surface, whether external or internal. Line profiles can be selected in any direction, or a surface roughness can be calculated from the whole profile.

In any LCSM as in any other optical microscope the results can be no better than the objective lens allows. The objectives for LCSM must have a very flat field and a large aperture if the focal plane is to be well defined [2, 4, 11, 12]. They must be well corrected for spherical aberration, but with a single wavelength laser source they need not be chromatically corrected. High power objectives used with immersion fluid give the highest apertures, but have a short working distance. This means that the high resolution in the axial or *z* direction can only be maintained for a short distance into a transparent sample.

The major use of LCSM up to the present has been as a fluorescence microscope in biology [4]. Its use in materials is growing [13] but many synthetic materials are opaque and can withstand vacuum, so their external surfaces can be viewed by SEM. Three dimensional imaging and metrology have been used to investigate water treeing and other dielectric defects in polymers [14, 15]. These specimens are transparent and have a low density of strongly scattering internal surfaces in complex shapes – ideal for LCSM. Other polymer materials studied by LCSM have included fiber reinforced composites, where a transparent epoxy resin matrix allowed internal interfaces to be seen [16] and latex suspensions [17]. In all these cases the authors emphasized the advantage of a nondescructive, non-contact method of three-dimensional observation.

6.2.1.1 *Raman imaging*

Replacing the detector in an LCSM (Figure 6.1) with a filter or spectrometer, or adding scanning optics to a micro-Raman spectrometer makes a new instrument which can form 'Raman images'. These recently available instruments and their applications to polymer science are reviewed by Meier and Kip [19]. They can distinguish between components on the basis of the chemical groups they contain, with the same advantages as the LCSM. That is, the technique is non-destructive, non-contact, and three-dimensional. Raman spectral lines are generally several orders of magnitude weaker than the incident light, and the power of the incident light is limited (or the sample will melt). This means that scanning a Raman image can be slow.

There are several different optical arrangements. Using 2-D scanning and a filter to select a particular Raman line produces an image where the signal depends on the concentration of a selected chemical group [19]. A 1 μm particle of PE in a 2 μm thick film of iPP was easily seen in this way [18], but normal optical microscopy could also distinguish such a particle. More subtle changes due to chemical degradation have been imaged in a system where the confocal aperture is a slit, and a 2-D CCD detector forms a full Raman spectrum for each point on the line [20]. Figure 6.2 shows an example where the line of illumination was stationary on a cross-section through a film of PVC. The intensity of a spectral peak at 1520 cm^{-1} is directly related to the local concentration of C=C bonds, which are a product of degradation in PVC. The control film was not degraded, and shows uniform peaks at 1436 and 1600 cm^{-1} that come from the plasticizer in the film. The degraded film has a very strong

1520 cm^{-1} peak, so shorter exposures suppress the plasticizer peaks. Degradation is greatest at the film surface, as expected, but the profile can be analyzed numerically to determine diffusion and reaction kinetics [20]. Clearly this method could be extended to give the spectrum and thus the chemistry of every point in a 2-D image, which can come from an internal plane in a transparent material.

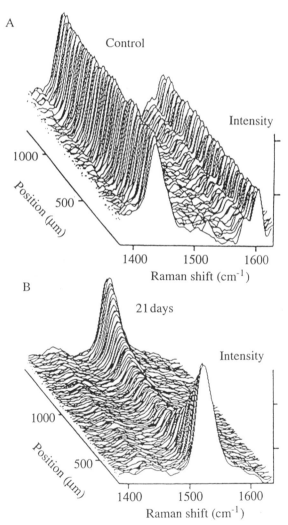

Fig. 6.2 Raman spectra from a line scan through the thickness of PVC film; (A) a control sample, (B) a film degraded with triethylenediamine for 21 days [20]. The band at 1520 cm^{-1} is due to C=C, from the dehydrochlorination of PVC. Its intensity as a function of depth can be analyzed to give degradation and diffusion kinetics. (From Bowden *et al.* [20]; reproduced with permission.)

6.2.2 Near field optical microscopy

Consider using a flexible optical fiber to focus the illumination on the object and collect the scattered light. If the fiber or the specimen is scanned, we have a scanning optical microscope, and if there is some depth selectivity, a kind of confocal scanning microscope [21]. The near-field optical microscope is similar to this scanning optical microscope, but with the extra requirement that the fiber is extremely small, and very close to the surface of the sample [22, 23]. Very close here means at much smaller distance than the wavelength of light – this is 'near-field' – and the advantage gained is that the resolution can be much better than in other optical microscopes. The resolution can be less than one tenth of the wavelength, about 30 nm for visible light. The equipment used to provide this close surface scanning is extensively borrowed from scanning probe microscopy (Section 6.5.2).

As the fine optical fiber can be inserted close to an object in water, the first applications of this instrument have been directed towards high resolution observation of tissue sections and living cells [24]. The associated photon tunneling microscope has been applied to polymer surfaces [25]. In principle all the contrast modes of normal optical microscopy can be used in the near-field and there may be other unique contrast mechanisms [26]. The light signal can be analyzed spectroscopically to give detectability limits down to a single molecule, or nano-aggregates in doped polymers, going beyond the realm of the 'micro-Raman' analyzer to the pico- or atto-Raman [27].

6.3 SCANNING ELECTRON MICROSCOPY

6.3.1 Low voltage SEM

'Low voltage' for an SEM means operating with a beam voltage below 5 kV. It may seem that low voltage SEM is not a special technique today, because it is unusual if a modern SEM instrument is not advertised as operating with a wide range of beam voltages. The lower limit may be 1 kV, 500 or even 200 V. Operation at low voltages is described in general texts on SEM [28, 29] and now there is a book entirely devoted to it [30]. The application of LVSEM to polymers has been reviewed [31, 32]. Many older instruments could not form images at very low voltages for practical reasons; contamination and e–m fields affected the image. The information in a low voltage image comes strictly from the sample surface, and is easily spoiled by contamination. Modern instruments have better vacuum; not only is the column at lower pressures, but the residual gas is cleaner, because of oil-free pumps. Unshielded small magnetic and electric fields have a negligible effect if the beam voltage is high, but can completely destroy the image at low beam voltages. Improved and simplified column design has reduced this problem, when the microscope is installed in a suitable location.

These improvements allow the microscopes to operate at low voltage, but as shown in Section 3.2.1 and Table 3.5, not all SEMs will produce the same resolution at low voltages. The high chromatic aberration and low brightness of tungsten filaments at low beam voltages severely restricts resolution. A field emission gun has lower energy spread and very high brightness, so that the resolution at low voltage in an FESEM is as good as or better than that of a normal SEM at high voltage. There are several types of electron guns that are better than the standard tungsten filament. These include the lanthanum hexaboride (LaB$_6$) rod and the Schottky emission gun as well as the cold and heated field emission guns [28, 30]. There are trade-offs in the choice of electron source. For example the cold field emission gun has the smallest beam energy spread and best resolution, but needs an extremely good vacuum in the gun chamber (which is expensive). Figure 5.49 is a direct comparison of images of a polymer blend obtained at 5 kV with a LaB$_6$ electron gun and with a field emission gun, and the image from the FESEM contains more information.

There are several important differences in image formation at low voltages in the SEM, and these differences are due to the small penetration of the incident beam. The range of electrons, R, can be expressed as $R = aE^n$, where E is the electron energy and a and n are constants. According to the general empirical expression of Kanaya and Okayama [33, 28] if R is in μm,

$$R = \frac{0.0276\,A}{Z^{0.889}\rho}\,E^{1.67}$$

where A is the atomic weight, Z the atomic number and ρ the relative density of the specimen. For a polymer modeled as carbon of relative density one, this gives $R = 70$ nm at 1 kV and 3 μm at 10 kV. Expressing R in terms of mass thickness, μg cm^{-2}, removes the dependence on density. Experimental data for carbon is fitted by $n = 1.4, a = 6$, when R is in μg cm^{-2} and E in kV [30]. Putting in a density of 1 g cm^{-3} gives the electron range in a hydrocarbon polymer as 60 nm at 1 kV, and 1.5 μm at 10 kV. This is in approximate agreement with the general expression. In higher density, higher atomic weight material the range will be even less, 15 nm in silicon at 1 kV, for example. Figure 6.3 shows this dramatically as a simulation of electron trajectories in the sample. The region containing the electron tracks is the interaction volume. As described in Section 2.3 and Fig. 2.4, the interaction volume controls the resolution in backscattering mode and x-ray analysis. This means that the resolution calculated in Table 3.5, which considered only the size of the incident beam, did not tell the whole story.

Figure 6.4 is a schematic diagram of the interaction volumes at high beam voltage and

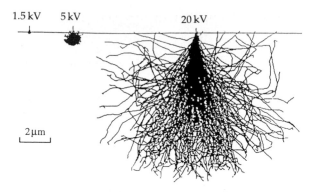

Fig. 6.3 Simulation of the interaction between incident electrons and the sample, as a function of incident beam voltage, by Monte Carlo calculation of the electron trajectories [49]. The volume is much less at low beam voltages. (From Joy and Pawley [49]; reproduced with permission.)

at low beam voltage. It shows a surface layer that is the source of the low energy secondary electrons. The thickness of this layer, Λ, depends on the specimen, and is about 3 nm for aluminium, 20 nm for polymers [30, 34, 35]. It is not really a sharply defined surface layer; the chance of an electron escaping the specimen falls exponentially with depth, as $\exp(-z/\Lambda)$ [36]. The depth distribution will depend on the number of electrons produced at each depth, and this can only be found by Monte Carlo simulation [35–37]. In any case, much of the interaction volume of a 1 kV beam is within the surface layer where secondary electrons can escape the specimen. Therefore a large fraction of the secondaries that are produced can escape. At high voltage, many more low energy electrons are produced (the number is proportional to the incident beam energy) but they are nearly all produced deep in the specimen and cannot escape. The yield of secondaries is therefore higher for the low beam voltage.

Secondary electrons produced by a high energy beam also have a broader distribution in space, because the emerging backscattered electrons (BSE) produce secondaries as they leave the specimen. These secondaries are called SE2, to distinguish them from SE1, which are second-

aries produced directly by the primary incident electrons [28–30]. Figure 6.4 shows how these SE2 electrons may emerge from a wide region of the specimen at high beam voltages. At low voltage, the backscattered electrons must emerge very close to the primary beam and the SE1 and SE2 cannot be distinguished. Figure 6.5 shows schematically the resulting spatial distribution of secondary electrons. At low energy things are simple, and there is a single sharp peak, if the incident beam is focused to a small diameter. At high energy the SE1 peak retains this sharpness, but there is a much broader peak of SE2, with a resolution more like that of the BSE. High

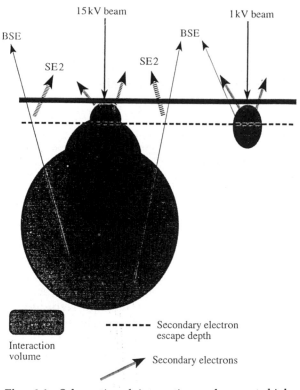

Fig. 6.4 Schematic of interaction volume at high beam voltage (15 kV) and at low beam voltage (1 kV), showing that at high beam voltage, most secondary electrons are produced too deep to emerge from the sample. Those secondary electrons that do escape at high beam voltage can arise from a large area, much larger than the size of the incident beam. SE2 are secondary electrons produced by backscattered electrons as they leave the surface.

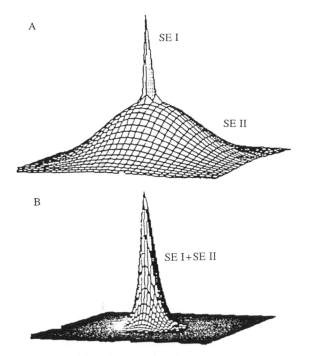

Fig. 6.5 Intensity profiles of secondary electrons leaving the specimen surface, when the surface is perpendicular to the beam. At high beam voltage (A) a sharp SE1 peak is superimposed on the very much broader SE2 peak. At low beam voltage (B) there is only a single peak. (From Vezie *et al.* [32]; reproduced with permission.)

resolution information can still be obtained, but the SE2 acts as a noise background, reducing contrast. At low resolution both SE1 and SE2 contain information. The profile at medium energy looks better, as the SE2 peak is much less broad, but it may be worse for high resolution. The SE2 signal is not varying slowly enough to be discarded as noise by simple image processing, yet does not have the highest resolution.

The secondary electron image appearance is different at low voltage. One reason is that the interactions are all at the surface, so that all of the image features are much more sensitive to the exact state of the surface, contamination and all. The other reason is that the effect of tilting the surface is different. At high voltage the secondary emission is strongly increased when the sample is tilted, because part of the interaction

volume is brought nearer to the surface. This effect is shown schematically in Fig. 2.5. An even larger increase in secondary emission comes if there is a sharp edge in the specimen surface. Figure 6.6 is modified from Fig. 6.4, with the sample surfaces each now having a sudden perpendicular edge near to the spot where the primary beam arrives. Evidently at high beam voltage the lateral leakage from a deep part of the interaction volume will drastically increase the secondary signal, even when the beam is well away from the edge. At low voltage there will be no effect until the beam is very close to the edge. It will be a small effect, since a large part of the interaction volume is already within a surface

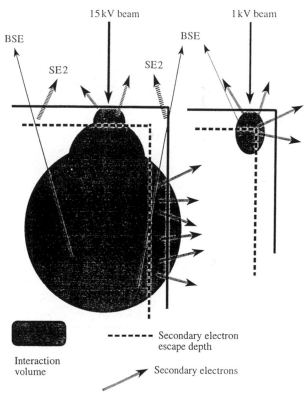

Fig. 6.6 Schematic of the effect of a step or edge on the emission of secondary electrons at high beam voltage (15 kV) and at low beam voltage (1 kV). The emission increases a lot at high beam voltage even when the beam not very close to the edge; the effect is smaller and limited to when the beam falls next to the edge at low beam voltages.

region where secondaries can escape. Figure 6.7 shows this effect very clearly, using a silicon sample produced by microlithography that has sharp perpendicular edges. At the 'normal' SEM operating voltages there is a lot of glare near the edges, and they are ill-defined. At 30 kV the glare spreads from one edge to the other, making the whole raised structure brighter. As the incident beam voltage is reduced, this effect goes away. At the same time, the detailed surface features become clearer; at the lowest beam voltage, thin surface layers become prominent.

If the detector is to one side of the sample, then its collection efficiency will be greater for surfaces tilted towards it than away from it. This is true both at high and at low beam voltages, even

though tilt of the sample surface has a small effect at low beam voltage. The topographic contrast in SEI is therefore altered at low voltage but not removed. Figure 6.7 shows that the low voltage image retains the appearance of a shadowed view of a three-dimensional object, which makes the SEM images so easily interpretable. The normal Thornley–Everhart secondary detector has a grid voltage of +250 V to attract the secondaries into it. When the beam voltage is less than a few kV, this electric field can affect the primary beam trajectory and thus distort the image. One way of overcoming this is to use two similar detectors, one on each side of the sample chamber. The primary beam is thus in a low field region, and topographic contrast

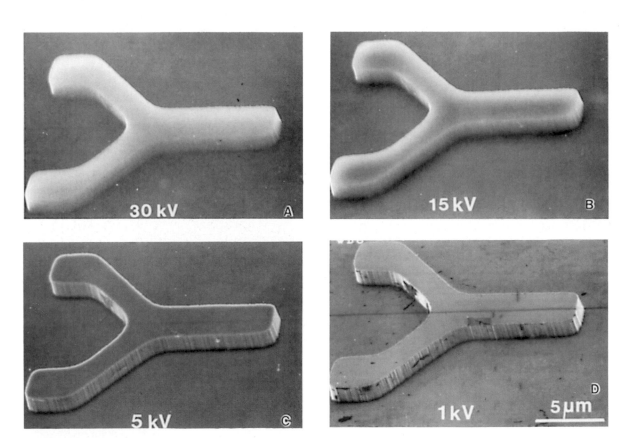

Fig. 6.7 Secondary electron images of a raised bar on silicon, tilted 45°, with the beam voltage set to (A) 30 kV, (B) 15 kV, (C) 5 kV and (D) 1 kV. The excess emission due to the edges spreads right across the bar at 30 kV. At 5 and 1 kV there is only a narrow, and a very narrow, bright line near the edge of the bar. (From Reimer [30]; reproduced with permission.)

can be manipulated (as for BSE in conventional SEM) by adding or subtracting the two signals [38, 39].

BSE are not much used in LVSEM, although the resolution and atomic number contrast are both good. There are many more secondaries than backscattered electrons at low voltage, and this is especially true for low atomic number samples such as polymers. It is also comparatively difficult to detect these few low energy backscattered electrons. A scintillator [40] or a multi-channel electron plate can detect them directly [41, 42]. Alternatively they can be detected indirectly, through the secondaries produced by their collision with the chamber walls (SE3) [30, 43].

So far we have not mentioned the most important feature of LVSEM for polymer samples. At lower beam voltages the number of backscattered electrons stays much the same, and the yield of secondaries increases, because all interactions are nearer the surface. At some beam energy, called $E2$, the total number of backscattered and secondary electrons becomes the same as the number of primary beam electrons. That is, the total electron yield is one at $E2$. In principle any insulating specimen can be observed without metal coating by operation at this beam voltage, which is typically in the range of 0.8 to 1.6 kV for synthetic polymers [28, 34].

Figure 6.8 is a schematic plot of total electron yield as a function of beam energy. At a high incident beam energy, $E \gg E2$, the number of electrons emitted from the specimen is much less than the number arriving. The specimen in the beam becomes negatively charged, and this potential repels and slows down the electrons in the incident beam. As a result, the incident energy of the beam falls until it reaches $E2$, when no more charging occurs. The irradiated specimen is then highly negatively charged, and will be much brighter than uncharged parts of the specimen. Conversely, at low beam energy $E < E2$ the yield is more than one, so the specimen becomes positively charged. The charge attracts the incoming primary beam electrons, so that they arrive with more energy.

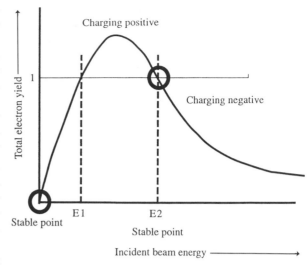

Fig. 6.8 Total electron yield as a function of the incident beam voltage. When the yield is less than one, the sample charges negatively. When it is greater than one, the sample charges positively. This makes E2 (and zero) a stable point with no further charging.

This process continues until the electrons arrive with energy $E2$, which is thus a stable point. As the beam energy goes to zero, so does the electron yield, so there is another energy $E1$ where the total electron yield is one (Fig. 6.8). If the incident beam energy $E < E1$ then the stable point is zero, not $E2$. $E1$ is a very low energy, 100–200V, which is not accessible in the SEM.

In practice it is found that operation at or near to $E2$ reduces charging considerably, but for an entirely stable image it is still necessary to apply a thin conductive coating. The reason is simple; the exact value of $E2$ depends on the chemical nature of the sample and on the local tilt of the surface [30, 44–46]. The effect of tilt on secondary emission is still present, although smaller, at low voltages. Thus a tilted specimen releases more secondaries, and will have a higher value of $E2$ than a specimen with no tilt. A specimen without local variations of tilt or chemistry would be boring indeed, so all real samples will have local variations in $E2$, and cannot maintain exact charge balance at all points.

Charging is a time dependent phenomenon; the charged spot on the specimen will act as a charged circuit with capacitance C and leakage resistance R. The very high value of R means that the time constant for charge decay once the beam moves off the charged spot is long. This provides a simple method for finding $E2$ for a particular material and tilt. Select an area at low magnification, then switch to a higher magnification (say five times higher) for a few seconds, and return to the lower magnification. The more highly irradiated central area that was the field of view at high magnification will still be charged more than the rest of the field of view. If it is brighter, then it is negatively charged and the beam voltage is too high, above $E2$. If it is darker, the beam voltage is below $E2$. Repeat until $E2$ is found [47]. This is one of the nine methods for determining $E2$ listed by Reimer [ref. 30, p. 78]. If a specimen is made especially for the purpose of this test, it should be metal coated except for narrow strips [46, 48].

The time dependence of charging makes slow scans more sensitive than fast scans. In fast scanning mode, none of the image points have much time for the charge to leak away, so all are at about the same potential. The uniform charge may subtly alter the image, but the effects are much less than at slow scan rates, where there are large potential differences between different points on the image. These will distort or destroy the image. The effect may be apparent even in normal (high voltage) SEM operation, when a good image on the viewing screen gives a bad one in slow scan photographic mode. The fast scans have too much noise, so the answer is to make many fast scans and sum the images. This is very easy with digital image processing.

It is not easy to predict how radiation damage varies with beam voltage in the SEM. Because the secondary emission is greater at low voltages, fewer low energy incident beam electrons are needed to form a given image, and each deposits less energy in the specimen [49]. However, the electron range decreases as about $E^{1.5}$, so the energy deposited per unit mass (the radiation dose, Section 3.4.2) is much the same or slightly greater at low voltages. Radiation affects only a thin surface layer, but this is exactly the surface layer that provides all the information in the image. In the SEM the unirradiated bulk of the sample serves to stabilize the damaged surface layer (Section 3.6). The stabilization should be better if the damaged layer is thinner, and if mass is lost, the mass loss will be lower [30, 50]. If the beam is stationary (or nearly so at high magnifications), the irradiated mass depends on the interaction volume $\approx$ range^3 and the radiation dose will be much greater at low voltages, and concentrated in a very small spot. Nevertheless visible changes to the image are less at low beam voltages [49, 51].

X-ray microanalysis is still possible at low beam voltages, but may not be easy. There are the advantages of better spatial resolution and reduced absorption correction, as the volume generating the x-rays is small and near the surface. The disadvantages are that the x-ray intensity produced is small, and not all elements can be analyzed. In principle elements of higher atomic number can be identified using L and M shell x-rays, but these are much more complicated than the simple K shell emissions. The surface sensitivity may be useful sometimes, but is more likely to be a problem; a thin conducting coat may be a significant part of the electron range, and without it the specimen may charge up. (Operation at voltages as low as $E2$ is not likely for x-ray microanalysis). Having a beam stationary on the sample for a long time to integrate the weak signals is a perfect way to produce contamination. Contamination buildup can stop the electrons in a low voltage beam before they reach the specimen surface. Then the analysis does not relate to the specimen at all.

Among many examples of the use of the LVSEM on polymers, Berry [52] characterized the phase structure of a wide range of polymer blends. Price and McCarthy [53] found the reduced interaction depth particularly valuable for imaging low-density polymer foams. Himelfarb and Labat [54] compared images of polymer blends at low voltage and at conventional voltages. The contrast was greater at higher voltages when one component of the blend was stained with a heavy metal, but otherwise visibility of small regions of either phase was better at low voltage. Other practical examples are described in detail in Chapters 4 and 5.

Many other examples of the application of LVSEM to real problems in polymer characterization are given in Chapters 4 and 5. These include images of the free surface of membranes (Figs 4.26, 4.32 and 5.32), and the fracture surfaces of isotropic blends (Fig. 5.80) and fibers (Fig. 5.113). The membrane surface micrographs were taken with magnifications of about 80–200,000 times, and some show details well below 10 nm in size. Molded grooves in a CD substrate are also shown as part of a comparison of FESEM, TEM and AFM in Fig. 5.43. Figure 5.85 [55] is another comparison of these techniques; it shows that excellent quality medium resolution images of stained block copolymer domains could be obtained with both FESEM and AFM.

6.3.2 High resolution SEM

The high resolution SEM (HRSEM) is an instrument that combines a field emission gun with a short focal length final condenser lens [56, 49]. A final lens with a focal length of only one or two mm gives similarly small values of the aberration coefficients C_s and C_c. As described in Section 3.2.1, this gives a very small probe size at high beam voltages. Such a lens combined with the small energy spread δE of a field emission gun gives a small probe size even at low beam voltages. A disadvantage of this short focal length is that the depth of field is small, so that a rough surface cannot be brought into focus all at once. The focal length is similar to that of a TEM objective lens and requires a geometry similar to the TEM, with the specimen inside the lens. This limits the maximum specimen size to one only a few mm across, another disadvantage.

The HRSEM is very often operated at low beam voltage, and the technique may be specifically referred to as 'high resolution low voltage SEM'. As discussed in the previous section, the small interaction volume is one reason for low voltage operation. For polymers, the second and more important reason discussed above is the reduction or elimination of charging artifacts. High resolution information cannot be imaged correctly through a thick metal coating. A very thin coating (or even none at all) may suffice at low beam voltages [32]. More directly, users of such microscopes try a wide range of beam voltages and may often find that a low voltage in the range 1 to 3 kV gives the 'best' image [32, 49]. Sometimes a SEM with a field emission gun and a normal large specimen chamber, which we would call a FESEM, may be described as 'high resolution low voltage SEM'. This is confusing but reasonable, as the resolution at low voltage *is* high compared to that obtained with a 'normal' tungsten filament instrument.

Since the space around the specimen is very limited in an HRSEM, a standard secondary electron detector cannot be put close to it. Luckily the strong solenoidal magnetic field of the final condenser lens traps low energy electrons, forcing them to move in spirals along the axis of the lens. A small electric field draws the spiralling electrons up the column and to a detector above the upper pole piece of the lens. In the normal large specimen chamber arrangement the secondary electron detector picks up electrons from many sources. They are produced in the specimen (SE1 and SE2), when backscattered electrons hit the chamber walls (SE3) and produced in the column (SE4). SE3 electrons carry only low resolution information, SE4 electrons no information at all and they may be up to 60% of the signal [57]. In the 'through-lens' detector, only the secondaries produced near the optic axis of the microscope are efficiently transferred to the detector, so this detector system has less background noise, better contrast at high resolution, than the normal arrangement. The 'through-lens' detector gives the normal topographic contrast [58], but higher features also appear brighter, presumably due to better collection efficiency [32].

Theoretical prediction of the exact resolution achieved in the HRSEM is difficult. The thicker metallic coating needed for an insulating specimen and the extra noise due to the SE2 electrons make the resolution worse at higher beam voltage, cancelling the effect of the smaller probe diameter that can be produced. The excitation of secondary electrons can occur some distance from the track of the primary electron. This delocalization [59] has been estimated as 0.5–5 nm, so it is not clear if it is important [49, 60]. It does not apply to backscattering, so in principle (though not yet in practice) BEI will have a better resolution than SEI [60]. With a test specimen of carbon coated with tungsten particles – an ideal high contrast specimen – Joy and Pawley show that 1 nm particles can be imaged in SE1 from 1.5 to 30 kV, and in BEI from 4 to 30 kV [49].

The same authors have simulated images of simple structures such as lines of square cross-section using Monte Carlo techniques to determine resolution [49]. The edges of the bars are bright, and the resolution is the limiting size when the two edges can just be distinguished. For gold this is generally near to a 2×2 nm bar, and for carbon, a 5×5 nm bar. This resolution is predicted from the finite size of the diffusion region, where secondary electrons can escape [60]. The highest and the lowest beam energies gave the best visibility of the edges, but the low beam energy images had much better signal to noise ratios because of the larger secondary signal. At 1 kV a 2×2 nm carbon bar could be 'partially resolved' in this simulation. This resolution limit can be reduced by coating the sample with a thin uniform film of metal, if the sample is a biological or polymeric material of low density and low atomic number. Up to now it has been assumed that any metal coating must degrade the image, by changing the feature sizes. But if a uniform 1–1.5 nm metal coating can be applied the feature distortion will be small while the secondary emission from it can be significant. This increases the signal to noise ratio, the contrast of steps and improves the resolution because of the smaller mean free path of electrons in the high density material [49]. Such thin metal films need special tools to prepare (Section 4.7) and are not entirely stable [32].

6.3.3 High pressure (environmental) SEM

The high pressure SEM (HPSEM) operates with a gas pressure in the specimen chamber in the range 0.1 to 30 torr (10–4000 Pa) [61–63]. The 'high pressure' in the HPSEM is high only in comparison to the vacuum inside a normal SEM, which has a pressure of 10^{-6} torr or less in the specimen chamber. As the normal micro-torr pressures are called 'high vacuum' some authors call the HPSEM a 'low vacuum SEM', which is fine, except that its abbreviation cannot be distinguished from low voltage SEM. An important pressure for biological, polymer or mineral samples that must be observed wet and not frozen is the vapor pressure of water at room temperature, 25 torr (3.2 kPa or 0.46 psi). Specimens can be maintained at their normal water content in the specimen chamber when there is this pressure of water vapor, independent of the pressure of any other gases.

Any increase of pressure in the gun chamber from its normal very low value will cause the electron gun to fail after a short time. To protect it the HPSEM contains some limiting apertures that restrict the flow of gas up the optical column from specimen to gun chamber. In the ESEM (Electroscan Inc.) there are four apertures. Two are conventionally placed, separating the gun chamber from the column, and the column from the specimen chamber (the final aperture). Two more are placed just above the final aperture, dividing the vacuum system into five separate regions in all. Each of these has its own vacuum pumping system, and the result is a mechanically complex microscope with a sequence of pressures that the electrons experience as they move from gun to specimen: 10^{-7}, 10^{-6}, 10^{-4}, 10^{-1} and 10^{1} torr.

The working distance – the separation of specimen and final aperture – must be kept small to limit the distance that the beam travels through the highest pressure region. Multiple scattering of electrons by the gas will smear out the fine focus of the probe if the pressure is too high or the working distance too long. The beam voltage must be kept in the 'normal' SEM operating range of 15 kV and above. The gas in the specimen chamber has some effect on the primary beam electrons and the backscattered electrons, but a greater one on the low energy secondaries, as they have a very small mean free path. The secondaries ionize the gas as they collide with the gas atoms producing positive ions and more electrons. An attractive positive voltage on a detector will make all the electrons drift rapidly towards it, with much the same result as if they had followed a free trajectory as they do in the regular SEM. If the attractive electric field is sufficiently large, each drifting electron will be accelerated enough to gain the energy to cause more ionization at its next collision with a gas atom. As each collision can release more than one electron, the result is an amplification of the secondary electronc current [62, 64].

The detector system just described is analogous to a gas-filled detector used for x-rays. At low applied voltage it is like an ionization chamber, and when there is amplification, like a proportional counter. An x-ray proportional counter has a limited response rate, of about 10 kHz, but the x-ray systems are normally operated at pressures close to atmospheric, with local fields up to $100 \text{ MV}^{-1} \text{ m}$ that give amplifications of several thousand times. These conditions can produce very high concentrations of positive ions by the central wire. The ions reduce the applied field, and the response rate is limited by the need for them to diffuse away. In the case of the HPSEM, the field is more uniform, the pressure and the field are lower. All these factors tend to reduce local ion concentrations and thus increase the response frequency.

The amplification will work best in only a limited range of pressures. Too high a chamber pressure makes the mean free path of electrons very small; a high field is then needed to accelerate them, and the high voltage required may not be practicable. Too low a pressure and the effect will be small because the electrons will have few collisions as they travel from specimen to detector. The exact form of the signal contrast is difficult to predict because of the complicated

effects that local topography may have on the electric field. Nevertheless, the images look like those of normal SEM secondary imaging.

Charging of the specimen is suppressed in the HPSEM, as the space above the specimen is full of charged particles of low energy. In particular, insulating specimens normally charge negatively at these incident beam voltages, and the positive gas ions are attracted to the specimen, neutralizing it. It is this combination of high surrounding gas pressure and no requirement for conductive coating that makes the instrument uniquely useful. Experiments that demand its use include the dynamic observation of reactions involving a solid and a gas or liquid, such as oxidation or other corrosion. However, if specimen charging is the only problem, a low voltage SEM is a more practical solution. The range of applications for the HPSEM is wide [61, 65]. Dynamic observations at elevated temperature have mostly concerned the oxidation of metals [66, 67] but have recently been made on the melt processing of polymers [68]. Observation of wet samples includes minerals [69], biological samples [70], and polymers such as hydrophilic sponges [71]. Oily as well as wet specimens can be imaged directly, and this is relevant to those involved in oil production [72] and to those planning to use polymer materials to clean up oil spills [73]. Metal coating can improve the image in the HPSEM, and may be applied to samples that are superficially dried, but would contaminate a high vacuum system (see Fig. 5.53, Section 5.3.3.3).

6.4 TRANSMISSION ELECTRON MICROSCOPY

6.4.1 High resolution transmission electron microscopy (HREM)

HREM is here taken to mean the resolution of the intermolecular or interatomic spacings in ordered materials. The theory and practice of this technique is well established [74–76] and its applications to materials science [77–79], to organic materials [80] and to polymers [81–83] have been reviewed. Polymers are extremely difficult materials for this technique. They are often poorly ordered, but a more important problem is that they are sensitive to radiation damage (Section 3.6). High resolution images require high electron doses. To obtain good results from polymers it is vital to use the low-dose techniques [84] that prevent any excess irradiation of the sample.

The basic principle of lattice resolution is explained schematically in Fig. 6.9. A thin specimen with a periodic electron density, spacing d, is illuminated with a plane wave, wavelength λ. The result is a transmitted plane wave with a periodic phase change. Far from the specimen, this wave becomes diffracted plane waves at angles 2θ given by Bragg's Law $n\lambda = 2d \sin\theta$. In electron diffraction, the wavelength is much less than d so the diffracted angles are very small and $2\theta = n\lambda/d$. Figure 6.9 shows only the waves with $n = 0$ and 1. To obtain lattice images, diffracted and transmitted waves are allowed to pass through the objective aperture and the imaging lenses, indicated as a single thin lens in the figure. This forms an image of the specimen with magnification M. As shown in Section 3.1, using ray diagrams, divergence angles α at the object are reduced to α/M at the image. Thus the diffracted waves recombine at an angle $2\theta/M$. The intensity is a maximum where the two waves are in phase, and the separation of these regions in the image plane is Md. A magnified image of the lattice spacing has been formed.

The spherical aberration of the objective lens and any defocus both add phase shifts to the diffracted beams that depend on the angle 2θ. From Fig. 6.9 it appears that phase shifts will move the image laterally. If the scattering is weak, the transmitted wave acts as a reference phase, and the phase of the diffracted beam is naturally $90°$ to this. Then, as for phase contrast generally, there will be no contrast in a perfectly formed image. As d decreases 2θ increases and the phase shift χ due to defocus and aberration increases. This will cause the image contrast to appear and strengthen, then disappear at

$\chi = 180°$, and reappear inverted. The resolution limit of the image is normally taken to be the point just before the first zero in this contrast transfer function [85]. At optimum focus this is close to the point to point resolution calculated in Section 3.1, at about $0.5(C_s \lambda^3)^{1/4}$. The appearance of the image will change in a complicated manner with very small changes in focus.

For the best resolution small values of spherical aberration and electron wavelength are required. Small wavelength corresponds to high accelerating voltage, so many modern high resolution microscopes operate at 200 to 400 kV.

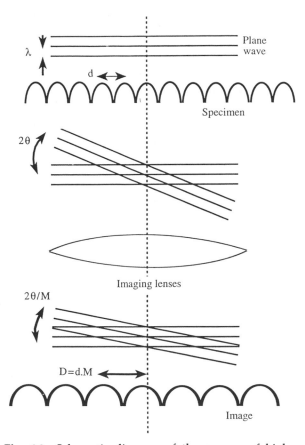

Fig. 6.9 Schematic diagram of the process of high resolution imaging of atomic lattice planes. The electron density variation in the specimen causes diffracted waves to form, at $2\theta=\lambda/d$. The imaging lenses reduce the off-axis angle by a factor of the magnification M, so that on recombining the waves, they are in phase every $d.M$ on the image.

Figure 6.10 is a map of theoretical resolution as a function of C_s and voltage, with current instruments placed on the map. There are other factors that affect image formation and limit resolution. These include the chromatic aberration of the lenses, the beam divergence and the magnification [75, 83]. A radiation sensitive sample might require the use of a low magnification where the film cannot record all the detail that would be in the image. A practical resolution of 0.17 nm in poly(*para*-xylylene) has been reported at 200 kV [83].

There are two levels of information normally used in HREM. When many diffracted beams are captured within the resolution limit, the image will contain detailed information on atomic positions within the unit cell. These images have been used to determine the structure of complex oxides with large unit cells [74]. At the other extreme, if there is only one diffracted beam represented in the image, it can only contain sinusoidal intensity fringes. A few beams will give a lattice of crossed fringes. In this case, the spacing information in the image is little more than that obtained by diffraction. However, images of both types show crystal interfaces and defects. The normal procedure in HREM includes accurate determination of all the operating parameters, and numerical simulation of the expected image from model structures [74, 75, 83]. Agreement between simulation and experiment is taken as evidence for the existence of the model structure. The calculations are complicated, so when dealing with details, more than one structure could cause a particular image. Obtaining sufficiently accurate parameters as inputs may be difficult.

The discussion so far is a general one for all materials. It assumes that the structure extends through the sample, and takes no account of radiation sensitivity. Polymers can have very small crystals and can be very radiation sensitive. Simulation of a small crystal embedded in disordered material indicates that the surroundings will reduce contrast and add noise, but not otherwise affect the periodic image. As described

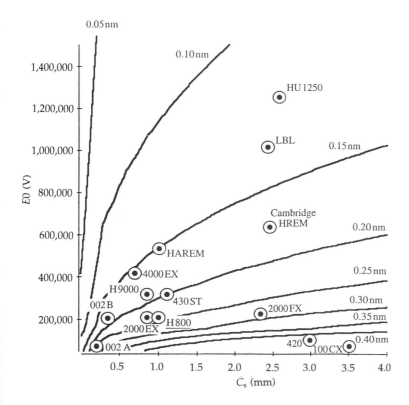

Fig. 6.10 Theoretical resolution of the TEM as a function of the operating voltage and the spherical aberration coefficient of the objective lens. The capabilities of various current generation instruments are indicated. (From Martin and Thomas [83]; reproduced with permission.)

in Section 3.6, radiation sensitivity will limit the number of electrons that can be used to form the image. The noise in the electron image that is due to the finite number of particles involved (shot noise) will limit the resolution through the minimum signal to noise ratio for a visible feature. The limited irradiation dose also limits the magnification for a film detector that requires a given dose to form a visible image. If this magnification is low, then the resolution of the film may become the resolution limit of the image.

Different polymers have a wide range of sensitivity to the electron beam. Aromatic, rigid rod and in general high melting point polymers [86] will be more resistant to beam damage, needing a flux J_a of over 100,000 electrons nm^{-2} to be made amorphous. Low melting point flexible chain aliphatic polymers may withstand only a few hundred electrons nm^{-2}. The quality of the image that can be obtained is as widely variable. The first experiments on the lattice resolution of polymers used highly resistant materials such as poly p-xylylene (PPX) [83, 87], PPTA [88, 89] and polydiacetylene [90]. This type of material continues to give the best images.

In PPX, for example, a view of the molecular chains projected down the chain axis using multiple diffracted beams helped understand the packing in the monoclinic and hexagonal crystal structures [91]. A similar molecular view was obtained for polydiacetylenes [92]. More recently twins, bending of crystal planes and arrays of edge dislocations making up a small angle grain boundary have all been imaged in this type of material [93, 94]. A result from this last paper which is quite as good as many images from inorganic materials is shown in Fig. 6.11. Electron irradiation causes the solid-state polymerization of the diacetylene monomer, and the electron irradiation of the beam has been used to study the process *in situ* [95]. This study turns the disadvantage of TEM work into an advantage.

Less radiation resistant flexible chain polymers that have been studied by HREM include iPS [96], PTFE [97, 98], PE [99, 100], poly(β-hydroxybutyrate) [101], poly(p-phenylene sulphide)

[102] and cellulose [103, 104]. In most of these cases the new information that can be obtained from the HREM is rather limited. Because of the very small electron dose that can be used, the images are noisy. Image processing can reduce the noise and emphasize the periodic part of the image, but crystal defects and crystal boundaries are not periodic. Information about them can be strongly affected by the image processing process; the apparent position and form of lattice defects changes if the image filtering is changed slightly [98]. The comparatively radiation resistant syndiotactic PS (s-PS) showed internal boundaries assigned to different packings of the helical molecules [105].

In partially ordered polymers, HREM of smectic liquid crystalline polymers show edge dislocations and other defect structures [81, 106] and the packing of discs in a discotic has been determined [107]. Much work has been done on nematic or poorly crystalline rod-like systems such as PBZT and PBO [108-111], where the results include the crystal size, perfection and orientation after thermal treatment [110] and after plastic deformation [111]. Figure 5.112B is an image from ref. 111 which shows the details of local deformation of molecular planes in a kink band produced by plastic deformation. Molecular scale defects relevant to the mechanical properties of the material can be imaged even in the partially ordered material. Pyromellitic dianhydride-oxydianisole (PMDA-ODA) polyimide normally has a low crystallinity, and HREM of a solution grown single crystal shows limited coherence of the lattice planes [112]. The necessity of using a very thin layer of material as a sample in the TEM was used to advantage in a study of the ordering of PMDA-ODA during imidization [113]. Figure 6.12 is an image of a droplet of PMDA-ODA with the crystals outlined; the HREM shows the crystal interfaces as well as the crystal size distribution and its relation to sample thickness.

These examples show that new information can be obtained by the HREM of polymers, but the experiments are not easy. What are the practical difficulties? They are those of regular

Fig. 6.11 HREM of a radiation resistant polymer, a polydiacetylene. This material contains large perfect crystals, with a comparatively large lattice spacing of 1.2 nm. Individual [010] edge dislocations making up a small angle (5°) grain boundary are clearly visible. A JEOL 4000 EX operating at 400 kV was used. (From Wilson and Martin [94]; reproduced with permission.)

TEM of polymers, taken to an extreme. Thus HREM needs even thinner specimens than regular TEM, preferably below 50 nm. High resolution needs an expensive TEM, so adding a 'low-dose' system to store the lens and alignment settings is only a small amount extra. This allows the preset conditions for a well focused high magnification image to be switched back in after searching for an interesting area at very low dose rate. See ref. 84 for details of how best to limit the exposure of the specimen in this case. Until computer controlled automatic alignment becomes available, only an experienced operator can align the TEM to the accuracy needed for HREM. Then the exact Gaussian or Scherzer focus positions must be recognized by their effect on the contrast of thin samples. This is because at the limited magnification forced by the radiation sensitive samples, the lattice detail may be only 25–250 μm at the viewing screen. These spacings are hardly visible in the focusing binoculars. Martin [84] points out that at the lower end of this range, it is difficult to see noisy fringes in the developed negative, even with a 10× magnifying lens. He recommends using a laser beam to get a light scattering pattern of the negative, to search for areas with a periodic image.

6.4.2 Structure determination by electron diffraction

Unknown crystal structures are generally determined by x-ray diffraction from single crystals, but this needs crystals that are tens and preferably hundreds of micrometers on a side. If the crystals are intrinsically smaller, then electron diffraction may be used. There are significant problems in using electrons. The sample must be very thin, but ideally should not be bent, and the effects of inelastic scattering must be taken into account. Recently, and largely due to the work of one group, structure determination by electron diffraction has been extended to organic molecular crystals and to polymers, overcoming the extra problems of radiation damage. Dorset summarizes the problems and the methods used to overcome them in an article titled, 'Is electron crystallography possible? The direct determination of organic crystal structures' [114]. The answer is yes, and the methods have been applied to polyethylene [115] and other linear polymers [116, 117]. The polymer studies are a continuation of extensive work on alkanes [118–120], cycloalkanes [121] and perfluoroalkanes [122]. These alkanes were grown epitaxially on organic substrates to have a view perpendicular to the chains.

Radiation damage is still a limiting factor [123] as higher resolution structural analysis can be performed on radiation resistant organic mole-

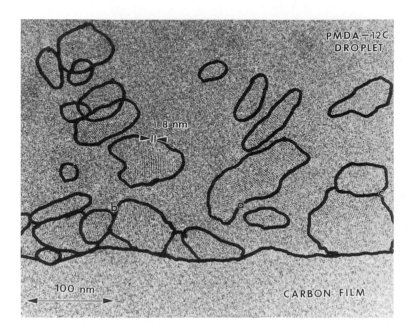

Fig. 6.12 HREM of the edge of a droplet of a PMDA-12C polyimide. Individual small crystallites are identified by their 1.8 nm (001) spacing, but are made clearer in this reproduction by a line drawn round them. A JEOL 2000 EX operating at 200 kV was used. (From Martin *et al.* [113]; reproduced with permission.)

cules [124]. The purpose of this short section is to point out that even for highly radiation sensitive materials, electron diffraction can be a fully quantitative tool for structural analysis, and not simply for the determination of orientation or texture.

6.5 SCANNING TUNNELING MICROSCOPY

A very basic description of the STM and other scanning probe microscopes was given in Section 2.5. A fine solid probe is mechanically scanned over a sample surface and interacts with it. In the STM the interaction is so localized that individual atoms can be resolved, and vertical resolution is even better. The sensitivity of the instrument has opened whole new areas in surface science, but for the same reason most of the stunning images of atomic arrangements have been obtained on stable conducting or semi-conducting materials in ultra-high vacuum. Discussion of the use of the instrument at lower resolutions, for measurement of surface features, is largely deferred to Section 6.6.3.

There is no shortage of books, reviews and general articles describing the STM and its applications. These range from a three-volume comprehensive treatise [125] to short readable surveys [126, 127] through many reviews. These may be general [128, 129], review instruments [130] or concentrate on applications for materials [131] or biology [132]. The size of the general recent literature is proved by a review that has over 400 references [133]. Any current issue of the journals, *Langmuir*, *Surface Science* or *Ultramicroscopy* is likely to have articles on the STM or AFM. The recent monograph by Chen [134] called *Introduction to Scanning Tunneling Microscopy* may be too detailed for most users. As the field as a whole and the instrumental capabilities are still changing rapidly, information from manufacturers can be very helpful [135].

6.5.1 Principles of scanning tunneling microscopy

The essence of the STM is the quantum mechanical tunneling current that passes across a gap between two conductors with a bias voltage between them. The current depends strongly on s, the size of the gap. This is an understatement; a simple expression for the current is $V \exp(-As)$ where V is the bias voltage and A is a constant, typically about 20 nm^{-1} [129, 136]. Thus for a 1 V bias the current is 2 nA at a gap of 1 nm and negligible at a gap of 2 nm. Reducing the gap from 1 nm to 0.99 nm, a change of 0.1 Å, will increase the current by 22%. This is the basis for the extreme vertical sensitivity of the instrument. Such sensitivity means that vibration and thermal drift must be almost eliminated for the STM to work at all. It also means that the positional control of the probe position must be extremely accurate.

The lateral resolution depends on the size of the region where current passes. If fine detail exists in the specimen, then its appearance depends on the form of the tip of the probe [137]. Considering the tip as a smooth hemisphere, the exponential dependence of current on the gap makes the total current completely dominated by the current passing through a small central patch where the gap is smallest. For example, using the above formula, if a tip of radius 5 nm has a minimum gap of 1 nm, essentially all of the tunneling current will come from a patch less than 2 nm in diameter. This is excellent though not atomic resolution. A blunt tip (and here a radius of more than a few nanometers is 'blunt') can have more complicated effects than limiting the resolution in this way. Figure 6.13 shows that if the sample has surface relief of even a few atomic steps, the small patch where the tunneling current is concentrated can move around the tip. This changes the image by smoothing out the signal, and measurement of surface roughness will be affected [138].

For atomic resolution, we must admit that the probe tip is not a smooth mathematical surface but is rough and has asperites. If such a high point is the nearest point to the sample surface, all the tunneling current will be concentrated there. The highest atomic cluster on this peak will act as a fine probe, perhaps with atomic resolution. If there are two atoms equally near the

specimen and a few Ångströms apart, then a double image will be formed. This simple example shows that the high resolution image will depend on the arrangement of the few atoms at the tip of the probe. If the tip is rough and has many points of contact the image will be due to the sum of all the interactions.

More detailed consideration of tunneling current shows that it depends on the density of electron states in the tip and the sample and non-linearly on the bias voltage as well as on the gap [139]. This is the basis for *scanning tunneling spectroscopy* (STS) where the variation of current with bias is used to measure the localized surface electronic states, particularly in semiconductors [140]. The less positive feature of this is that the STM image will depend on the composition and the geometrical arrangement of atoms at the tip. It will change contrast in some unpredictable way if an atom or molecule is adsorbed onto the active part of the tip [141].

The STM thus has the problem that the image is some convolution of the properties of the sample and those of the tip. Tips are commonly made of tungsten, for the high melting point

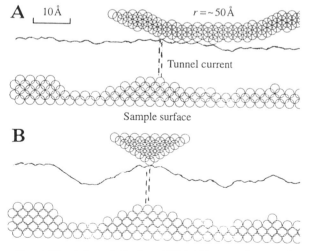

Fig. 6.13 A broad tip in a STM (A) may change the position where the localized current flows as it moves across a rough surface. The result is to smooth out the apparent surface. A sharp pyramidal tip (B) follows the surface much more accurately. (From Nishikawa *et al.* [137]; reproduced with permission.)

means that atoms are strongly held in position. Electrochemical etching produces a sharp tip [142], and 'sharp' is again an understatement as the tip radius may be only 2–3 nm. Ion milling produces sharp tips that are very reproducible in shape, suitable for metrology [143, 144], but for atomic resolution where only a small very flat area of the sample is scanned, these complications are not necessary. A sheared piece of fine wire will have some roughness at the end, and the atoms nearest the sample act as the tip. The shape of the rest of the probe is not important. In a high vacuum STM the tip may be flashed at high temperature after mounting in the microscope. This reduces adsorbates, and so increases the tip stability.

In one case at least the tip structure has been determined directly by field ion microscopy [145]. It was found to consist of six tungsten atoms, and gave a STM resolution of 0.4 nm, in agreement with theory. Most users cannot do this direct characterization, and the hourly changes in tip structure [141] make the idea highly inefficient. Some characterize the tip in an operational manner, ideally by forming an image of a known structure immediately before and after investigating a new one [146]. Again the lack of stability often makes this inappropriate. A tip that goes bad – producing obviously poor images – may be reformed by increasing the bias slightly (to 5 V) or with high voltage (100–1000 V) pulses [141].

6.5.2 Instrumentation and operation of the STM

The STM consists of a head, vibration isolation and electronics. The head contains the probe tip and its positioner. Fine control of position is always by piezoelectric actuator. Several geometries are possible, but a common design is a piezoelectric tube where a voltage applied across the thickness of the tube wall causes the tube to extend along its axis [147]. The tip is mounted on the end of the tube, and the sample surface (x, y) is perpendicular to the tube axis z (Fig. 6.14A). One electrode is on the inside of the tube, and

there are four sector electrodes on the outside. A voltage on the inner electrode causes the tube to extend or contract for z-axis control. Opposite pairs of outer electrodes with opposite voltages cause the tube to bend about the x or y-axis. For small deflections, bending the tube makes the tip move in the x, y plane. With low noise electronics, control approaches 1 pm – which is $< 1\%$ of an atomic diameter!

The head may be optimized for the highest resolution, or for a lower magnification and resolution. The most suitable tip will be different [138, 143, 144, 148] and so will the sensitivity of the piezoelectric actuator. For a particular tubular actuator, the transverse motion sensitivity is 3 nm V^{-1}. Thus at high resolution, for control better than 0.1 Å, the noise in the driving signal must be below 3 mV. To scan an area 3 μm across needs a driving voltage of 1 kV. If the display is 15 cm across, this scan area corresponds to a magnification of 50,000 times. To use the STM at a low magnification, a longer piezo tube would be required. A longer tube is also needed for the greater z range expected at larger scan areas. A long tube goes against the requirement for small stiff structures, and for large z ranges the tripod scanner (Fig. 6.14B) may have advantages. In this device the x, y and z piezoelectric elements are independent, and choice of the pivot point position and z piezo length allow separate control of z-range and x, y scan range.

Another level of position control moves the specimen towards the probe tip, to put it into the tunneling regime. This is generally integrated into the STM head and although it may be called coarse positioning, remember that in this instrument a motion of 0.1 μm is very 'coarse'. This position control may be piezoelectric, mechanical or some combination of these. Examples are the piezoelectric walker or inchworm, and using a fine pitch screw to push onto a soft spring in contact with a stiff specimen mount. The effectiveness of this coarse positioning and specimen alignment are very important for the ease of use of the STM. For selecting a specific area to view, the head may be integrated with an optical microscope.

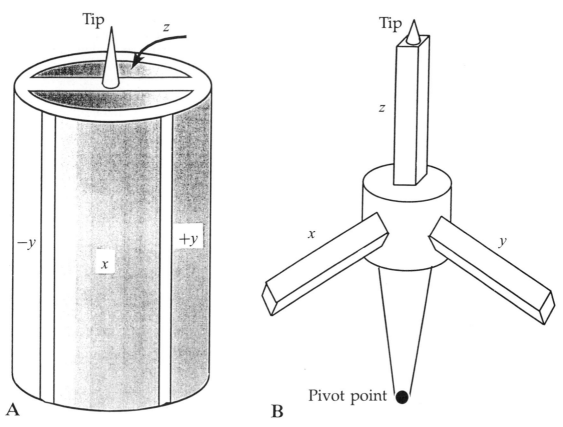

Fig. 6.14 Piezoelectric scanners used in STM and AFM. (A) Tube scanner. This is a monolithic tube of piezoelectric ceramic, with the tip attached at one end – attached to the tube, or mounted centrally as shown. A voltage applied across the tube wall causes it to lengthen. The outer electrode is divided into four segments, so opposite voltages applied to opposite segments make the tube bend and the tip scan. (B) Tripod scanner. Three stacked piezoelectric ceramics are placed orthogonally to give independent motion in *x*, *y* and *z*. The mechanical design is not so simple as the tube, but it allows the *z*-range to be chosen independently of the *xy* scan range.

Since the tip to sample distance must be kept stable at the sub-Ångström level, isolation from mechanical vibration and shock is vital. The first STM used superconducting levitation for vibration isolation [149]. Current designs use small and rigid structures for the STM, so that their resonance frequencies are high (100 kHz for the piezoelectric tube support of the tip [147]). This reduces the effect of low frequency noise on the system and allows rapid scanning. High frequency noise is reduced by mounting the STM on a heavy mass supported by a soft suspension with damping. Clearly when these strict mechan-

ical constraints are combined with other requirements such as operation at low temperatures or in ultra-high vacuum the mechanical design is not simple [129, 134]. Another constraint is that the tip must be shielded from the high voltage signals that drive the piezoelectric actuators.

The electronics of the STM must have low-noise high voltage amplifiers to drive the tip motions, and a sensitive low-noise amplifier chain for the tunneling current. These analogue signals are converted from and to digital signals for computer control. In both the scanner systems described above, the scan is not exactly planar, with errors increasing with scan range. Computer linking of scan with z-axis drive can correct for this, or it can be corrected by image processing. This correction is not simple and others are needed, because the piezoelectric elements are not ideal. The displacement is not exactly linear with applied voltage, and may show hysteresis even at slow scan rates [150, 151].

The basic imaging modes are the fast scan at constant height, where the signal is the tunneling current, and the slower scan at constant tunneling current. Tunneling current is kept constant by a feedback loop of probe height, and the signal is the probe position. These modes are shown in Fig. 2.7, which also shows schematically the problem with constant height operation. If the specimen is not very flat, the tip may crash into the sample. This is likely to disrupt the arrangement of atoms at the tip, and change the imaging. Tip crashing is more likely if the gap is small and a small gap is preferred for high current and thus good signal to noise ratio. The slow scan allows atomic resolution even when the specimen is not completely flat, as atoms can be resolved on both sides of an atomic step [152]. However, a slow scan is more sensitive to thermal drift and other instabilities.

The feedback loop used for constant current mode may be directly analogue, or through the digital system. The finite response time of the piezoelectric driver is important here. Generally the feedback system is operated near instability, to obtain accurate high frequency control, and thus good images at high scan rates without crashes. If there is instability, false periodicities and extra noise may appear and if the response is too slow, the image will be smoothed in the direction of faster scanning.

6.5.3 STM of insulators

There are three methods for the study of insulators in the STM:

(1) Coat the insulating samples with a thin layer of metal or carbon by sputtering [153] or evaporation, just as for the SEM at normal operating beam voltages. This procedure prevents atomic resolution, but provides a stable surface for metrology.

(2) Use a very thin layer of the material, or individual molecules, on conducting substrates. These may have different structures to that of the surface of bulk material, but are interesting in their own right.

(3) Use the insulator itself in the STM.

This last method seems very surprising, as an electric current through the specimen is the output signal, but it has been shown to work on some inorganic compounds. The tunneling current is so small that the bulk conductivity does not need to be very high to allow it to escape. Alternatively there may be sufficient surface leakage in surfaces affected by air, or in a surrounding electrolyte, to dissipate the transferred charge. This is the case for TiO_2 [154, 155], and off-stoichometric TiO_2 can be imaged in vacuum [156]. Single crystals of doped oxides can be sufficiently conducting to give clear atomic resolution [157]. If the sample has a large band gap, high bias voltages may be needed to get tunneling to occur, but occasionally the surface structures provide the necessary electron states at low bias. Polycrystalline materials are more difficult to deal with, as the conductivity changes from place to place. Although materials do not have to conduct as well as metals or silicon to be used in the STM, the insulating properties of most polymers are too good for this to work. Conducting polymers can be imaged in this way, but for non-conducting polymers and other organics, one of the other two methods must be used at present. Although it is paradoxical, we should add a fourth method to the above list.

(4) Do not use the STM on insulators, use the AFM instead.

Many of the early examples of the use of STM on polymers in the literature date from a time when the AFM was less available and less well developed. Today more of this work would be done with the AFM.

6.5.4 Adsorbed organic molecules

Many STM experiments on organics have used monolayers, multilayers or separate individual molecules adsorbed onto conducting substrates [129, 158]. Graphite (highly oriented pyrolytic graphite, HOPG) is a common substrate. It is flat, conducting, stable and easily prepared. Another reason for choosing HOPG has been that atomic resolution could easily be obtained in air [159]. However, resolution at the atomic level, some-times of anomalous structures, is obtained even under conditions where tunneling should not be effective. This includes when the tip and sample are in strong mechanical contact [160]. Explana-tions of this include elastic deformation of the sample [160, 161] and when the specimen is imaged in air, point contact conduction through contamination [162]. Many artifacts have now been identified in these samples. These include the appearance of regular bands or arrays that are not related to any real surface structure and molecular images where there are no adsorbates [162–165]. Observation of the graphite periodi-city has often been used as an indication that the instrument is in a condition to produce good tunneling images at high resolution [166]. Un-fortunately this is not always correct. Extra care is required, and some prefer the use of more difficult substrates, such as gold or platinum.

Generally, careful and systematic study of the image under a variety of conditions is needed for confidence in any observation of new or un-expected features [129, 146, 164]. Alternatively, there should be independent confirmation from another technique. This is not a particularly satisfactory situation, but common in high resolution microscopy techniques. Since polymer samples are much softer than graphite there may be severe distortion due to the mechanical effects of the tip. The force between tip and sample during tunneling microscopy has been found to be between 100 nN and 1 μN [167], which corresponds to very high stresses in the small contact area (over 100 GN m^{-2} at 1 nm^2). In the worst case, the tip plows into the adsorbate and drags it around the substrate. The electric fields around the tip can also apply large forces to any adsorbed molecules. At a bias of 1 V, the local field is 1 GV m^{-1}, and the field gradients are extremely large. This could easily rotate dipolar molecules, and attract or expel any molecule from the high field region, depending on its dielectric constant. Of course these effects can be put to use, turning the STM into a device for nanofabrication and surface modification of any material [168, 169], including polymers [170].

Convincing molecular images of smaller mo-lecules can be obtained on graphite and other substrates [171]. Aromatic molecules and those that adhere well to the substrate are the best [171, 172]. If the sample is a liquid or liquid crystal, there is no need to make a monolayer sample. The probe will scan in the liquid layer, and can clearly image the static adsorbed structure below. Figure 6.15A is an example of this. It shows individual alkane molecules adsorbed onto graphite from solution, and the image was taken in the presence of excess solvent. The image is unfiltered, and shows streaks and distortion typical of unprocessed images. Distortion may be due to drift or to hysteresis in the scanning system that can be corrected [149, 150, 173]. Since even pure noise can be processed into an apparently acceptable image, it is good practice to include some unprocessed images in publications.

Even when care is taken, the results may not be quantitatively reliable. A recent example of this is the new structure for an adsorbed layer of decylcyanobiphenyl seen in the STM and con-firmed by x-ray diffraction from the surface using a synchrotron source [174]. The spacings deter-mined by the two methods are 10% off, and the authors suggest that the calibration of the STM image may be inaccurate.

For multilayers of small molecules, thicker polymer specimens, or very large molecules such as proteins or DNA, there are extra problems [132, 164, 175]. These materials are normally non-conducting, and the way that the image is formed is not clear. Metal coating can be used on large individual molecules, and resolution of 1–2 nm is achieved with biological samples in this

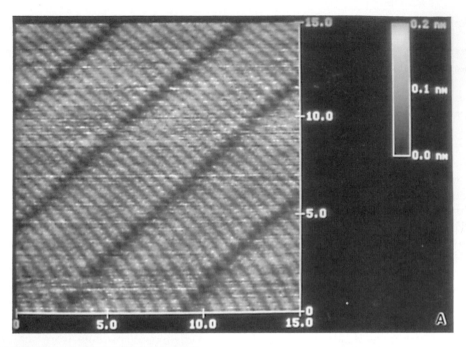

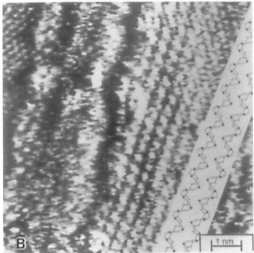

Fig. 6.15 (A) Unprocessed high resolution STM image of $C_{32}H_{66}$ adsorbed from solution onto highly oriented pyrolytic graphite (HOPG). The image was taken in the presence of excess solution at a bias of 0.3 V and a tunneling current of 100 pA. Individual molecules are seen arranged in rows. (From K. D. Jandt, M. J. Miles, J. Petermann and N. H. Thomson, unpublished.) (B) STM image of HOPG partially covered with an ultrathin highly oriented film of poly(butene-1). The atomic lattice of graphite is resolved in the upper left corner of the image. At the right are extended molecules of the polymer, with an intermolecular distance of 0.42 nm. A schematic arrangement of the polymer chains has been inserted at the right of the image; this sketch is not exactly to scale. (From Fuchs *et al.* [176]; reproduced with permission.)

way [175]. Salmerton *et al.* [165] suggest that the metal coated samples have the highest possible true resolution largely because the molecules do not move around. After a very careful study of uncoated samples using an UHV STM, Wilson *et al.* [172] say of DNA,

'It is unfortunate that we can obtain data similar to that of several highly optimistic publications if we overlook impurities, statistical significance, and sensible physical mechanisms.' Nawaz *et al.* [146] conclude with the following set of rules for reliable imaging of smaller molecular species.

(1) Extensive lateral motion must be prevented, most effectively by chemical bond to the substrate.

(2) A sufficiently flat perfectly conducting substrate must be used.

Nawaz *et al.* preferred sequential evaporation of silver and gold on freshly cleaved mica.

(3) A comparative reference sample should be readily available.

A substrate that has no adsorbates, or ideally Si(111) 7×7.

(4) Auxiliary checks with other techniques are helpful.

For example, spectroscopy to confirm the presence of the adsorbate on the imaged area.

These rules will also apply to polymers that can spread out on the surface. Figure 6.15B shows an image of an ultrathin polymer film adsorbed onto HOPG where the 'reference' sample, a region with no adsorbates, is visible on the same image [176]. A region at the upper left shows the graphite periodicity, and at the right are long extended molecules of polybutene-1, 0.42 nm apart, with a molecular model structure overlaid. There is good reason to be optimistic about the prospects of STM using adsorbed organic materials [171], but obtaining high resolution information is hardly routine, and artifacts are easy to obtain. Some of the other non-conducting polymeric materials that these techniques have been applied to are polyethers [177, 178], polyamides [179], polyimides [180] and polydiacetylenes [181].

6.5.5 Other polymer applications

There is a continuous range of polymer specimen thickness possible from sub-monolayer coverage by deposition from suspension, through ultrathin films to bulk samples. Samples that are thick enough and non-conducting enough to block any conduction to the substrate must be metal coated. The STM has been applied to the study of a wide range of polymer samples using metal coating.

These include fracture surfaces in a range of amorphous materials [182, 183], single crystals [184], samples made by evaporation from an entangled solution [185] or by spin casting [186], microtomed sections [187] and fabricated structures [188, 189]. In most of these applications the STM images are qualitatively similar to those that can be obtained in the SEM or from surface replicas in the TEM, except when there are very small height differences involved. Then the extraordinary vertical sensitivity of the STM can add new quantitative information [185] (see also Figs 5.43 and 5.44). SPM instruments have a great advantage in accurate measurement of height, and quantitative metrology with a probe microscope will be discussed in Section 6.6.3.

Thin films for STM study have been made by drawing [176] and frictional transfer [190]. They show interesting structures at the meso-scale level, small crystals and surface features, which correlate well with details visible in the TEM. The STM (and AFM, see Fig. 6.19) images can be complementary to the TEM, as they show only one surface. The TEM gives a projection through the thin film, a superposition of detail from both surfaces that is more difficult to interpret. In the PTFE film formed by rubbing [190] there are rods visible parallel to the rubbing direction, separated by 0.6 nm. This agrees with AFM results and the expected separation of the rod-like helical molecules of PTFE.

Even thick samples of conducting polymers do not need metal coating for the STM, but most of the literature refers to thin films or isolated molecules [191, 192]. The larger scale morphology and the local conductivity of polymer films are also studied [193, 194]. An important use of the STM is to study the growth processes of conducting films by electrodeposition or electrochemical polymerization [194, 195]. The scanning electrochemical microscope is a more specialist tool for these investigations [196].

6.6 SCANNING FORCE MICROSCOPY

'Scanning force microscope' is sometimes used as an alternative name for the atomic force microscope, particularly when individual atoms are not resolved. Here we use it to include the AFM and its close relative, the frictional force microscope (FFM). Both share much of the construction and features of the scanning tunneling microscope, but the interaction signal between probe and specimen is the force on the probe. This is the vertical force in the AFM, and the horizontal (and vertical) force in the FFM. If the interaction is as localized as it can be in the STM, then the microscopes can map out atomic positions in conductors and insulators. If the interaction is not quite so localized, the AFM is still a very high resolution probe of surface topography and surface properties.

As with the STM there is no shortage of modern literature on the subject, and several previously referenced works deal with both STM and AFM [128, 129, 197]. There are excellent reviews [198–201], summaries [202, 203] and at least one very detailed book for designers [204]. Again as with the STM, manufacturers' literature can be very helpful [135]. Frictional force microscopy is described in refs 128, 129, 201 and 204.

6.6.1 Principles of atomic force microscopy

The basic construction and electronics of the AFM are the same as that of the STM, as described in the previous section. The probe need not be a conductor, and it is mounted on the end of a cantilever. In a simple operating mode, the detected signal is the deflection of this lever caused by a force due to the interaction of tip and specimen. If forces caused by single atoms (or small groups of atoms) are to produce detectable motion of the lever, it must be soft. A spring constant of 0.5 N m^{-1} is typical for such a soft lever. An apple placed on such a spring would deflect it by 5 m, and a possible interatomic force for a single strong bond of 10 nN by 50 nm [129, 205]. The necessary presence of a soft structure goes against the design rules given above for the STM. To keep the resonance frequency of the probe support high, the lever must be very small and light. Levers are now made by the micro-lithographic methods devel-

oped for integrated circuit manufacture [206]. A lever of silicon or silicon nitride might be 100 μm long, 70 μm wide and less than 1 μm thick, with a resonant frequency of above 80 kHz [129]. This is high enough to allow a relatively simple vibration isolation arrangement.

The deflection of the lever is very small. It can be detected by reflecting light off the back of the lever into a split photodiode, which gives a signal proportional to the deflection of the light beam. This is a sensitive non-contact detector, but the optical alignment may be difficult to maintain while the probe is scanning a large area. In these instruments it is simpler if a small specimen is scanned under the tip. Alternatively a capacitance gauge on the back of the lever or a strain sensor can be used to detect the deflection of the lever. These can be integrated into the device by microfabrication. Typically the ideal (noise free) resolution of these detectors is 0.01 nm [200]. With a spring constant of 0.5 N m^{-1} the resolved force would be 5 pN.

The instrument can be operated in the two basic modes of constant height and constant signal – here constant deflection or force – as described for the STM and shown in Fig. 2.7. Normally the AFM is operated with feedback to give constant signal unless a small flat area is being scanned. However, because the force is a complex function of the probe position, there are several quite different imaging modes used for the AFM.

6.6.1.1 *Forces acting on the probe*

Idealized curves of interatomic potential and force are shown in Fig. 6.16A. The equilibrium separation is at A. Figure 6.16B is a closer look at the forces near this point. Forces above the horizontal zero axis are repulsive, and this occurs at separations less than A, when the two atoms are in contact. Long range forces such as van der Waal forces are attractive, and occur at separation greater than A. When the force measured is attractive, the imaging mode is said to be non-contact. Non-contact modes involve small forces, and if they are tested at some distance from the surface, long range forces are involved. The resolution is not so good as that of the contact modes. Electrostatic charges and the influence of the surrounding medium can alter the real force separation curve [207].

This extremely simple picture is complicated by three effects. One is that in air, the environment of many AFM experiments, there will be a layer of contamination (usually including water) on the sample surface. This layer may be 2–50 nm thick, or may condense in the gap between tip and surface even if it is absent elsewhere. As the tip approaches the surface it touches this contamination layer, is wetted, and capillary forces pull it towards the surface. These forces can be much greater than the van der Waal forces, and will depend on the sample, the humidity and the tip shape. At the apparent contact separation, with zero total force, there

will be attractive capillary forces around the surface of the probe, and a large repulsive force at the tip. The force distance curve will show hysteresis when there is contamination present, as the tip will remain wetted up to larger separations as it is withdrawn from the surface.

The second complication is that the position of the tip is not under direct control. At any image point the signal is the deflection (force) for a given position of the cantilever support, not for a given separation between tip and surface. Figure 6.16C is a plot of force versus position, created using the force – interatomic separation plot of Fig. 6.16B. The cantilever beam may have a low spring constant to make it sensitive to small forces. If the slope of the force – separation curve is greater than the value of the cantilever spring constant, then the deflection will not be stable. As the probe approaches the surface it will suddenly

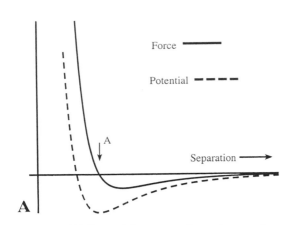

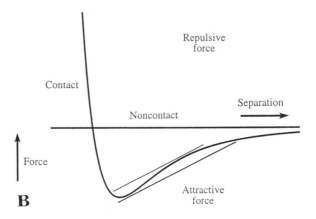

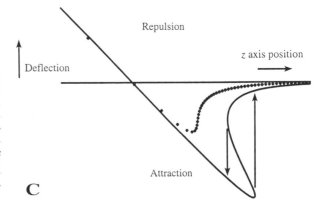

Fig. 6.16 (A) Potential energy of an atom probe approaching a surface, as a function of separation, and the resulting normal force, the differential of the energy. Point A is the position of equilibrium contact. (B) Enlarged view of the force versus separation curve near the equilibrium contact point. Smaller separations give a repulsive force, and the probe is said to be in contact with the surface. The straight lines correspond to the stiffness of the cantilever, and greater slopes make the probe position unstable. (C) Cantilever deflection as a function of the distance of the cantilever support above the surface. The vertical lines show how the probe will be unstable and jump into and out of contact with the surface. A stiffer cantilever would produce the smaller deflection shown as a dotted line, and suppress the instability.

jump towards the surface. There will be a reverse jump at larger separations as the tip is withdrawn from the surface. The tangent lines in Fig. 6.16B represent a spring constant (force/distance) of a cantilever and show schematically where this would happen. In Fig. 6.16C the jumps occur when the curve becomes vertical, as shown by the arrows. A stiffer spring, as suggested by the dotted line in Fig. 6.16C, suppresses the instability but of course the deflections are smaller.

A jump towards the surface is even more likely when contamination is present, because then the probe and surface are joined by a liquid layer and surface tension pulls them together. In this case, the probe jumps to a position where there is a net attractive force, but the tip is in repulsive contact with the surface. Operating the microscope in a good vacuum, in dry gas, or in a liquid reduces or eliminates the effects of contamination. Water is often quoted as the liquid, but if there is an oily contamination film on a hydrophobic surface, iso-propanol or other organic solvents may be better than water. The primary aim of this is to have lower forces on the surface. Detailed interpretation of the measured forces is complex [207], and contamination layers makes it even more difficult.

The third complication is that the measured force – position curve relates to the distance from sample support to probe support. There are several compliant objects in series here. These are the cantilever, the tip, the tip–surface contact region, and the rest of the sample. The softest of these objects is the cantilever, and it deforms the most. However, the sample will also deform under the load applied by the tip. If the sample is soft, its mechanical properties will affect the results. These are elastic properties if the deflections are small, plastic properties if the local forces are large.

6.6.1.2 *Imaging modes*

The imaging modes of the AFM can be characterized by the region of the force – separation curve that is being used and by whether the probe is vibrating. These modes may be described as 'static' or 'dc' and 'modulated' 'dynamic' or 'ac' but the 'static' probe is still moving in all directions at frequencies up to several kHz. The static or non-vibrating probe is deflected by the force of interaction of the tip with the specimen, and this deflection is used to adjust the z-axis piezoelectric driver to give a constant deflection. The tip may be in contact with the surface, so that the deflecting force is repulsive, or it may be out of contact, with an attractive force on the cantilever.

If there is a contamination layer, there is a third possibility; the tip may be held in contact with the surface by the capillary forces, even when the force on the cantilever is attractive. This causes confusion, since attractive force on the cantilever can be taken to mean non-contact mode, but here high resolution is possible because the tip is in contact with the surface. A true non-contact mode with a static probe is difficult to maintain in a stable manner, as the soft lever is near instability.

The vibrating or dynamic probe has another piezoelectric driver at the base of the cantilever. This is driven with a high frequency alternating voltage, to set the lever into vibration at a frequency at or near to resonance. A force or more accurately a force gradient in the region where the tip is oscillating causes the resonant frequency to change, and at a fixed drive frequency this changes the amplitude and the phase of the oscillations [198]. The sensitivity is high even when the lever is comparatively stiff, and therefore not subject to instability. Meyer [200] quotes a detectable force gradient of $10 \ \mu N \ m^{-1}$. Depending on the form of the force–distance curve, this may correspond to a force as small as 0.1 pN at a probe gap of 1 nm.

If the amplitude of the oscillations is kept low (about 1 nm) then the force gradient can be probed at a range of distances from the surface. At larger distances, long range forces are involved and so the highest molecular resolution is not possible. A large amplitude oscillation can move the probe into the contamination layer, to close contact with the surface and all the way out again in each cycle. There are large force gradients at contact, so the vibration is sensitive to the local interaction, even when contact with the surface is only for a small part of the cycle. Thus the image can have high resolution and a limited contact with the specimen. Modes such as this may be called 'tapping' or 'hopping' modes, and can give high resolution even through a contamination layer. The mode prevents frictional forces from dragging the sample surface, but care is needed to be sure that the tip does not apply forces which are too large to the specimen at contact (that would be 'hammering' mode!).

There are yet more modes, where the aim is to investigate the surface compliance and adhesion of the sample. If the x–y scan is stopped, the force – distance curve can be used to estimate the elastic and plastic properties of the sample [208]. Hysteresis normally indicates contamination, but when the sample is in a water solution of known ion concentration and pH this is not so. Changing adhesive hysteresis now indicates chemical interaction between tip and specimen [209]. An image of the surface modulus can be obtained with a small amplitude low frequency modulation applied to the cantilever support during normal (static, contact) AFM imaging [210]. Since the amplitude is low, the total force is always repulsive and the result is an oscillating force on the sample, not a vibrating probe. The detector output goes to an amplifier sensitive only at the applied frequency, as well as to the usual feedback loop that keeps the mean force constant. The oscillating cantilever displacement gives the local force – displacement curve. As the cantilever and the sample are mechanically in series, a stiff cantilever will be more affected by changes in the stiffness of the sample. A very stiff cantilever and a large force of 1 mN was used to distinguish carbon fibers in a composite, with a resolution of about 60 nm [210]. To obtain numbers for the elastic properties of the solid is more difficult; the geometry of the tip and the surface must be well known, and the effects of surface energy considered [207, 211].

6.6.2 Atomic resolution in the AFM

The AFM, like the STM, can form images of surfaces with sufficiently high lateral and vertical resolution to see atoms. However, as for the STM, the high resolution image is formed by the interaction of a poorly defined tip structure with the surface. The basal layer interatomic repeat structure of graphite and other layered materials is comparatively easy to obtain. Once again it has been found that such images can be obtained when they should not be. In particular, 'atomic resolution' images have been formed when the force between tip and surface was so great that hundreds of atoms must have been involved in the contact. Explanations for this have been given in terms of sliding of layers, frictional forces on tilted tips [197], and the addition of many signals from multiple contact sites which all have the same periodicity.

Atomic resolution in ionic crystals such as lithium fluoride and sodium chloride was more convincing, as there are no layers to slide, and the height corrugations measured by AFM matched simulations. Nevertheless some users were still skeptical, because almost all the early results seemed to show ideal regular lattices, with no clearly distinguishable defects. Also, such images can be obtained even with a large tip that does not properly reproduce the surface features at lower resolution [212]. In contrast, the STM of semiconductors shows adatoms, missing atoms, steps and other point defects, all clearly imaged.

These concerns have not stopped researchers applying the AFM to many polymer samples, and obtaining sub-nanometer resolution. A review showing the surface lattice structure of *n*-alkanes, individual polyethylene chains and polypropylene helices with imaged methyl groups [213] appeared in the same volume as a survey by Binnig [199] (the inventor of the instrument), which suggested 5 nm as a reasonable lower limit to useful resolution at that time. A recent paper by Ohnesorge and Binnig [214] called, 'True atomic resolution by atomic force microscopy ...' showed that individual atom sites on a crystal surface (of calcite) can be resolved. But steps and defects are only seen when the contact force is kept below 100 pN. At higher contact forces, steps and other features were wiped away leaving the perfectly ordered surface [214]. In most of the literature of the AFM of polymers the force applied to the surface is not given, but it will normally have been between 1 and 10 nN. Most AFM instruments cannot now operate with forces below 100 pN.

Calcite is not a particularly strong material, but most polymers are no stronger and less rigid. It thus seems likely that unless careful precautions are taken to minimize the applied forces, the surface will be affected by the probe in some way, even when the correct lattice periodicities appear in the image. A very simple calculation gives a local stress under the probe; for a force of 1 nN and a contact dimension of 0.5 nm, the stress is 4 GPa. It is difficult to imagine all the atoms on the surface of a bulk polymer staying in the same place under this stress. Following the rules given for STM images in the previous section, the most reliable atomic resolution AFM images of organic materials should come from monolayers of reasonably small molecules adsorbed onto a rigid substrate. Even with such Langmuir–Blodgett thin films there is evidence of the strong interaction between probe and specimen [215]. The measured lattice periodicity varies with the scanning direction, as molecules are pushed out of the way by the scanning probe, then spring back as it passes. While deformation of the sample is a problem for imaging, the manipulation of extremely small volumes of polymer is of great interest for nano-technology and for the study of flow and failure on the extremely small scale [216, 217].

These problems and the lack of a good theoretical basis for the image formation at high resolution makes single images of polymer surfaces not very reliable if they are unsupported by other techniques. On the other hand, the AFM structure may re-appear during imaging under different conditions and with different tips, or it may be supported by other experiments. Then the imaged structure cannot be discarded just because the image formation is not understood.

Take for example, the case of thin films of PTFE made by rubbing the polymer on a substrate. As already mentioned, STM of very thin regions shows long objects lying parallel to the rubbing direction, and about 0.6 nm apart [190]. AFM investigations show very similar structures, though the spacing may be slightly different [218, 219]. It is known that PTFE molecules are largely extended rods, and diffraction shows that the orientation of the molecules in the film is extremely good, with an intermolecular spacing of 0.55 nm [220]. The AFM images show some detail of the internal structure of the helical molecules, but it is difficult to judge the reliability of new information at the highest resolution.

There are many publications showing images with atomic or molecular resolution in polymeric materials. Some specimens are very thin films, epitaxially grown [221], or otherwise placed on a rigid substrate [222]. Some are of comparatively rigid materials [223, 224], while others are the lateral surfaces of flexible chain polymer crystals in oriented fibers [225, 212]. Patil and Reneker show an image of the fold surface of a single crystal of polyethylene, with the correct lattice, and bumps that look like sharp folds [226]. From studies on adsorbed materials, it is evident that a scanning probe can penetrate liquid and show the structure of an inner ordered layer. So when the sample is a semi-crystalline polymer with T_g below the temperature of observation, an ordered AFM image does not preclude an outer disordered layer.

Jandt *et al.* [222] studied highly oriented thin films of polybutene-1 on which a little tin had been evaporated. TEM of these samples showed isolated small crystals of tin epitaxially related to needle-shaped crystals of the polymer [227]. They used a liquid environment to minimize the surface forces, but the regular contact mode removed the crystals of tin from the surface even when the contact force was 2 nN or less. Figure 6.17 shows AFM images of the polymer film at medium and high resolution [222]. The needle crystals are 20–35 nm across in agreement with TEM data. Figure 6.17A shows crystal branching at the upper right, and fine striations parallel to the molecular chain axis; these are surface features not visible in the TEM. At higher magnification, Fig. 6.17B and C, the fine parallel striations are still visible, and a 0.7 nm repeat appears along the molecular chain axis. This structure is reproducible, and the repeat distance is fairly close to the value obtained by x-ray diffraction from the bulk material (unlike the strongly adsorbed layer of the same material shown in Fig. 6.15).

Small contact forces are desirable for soft materials even if they are not crystalline. An AFM study of DNA using commercial tips, obtained a resolution of 3 nm in iso-propanol [228]. This was with a peak attractive force of 1 nN, and when this force was larger the resolution was worse. The size of the resolved detail increased with the square root of the peak attractive force in the range 1–10 nm [228]. An AFM study of the mechanical properties of elastomers and cells using the force – displacement curve [211] gives local Young's moduli of 0.6–2.4 MPa for rubber, but only 0.013–0.15 MPa for a living cell. This very low value suggests that

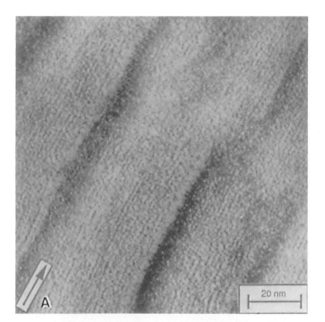

Fig. 6.17 AFM images of poly(butene-1) thin films [222]. (A) Medium magnification, showing a small part of four long needle-like crystals about 35 nm across. There are fine striations along all the crystals, parallel to the molecular chain axis. At the upper right, a crystal branch is occurring at the surface of the film. (B) At high magnification of the same material the striations remain visible, and finer molecular scale features appear. These are reproducible under a range of scanning conditions. The repeat distance along the chain axis agrees with that expected from the bulk crystalline structure. (C) is a processed version of (B), the original image. (From Jandt *et al.* [222]; reproduced with permission.)

the applied force must be kept at 1–10 pN for a realistic height determination of a cell at nanometer resolution.

Molecular dynamics modeling of the interaction between a tip and the sample surface will provide a path to a better theoretical understanding of the imaging process [229]. At the present, most modeling deals with the interaction of well-defined surfaces of metals or diamond [230–232]. We look forward to such methods being applied to polymer surfaces.

6.6.3 Metrology using scanning probe microscopy

The use of the AFM as a probe of surface topography at a resolution of several nanometers

is well established, in contrast to doubts that have been expressed about some images with atomic resolution. The STM can be used in the same way for conducting materials [143, 144, 148]. In both instruments the spreading of localized interaction forces at lower resolution reduces the concerns of distortion and damage to the surface by the probe. Many areas of polymer science can benefit from information of surface structure on this less than atomic scale. For example, Fig. 6.18 shows the complex structure of a side-chain liquid crystal polymer which contains lamellae and fine scale banding running across the lamellae [233].

There are several image artifacts that may make the image an inaccurate representation of the surface [151]. The most important of these is

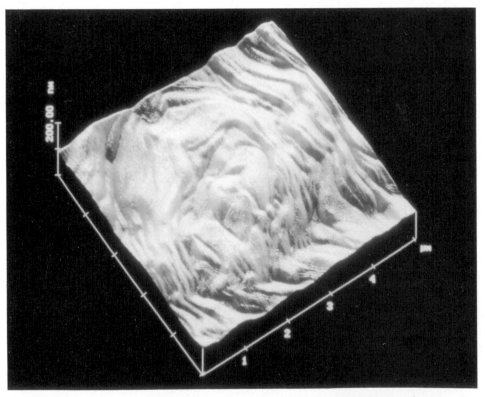

Fig. 6.18 AFM image of the free surface of a side chain liquid crystal polymer, showing three levels of organization. The polymer contains lamellae 50–300 nm thick. These lamellae are arranged in concentric layers giving a larger scale structure 7 to 10 μm across which is similar to hedrites that are found in crystalline polymers. Fine banding with a periodicity of about 30 nm appears within the lamellae and runs perpendicular to the long axis of the lamellae. (From Jandt *et al.* [233]; reproduced with permission.)

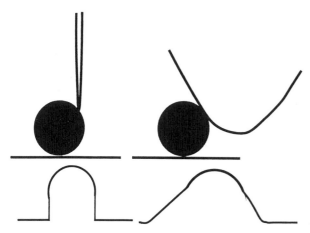

Fig. 6.19 Schematic of an extremely sharp tip probing the shape of a test sphere, and a much broader tip probing the same structure. The apparent size of the object and the slope of its size depends on the shape of the tip. If the object is known to be a sphere, the tip shape can be calculated from the images it produces.

due to the tip shape. As Fig. 6.13 showed for high resolution, a tip cannot form an accurate image of a surface that has steep slopes. The effect is most clearly seen when a blunt tip is used to image an abrupt step, or a sphere adsorbed on the surface [144, 148, 151, 234] as shown schematically in Fig. 6.19. The image at the step is an image of the profile of the tip. The broader the tip, the wider such features appear to be. This is well known, and easily recognized as it causes any sharp asperity to have the same apparent shape – the shape of the tip.

This problem exists at any resolution, but at these 'low' resolutions of more than a few nanometers the tip shape is stable and measurable. Tip shape can be measured by observing the image of a known shape, and can be relied on to stay the same for a reasonable time. Latex

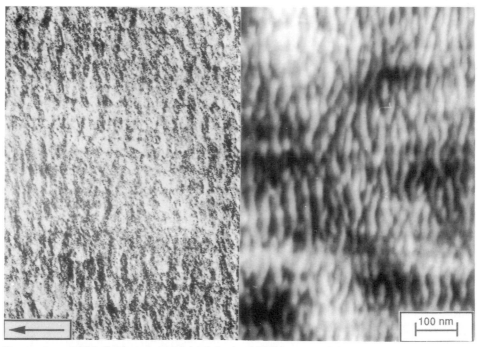

Fig. 6.20 Comparison TEM (left) and AFM (right) of a highly oriented PE film with lamellar structure. The arrow shows the molecular orientation. The TEM sample was shadowed with platinum and the AFM sample was not; they were otherwise identical. Both images can be analyzed to give the same values for lamellar thickness and their height above the surrounding film, but the AFM image is clearer. It also shows how the lamellae interlock and branch. (From Jandt [227]; reproduced with permission.)

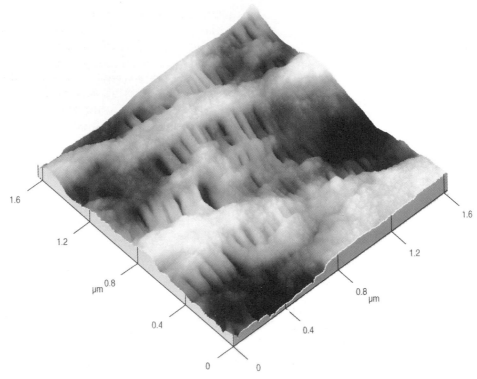

Fig. 6.21 AFM of a microporous membrane of stretched i-polypropylene, taken in a non-contact mode. The profiles at the edges of the image show how a quantitative surface measurement of any feature is easily obtained. (Supplied by P. Marella, Park Scientific Instruments Analytical Services Lab., 1171 Borregas Ave, Sunnyvale, CA 94089.)

spheres can calibrate horizontal and vertical magnification simultaneously [235]. The effects of tip shape on the image can then be calculated, and algorithms and software to correct the image distortion are becoming available [236]. Naturally, information on the parts of the surface that the probe never contacts cannot be recovered, but these unreliable parts can be flagged and ignored.

If rough surfaces are to be measured, then a tip shaped like a needle will have fewer problems than a pyramid or cone, even if the cone has a very small radius at the tip. Tips of this form are made by etching away material from a standard tip, or adding a thin spire to it by electron beam assisted chemical vapor deposition (CVD), and are available commercially. These can probe deep and narrow trenches, but they are more easily

bent. Bending by frictional forces shifts the image, in a direction controlled by the scan direction. A steep feature will increase the bend, and the recorded slope will not be correct. A sudden drop in surface height may cause bending by axial forces when the tip hits the surface again. This will give a 'shadow' behind a wall, which can be eliminated by adjusting the feedback [148].

These problems, though important for accurate measurement of feature size, should not be allowed to hide the fact that the AFM can show detail and make accurate measurements at the 5 nm level with little or no specimen preparation. The results compare well with high resolution LVSEM, as shown in Fig. 5.58. The AFM is not only cheaper, but allows *in situ* environmental studies without the problem of vacuum or

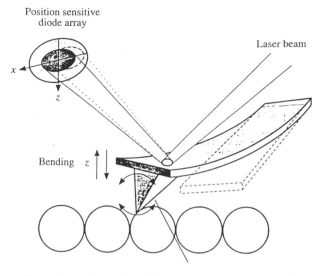

Position sensitive
diode array

Laser beam

Bending

Fig. 6.22 Schematic of the detector system in a frictional force microscope. A laser beam is reflected off the back of the cantilever and onto the center of a four segment light sensor. When the cantilever bends, due to a normal force, there is an intensity difference between the top and bottom halves. When it twists, due to frictional forces, there is an intensity difference between the left and right halves. (From Overney and Meyer [201]; reproduced with permission.)

radiation damage of the surface. A direct comparison with TEM is shown in Fig. 6.20, for a thin film of PE. The film was shadowed with Pt for the TEM, so that the topography of one surface contributes most contrast to the image, and the height of the surface features can be measured in both techniques [227]. The AFM image is superior and much easier to interpret.

Instruments continue to improve and everyone realizes that small interaction forces are better for imaging soft and weak samples. Of all the imaging modes, the static contact (or repulsive) mode applies the greatest local forces. Thus as other imaging modes improve in resolution and ease of use, they are more often used for polymer studies. These modes include static and dynamic non-contact (attractive) modes, and modes described as 'tapping' or 'hopping' where the tip is in repulsive contact with the sample for a small fraction of the time. Figure 5.79 showed a low resolution example of non-contact AFM imaging of a polymer latex. Figure 6.21 is a higher resolution non-contact example, of a iPP microporous membrane; the image quality is very

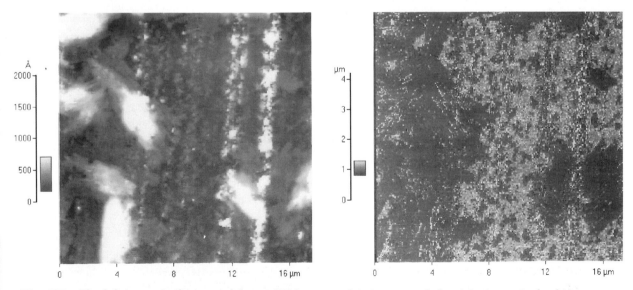

Fig. 6.23 The left image is the normal force, AFM topographic image, and the right image is the friction or lateral force image obtained during the same scan. The sample is a polyacetate film. In the FFM image bright areas correspond to high frictional forces. The low friction regions may be due to local surface segregation of low molecular weight material. (Supplied by P. Marella, Park Scientific Instruments Analytical Services Lab., 1171 Borregas Ave, Sunnyvale, CA 94089.)

similar to the images of membranes in Chapter 5: Figure 5.25C using contact mode AFM, and Figure 5.32C using the FESEM.

6.6.4 Frictional force microscopy

As an AFM tip is dragged along the surface in its contact mode of operation, there will be frictional forces that deflect the lever to one side. A frictional force microscope (FFM) forms an image using this interaction [237]. A normal AFM cantilever is wide, or has a broad base to resist twisting. If the lever is narrow, it will twist due to sideways frictional forces, and a detector can be designed to detect lateral motion [237]. A four-segment photodiode or a 2-D capacitance probe can detect vertical and lateral deflection of the lever at the same time (Fig. 6.22). The torsional spring constant can be as low as 0.1 N m^{-1}, giving a lateral force sensitivity of 10 pN. The lateral deflection can form the frictional force image signal, while the vertical deflection acts as a feedback input to keep the probe at constant vertical force, as well as giving the topographic signal.

Frictional forces have a chemical adhesion component, and one due to the physical interaction of rough surfaces. The FFM, with its very local determination of topography and lateral force, can distinguish these two components and thus distinguish regions of different surface chemistry. This begins to be micro-analysis, a feature that has been missing from the regular AFM. As the lever is not exactly parallel to the surface, and the surface is not exactly flat, there will be some cross-talk between the two signals. That is, the topographical signal is really a force perpendicular to the local surface, and this will have a lateral component on a slope [237–239]. Similarly frictional forces due to any motion not perpendicular to the long axis of the lever may cause it to bend or buckle and this will appear as a vertical deflection of the lever [197, 240, 241]. Thus corrections are required, particularly on rough surfaces, to make sure that part of the lateral or frictional force signal is not caused by the topography.

The friction force microscope is still in an early stage of development and use, and relatively few papers have been published as yet. Some deal with friction at high resolution on a standard substrate for AFM such as mica [242, 243]. Others concern lubricating films or their models, such as Langmuir–Blodgett films [210, 244], phase separated films [245] and surfactants such as dodecanol [246]. From such organic films it is a short step to polymer surfaces and polymer blends [247]. Figure 6.23 shows as an example a pair of images, AFM and FFM, acquired simultaneously, of a single phase polyacetate film. The low friction regions, dark in the FFM image, may be associated with low molecular weight material on the film surface.

REFERENCES

1. M. B. Rhodes and R. P. Nathhorst in *Computer Applications in the Polymer Laboratory*, ACS Symposium Series 313 (ACS, Washington DC, 1986) p. 155.
2. D. L. Taylor, M. Nederlof, F. Lanni and A. S. Waggoner, *Amer. Scientist* **80** (1992) 322.
3. T. Wilson, *Theory and Practice of Scanning Optical Microscopy* (Academic Press, London, 1984).
4. J. B. Pawley, Ed., *Handbook of Biological Confocal Microscopy*, edited by J. B. Pawley (Plenum Press, New York, 1990).
5. T. Wilson, Ed., *Confocal Microscopy* (Academic Press, London, 1990).
6. P. C. Cheng, T. H. Lin, W. L. Wu and J. L. Wu, Eds, *Multidimensional Microscopy* (Springer Verlag, New York, 1994).
7. C. J. Cogswell and J. W. O'Byrne, *SPIE* **1660** (1992) 503.
8. D. H. Burns, R. B. Hataganagadi and F. A. Spelman, *Scanning* **12** (1990) 156.
9. J. Pawley, *Trans. Roy. Microsc. Soc.* **1** (1990) 381.
10. A. Boyde, in *Proc. 13th Int. Congr. on X-Ray Optics and Microanalysis, 1992*, edited by P. B. Kenway, P. J. Duke, G. W. Lorimer, T. Mulvey, I. W. Drummond, G. Love, A. G. Michette and M. Stedman (IOP, Bristol, 1993) p. 353.
11. M. L. Scott, *Amer. Lab.* **24** (1992) 34.
12. H. Kitagawa, in *Multidimensional Microscopy*, edited by P. C. Cheng, T. H. Lin, W. L. Wu and J. L. Wu (Springer Verlag, New York, 1994).

13. A. Knoester and G. J. Brakenhoff, *J. Microsc.* **157** (1990) 105.

14. J. Suhar, J. R. Laghari and P. C. Cheng, *IEEE Electrical Insul. Mag.* **8** (1992) 20.

15. E. Morueau, A. Laurent and C. Mayoux, in *Sixth Int. Conf. on Dielectric Materials, Measurements and Applications*, IEE Conference Publ. 363 (IEE, Stevenage, England, 1992) p. 488.

16. J. L. Thomason and A. Knoester, *J. Mater. Sci. Lett.* **9** (1990) 258.

17. H. Yoshida, K. Ito and N. Ise, *Phys. Rev.* **B4** (1991) 435.

18. D. N. Batchelder, C. Cheng and G. D. Pitt, *Adv. Mater.* **3** (1991) 566.

19. R. J. Meier and B. J. Kip, *Microbeam Anal.* **3** (1994) 61.

20. M. Bowden, P. Donaldson, D. J. Gardiner, G. Rice and D. L. Gerrard, *Anal. Chem.* **63** (1991) 2915.

21. K. P. Ghiggino, M. R. Harris and P. G. Spizzirri, *Rev. Sci. Instr.* **63** (1992) 2999.

22. M. Isaacson, J. A. Cline and H. Barshatzky, *J. Vac. Sci. Technol.* **B9** (1991) 3103.

23. D. W. Pohl, in *Scanning Tunneling Microscopy II*, edited by M. S. Guntherodt and R. Wiesendanger, Springer Series in Surf. Sci. Vol. 28 (Springer Verlag, Berlin, 1992).

24. E. Betzig and J. K. Trautman, *Science* **257** (1992) 189.

25. J. M. Guerra, M. Srinivasarao and R. S. Stein, *Science* **262** (1993) 1395.

26. J. K. Trautman, E. Betzig, J. S. Weiner, D. J. DiGiovanni, T. D. Harris, F. Hellman and E. M. Gyorgy, *J. Appl. Phys.* **71** (1992) 4659.

27. D. Birnbaum, S.-K. Kook and R. Kopelman, *J. Phys. Chem.* **97** (1993) 3091.

28. J. I. Goldstein, D. E. Newbury, P. Echlin, D. C. Joy, A. D. Romig Jr., C. E. Lyman, C. Fiori and E. Lifshin, *Scanning Electron Microscopy and X-ray Microanalysis*, 2nd Edn (Plenum, New York, 1992).

29. L. Reimer, *Scanning Electron Microscopy, Physics of Image Formation and Microanalysis* (Springer Verlag, Berlin, 1985).

30. L. Reimer, *Image Formation in Low-Voltage Scanning Electron Microscopy* (SPIE, Bellingham, 1993).

31. J. H. Butler, in *Microbeam Analysis – 1991*, edited by D. G. Howett (San Francisco Press, San Francisco, 1991) p. 565.

32. D. L. Vezie, W. W. Adams and E. L. Thomas, *Polymer*, **36** (1995) 1761.

33. K. Kanaya and S. Okayama, *J. Phys. D, Appl. Phys.* **5** (1972) 43.

34. D. C. Joy, *Scanning Microsc.* **2** (1988) 57.

35. D. C. Joy, *J. Microsc.* **147** (1987) 51.

36. D. C. Joy, *J. Microsc.* **140** (1985) 283.

37. K. Murata, H. Kawata and K. Nagami, *Scanning Microsc. Suppl.* **1** (1987) 83.

38. B. Volbert and L. Reimer, in *Scanning Electron Microscopy 1980* (SEM Inc., AMF O'Hare, 1980) p. 1.

39. M. Brunner and R. Schmid, *Scanning Microsc.* **1** (1987) 1501.

40. R. Autrata and J. Hejna, *Scanning* **13** (1991) 275.

41. P. E. Russell and J. F. Mancuso, *J. Microsc.* **140** (1985) 323.

42. M. T. Postek, W. J. Keery and N. V. Frederick, *Rev. Sci. Instrum.* **61** (1990) 1648.

43. L. Reimer and B. Volbert, *Scanning* **2** (1979) 238.

44. J. B. Pawley, *J. Microsc.* **136** (1984) 45.

45. D. C. Joy, *Scanning* **11** (1989) 1.

46. L. Reimer, U. Golla, R. Böngeler, M. Kässens, B. Schindler and R. Senkel, *Optik* **92** (1992) 14.

47. J. Chang, S. Krause and R. Gorur, in *Proc. XIIth Int. Congr. Electr. Microsc.*, Vol. 4 (San Francisco Press, San Francisco, 1990) p. 1108.

48. T. Ichinokawa, M. Iijama, A. Onoguchi and T. Kobayashi, *Jap. J. Appl. Phys.* **13** (1974) 1272.

49. D. C. Joy and J. B. Pawley, *Ultramicroscopy* **47** (1992) 80.

50. L. Reimer and A. Schmidt, *Scanning* **7** (1985) 47.

51. Ya Chen and J. B. Pawley, in *Multidimensional Microscopy*, edited by P. C. Cheng, T. H. Lin, W. L. Wu and J. L. Wu (Springer Verlag, New York, 1994).

52. V. K. Berry, *Scanning* **10** (1988) 19.

53. C. W. Price and P. L. McCarthy, *Scanning* **10** (1988) 29.

54. P. B. Himelfarb and K. B. Labat, *Scanning* **12** (1990) 148.

55. D. W. Schwark, D. L. Vezie, J. R. Reffner, E. L. Thomas and B. K. Annis, *J. Mater. Sci. Lett.* **11** (1992) 352.

56. T. Nagatani, S. Saito, M. Sato and M. Yamada, *Scanning Microsc.* **1** (1987) 901.

57. H. Seiler, *J. Appl. Phys.* **54** (1983) R1.

58. H. Kawamoto, S. Yamazaki, A. Ishikawa and R. Buchanan, *Scanning Electr. Microsc.* Part 1 (1984) 15.

59. M. Isaacson and J. P. Langmore, *Optik* **41** (1974) 92.

60. D. C. Joy, *J. Microsc.* **161** (1991) 343.
61. G. D. Danilatos and R. Postle, *Scanning Electron Microsc.* Part 1 (1982) 1.
62. G. D. Danilatos, *Adv. in Electronics Electron Phys.* **71** (1988) 109.
63. V. Robinson, *J. Comp. Assisted Microsc.* **4** (1992) 247.
64. R. Durkin and J. S. Shah, *J. Microsc.* **169** (1993) 33.
65. E. Doehne and D. C. Stulik, *Scanning Microsc.* **4** (1990) 275.
66. R. A. Rapp, *Pure Appl. Chem.* **56** (1984) 1715.
67. T. A. Ramanarayanan and J. Alonzo, *Oxidation of Metals* **24** (1985) 17.
68. K. Ramani, C. J. Hoyle and N. C. Parasnis, in *Use of Plastics and Plastic Composites: Materials and Mechanics*, ASME MD Vol. 46 (ASME, New York, 1993) p. 633.
69. S. L. Sarkar and Xu Aimin, *Cement and Concrete Res.* **22** (1992) 605.
70. B. Little, P. Wagner, R. Ray, R. Pope and R. Scheetz, *J. Indust. Microbiology* **8** (1991) 213.
71. T. V. Chirila, Yi-Chi Chen, B. J. Griffin and I. J. Constable, *Polym. Internat.* **32** (1993) 221.
72. S. Mehta, in *Proc. 1991 SPE Annual Technical Conference* (SPE, Richardson TX, 1991) p. 445.
73. H.-M. Choi and H.-J. Kwon, *Text. Res. J.* **63** (1993) 211.
74. P. Buseck, J. Cowley, L. Eyring, Eds, *High-Resolution Transmission Electron Microscopy and Associated Techniques* (Oxford University Press, Oxford, 1988).
75. J. C. H. Spence, *Experimental High Resolution Transmission Electron Microscopy*, 2nd Edn (Oxford University Press, Oxford, 1988).
76. S. Amelinckx, in *Materials Science and Technology*, Vol. 2A, edited by E. Lifshin (VCH Publishers, Weinheim, 1993).
77. J. M. Gibson, *MRS Bulletin* **16**(3) (1991) 27.
78. S. Amelinckx, in *Examining the Submicron World*, edited by R. Feder, J. W. McGowan and D. M. Shinozaki, NATO ASI Series B 137 (Plenum Press, New York, 1986).
79. L. Kihlborg, *Progr. Solid State Chem.* **20** (1990) 101.
80. T. Kobayashi and S. Isoda, *J. Mater. Chem.* **3** (1993) 1.
81. I. G. Voigt-Martin, H. Krug and D. Van Dyck, *J. Phys. (Paris)* **51** (1990) 2347.
82. D. Vesely, in *Proc. of the IOP Electron Microscopy and Analysis Group Conference 1991*, edited by F. J. Humphreys (IOP, Bristol, 1991) p. 329.
83. D. C. Martin and E. L. Thomas, *Polymer* (1995) In press.
84. D. C. Martin, P. M. Wilson and J. Liao, *Polym. Prepr.* **33** (1992) 245.
85. M. A. O'Keefe, *Ultramicroscopy* **47** (1992) 282.
86. S. Kumar and W. W. Adams, *Polymer* **31** (1990) 15.
87. A. Keller, *Koll Z. u Z. Polymere* **231** (1969) 389.
88. M. G. Dobb, A. M. Hindeleh, D. J. Johnson and B. P. Saville, *Nature* **253** (1975) 189.
89. M. G. Dobb, D. J. Johnson and B. P. Saville, *J. Polym. Sci.* **C58** (1977) 237.
90. R. T. Read and R. J. Young, *J. Mater. Sci.* **19** (1984) 327.
91. M. Tsuji, S. Isoda, M. Ohara, A. Kawaguchi and K. Katayama, *Polymer* **23** (1982) 1568.
92. P. H. J. Leung and R. J. Young, *Polymer* **27** (1986) 202.
93. P. Pradere and E. L. Thomas, *Phil. Mag.* **A60** (1990) 177.
94. P. M. Wilson and D. L. Martin, *J. Mater. Res.* **7** (1992) 3150.
95. J. Liao and D. L. Martin, *Science* **260** (1993) 1489.
96. M. Tsuji, S. K. Roy and R. St. John Manley, *Polymer* **25** (1984) 1573.
97. H. D. Chanzy, P. Smith and J. F. Revol, *J. Polym. Sci.* **C24** (1986) 557.
98. H. D. Chanzy, T. Folda, P. Smith, K. Gardner and J. F. Revol, *J. Mater. Sci. Lett.* **5** (1986) 1045.
99. J. F. Revol and R. St. J. Manley, *J. Mater. Sci. Lett.* **5** (1986) 249.
100. H. D. Chanzy, P. Smith, J. F. Revol and R. St. John Manley, *Polymer Commun.* **28** (1987) 133.
101. J. F. Revol, H. D. Chanzy, Y. Deslandes and R. H. Marchessault, *Polymer* **30** (1989) 1973.
102. A. Uemura, M. Tsuji, A. Kawaguchi and K.-I. Katayama, *J. Mater. Sci.* **23** (1988) 1506.
103. J. Sugiyama, H. Harada, Y. Fujioyoshi and N. Uyeda, *Planta* **166** (1985) 16.
104. J.-F. Revol, *J. Mater. Sci. Lett.* **4** (1985) 1347.
105. P. Pradere and E. Thomas, *Macromolecules* **23** (1990) 4954.
106. I. G. Voigt-Martin and H. Durst, *Macromolecules* **22** (1989) 186.
107. I. G. Voigt-Martin, R. W. Garbella and M. Schumacher, *Macromolecules* **25** (1992) 961.
108. K. Shimamura, J. R. Minter and E. L. Thomas, *J. Mater. Sci. Lett.* **18** (1983) 54.
109. W. W. Adams, S. Kumar, D. C. Martin and K. Shimamura, *Polymer Commun.* **30** (1989) 285.

110. D. C. Martin and E. L. Thomas, *Macromolecules* **24** (1991) 2460.

111. D. C. Martin and E. L. Thomas, *J. Mater. Sci.* **26** (1991) 5171.

112. J. R. Oyeda and D. C. Martin, *Macromolecules* **26** (1993) 6557.

113. D. C. Martin, L. L. Berger and K. H. Gardner, *Macromolecules* **24** (1991) 3921.

114. D. L. Dorset, *Ultramicroscopy* **38** (1991) 23.

115. D. L. Dorset, *Macromolecules* **24** (1991) 1175.

116. H. Hu and D. L. Dorset, *Macromolecules* **23** (1990) 4604.

117. D. L. Dorset, *Polym. Prepr.* **33** (1992) 243.

118. B. Moss, D. L. Dorset, J. C. Wittmann and B. Lotz, *J. Macromol. Sci. Phys.* **B24** (1985) 99.

119. D. L. Dorset, *J. Polym. Sci., Part B: Polym. Phys.* **24** (1986) 79.

120. D. L. Dorset and F. Zemlin, *Ultramicroscopy* **33** (1990) 227.

121. D. L. Dorset and S.-L. Hsu, *Polymer* **30** (1989) 1596.

122. W. P. Zhang and D. L. Dorset, *Macromolecules* **23** (1990) 4322.

123. R. M. Glaeser and K. H. Downing, *Ultramicroscopy* **47** (1992) 256.

124. D. L. Dorset, W. F. Tivol and J. N. Turner, *Ultramicroscopy* **38** (1991) 41.

125. M. S. Guntherodt and R. Wiesendanger, Eds, *Scanning Tunneling Microscopy I, II and III*, Springer Series in Surf. Sci. Vol. 20, 28 and 29 (Springer, Berlin, 1992, 1993).

126. R. J. Hamers, *MRS Bull.* **16**(3) (1991) 22.

127. J. Shen, R. G. Pritchard and R. E. Thurstans, *Contemp. Phys.* **32** (1991) 11.

128. J. Jahanmir, B. G. Haggar and J. B. Hayes, *Scanning Microsc.* **6** (1992) 625.

129. N. J. DiNardo, in *Materials Science and Technology*, Vol. 2B, edited by E. Lifshin (VCH Publishers, Weinheim, 1994).

130. H. Rohrer, *Ultramicroscopy* **42–44** (1992) 1. (The rest of this issue contains many articles on STM).

131. J. E. Griffith and G. P. Kochanski, in *Annual Review of Materials Science*, Vol. 20, edited by R. A. Huggins, J. A. Giordmaine and J. B. Wachtman Jr. (Annual Reviews, Palo Alto, 1990) p. 219.

132. B. L. Blackford, M. H. Jericho and P. J. Mulhern, *Scanning Microsc.* **5** (1991) 907.

133. L. E. C. van de Leemput and H. van Kempen, *Rep. Progr. in Phys.* **55** (1992) 1165.

134. C. J. Chen, *Introduction to Scanning Tunneling Microscopy* (Oxford University Press, Oxford, 1993).

135. Some examples are the *Application and Technical Notes* from TopoMetrix Inc. and the *PSI Probe* technical newsletter from Park Instruments Inc. Addresses for these and other manufacturers of SPMs can be found in Appendix VIII.

136. G. Binnig, H. Rohrer, C. Gerber and H. Weibel, *Appl. Phys. Lett.* **40** (1982) 178.

137. O. Nishikawa, M. Tomitori and F. Iwawaki, *Mater. Sci. Engn.* **B8** (1991) 81.

138. R. Hiesgen and D. Meissner, *Ultramicroscopy* **42–44** (1992) 1403.

139. M. Tsukada, K. Kobayashi, N. Isshiki and H. Kageshima, *Surf. Sci. Reports* **13** (1991) 265.

140. R. J. Hamers, *Ann. Rev. Phys. Chem.* (1989) 521.

141. R. J. Wilson, P. H. Lippel, S. Chiang, D. D. Chambliss and V. M. Hallmark, *Ultramicroscopy* **47** (1992) 212.

142. A. J. Melmed, *J. Vac. Sci. Technol.* **B9** (1991) 601.

143. J. E. Griffith, D. A. Grigg, M. J. Vasile, P. E. Russell and E. A. Fitzgerald, *J. Vac. Sci. Technol.* **B9** (1991) 3586.

144. J. E. Griffith, D. A. Grigg, M. J. Vasile, P. E. Russell and E. A. Fitzgerald, *J. Vac. Sci. Technol.* **A10** (1992) 674.

145. Y. Kuk and P. J. Silverman, *Appl. Phys. Lett.* **48** (1986) 1597.

146. Z. Nawaz, T. R. I. Cataldi, J. Knall, R. Somekh and J. B. Pethica, *Surf. Sci.* **265** (1992) 139.

147. G. Binnig and D. P. E. Smith, *Rev. Sci. Instrum.* **57** (1986) 1688.

148. M. G. Walls, *Annales de Chimie* **17** (1992) 191.

149. G. Binnig, H. Rohrer, C. Gerber and H. Weibel, *Phys. Rev. Lett.* **49** (1982) 57.

150. R. C. Barrett and C. F. Quate, *Rev. Sci. Instrum.* **62** (1991) 1393.

151. *Artifacts in SPM*, TopoMetrix Technical Report (TopoMetrix Corp. Santa Clara CA, 1993).

152. R. Weisendanger, G. Tarrach, D. Burgler and H.-J. Guntherodt, *Europhys. Lett.* **12** (1990) 57.

153. C. M. MacRae, *Ultramicroscopy* **42–44** (1992) 1337.

154. K. Itaya and E. Tomita, *Chem. Lett.* **1989** (1989) 285.

155. F.-R. Fan and A. J. Bard, *J. Phys. Chem.* **94** (1990) 3761.

156. G. S. Rohrer, V. E. Heinrich and D. A. Bonnell, *Science* **250** (1990) 1239.

157. T. Matsumoto, H. Tanaka, T. Kawai and S. Tawai, *Surf. Sci. Lett.* **278** (1992) L153.

158. S. Chiang, in *Scanning Tunneling Microscopy I*, edited by M. S. Guntherodt and R. Wiesendanger, Springer Series in Surf. Sci. Vol. 20 (Springer Verlag, Berlin, 1992).

159. J. Schneir, R. Sonnenfeld, P. K. Hansma and J. Tersoff, *Phys. Rev.* **B34** (1986) 4979.

160. J. B. Pethica, *Phys. Rev. Lett.* **57** (1986) 3235.

161. J. M. Soler, A. M. Baro, N. Garcia and H. Rohrer, *Phys. Rev. Lett.* **57** (1986) 444.

162. R. J. Colton, S. M. Baker, R. J. Driscoll, M. G. Youngquist, J. D. Baldeschwieler and W. J. Kaiser, *J. Vac. Sci. Technol.* **A6** (1988) 349.

163. C. R. Clemmer and T. P. Beebe, Jr., *Science* **251** (1991) 640.

164. C. R. Clemmer and T. P. Beebe, Jr., *Scanning Microsc.* **6** (1992) 312.

165. M. Salmeron, T. Beebe, J. Odriozola, T. Wilson, D. F. Ogletree and W. Siekhaus, *J. Vac. Sci. Technol.* **A8** (1990) 635.

166. R. Yang, D. F. Evans, W. A. Hendrickson and J. Baker, *J. Phys. Chem.* **95** (1991) 3765.

167. S. N. Magonov and M.-H. Whangbo, *Adv. Mater.* **6** (1994) 355.

168. R. Wiesendanger, *Appl. Surf. Sci.* **54** (1992) 271.

169. T. T. Tsong, *Mater. Chem. Phys.* **33** (1993) 1.

170. C. J. Roberts, M. J. Wilkins, G. Beamson, M. C. Davies, D. E. Jackson, P. D. Scholes, S. L. B. Tendler and P. M. Williams, *Nanotechnology* **3** (1992) 98.

171. J. P. Rabe, *Ultramicroscopy* **42–44** (1992) 41.

172. R. J. Wilson, K. E. Johnson, D. D. Chambliss and B. Melior, *Langmuir* **9** (1993) 3478.

173. J. F. Jorgensen, K. Carneiro and L. L. Madsen, *Nanotechnology* **4** (1993) 152.

174. P. Dai, S.-K. Wang, H. Taub, J. E. Buckley, S. N. Ehrlich, J. Z. Larese, G. Binnig and D. P. E. Smith, *Phys. Rev.* **B47** (1993) 7401.

175. M. Amrein and H. Gross, *Scanning Microsc.* **6** (1992) 335.

176. H. L. Fuchs, M. Eng, R. Sander, J. Petermann, K. D. Jandt and T. Hoffmann, *Polym. Bull.* **26** (1991) 95.

177. R. Yang, X. R. Yang, D. F. Evans, W. A. Hendrickson and J. Baker, *J. Phys. Chem.* **94** (1990) 6123.

178. M. Sano, D. Y. Sasaki and T. Kunitake, *Macromolecules* **25** (1992) 6961.

179. M. E. Hawley and B. C. Benicewicz, *J. Vac. Sci. Technol.* **B9** (1991) 1141.

180. H. Sotobayashi, T. Schilling and B. Tesche, *Langmuir* **6** (1990) 1246.

181. T. E. Wilson, D. F. Ogletree, M. B. Salmeron and M. D. Bednarski, *Langmuir* **8** (1992) 2588.

182. C. M. Agrawal, K. Hunter, G. W. Pearsall and R. W. Henkens, *J. Mater. Sci.* **27** (1992) 2606.

183. D. M. Kulawansa, S. C. Langford and J. T. Dickinson, *J. Mater. Res.* **7** (1992) 1292.

184. R. Piner, R. Reifenberger, D. C. Martin, E. L. Thomas and R. P. Apkarian, *J. Polym. Sci., Part C: Lett.* **28** (1990) 399.

185. D. H. Reneker, J. Schneir, B. Howell and H. Harary, *Polymer Commun.* **31** (1990) 167.

186. T. G. Stange, R. Mathew, D. F. Evans and W. A. Hendrickson, *Langmuir* **8** (1992) 920.

187. G. W. Zajac, M. Q. Patterson, P. M. Burrell and C. Metaxas, *Ultramicroscopy* **42–44** (1992) 998.

188. L. Wefers and E. Schollmeyer, *J. Polym. Sci., Part B* **31** (1993) 23.

189. A. M. Baro, L. Vazquez, A. Bartolame, J. Gomez, N. Garcia, H. A. Goldberg, L. C. Sawyer, R. T. Chen, R. S. Kohn and R. Reifenberger, *J. Mater. Sci.* **24** (1989) 1739.

190. P. Bodo, C. Ziegler, J. R. Rasmusson, W. R. Salaneck and D. T. Clark, *Synth. Met.* **55** (1993) 329.

191. Y. Oka and A. Takahashi, *Polymer J.* **23** (1991) 805.

192. E. Lacaze, K. Uvdal, P. Bodo, J. Garbarz, W. R. Salaneck and M. Schott, *J. Polym. Sci.* **B31** (1993) 111.

193. S. J. Kamrava, M. Zagorska, B. Krische and S. Soderholm, *Phys. Script.* **44** (1991) 112.

194. H. Yang, F.-R. F. Fan, S.-L. Yau and A. J. Bard, *J. Electrochem. Soc.* **139** (1992) 2182.

195. M. P. Everson and J. H. Helms, *Synth. Met.* **40** (1991) 97.

196. M. V. Mirkin, F.-R. Fan and A. J. Bard, *Science* **257** (1992) 364.

197. E. Meyer and H. Heinzelmann, in *Scanning Tunneling Microscopy II*, edited by M. S. Guntherodt and R. Wiesendanger, Springer Series in Surf. Sci. Vol. 28 (Springer, Berlin, 1992).

198. D. Sarid and V. Elings, *J. Vac. Sci. Technol.* **9B** (1991) 431.

199. G. Binnig, *Ultramicroscopy* **42–44** (1992) 7.

200. E. Meyer, *Progr. Surf. Sci.* **41** (1992) 3.

201. R. Overney and E. Meyer, *MRS Bull.* **18**(5) (1993) 26.

202. D. Rugar and P. Hansma, *Physics Today* **43** (1990) 23.

203. S. M. Hues, R. J. Colton, E. Meyer and H.-J. Guntherodt, *MRS Bull.* **18**(1) (1993) 41.

204. D. Sarid, *Scanning Force Microscopy* (Oxford University Press, Oxford, 1991).

205. G. Binnig, C. F. Quate and C. Gerber, *Phys. Rev. Lett.* **56** (1986) 930.

206. S. Akamine, R. C. Barrett and C. F. Quate, *Appl. Phys. Lett.* **57** (1990) 316.

207. N. A. Burnham, R. J. Colton and H. M. Pollock, *Nanotechnology* **4** (1993) 64.

208. N. A. Burnham and R. J. Colton, *J. Vac. Sci. Technol.* **A7** (1989) 2906.

209. J. H. Hoh, J.-P. Revel and P. K. Hansma, *Nanotechnology* **2** (1991) 119.

210. P. Maivald, H. J. Butt, S. A. C. Gould, C. B. Prater, B. Drake, J. A. Gurley, V. B. Elings and P. K. Hansma, *Nanotechnology* **2** (1991) 103.

211. A. L. Weisenhorn, M. Khorsandi, M. Kasas, V. Gotzos and H. J. Butt, *Nanotechnology* **4** (1993) 106.

212. S. S. Sheiko, M. Möller, E. M. C. M. Reuvekaml and H. W. Zandbergen, *Ultramicroscopy* **53** (1994) 371.

213. W. Stocker, B. Bickman, S. N. Magonov, H. J. Cantow, B. Lotz, J. C. Wittman and M. Moller, *Ultramicroscopy* **42–44** (1992) 1141.

214. F. Ohnesorge and G. Binnig, *Science* **260** (1993) 1451.

215. M. Florsheimer, A. J. Steinford and P. Gunter, *Surf. Sci.* **297** (1993) L39.

216. O. M. Leung and M. C. Goh, *Science* **255** (1992) 64.

217. E. Hamada and R. Kaneko, *Ultramicroscopy* **42–44** (1992) 184.

218. P. Dietz, P. K. Hansma, K. J. Ihn, F. Motamedi and P. Smith, *J. Mater. Sci.* **28** (1993) 1372.

219. S. N. Magonov, S. Kempf, M. Kimmig and H.-J. Cantow, *Polymer Bull.* **26** (1991) 715.

220. J. C. Wittman and P. Smith, *Nature* **352** (1991) 414.

221. B. Lotz, J. C. Wittman, W. Stocker, S. N. Magonov and H. J. Cantow, *Polymer Bull.* **26** (1991) 209.

222. K. D. Jandt, T. J. McMaster, M. J. Miles and J. Petermann, *Macromolecules* **26** (1993) 6552.

223. D. Snetivy, G. J. Vancso and G. C. Rutledge, *Macromolecules* **25** (1992) 7037.

224. G. Bar, S. N. Magonov, A. Gorenberg, H. D. Bauer and H. J. Cantow, *Polym. Prepr.* **33** (1992) 792.

225. D. Snetivy and G. V. Vansco, *Polymer* **35** (1994) 461.

226. R. Patil and D. H. Reneker, *Polymer* **35** (1994) 1909.

227. K. D. Jandt, *Untersuchungen zur Polymer-Metall-Epitaxie in Computersimulation und Experiment*, VDI Reihe 5, 302 (VDI-Verlag, Dusseldorf, 1993).

228. J. Yang and Z. Shao, *Ultramicroscopy* **50** (1993) 157.

229. U. Landman and W. D. Luedtke, in *Scanning Tunneling Microscopy III*, edited by M. S. Guntherodt and R. Wiesendanger, Springer Series in Surf. Sci. Vol. 29 (Springer, Berlin, 1993).

230. U. Landman, W. D. Luedtke, N. A. Burnham and R. J. Colton, *Science* **248** (1990) 454.

231. J. A. Harrison, C. T. White, R. J. Colton and D. W. Brenner, *Surf. Sci.* **271** (1992) 57.

232. J. A. Harrison, C. T. White, R. J. Colton and D. W. Brenner, *Phys. Rev.* **B46** (1992) 9700.

233. K. D. Jandt, D. G. McDonnell, J. M. Blackmore and M. J. Miles, *Polym. Bull.* **4** (1994) 487.

234. D. C. Martin, J. R. Ojeda, J. P. Anderson and G. Pingali, AFM/STM Symposium, US Army Natick Research Center, Natick MA, June 1993.

235. Y. Li and S. M. Lindsay, *Rev. Sci. Instrum.* **62** (1991) 2630.

236. G. S. Pingali and R. Jain, in *Proc. IEEE Workshop on Applications of Computer Vision, Palm Springs, CA* (IEEE, 1992) p. 282.

237. C. M. Mate, G. M. McClelland, R. Erlandsson and S. Chiang, *Phys. Rev. Lett.* **59** (1987) 1942.

238. S. Grafstrom, M. Neitzert, T. Hagen, J. Ackermann, R. Neumann, O. Probst and M. Wortge, *Nanotechnology* **4** (1993) 143.

239. G. Haugsted, W. L. Gladfelter and E. B. Weberg, *Langmuir* **9** (1993) 3717.

240. B. K. Annis and D. F. Pedraza, *J. Vac. Sci. Technol.* **B11** (1993) 1759.

241. R. J. Warmack, X. Y. Zheng, T. Thundat and D. P. Allison, *Rev. Sci. Instrum.* **65** (1994) 394.

242. O. Marti, J. Colchero and J. Mlynek, *Nanotechnology* **1** (1990) 141.

243. S. Fujisawa, Y. Sugawara, S. Ito, S. Mishima, T. Okada and S. Morita, *Nanotechnology* **4** (1993) 138.

244. E. Meyer, R. Overney, D. Brodbeck, L. Howald, R. Luthi, J. Frommer and H. J. Guntherodt, *Phys. Rev. Lett.* **69** (1992) 1778.

245. R. Overney, E. Meyer, J. Frommer, D. Brodbeck, L. Howald, R. Luthi, H. J. Guntherodt, M. Fujihara, H. Takano and Y. Gotoh, *Nature* **359** (1992) 335.

246. S. J. O'Shea, M. E. Welland and T. Rayment, *Appl. Phys. Lett.* **61** (1992) 2240.

247. E. Hamada and R. Kaneko, *Ultramicroscopy* **42–44** (1992) 184.

CHAPTER SEVEN

Problem solving summary

The preceding chapters have provided a description of microscopy techniques, imaging theory and the specimen preparation methods required to investigate polymer structures. The theme of this chapter is to put all of this together within a useful framework. This framework might be a review to experienced microscopists (who likely have developed their own protocols), but it will provide useful information regarding problem solving ideas. A problem solving protocol will be developed that permits microscopy characterizations to follow an easy and short path to a solution. These characterizations will all be classified as 'problems' that require a solution. Problems can range from simple to complex and include, for example, determination of the phase structure in a polymer blend, the cause of failure of a composite or the complete and fundamental characterization of a new membrane, fiber, film, etc. Clearly, such problem solving will require a range of time and effort, but the protocols used to begin the characterization and to know when the problem is solved are similar overall. Generally more than one technique is required to solve problems relating to polymer morphology and thus complementary multidisciplinary techniques are important in conducting problem solving analyses. Interpretation of the images produced is of critical importance in evaluating polymer structures and thus the topic of artifacts will be included in this discussion. Finally, although structural characterizations cannot generally be accomplished without microscopy

methods and techniques, there are other complementary analytical techniques that are often quite important in understanding polymer structures. The last section will be devoted to a short description of these techniques, including x-ray diffraction, thermal analysis, electron spectroscopy and others.

7.1 WHERE TO START

One of the most difficult decisions that must be made in the microscopy laboratory is how to *start* solving a problem. The difficulty is that it is not always obvious where to start and to know ahead of time the full range of techniques that will be required. It is also difficult to deal with this question when considering that there is a wide range of techniques that can be used for problem solving. However, there are some simple concepts to consider before beginning the microscopy characterization. The protocol that will be discussed here is not necessary in all cases. If a measure of the orientation of a fiber is needed, for instance, a problem solving protocol might not be needed if the investigator is aware that the birefringence can be measured using a polarizing optical microscope and a compensator. But, if the dispersed phase distribution and particle size of a polymer blend are to be correlated with impact strength, then it might not be as obvious where to start solving the problem.

7.1.1 Problem solving protocol

The steps involved in the problem solving protocol are outlined in Table 7.1. They are rather simple and do not take much time to consider and such a protocol can save time in the long run. The protocol involves steps typical of scientific inquiry: collect all the currently known facts, determine the nature of the problem, state the objective of the study, obtain the correct specimen, be sure to have experimental controls, look at the sample with the naked eye and then with a stereo microscope. These provide an aid to selection of the specific microscopy techniques and preparation methods needed to begin to address the objectives. The result should be that clearly defined analyses are conducted.

It is essential to know all the facts relating to the problem to be solved. In the worst case someone simply asks you to take a picture and not ask any questions. This sounds and is very simple. However, then you have a picture, you might even have the right picture, but you do not have a solution to the problem! The time used to take the picture is likely wasted because the problem still must be discovered and solved. It is important to gather all the relevant facts regarding the problem and the specimens that are to be characterized. It is useful to write down the objective of the experiment in order to focus on the problem itself rather than conducting a complete characterization that reveals the structure but provides more information than is needed. The next steps involve consideration of the specimens required to solve the problem. In the case of the multiphase blend described above the correct specimens might be those that have

Table 7.1 Problem solving protocol

(1) Collect all the facts/data about the problem
(2) Clearly define the problem and objectives
(3) Select the specimen to solve the problem
(4) Define needed controls to aid interpretation
(5) Examine specimens with the naked eye
(6) Examine specimens with a stereo-binocular microscope

Table 7.2 Polymer structures

Structure types	Characteristic sizes
Crystal (unit cell)	0.2–2 nm
Chain (sequence length)	2–100 nm
Lamellar crystal thickness	5–50 nm
Fibrils	5–50 nm
Spherulites	1–100 μm
Copolymer and blend phase domains	2 nm–100 μm

been tested and exhibited a range of impact strength properties. Only the tested samples will provide the necessary data and permit comparison with specific property values whereas using the average value from a physical test with a sample taken at random is not nearly as useful. Controls are important in any scientific study; this might involve the assessment of a specimen before and after treatment, or under various conditions.

Specimens should be observed with the naked eye as the size, shape, color and gross morphology are all important to consider when choosing microscopy techniques and preparation methods. A stereo microscope is useful to observe the specimen at low magnification as it often provides an overview of the specimen which can be critical in determining the area to be studied in more detail. One example will show the utility of this low power, inexpensive microscope. A plastic key cap for a computer terminal was brought into the laboratory one day with black specks on the part that required assessment via SEM/EDS. Looking at the key cap by eye it looked like black specks were present. Observation in a stereo microscope showed the specks were on the surface and easily brushed off, saving the time it would have taken to conduct x-ray experiments.

7.1.2 Polymer structures

The first step in the selection of a microscopy technique is to know the size of the polymer structures to be characterized. In fundamental studies the answer might be that all of the structures present must be understood whereas in more routine studies a specific structure must be evaluated as part of the problem solving process. An example of a specific structure that is often evaluated is the spherulite. Spherulite sizes in semicrystalline polymers often determine the properties of the material. Monitoring this structure can be important in structure–property determinations. A listing of the most common polymer structure types, described in Chapter 1, and their characteristic sizes are shown in Table 7.2. Clearly, different microscopy techniques must be used to characterize these different structures.

7.2 INSTRUMENTAL TECHNIQUES

Once the objective of the experiment is known and the specimens selected for study, the next major step is the selection of the microscopy techniques and the specimen preparation methods required to image the polymer structures of interest (Table 7.2). If lamellar crystals must be evaluated, for instance, there is no point in considering most optical techniques as they will only provide an overview of these structures. Comparisons are made in this section regarding the various techniques, in both the text and tables, as an aid in this selection process. Observations of polymer structures are limited by their size, the specimen preparations required and the imaging techniques. If similar structures are observed by several different methods and techniques they are more likely to be representative of the material. In this section the advantages and disadvantages or limitations of the techniques will be compared.

Table 7.3 shows the relation of the polymer structures and their sizes, superimposed on the range of structural sizes observable by the various microscopy and scattering techniques. An important point is the overlap among the various techniques, which makes complementary analyses possible. For example, study of spherulitic structures is shown to be possible by optical, SEM or TEM methods. A flow chart (Fig. 7.1) is provided at the end of this section as an aid in the final selection of a characterization technique.

Table 7.3 Structural characterization

Structure Size	Macroscopic		Spherulitic		Domains		Lamellar	Crystal lattice	
	10 mm	1.0 mm	0.1 mm	10 μm	1 μm	0.1 μm	10 nm	1 nm	0.1 nm
Technique		Eye							
		Stereo-binocular							
			Optical microscope						
				SEM					
					TEM				
				STM / AFM / FFM					
						SAXS			
								WAXS	
								SAED	

7.2.1 Comparison of techniques

Within the ranges of the general techniques shown in Table 7.3 there are many specific techniques that provide useful information relating to polymer characterization. A listing of the more commonly employed microscopy techniques is shown in Table 7.4 with the type of features that are commonly imaged, the typical size range of the structures and magnifications. This table is meant to summarize the application of these techniques, as more detailed information is found in Chapters 2, 3 and 6. Polarized light microscopy is very important in characterizing the spherulitic textures of crystalline polymers. The nature of the orientation in extruded and molded articles and the size and distribution of the spherulites have all been shown to relate to mechanical properties. An underutilized technique is phase contrast optical microscopy which enhances the observation of small differences in refractive index between polymers, permitting the imaging of the dispersed phase domains. Scanning electron microscope techniques permit imaging of surface topography, by SEI, and imaging with atomic number contrast, by BEI

imaging. Comparison of these two imaging modes has been shown in Chapter 5.

Various scattering, or diffraction, techniques also important in polymer characterization are listed in Table 7.5. The major differences among the various scattering techniques are the volume sampled, and the spatial resolution of the techniques. The x-ray diffraction techniques sample regions much larger than 100 μm, generally several millimeters across, although micro-XRD techniques have been used to sample 50–100 μm diameter areas. Selected area electron diffraction techniques sample areas about 1–3 μm in diameter and thus provide crystal lattice information on very small crystallites that are difficult to consider by x-ray scattering techniques. Areas on the order of 50–100 nm are sampled by electron microdiffraction which permits study of phase separated materials and very small local differences in the crystallinity of materials.

The various general microscopy techniques are listed and compared in Tables 7.6 and 7.7. Table 7.6 compares optical, electron and scanning probe microscope techniques, with the magnification, resolution, field of view and imaging

Table 7.4 Microscopy techniques

Type	Features	Size range	Magnification
Optical			
Bright field	Macro-, microstructures, color, homogeneity	1 cm – 0.3 μm	1–1000×
Polarized light	Spherulitic textures	1 cm – 0.5 μm	1–1000×
Phase contrast	Phase variations, refractive index differences	100 μm – 0.2 μm	50–1200×
Electron			
Scanning (SEI)	Surface topography	1 mm – 5 nm	10–50 000×
Scanning (BEI)	Atomic number contrast	1 mm – 20 nm	10–10 000×
Transmission	Internal morphology, lamellar and crystalline structures	10 μm – 0.2 nm	2000–5 × 10^6×
STEM	Internal morphology, lamellar and crystalline structures	100 μm – 1 nm	300–0.3 × 10^6×
Scanning probe			
STM	Surface topography	10 μm – 0.2 nm	2000–5 × 10^6×
AFM	Surface topography of insulators	10 μm – 0.4 nm	2000–1 × 10^6×
FFM	Friction and surface chemistry	10 μm – 1 nm	2000–1 × 10^6×

Table 7.5 Diffraction techniques

Technique	Acronym	Information	Sampled region	Sample thickness
X-ray diffraction	XRD			
Wide angle	WAXS (WAXD)	Crystal lattice Crystal size, Crystal or molecular orientation	1 mm 0.1–10 mm	1 mm
Small angle	SAXS	Lamellar thickness Domain size Fibril size and orientation	300 μm 100–600 μm	1 mm
Electron diffraction				
Selected area	SAED		5 μm	10–200 nm
		as for WAXS above, but local information	2–10 μm	
Micro-diffraction	μdiff		20 nm 5–100 nm	10–200 nm

system listed. Useful magnifications and typical resolutions given are not the values on the knobs of the instrument, or in the instrument brochure, but these are approximate values typical of routine performance that should be considered when choosing a technique to solve a problem. The difficulty with the 'best' resolution values is that often these cannot be obtained when imaging polymers due to specimen preparation and beam damage limitations. Consideration should be given to the size of the field of view of the technique (Table 7.6). One advantage of the light microscope is that large fields may be imaged and much of the specimen is observed rather than just very small areas, as in the TEM. This means that fewer samples are required to ensure that the analysis of the area studied gives information about the whole specimen. Complementary analyses provide important data regarding the uniformity of the structures in the material by analyzing both larger areas, for an overview of the structure, and smaller areas, in greater detail.

A summary and comparison of the performance of electron microscopes is found in Table 7.7. This lists typical specimen thickness, viewing mode, accelerating voltage, image resolution and x-ray analysis spatial resolution by either energy

Table 7.6 Comparison of various microscopies

	Stereo-binocular	Compound	SEM	TEM	AFM
Useful magnifications	5–100×	30–1500×	20–60,000×	3000–250,000×	3000–250,000×
Typical resolution	10 μm	1 μm	10 nm	1 nm	3 nm
Best resolution	2 μm	0.2 μm	4 nm	0.2 nm	0.3 nm
Field of view	Very large 5 mm, 50×	Large 2 mm, 50×	Large 20 μm, 5000×	Small 2 μm, 50,000×	Small 2 μm, 50,000×
Imaging system	Light optical	Light optical	Scanning electron beam	Electron optical	Scanning solid probe
	Glass lenses	Glass lenses	No imaging lenses	Magnetic lenses	No lenses

Table 7.7 Electron microscopy techniques

Instrument	'Regular' SEM	LVSEM	STEM	TEM
Specimen type	Thick	Thick	Thin	Ultrathin
Beam energy (kV)	10–40	1–5	20–100	20–400
Useful magnifications	20–50,000×	20–100,000×	200–200,000×	3000–250,000×
Image resolution	4 nm	3 nm	< 1 nm	0.2 nm
X-ray spatial resolution	1 μm	(0.1 μm; few x-rays produced)	0.1 μm	0.1 μm

or wavelength spectroscopy. The tables provide a broad view of the techniques available in the microscopy laboratory and they should be an aid in selection of the specific techniques required to solve materials problems.

The relatively low cost and wide availability of SEMs and more recently AFMs has led to concentration on these techniques, with less use of the TEM and optical microscopes. SEMs have great advantages; they produce easy to interpret images with excellent depth of field at high resolution. Similarly the AFMs are opening up new areas of surface studies in polymers. However, while the resolution and depth of field in an OM is limited, a wealth of information becomes available when optical microscopy is used. The size of the specimen that can be examined is tens of times the size of an SEM specimen and hundreds of times the size of a TEM specimen. Instrumental cost is much less for optical microscopes, although a research grade polarizing microscope, fully equipped for phase contrast and reflected light, can cost well above the price of a small SEM. The sections that follow consider each of the major microscopy techniques.

7.2.2 Optical techniques

Optical microscopy is extremely useful in providing a rapid view of a relatively large area of the specimen. It should be used as a starting point in most microscopic problem solving, to show the general appearance of the sample, and its structure at low resolution. Specialists solving problems that require the use of electron microscopy or atomic force microscopy regularly use the optical microscope to define the location of the very small area that can be imaged by these high resolution techniques, and to show the relation of the high magnification images to larger scale features. Examples include using polarizing microscopy to show levels of heterogeneity in oriented films and fibers before sectioning for the TEM, reflected light microscopy of molded surfaces before profiling with the AFM, and phase contrast microscopy of multiphase polymers before study of fracture in the SEM. In the examination of spherulites there is less chance of artifacts in the preparation of a thin section for optical study than in the preparation of an ultrathin section for TEM or an etched film for SEM. In a similar manner, dispersed phase particles can be imaged by phase contrast optical microscopy of thin sections, TEM of stained ultrathin sections or fractured bulk samples examined in the SEM. In both examples optical techniques are useful, rapidly providing an overview of the structures in relation to the whole specimen.

A major advantage of optical microscopy techniques is the ease of sample preparation. Thin fibers, films, or membranes can be placed directly in an appropriate immersion oil on a glass slide and information regarding the crystalline or dispersed phases, orientation, birefringence, etc., can be readily determined (Section

4.1.1). Sectioning (Section 4.3.2) of thicker materials is routinely accomplished in very short times, on the order of 30 min or so, with steel or glass knives. Observations and measurements of spherulite sizes, local orientation in molded parts and fiber orientation are also conducted with the optical microscope. Phase contrast and Nomarski techniques provide contrast in multiphase polymers. Small differences in refractive index are enough to make the dispersed phases distinct, so the dispersed phase size and distribution can be measured. Even hard composites, such as glass fiber reinforced thermoplastics, can be thin sectioned by grinding and polishing methods (Section 4.2.2) for reflected light microscopy. Samples for optical study are not placed in a vacuum, as with electron microscopy, and thus volatiles are not removed. Furthermore, beam damage, which is common in electron microscopy of polymers, does not occur in an optical microscope. Finally, there are many optical techniques that permit quantitative measurement of thickness, refractive indices, roughness and orientation. These cannot be conducted in the SEM.

Optical microscopy has limited resolution and a decreasing depth of field with increasing magnification. The resolution limit is on the order of 0.2 μm, although 1 μm is typical for routine analyses. Magnifications used are about 150–1000× with a limit at about 2000×. The poor depth of field is not a major problem if the specimen is flat, but for round specimens such as fibers this is a major problem as little of the specimen is in focus at one time. Confocal laser scanning optical microscopy (Section 6.2.1) removes this problem, but the instrument is not simple or inexpensive. Overall, optical techniques are generally applied to the characterization of polymers as important information can be obtained in short times with minimum capital expenditure and relatively easy sample preparation methods. For many structural studies optical techniques provide a solution to the problem. In more complex problems these techniques provide key information that can lead to other appropriate techniques.

7.2.3 SEM techniques

The advantages of the SEM are well known: images with a three dimensional appearance, great depth of field, ease of operation and ease of specimen preparation. These advantages translate into micrographs that are easier to understand than the micrographs obtained by most optical and TEM techniques. The surface of even rather large samples can be imaged directly in the SEM rather than indirectly, as for TEM of replicas. Images can be formed by combinations of signals (secondary electrons, backscattered electrons and x-rays) and the electronic signal processed to form a variety of micrographs with exceptional detail. Micrographs showing surface topography and chemical contrast are readily obtained in a short time by SEM and they are often easy to interpret. Additionally, benchtop microscopes are relatively low in cost and even the larger SEMs are available at half to one third the cost of a TEM. Research SEMs are also now available that are analytical instruments with combined high resolution imaging and elemental analysis by energy and wavelength dispersive x-ray techniques.

Limitations in SEM imaging include: changes in the specimen caused by the vacuum and the electron beam, lack of internal detail, limited resolution and difficulties in interpreting image details. Volatiles present in a sample are removed in the vacuum, causing changes in the specimen surface as a function of time, often leaving a residue that is unrelated to the original structure. The removal of volatile components in surface finishes on textile fibers led to the need for a replication method (Section 4.6) so the structures could be imaged without the specimen being placed in a vacuum. Alternatively an HPSEM (Section 6.3.3) could be used. Radiation effects cause the most concern in relation to specimen damage and, in an earlier discussion (Sections 2.5 and 3.4), radiation damage was shown to change the structure of some polymers and to affect imaging and resolution. Irreversible radiation effects are often responsible for the formation of structures which are easily misinterpreted. SEM images provide only surface detail, although samples for internal or bulk study at good resolution can be specially prepared, e.g. by fracturing or sectioning. A modern SEM with a field emission gun can be operated at low beam voltage. This limits radiation damage to a thin surface layer, and reduces the need for metal coating of polymer specimens.

Interpretation of SEM images requires assessment of the data in the micrograph in light of both potential artifacts in imaging and specimen preparation and what is known regarding specimen properties. At the risk of being controversial, SEM imaging can be too easy. With reasonable capital expenditure, rapid sample preparation and minimum training almost anyone can be taught to take a picture with the SEM. However, the esthetically pleasing SEM micrographs might have little to do with the structural problem under study. Many experienced and talented microscopists have shown that the SEM is useful for the study of polymers and as an aid to problem solving. It is important to examine specimens for structures that are representative and relate to the problem and then correlate them with the relevant properties of the material. These structure–property applications are a major contribution made possible by SEM studies during the last 30 years.

7.2.4 TEM techniques

Transmission electron microscopy offers excellent resolution, down to the atomic level. It can provide information about molecular orientation and molecular ordering in crystals or liquid crystals, even when the ordered regions are extremely small. Combination of bright field and dark field electron microscopy with electron diffraction permits the identification of the structure of ordered regions and measurement of their orientation, perfection and size. Crystals only a few nanometers across can be detected and identified. In multiphase polymers, the dispersed phase structures can often be imaged and domains observed and quantified over a size range from less than 10 nm up to 1 μm. Attachment of EDS detectors permits the identification of local elemental composition variations at a spatial resolution of about 0.1 μm. Electron energy loss analysis theoretically can provide analysis of light elements at a spatial resolution of about 20 nm. Important details of the fibrillar nature of highly oriented fibers and the transformation upon deformation of spherulites into lamellar structures have been shown by TEM characterization.

The disadvantages associated with obtaining TEM images are high capital expenditure, tedious and time consuming specimen preparation methods, the need for highly trained personnel and 'two dimensional' images which are difficult to interpret. Another serious problem for polymers is that radiation damage is often severe in the TEM, so that for many materials the high resolution structural information is very difficult to obtain. There are two major reasons for the time consuming nature of specimen preparation for TEM: the specimens must be extremely thin, on the order of 50 nm thick, and extra steps are often required to increase the contrast in polymer specimens. Most of the methods for producing ultrathin sections are slow and require major capital acquisitions themselves, such as ultramicrotomes with diamond knives for sectioning. Specimen preparation methods involve replication, staining or etching due to the lack of inherent contrast. TEM imaging requires an understanding of image formation, the effect of the electron beam on the specimen and knowledge of the instrument itself. Image interpretation is difficult due to changes in the specimen caused by radiation damage or exposure to vacuum. Artifacts are often caused by the specimen preparation methods. The observed sample volume is very small, and the image is not intuitively understandable in the way that an SEM image is. These difficulties notwithstanding, fundamental polymer characterization generally involve the application of TEM techniques.

7.2.4.1 *STEM techniques*

Scanning transmission electron microscopy gives essentially the same type of results and has the same type of difficulties as the conventional TEM. There are two types of instruments, the 'dedicated' STEMs, which generally have a UHV column, and the TEM based instruments mostly known as AEMs (analytical electron microscopes). A detailed comparison of STEM and TEM was given in Section 2.4.1.3. There are some advantages in using the STEM on polymer samples; in particular it seems that thicker samples can be used. However, the added complexity and cost, combined with lower resolution in the AEM STEM mode, make it unlikely that either kind of instrument would be purchased for polymer studies.

7.2.5 SPM techniques

Scanning probe microscopy includes scanning tunneling microscopy, atomic force microscopy and frictional force microscopy. At the time of writing it is not yet clear how large an area of application these new techniques will find in polymers, but it is clear that they will be very important. AFM in particular, with its ability to form very high resolution images of the surface of non-conductors, in air or in water, is already producing images superior to those of the SEM, at a lower capital cost for the instrument. The AFM surface images are like the SEM images, easy to understand, and they are directly quantitative, giving the absolute height profiles and roughness of the imaged surface. Specimen preparation is trivial, there is no radiation damage, and image resolution can extend to the atomic level. A major problem at present is that theoretical understanding of the image formation is weak, and this reduces confidence in the results.

As a solid scanning probe is used, the specimen can be affected by the imaging process, but the damage is now mechanical instead of radiation damage. It is comparatively easy to make pits and holes in polymer samples with the SPM, so that care still has to be taken in imaging.

7.2.6 Technique selection

The specific microscopy techniques and appropriate preparation methods required to solve structural problems must be selected now that the techniques and the problem solving protocol (Table 7.1) have been considered. The advantages and limitations of these techniques have already been considered but questions still remain. When should optical techniques be utilized in solving polymer structural problems? How can experiments be conducted so that it is clear if further study is required? Are there any simple 'formulas' for successfully conducting microscopy studies of diverse materials in a manner that is time effective, cost effective and really provides structural answers? The flow chart in Fig. 7.1 is included as an aid in selection of a technique to solve structural problems. Conducting the actual studies is an iterative process where an experimental plan is developed, the studies conducted and further experiments planned as required. Experience has shown that certain techniques are the most likely to provide meaningful answers to certain structural problems. The structure–property applications, in Chapter 5, are examples of the application of both single techniques and complementary techniques to problem solving.

7.3 INTERPRETATION

There are many facets to the interpretation of images as part of structure determination and problem solving. First, it is necessary to know the effect of specimen interactions with the microscope and to understand the image formation process. Next, the effects of specimen preparation must be understood. Many of the methods of preparing specimens suitable for microscopy can deform all or part of the sample, and can produce a wide range of artifacts. For the present discussion artifacts are defined as any features present in a micrograph that do not correspond to structural detail in the original material. The artifacts may be introduced during specimen preparation, by radiation, thermal or mechanical damage in the microscope, or by some imperfec-

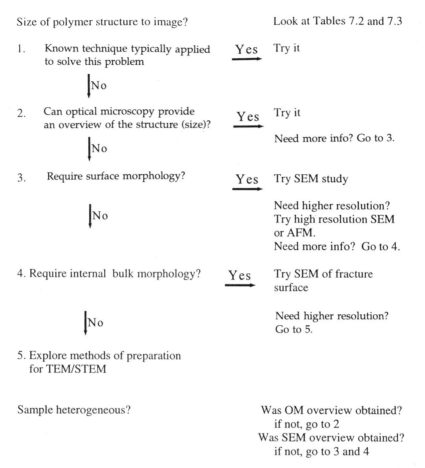

Fig. 7.1 Problem solving flow chart. Questions to consider when selecting a microscopy technique.

tion in the imaging process. In problem solving by microscopy, the only important artifacts are those that are not recognized as such, and are erroneously attributed to the structure of the material. Finally, the nature of the material and its physical and mechanical properties must be considered as part of the image interpretation process. The formation of artifacts will be reviewed as the final consideration in the image interpretation process. For examples of structure determination, and image interpretation, the reader is directed to the examples of structure–property applications at the end of each subsection of Chapter 5.

7.3.1 Artifacts

A major consideration in the selection of preparation methods for microscopy study is the nature of the potential artifacts formed, although time, cost and the capital equipment required are also important factors. In a busy laboratory, time considerations are very important, especially if time consuming methods also have potential artifacts. The accessory equipment available must also be considered, although for this discussion it will be assumed that the laboratory has the equipment required for most general preparations. A complete discussion of specimen preparation methods can be found in Chapter 4. Typical preparations for microscopy will be outlined here with emphasis on the nature of potential artifacts.

7.3.1.1 *Artifacts in optical microscopy*

Typical preparations for optical microscopy include simple preparations, microtomy and polishing. Placement of a thin specimen (fiber, film, etc.), as is or in an immersion oil, is a simple preparation (Section 4.1.1) which is rapid and inexpensive. The major potential artifacts relate to the thickness of the sample or damage by the oil. Samples thicker than 10–50 μm generally do not reveal much structural detail by transmitted light techniques due to overlapping textures which may be misinterpreted or appear as a lack of structure in the specimen. Microtomy (Section 4.3.2) is the most popular method for the preparation of specimens from fibers, films, membranes and engineering resins. Sectioning takes reasonable times, on the order of minutes to several hours per specimen, and microtomes are rather inexpensive. Potential artifacts are the deformation of the specimen, which can cause changes in the shape or the phases within the specimen, or produce stress induced structures which can be interpreted as being present in the polymer. Knife marks are found on the surface of the section although this may be minimized by immersion of the sections in an oil with a refractive index similar to that of the embedding medium. In grinding and polishing (Section 4.2) of tough resins or filled composites, potential artifacts result from the deformation of the specimen, including pullout of the filler fibers or particles, undercutting of the resin, and the addition of a directionality to the specimen structures. Selection of the best polishing cloths and polishing media for a given type of specimen will minimize these effects. Polishing is an art, as is true of many microscopy preparations, and both care and experience are required for artifact-free preparations.

7.3.1.2 *Artifacts in SEM*

SEM preparations are generally direct and rapid, yet there are quite a number of potential artifacts. Paints, glues and tapes used to attach the specimen to the SEM stub can wick up or contaminate the specimen surfaces, adding false structures to the specimen. The preparation of specimens from the bulk often involves deformation of the specimen, e.g. peeling, fracturing and sectioning. Peeling (Section 4.3.1) results in specimen fibrillation, which is a function of the force used, which is not controlled, and the sample history. Fracturing (Section 4.8) at room temperature often causes deformation of soft or rubbery phases whereas fracture in liquid nitrogen causes less deformation but tends to be less reproducible. Hand fracturing is not reproducible whereas standardized testing more commonly provides fracture surfaces whose structures can be related to mechanical properties. Sections (Section 4.3) and sectioned block faces exhibit knife damage, obvious by surface topography imaging. Etching (Section 4.5) is one of the methods with the most potential artifacts. Chemical etching is known to remove polymer material, often redepositing it on the surface and creating new structures which might not have much to do with the structure of the specimen. Ion etching (Section 4.5.2) is known to cause heating and melting and to create structures with directionality if great care is not taken with two water cooled guns at low glancing angles, whereas plasma etching has much less chance of artifact formation. Replication (Section 4.6) is replete with artifacts as much fine detail can be lost and new textures added, resulting in images which are often quite difficult to interpret. The application of conductive coatings (Section 4.7) to polymers can also provide artifacts, such as the formation of grain structures. Finally, irreversible radiation damage and reversible charging effects also can cause the formation of artifacts. The picture is not as gloomy as it may seem and if care is taken important structural information can be obtained rather rapidly. Controls must be used to ensure the validity of the observations and more than one specimen preparation method should be used if etching or replication are applied.

7.3.1.3 *Artifacts in TEM*

Ultrathin films and sections, required for TEM and STEM techniques, are produced by methods such as the formation of single crystals (Section 4.1.3.3), dispersion (Section 4.1.3.2), film casting (Section 4.1.3.5), replication (Section 4.6) and ultramicrotomy (Section 4.3.4). The last two methods are most commonly applied to the study of industrial materials although dispersion and sonication provide thin fragments of the specimen quite rapidly. The disadvantage of sonication is that the location of the fragment in the original specimen is unknown whereas in ultrathin sectioning such information is known although the method is tedious. Often, ultrathin sections have no contrast, and enhancement techniques such as staining (Section 4.4) and shadowing (Section 4.7) are required to image the structures of interest. Controls must be examined as artifacts can result from stain deposits and false enhancement of shadowed structures. Replication is an alternative to sectioning a bulk material, although there are disadvantages in examining a replica when the specimen itself might be examined by a scanning technique at similar resolution (SEI). Additionally, an SEI image is much easier to interpret than a TEM image of a replica. Pretreatment of the specimen for replication can involve etching (Section 4.5) which can also produce artifacts. Special methods (Section 4.9) may be required for the preparation of soft or deformable microemulsion, latex or adhesive materials which require equipment for freeze drying (Section 4.9.2), critical point drying (Section 4.9.3) or freeze fracture (Section 4.9.4). These processes must be controlled and the resulting images carefully interpreted.

Radiation and thermal effects in the TEM and STEM can produce large scale changes in polymer specimens. These changes may be as severe as the disappearance of the specimen due to depolymerization and evaporation. Unless previous studies or experience show that TEM or STEM imaging is required and no other techniques provide the required information, the best approach is to conduct an optical or SEM experiment, prior to the higher resolution technique, in order to evaluate the structure.

7.3.2 Summary

The most important issues in the solution of structural problems are image interpretation and development of structure–property relations. Imaging techniques and preparative methods must be chosen that provide images of the needed structures by the most efficient experiments. Several major principles have been emphasized for the imaging of structures. First, the problem solving protocol (Table 7.1) should be considered prior to developing an experimental plan. As part of this protocol the important properties of the material to be studied should be determined and the overall objective of the study developed. The size of the polymer structures required should be determined (Tables 7.2 and 7.3) as an aid to the selection of the appropriate microscopy techniques. Specimen preparation methods should be selected after considering the nature of the specimen itself, the types of structures to be determined and the potential artifacts. If a specimen can be examined directly, that is preferred over a less direct specimen preparation method, especially etching and replication. A key concept to keep in mind is that specimens are prepared for microscopy using deformation methods and that deformation and the other features of the preparation method, such as metal grain sizes, staining deposits and etchants, must be carefully controlled as part of image interpretation and problem solving. Finally, when more indirect and artifact prone methods are used multidisciplinary methods and techniques should be employed to confirm the nature of the polymer structures.

7.4 SUPPORTING CHARACTERIZATIONS

Microscopy techniques provide important information about the structure of polymers. This information is often necessary in order to develop structure–property relations, but it is rarely sufficient. For example, the interpretation of anisotropic mechanical properties requires assessment of orientation and morphology as they influence these properties [1]. Morphology is described using microscopy techniques whereas the nature of the orientation and the molecular structure must be determined by other analytical techniques. These techniques include x-ray scattering or diffraction (XRD), thermal analysis, electron spectroscopy for chemical analysis (ESCA), nuclear magnetic resonance (NMR), x-ray fluorescence (XRF), small angle light scattering (SALS), small angle neutron scattering (SANS), infrared and Raman spectroscopies and the wide range of chromatographic and wet chemical analyses. In another example, the morphology and properties of block copolymers have been described by Gibson *et al.* [2]. The investigative techniques used include microscopy and some of those described in this section: x-ray diffraction, infrared and thermal analysis. Techniques that are most frequently used to complement microscopy observations will be outlined here in short summary paragraphs which briefly describe the nature of the information available, the principle of the technique and several relevant references. The bulk of the present summary has been compiled from a book, *Modern Methods of Analysis*, produced by the Analytical Research Department, Summit Technical Center of Hoechst Celanese Corp. [3], and my colleagues are duly and gratefully acknowledged for this useful compilation.

7.4.1 X-ray diffraction

X-ray scattering techniques are the most commonly applied complementary discipline to microscopy for structural studies. The type of information that is obtained by x-ray scattering experiments includes phase identification and quantification, crystallinity, crystallite size, lattice constants, molecular orientation and structure, molecular packing and order and amorphous structure [4, 5]. Diffraction techniques that will be described include powder diffraction, wide angle x-ray scattering (WAXS) and fiber diffraction. Small angle x-ray scattering (SAXS) will be described in Section 7.4.4.

Crystalline materials can be identified by rapid computerized powder diffraction techniques. The principle of this technique [6,7] is that the crystallites within a sample, placed in a collimated x-ray beam, reflect x-rays at specific angles and intensities. The diffraction pattern can be recorded photographically, using a camera, e.g. a Debye–Scherrer camera, or using a powder diffractometer. Chemical analysis depends on the fact that each chemical composition and crystallographic structure produces a unique angular distribution of diffracted intensity. Analysis is based on comparison of the diffractometer scan with known standards. Typical applications of the powder diffraction technique to polymers would be the identification of mineral fillers in engineering resins, the nature of crystalline contaminants and determination of crystalline phases in a material.

Wide angle x-ray scattering [4, 5, 8] is used to identify the nature of crystalline phases on an atomic scale, the degree of crystallinity, the size of crystallites and the degree of perfection and orientation of crystallites. Crystalline materials give rise to sharp diffraction rings or peaks whereas amorphous materials produce broad diffuse scattering of x-rays. The crystallinity of a material is a function of the amount of sharp to diffuse diffracted intensity. Analysis of the width of reflections along any crystallographic axis provides data about crystallite sizes and the degree of disorder within the crystal in that direction. Smaller crystallites result in broader reflections than larger crystallites.

The degree of orientation of crystallites can be computed from the arc lengths, or the angular spread of a chosen reflection. Molded specimens are known to have structural heterogeneities, including layers that can be seen visually. WAXS techniques can provide information about the degree and direction of molecular orientation within these different layers as long as the layer, or section of the specimen, is smaller than the x-ray beam. Polymer fiber bundles, aligned in the x-ray beam, provide patterns which can be analyzed for the degree of alignment of the molecules along the fiber axis [5, 9]. The uniaxial nature of fibers makes the WAXS experiment straightforward and useful in polymer structure determination. This technique has been used to determine the crystalline content of fibers as a function of process parameters, such as spinning speed.

Recent trends in x-ray diffraction are the increased use of electronic data collection and computer analysis, and *in situ* or dynamic experiments. These include experiments where the x-ray patterns are obtained while the sample is stretched, heated to its melting point, or aligned with an electric field. Synchrotron radiation is several orders of magnitude more intense than regular laboratory x-ray sources, and its use allows real-time study of deformation or fiber spinning, for example. More synchrotron sources usable for polymer science are currently being built, and although they are limited to national facilities, access should improve in the future. The increased power of computer data analysis permits 'whole pattern analysis' where the crystal structure, orientation and crystallinity are simultaneously determined [10]. More detailed numerical analysis has led to interpretation of x-ray patterns in terms of three phases in semi-crystalline polymers instead of the usual two.

The ordering of polymers, determined by x-ray diffraction techniques, has been reported by many investigators. Classical methods are available [11] to determine the crystallinity in terms of the ratio of integrated peak intensity to the integrated intensity of the entire trace, although these methods depend upon good separation of crystalline peaks from the amorphous background. Hindeleh and Johnson [12] developed sophisticated calculation and peak separation techniques for the estimation of crystallinity and crystallite size in polyamide and polyester

fibers. They showed that the apparent crystallite sizes increase as a function of annealing. The x-ray calculations are indirect compared with direct high resolution electron microscopy observations of true mean crystallite sizes. X-ray diffraction provides peaks which are evaluated for peak broadening, with inherent complications due to the effects that crystallite size distribution and lattice distortion have on the calculation of the average, apparent crystallite size [13]. Electron microscopy can provide lattice fringe images or replicas which are used to determine the true crystallite size distribution if unaffected by specimen preparation [14, 15]. The structure of newly developed liquid crystalline copolyester fibers has also been determined by x-ray techniques. Variations in the positions and intensities of the peaks have been used to study the randomness of the polymers [16, 17]. Polymer structure determination using electron diffraction has been described in Section 6.4.2. Although it is less precise than XRD, valuable data are available from much smaller crystals than by XRD. X-ray diffraction is clearly the better technique for routine structural investigations on regions of 100 μm diameter or greater. Electron diffraction and imaging is performed on samples several orders of magnitude smaller.

An example of the use of complementary techniques is a paper by Frye *et al.* [18]. These authors compared the structural investigation of ultrahigh modulus linear polyethylene by electron microscopy, WAXS and combined nitric acid etching and low frequency Raman spectroscopy. Calculations of the integral breadth of the (002) reflection and dark field TEM gave the same crystal lengths. Additionally, nitric acid etching followed by Raman spectroscopy suggested that there is a broad crystal length distribution for high draw ratio, also consistent with the WAXS and TEM data. These techniques all show an increase in the crystal length in linear PE with increased draw ratio. Hall [19] has several chapters on the application of x-ray and neutron scattering techniques to the study of polymers. This includes chapters on the determination of the structures of aromatic polyesters by WAXS, computer analysis of diffraction patterns and the ability of SAXS to distinguish the morphology of crystalline polymers.

7.4.2 Thermal analysis

There are a variety of thermal techniques which are applied to the understanding of polymer materials. These include differential thermal analysis (DTA), differential scanning calorimetry (DSC), thermomechanical analysis (TMA), thermogravimetric analysis (TGA) and dynamic mechanical analysis (DMA) [20–24]. DTA techniques permit study of the thermal behavior of materials as they undergo transformations as a function of temperature. This permits evaluation of melting points, crystallinity, purity, heat of fusion, specific heat, reaction kinetics, etc. The principle of the technique is that a sample and a reference material are heated while both are monitored by thermocouples. The output of the instrument is the difference between the two thermocouple voltages. When there are no thermal transformations this output voltage is zero. If the output is positive there is an exothermic reaction whereas a negative voltage shows an endothermic reaction. DTA thermograms, plots of this output as a function of the reference temperature, provide data regarding glass transition, crystallization and melting parameters.

Differential scanning calorimetry is another thermal technique similar to DTA in the type of information available, although the experiment is more reproducible due to the nature of the instrument. Typically, a small sample and a reference material are heated at a constant rate and the power consumption or heat flow is measured as a function of temperature or time. The difference between the heat required by the sample and the reference is a direct measure of the thermal properties of the sample. The DSC thermogram is a plot of the differential heat flow versus the temperature or time. Integration of peaks gives the enthalpy change of the specimen. Again, glass transitions, crystallization and melt-

ing points are determined by this important and useful technique. In the case of liquid crystalline compounds the multiple thermal transitions often require complementary hot stage microscopy and x-ray diffraction techniques to identify the phases. One application of the technique is in the determination of the effect of the composition of copolymers on the glass transition temperature as an indicator of the degree of phase separation [2].

Thermomechanical analysis permits measurement of the dimensional changes of materials as a function of heating or cooling. TMA measurements include expansion and contraction, degree of crosslinking, glass transitions, crystallization temperatures, etc. A movable core differential transformer in the TMA, in combination with specific probes, permits the various measurements to be made as a function of temperature. Thermogravimetric analysis is a technique that provides a measure of the weight change of a material as a function of temperature. TGA measurements include thermal stability of polymers, determination of volatiles, additives, or solvents, decomposition temperatures and kinetics. TGA operates on a null balancing principle with a sensitive balance maintaining a reference position for comparison with the weight of the sample. A current flow is produced to balance variations in weight between the reference and the sample and this current is proportional to the change in sample weight. The relative thermal stability of polymers is quite important in end use properties.

Dynamic mechanical analysis techniques permit measurement of the ability of materials to store and dissipate mechanical energy during deformation. DMA is used to determine the modulus, glass transition, mechanical damping and impact resistance, etc., of thermoplastics, thermosets, elastomers and other polymer materials. Information regarding the phase separation of polymers is also available by DMA [2]. In DMA, viscoelastic materials are deformed in a sinusoidal, low strain displacement and their responses are measured. Elastic modulus and energy dissipation are the measured properties.

Stress-strain relationships are determined by DMA and temperature scans reveal glass transitions, crystallization and melting information. Blends of polypropylene and rubber have been studied by DSC where the intensity of one of the two crystallization exotherms was used as a measure of the polypropylene domains and compared to the size determined by TEM cryomicrotomy and osmium tetroxide staining methods [25]. Isothermal annealing of PET above the crystallization temperature was shown to influence the morphology and increase thermal stability by combined SAXS and DSC analysis [26]. An excellent text edited by Turi [21] described the instrumentation and theory of thermal analysis and its application to thermoplastics, copolymers, thermosets, elastomers, additives and fibers.

7.4.3 Spectroscopy

7.4.3.1 *X-ray fluorescence*

X-ray fluorescence spectroscopy (XRF) is a technique for the determination of elemental composition of materials, for elements greater than atomic number 11, present above 0.05% concentration [27–29]. The technique is similar to the EPMA, and x-ray analysis in the electron microscope (Section 2.6.1), except that the EPMA is used for local analysis whereas XRF is a bulk technique. Exposure of the sample to an x-ray beam causes electrons to be ejected and outer shell electrons to fall into the vacancies, emitting x-rays of discrete energy. Characteristic energies are associated with specific elements and the x-ray intensities are related to the concentration of the element in the sample. There are problems with this direct association of x-ray intensity and concentration, due to absorption by the matrix but standards and software programs are available to calculate elemental composition. XRF experiments have the advantage of being rapid and nondestructive. These techniques are usefully applied to the assessment of fillers, additives and contaminants in polymers, and software is available for quantitation.

7.4.3.2 *Infrared and Raman*

Infrared and Raman are complementary vibrational spectroscopies and both will be considered here. The absorption versus frequency characteristics of light transmitted through a specimen irradiated with a beam of infrared radiation provide a fingerprint of molecular structure. IR is absorbed when a dipole vibrates naturally at the same frequency in the absorber. The pattern of vibrations is unique for a given molecule and the intensity of absorption is related to the quantity of absorber. Thus, infrared spectroscopy permits the determination of components or groups of atoms which absorb in the infrared at specific frequencies, permitting identification of the molecular structure [30–35]. These techniques are not limited to chemical analysis. With instruments of high spectral resolution, the tacticity, crystallinity and molecular strain can also be measured. Copolymer dispersions can be determined as block copolymers absorb additively and alternating copolymers deviate from this additivity due to interaction of neighboring groups.

The conventional spectrometer with a dispersive prism or grating has been largely superseded by the Fourier transform (FTIR) technique. This uses a moving mirror in an interferometer to produce an optical transform of the infrared signal. Numerical Fourier analysis gives the relation of intensity and frequency, that is, the IR spectrum. FTIR can be used to analyze gases, liquids and solids with minimal preparation in short times. FTIR has been applied to the study of many systems, including adsorption on polymer surfaces, chemical modification and irradiation of polymers and oxidation of rubbers [36]. The application of infrared spectroscopy to the study of polymers has been reviewed by Bower and Maddams [35].

Infrared dichroism is a phenomenon which is used to measure the degree of orientation of the polymer chain. In this case, aligned groups, such as in a stretched polymer film, exhibit absorbance in the infrared which varies depending upon the alignment of the transition moments with respect to the polarization direction of the incident radiation [35]. Gibson *et al.* [2] discussed obtaining infrared dichroism measurements by uniaxially orienting a film sample and determining the absorbance of selected bands with radiation polarized parallel and perpendicular to the stretch direction. A dichroic ratio of these two absorbances is then related to an orientation function.

Laser Raman spectroscopy [30, 33–38] is a light scattering process where a sample is irradiated with a laser and the inelastically scattered light collected and analyzed. Functional groups, e.g. carbon–carbon double and triple bonds and carbon bonded to sulfur and chlorine, scatter incident radiation at characteristic frequency shifts in Raman spectroscopy [37]. The vibrational frequency of the group is the amount of shift from the exciting radiation. Raman spectra depend upon polarizability whereas infrared absorption relates to dipole moment changes so that for most groups either the vibrational band is active in the infrared or it is observed in the Raman. The techniques are complementary for molecular structure determinations.

An important development in Raman spectroscopy has been the coupling of the spectrometer to an optical microscope. This allows the chemical and structural analysis described above to be applied to sample volumes only 1 μm across [38]. No more sample preparation is required than that for optical microscopy, and the microscope itself can be used to locate and record the area which is analyzed. This has obvious practical application to the characterization of small impurities or dispersed phases in polymer samples. This instrument, which may be called the 'micro-Raman spectrometer', the 'Raman microprobe' or the 'Molecular Optics Laser Examiner' [39] has also been applied to the study of mechanical properties in polymer fibers and composites. It can act as a non-invasive strain gauge with 1 μm resolution, and this type of work has recently been reviewed by Meier and Kip [40]. Even if the sample is large and homogeneous, there may be advantages in using the micro-Raman instrument. The microscope

lenses are very efficient at collecting the scattered light, and because only a small volume is in the laser beam, a higher power density can be used without overheating and destroying the sample [40]. The main disadvantage is that the micro-Raman instrument is expensive, at up to $250,000 'fully loaded'. If the laser beam can be scanned, a Raman image can be formed; this is a kind of microscopy, and was described in Section 6.2.1.

7.4.3.3 *Nuclear magnetic resonance*

Nuclear magnetic resonance (NMR) uses the magnetic properties of nuclei, particularly the proton ^{1}H and ^{13}C, to observe structural features in polymer chains. A radio frequency field and a magnetic field are applied to the sample and at the resonance condition energy is absorbed. The exact resonance frequency depends on the local chemical environment of the atom. Measuring the small difference in resonance frequency, the *chemical shift*, and using model compounds allows the exact chemical structure to be determined [41]. Traditional NMR was a chemical technique applied to solutions to determine the chemical structure and tacticity. With Fourier transform and other techniques and today's more powerful magnets NMR can be applied to a wider range of nuclei in solution or in the solid state [42–44]. Chemical information can be obtained not only about the average structure, but also about defects such as copolymer structure [45] chain ends and branch point density [46]. There is also a wide range of structural information available. For example the linewidth of the resonances gives information about the mobility of the chain, or of specific side groups, and thus can give T_g or crystallinity.

7.4.3.4 *X-ray photon spectroscopy*

X-ray photon spectroscopy (XPS) was also known as electron spectroscopy for chemical analysis (ESCA). Both names are descriptive, as the essential feature of the technique is bombardment of a specimen with monochromatic x-ray photons, and the energy spectrum of electrons that are emitted is measured. The binding energy of the core electrons that are emitted and the kinetic energy that they possess sum to the x-ray energy. Peaks in the plots of electron intensity versus binding energy correspond to the core energy levels that are characteristic for a given element. Small shifts in the binding energy are caused by the state of the valence electrons, that is by the local chemical environment. The spectrum thus allows elemental and chemical analysis of the top 2–10 nm of the surface [47–49]. The specimen must be in a high vacuum, but apart from that the method is non-destructive. Other advantages are that the data interpretation is relatively straightforward, and sample preparation is simple. The spectra can sometimes be unclear because the electron spectrometer resolution is not sufficient to prevent peaks with different chemical shifts from overlapping. A data base of many spectra has been built up to aid identifications [50]. XPS has been applied to many polymer problems, particularly for the investigation of surface modification processes [36, 51] and the interaction of polymers and metal layers deposited on them [52]. Most XPS instruments have a spatial resolution of about 1 mm. Higher resolution is possible, but a very high flux of x-rays is needed to keep the signal high, and this may destroy most polymer specimens.

7.4.3.5 *Auger spectroscopy*

Auger spectroscopy [48, 49, 53] is a surface elemental analysis technique similar to XPS although with higher spatial resolution and lower detection limits. Auger analysis is difficult to apply to polymers due to the severe sample degradation which occurs during the analysis. Scanning Auger microscopy (SAM) is a method for the elemental mapping of a surface with a rastered electron beam. It gives 5 μm spatial resolution with depth profiling possible by ion etching. An energetic beam of electrons strikes the atoms of the material in a vacuum environment and electrons with binding energies less than the incident beam energy may be ejected

from the inner atomic levels, creating a singly ionized excited atom. This inner level vacancy is filled by de-excitation of electrons from other electron energy states. The energy released can be emitted as an x-ray (fluorescence) or transferred to an electron in any atom. If this latter electron has lower binding energy than the energy from the de-excitation then it will be ejected, with its energy related to the energy level separations in the atoms. Auger electrons are the result of de-excitation processes of these vacancies and electrons from other shells and a re-emission of an electron to carry away excess energy. The electrons emitted have a short mean free path, and thus all Auger electrons are from the first few atomic surface layers. The kinetic energy of the freed electron is detected and these energies reflect the variations in binding energies of the levels involved in the process. The spatial resolution of Auger spectroscopy is about 30–50 nm, similar to energy dispersive spectroscopy of thin sections. Auger analysis is used in the study of adhesion on metal surfaces, in adsorption, corrosion and oxidation studies.

7.4.4 Small angle scattering

Small angle scattering is a technique for the determination of morphological structures on a scale larger than that of the wavelength of the radiation used. Light, x-rays and neutrons are used for small angle scattering, and the experimental details of these techniques are very different. They share the property of forming good numerical averages of feature size over a comparatively large volume of sample, without giving local details of the morphology. They are therefore complementary to microscopy, which can give the local details, but usually with a restricted field of view.

7.4.4.1 *Small angle light scattering*

Small angle light scattering (SALS) uses light of wavelength 0.5 μm, so can investigate structures in the range 5–100 μm. Spherulites are structures of semicrystalline polymers which are in this size range and are studied by SALS techniques. In SALS a monochromatic, collimated and plane polarized laser beam passes through a thin polymer film. The scattered radiation is analyzed with a second polarizer, aligned with the first polarizer, and the scattering pattern is recorded on photographic film or by electronic detectors. The scattering of visible light is related to variations in the anisotropy and refractive index or polarizability of the specimen and this polarizability is affected by the molecular structure [54]. Thus, light scattering techniques provide information about molecular structure and orientation. Spherulites in crystalline polymers have been found to be anisotropic scatterers and theoretical scattering patterns have been calculated [55]. Typically, spherulitic structures are characterized by complementary SALS and polarized light microscopy techniques [54–56] where the scattering angle in the SALS pattern is used to determine the size of the spherulite. Stein *et al.* [54] have reviewed the theory and applications of light scattering techniques, including a comparison with x-ray and neutron scattering.

7.4.4.2 *Small angle x-ray scattering*

Small angle x-ray scattering experiments are used to analyze the macrostructure of materials on a scale of about 1–200 nm [57–60]. The SAXS technique provides information regarding the electron density distribution of the material and analysis of the angular distribution of the peak intensities (if there are peaks) reveals the periodicity and magnitude of that electron distribution. Thus, average morphological information is provided which is useful for the determination of the nature of voids or crystalline regions. Periodic structures, such as crystalline lamellae in polymers, produce small angle diffraction peaks whose measurements reflect the lamellar periodicity (i.e. thickness plus spacing). The principle of the technique is that x-rays are scattered by regions of varied electron density, such as voids or local crystalline regions and the intensity is related to the number of such regions and their contrast. SAXS is often used to define

the size and shape of voids or fibrils in fibers and to measure the lamellar spacing in crystalline polymers. Complementary studies involving SAXS and microscopy are common in polymers. They include studies of lamellar structures in polyethylene [61] and of fibrillar structures in rigid rod polymer fibers [62, 63]. Detailed analysis of the SAXS streak from high modulus fibers has been used to determine the length and orientation distribution of the fibrils [64]. To cover a wide range of feature size during phase segregation of polyethylene blends, SAXS was combined with SALS [65].

7.4.4.3 *Small angle neutron scattering*

Neutrons are uncharged particles which may interact with a specimen by nuclear interactions resulting in a transfer of energy and momentum between the neutron and target materials [66]. Neutron scattering from hydrogen is very different to scattering from deuterium, so chemically similar deuterated molecules can be distinguished in a matrix of normal hydrogenated polymer. This allows the shape of individual molecules to be derived [60, 66, 67]. An early triumph of small angle neutron scattering (SANS) was to show that the molecules in an amorphous polymer have a random coil shape as predicted by Flory. Since then there have been many studies of copolymers [68–70], blends [71–73], semi-crystalline polymers [74, 75] and drawn fibers [76]. The emphasis in these studies has been the shape of individual molecules in these different circumstances, their diffusion and segregation to surfaces. Comparison of the radius of polybutadiene in diblock copolymers showed that the TEM values were significantly smaller than those obtained by SANS [69]. Domain boundary thicknesses, domain sizes and domain packing order have also been determined by SANS measurements of solvent cast diblock copolymers and blends [70].

7.4.5 Summary

As was described in the earlier part of this text (Section 1.4.1), a very wide range of analytical techniques is used to determine the structure of polymeric materials. The chemical regularity, stereochemical configuration and molecular weight distribution make up the basic molecular structure. These are unaffected by physical processing but they define the starting material for such processing. The techniques used to determine this level of structure are essentially chemical ones.

The physical structure can be affected by processing and involves many variables, for example the molecular orientation and the distribution of dispersed phases in multiphase systems. If crystals are present, other important variables are the crystal structure, the degree of crystallinity, the crystal sizes and their arrangement into spherulites or other structures. This type of structure is determined by microscopy, and by a range of scattering and spectroscopic techniques, as has been discussed in this chapter. Second phase distribution is commonly determined by electron microscopy and by x-ray or neutron scattering. The degree of crystallinity may be obtained from the density of the sample, from wide angle x-ray diffraction, from thermal analysis and from NMR. In some cases the two phase model of amorphous and crystalline material is insufficient, and the quantity of interfacial material must be considered. The interfacial material is associated with the fold surface of the lamellae, and the quantity of such material in PE is determined by Raman, WAXD and NMR. Lamellar thickness is determined by small angle x-ray scattering, and in some materials by Raman (low frequency longitudinal acoustic mode) and by TEM. If all are available, the combined results give a much clearer picture than is available from a single technique as each method has its advantages, but to do all of these well is time consuming.

To some polymer scientists the word 'morphology' means the assembly and relative arrangement of crystals or of second phase particles, while others use 'supermolecular structure' to describe the same thing. Whatever the name, optical and electron microscopy and the complementary techniques of light and x-ray scattering are used to determine such structure. Common arrangements in crystalline polymers solidified from the melt are spherulites, row structures, stacks or bundles of lamellae, rods and fibrils. Poorly ordered materials may have randomly placed single lamellae. In general, microscopy is used to define the type of arrangement, and a scattering technique, which samples a much larger volume, is used to measure average dimensions of the various structures. Without the microscopy, a model must be assumed to interpret the scattering data; without the scattering, a great deal of quantitative microscopy must be done to ensure statistical sampling of the specimen structure. Thus, although our subject here is microscopy, it is apparent that at every point complementary techniques are vital for problem solving.

REFERENCES

1. I. M. Ward, *Mechanical Properties of Solid Polymers* (Wiley, New York, 1983).
2. P. E. Gibson, M. A. Vallance and S. L. Cooper, in *Dev. Block Copolym.* **1** (1982) 217.
3. D. G. Vickroy, Ed., *Modern Methods of Analysis* (Analytical Research Dept., Hoechst Celanese, Summit Technical Center, Summit, 1983).
4. H. P. Klug and L. E. Alexander, *X-ray Diffraction Procedures* (Wiley, New York, 1974).
5. L. E. Alexander, *X-ray Diffraction Methods in Polymer Science* (Krieger, Huntington, NY, 1979).
6. R. L. Snyder, in *Materials Science and Technology*, Vol. 2A, edited by E. Lifshin (VCH Publishers, Weinheim, 1993).
7. B. D. Cullity, *Elements of X-ray Diffraction* (Addison Wesley, Reading, MA, 1959).
8. D. L. Bish and J. E. Post, Eds, *Reviews in Mineralogy, Vol. 20, Modern Powder Diffraction* (Mineralogical Soc. of America, Washington, DC, 1989).
9. H. Tadokoro, *Structure of Crystalline Polymers* (Wiley–Interscience, New York, 1979).
10. W. R. Busing, *Macromolecules* **23** (1990) 4608.
11. P. H. Hermans and A. Weidinger, *Text. Res. J.* **31** (1961) 551.
12. A. M. Hindeleh and D. J. Johnson, *Polymer* **19** (1978) 27.

13. I. H. Hall and R. Somashekar, *J. Appl. Cryst.* **24** (1991) 1051.

14. A. M. Hindeleh and D. J. Johnson, *Polymer* **21** (1980) 929.

15. D. T. Grubb and D. Y. Yoon, *Polymer Commun.* **27** (1986) 84.

16. J. Blackwell and G. Gutierrez, *Polymer* **23** (1982) 671.

17. J. B. Stamatoff, *Mol. Cryst. Liq. Cryst.* **110** (1984) 75.

18. C. J. Frye, I. M. Ward, M. G. Dobb and D. J. Johnson, *J. Polym. Sci., Polym. Phys. Edn.* **20** (1982) 1677.

19. I. H. Hall, Ed., *Structure of Crystalline Polymers* (Elsevier–Applied Science, London, 1984).

20. W. W. Wendlandt, *Thermal Methods of Analysis* (Wiley–Interscience, New York, 1974).

21. E. Turi, Ed., *Thermal Characterization of Polymeric Materials* (Academic Press, New York, 1981).

22. M. E. Brown, *Introduction to Thermal Analysis* (Chapman and Hall, New York, 1988).

23. B. Wunderlich, *Thermal Analysis* (Academic Press, New York, 1990).

24. P. K. Gallagher, in *Materials Science and Technology*, Vol. 2A, edited by E. Lifshin (VCH Publishers, Weinheim, 1993).

25. A. Ghijels, N. Groesbeck and C. W. Yip, *Polymer* **23** (1982) 1913.

26. G. Groeninckx and H. Reynaers, *J. Polym. Sci., Polym. Phys. Edn.* **18** (1980) 1325.

27. E. Bertin, *Principles and Practice of X-ray Spectrometric Analysis* (Plenum, New York, 1970).

28. R. Jenkins, *X-ray Fluorescence Spectrometry* (Wiley–Interscience, New York, 1988).

29. R. Jenkins, in *Materials Science and Technology*, Vol. 2A, edited by E. Lifshin (VCH Publishers, Weinheim, 1993).

30. D. J. Cutler, P. J. Hendra and G. Fraser, in *Developments in Polymer Characterization*, edited by J. V. Dawkins (Applied Science, London, 1980) p. 71.

31. R. Zbinden, *Infrared Spectroscopy of High Polymers* (Academic Press, New York, 1964).

32. P. R. Griffiths, *Chemical Infrared Fourier Transform Spectroscopy*, Vol. 43 (Wiley, New York, 1975).

33. H. W. Siesler and K. Holland-Moritz, *Infrared and Raman Spectroscopy of Polymers* (Marcel Dekker, New York, 1980).

34. P. C. Painter, M. M. Coleman and J. L. Koenig, *The Theory of Vibrational Spectroscopy and its Application to Polymers* (John Wiley, New York, 1982).

35. D. I. Bower and W. F. Maddams, *The Vibrational Spectroscopy of Polymers* (Cambridge University Press, Cambridge, 1989).

36. L. H. Lee, Ed., *Characterization of Metal and Polymer Surfaces, Vol. 2, Polymer Surfaces*, edited by L. H. Lee (Academic Press, New York, 1977).

37. J. G. Grasselli, M. K. Snavely and B. J. Bulkin, *Chemical Applications of Raman Spectroscopy* (Wiley, New York, 1981).

38. J. G. Grasselli and B. J. Bulkin, *Analytical Raman Spectroscopy* (John Wiley, New York, 1991).

39. F. Adar and H. Noether, *Polymer* **26** (1985) 1935.

40. R. J. Meier and B. J. Kip, *Microbeam Analysis* **3** (1994) 61.

41. E. D. Becker, *High Resolution NMR – Theory and Chemical Applications* (Academic Press, New York, 1980).

42. E. A. Williams, in *Materials Science and Technology*, Vol. 2A, edited by E. Lifshin (VCH Publishers, Weinheim, 1993).

43. C. A. Fyfe, *Solid State NMR for Chemists* (CFC Press, Guelph, 1983).

44. R. A. Komoroski, Ed., *High Resolution NMR Spectroscopy of Synthetic Polymers in Bulk* (VCH Publishers, Deerfield Beach, 1986).

45. J. R. Ebdon, in *Developments in Polymer Characterization – 2*, edited by J. V. Dawkins (Applied Science, London, 1980) p. 1.

46. A. V. Cunliffe, in *Developments in Polymer Characterization – 1*, edited by J. V. Dawkins (Applied Science, London, 1978) p. 1.

47. V. D. Nefedov, *X-ray Photoelectron Spectroscopy of Solid Surfaces* (VSP, Utrecht, 1988).

48. D. Briggs and M. P. Seah, Eds, *Practical Surface Analysis by Auger and X-ray Photoelectron Spectroscopy* (Wiley, New York, 1990).

49. D. L. Allara and P. Zheng, in *Materials Science and Technology*, Vol. 2B, edited by E. Lifshin (VCH Publishers, Weinheim, 1993).

50. NIST, *X-ray Photoelectron Database, NIST Standard Reference Database 20* (NIST, Gaithersburg, 1989).

51. N. H. Turner and J. A. Schreifels, *Anal. Chem.* **64** (1992) 302R.

52. S. G. Anderson, J. Leu, B. D. Silverman and P. S. Ho, *J. Vac. Sci. Technol.* **A11** (1993) 368.

53. D. E. Ramaker, *Crit. Rev. Solid State Mater. Sci.* **17** (1991) 211.

54. R. S. Stein and G. P. Hadzhoannou, in *Polymer Characterization*, ACS Adv. in Chem. Series 203, edited by C. C. Craver (American Chemical Society, Washington, DC, 1983) p. 721.

55. R. J. Samuels, *J. Polym. Sci.* **9** (1971) 2165.

56. G. E. Wissler and B. Crist, *J. Polym. Sci., Polym. Phys. Edn.* **23** (1985) 2395.

57. A. Guinier and G. Fournet, *Small-angle Scattering of X-rays* (Wiley, New York, 1955).

58. O. Glatter and O. Kratky, Eds, *Small Angle X-ray Scattering* (Academic Press, London, 1982).

59. F. J. Balta-Calleja and C. G. Vonk, *X-ray Scattering by Synthetic Polymers* (Elsevier, Amsterdam, 1989).

60. C. Williams and R. P. May, in *Materials Science and Technology*, Vol. 2B, edited by E. Lifshin (VCH Publishers, Weinheim, 1993).

61. G. Blochl and A. J. Owen, *Coll. Polymer Sci.* **262** (1984) 793.

62. Y. Cohen and E. L. Thomas, *Macromolecules* **21** (1988) 433.

63. Y. Cohen and E. L. Thomas, *Macromolecules* **21** (1988) 436.

64. D. T. Grubb and K. Prasad, *Macromolecules* **25** (1992) 25.

65. K. Tashiro, M. M. Satkowski, R. S. Stein, Y. Li, B. Chu and S. L. Hsu, *Macromolecules* **25** (1992) 1809.

66. R. W. Richards, in *Developments in Polymer Characterization – 1*, edited by J. V. Dawkins (Applied Science, London, 1978) p. 117.

67. J. S. Higgins and H. Benoit, *Neutron Scattering of Polymers* (Oxford University Press, Oxford, 1993).

68. G. D. Wignall, H. R. Child, F. S. Bates, R. S. Cohen, C. Berney and R. J. Samuels, Proc. IUPAC 28th Macromol. Symp. (Oxford, 1982) p. 654.

69. C. V. Berney, R. E. Cohen and F. S. Bates, *Polymer* **23** (1982) 1222.

70. F. S. Bates, C. V. Berney and R. E. Cohen, *Macromolecules* **16** (1982) 1101.

71. W. Wu and G. D. Wignall, *Polymer* **26** (1985) 661.

72. B. J. Bauer, R. M. Briber and C. C. Han, *Macromolecules* **22** (1989) 940.

73. E. J. Kramer and H. Sillescu, *Macromolecules* **22** (1989) 414.

74. B. Crist, J. D. Tanzer and T. M. Finerman, *J. Polym. Sci., Part B: Polym. Phys.* **27** (1989) 875.

75. D. M. Sadler, in *Structure of Crystalline Polymers*, edited by I. H. Hall (Elsevier–Applied Science, London, 1984) p. 125.

76. D. M. Sadler and P. J. Barham, *Polymer* **31** (1990) 36, 43 and 46.

Appendices

APPENDIX I. ABBREVIATIONS OF POLYMER NAMES

Acrylonitrile–butadiene–styrene	ABS	Polyimide	PI
Acrylonitrile–butadiene rubber	ABR	Poly(methyl methacrylate)	PMMA
Acrylic–styrene–acrylonitrile	ASA	Polyoxymethylene	POM
Cellulose acetate	CA	Polypropylene	PP
Cellulose nitrate	CN	Poly(p-phenylene benzobisthiazole)	PBZT
Chlorinated polyethylene	CPE	Poly(p-phenylene benzobisoxazole)	PBZO
Diethyl triamine	DETA		(or PBO)
Ethylene–propylene–diene monomer rubber	EPDM	Poly(p-phenylene terephthalamide)	PPTA
		Poly(phenylene oxide)	PPO
Ethylene–vinyl acetate	EVA	Poly(phenylene sulfide)	PPS
High density polyethylene	HDPE	Polystyrene	PS
High modulus polyethylene	HMPE	Polysulfone	PSO
High impact polystyrene	HIPS	Polytetrafluoroethylene	PTFE
Hydroxy propyl cellulose	HPC	Polyurethane	PUR
Low density polyethylene	LDPE	Poly(vinyl acetate)	PVAC
Polyacetal (see Polyoxymethylene below)		Poly(vinyl alcohol)	PVA
Polyacrylonitrile	PAN	Poly(vinyl chloride)	PVC
Polyamide (nylon)	PA	Poly(vinylidene chloride)	PVDC
Polybutadiene	PB	Poly(vinylidene fluoride)	PVDF
Poly(butylene terephthalate)	PBT	Poly(p-xyxylene)	PPX
Polycarbonate	PC	Resorcinol–formaldehyde–latex	RFL
Polyethylene	PE	Styrene–acrylonitrile copolymer	SAN
Poly(ethylene oxide)	PEO	Styrene–butadiene rubber	SBR
Poly(ether ether ketone)	PEEK	Styrene–butadiene–styrene	SBS
Polyetherimide	PEI		
Polyethersulfone	PES		
Poly(ethylene terephthalate)	PET		

APPENDIX II. LIST OF ACRONYMS – TECHNIQUES

Analytical electron microscope	AEM
Atomic force microscopy	AFM
Backscattered electron imaging	BEI
Confocal scanning laser microscopy (or laser confocal scanning)	CSLM (or LCSM)
Confocal scanning optical microscopy	CSOM
Differential interference contrast	DIC
Energy dispersive x-ray spectroscopy	EDS
Field emission scanning electron microscopy	FESEM
Frictional force microscope	FFM
High pressure scanning electron microscopy	HPSEM
High resolution scanning electron microscopy	HRSEM
High resolution transmission electron microscopy	HRTEM (or HREM)
Infrared spectroscopy	IR
Lateral force microscope	LFM
Magnetic force microscope	MFM
Microdiffraction	μDiff
Near field optical microscope	NFOM
Nuclear magnetic resonance spectroscopy	NMR
Optical microscopy	OM
Phase contrast microscopy	PC
Polarized light microscopy	PLM
Scanning electron microscopy	SEM
Scanning ion conductance microscope	SICM
Scanning probe microscopy	SPM
Scanning thermal profiler	STP
Scanning transmission electron microscopy	STEM
Scanning tunneling microscopy	STM
Scanning tunneling spectroscopy	STS
Secondary electron imaging	SEI
Selected area electron diffraction	SAED
Small angle neutron scattering	SANS
Small angle x-ray scattering	SAXS

Transmission electron microscopy	TEM
Conventional transmission electron microsopy	CTEM
Wavelength dispersive x-ray spectroscopy	WDS
Wide angle x-ray scattering	WAXS

APPENDIX III. MANMADE POLYMERIC FIBERS

Fiber type	Generic name	Trademark	Manufacturer
Cellulosic	Acetate		
	Rayon	Coloray	Courtaulds
	Triacetate	Arnel	Hoechst Celanese
Non-cellulosic	Acrylic	Acrilan	Monsanto
		Creslan	Am. Cyanamid
		Orlon	duPont
	Aramid	Kevlar	duPont
	Copolyester	Vectran	Hoechst Celanese
	Fluorocarbon	Teflon	duPont
	Modacrylic	Sep	Monsanto
	Nylon	Antron	duPont
		Ultron	Monsanto
	Polybenzimidazole		Hoechst Celanese
	Polyester	Dacron	duPont
		Fortrel	Hoechst Celanese
	Polyethylene	Spectra	Allied
	Polypropylene	Herculon	Hercules
	Spandex	Lycra	duPont

APPENDIX IV. COMMON COMMERCIAL POLYMERS AND TRADENAMES FOR PLASTICS, FILMS AND ENGINEERING RESINS

Generic name	Tradename	Manufacturer	Typical end uses
Acrylonitrile–butadiene– styrene (ABS)	Absom	Mobay	Automotive, appliance housings, furniture, construction
Epoxy			Paints, coatings, adhesives, pipes, circuit boards
High impact polystyrene (HIPS)			Automotive, appliance housings, furniture
High density PE (HDPE)			Containers, pipes, fabricated parts
Low density PE (LDPE)			Packaging, films for bags, stretch wrap
Nylon polymer and resin	Vydyne Zytel	Monsanto duPont	Carpet yarns, tirecords, cigarette lighters, sporting goods, brushes
Polybutadiene in copolymers and blends			Tires, rubber articles, encapsulation
Poly(butylene terephthalate) (thermoplastic polyester) (PBT)	Celanex Valox	Hoechst Celanese General Electric	Automotive and other fabricated parts, bearings, housings
Polycarbonate	Lexan	General Electric	Bottles, safety glass, auto lenses, helmets, aircraft interiors
Poly(ether ether ketone) (PEEK)	Victrex	ICI AM.	Cable insulation, coatings, composites
Polyetherimide (PEI)	Ultem	General Electric	Aerospace seats, lights, wiring, films/ tapes
Poly(ethylene terephthalate) (PET)	Mylar	duPont	Films for packaging, coatings
PET engineering resins	Petlon Rynite	Mobay duPont	Extrudates and moldings, bottles, recording tapes, electrical insulation
Polyimide (PI)	Tulon Kapton Skyboard	Amoco duPont Monsanto	Printed circuit boards, insulation/films for motors, adhesives, electronics
Poly(methyl methacrylate) (PMMA)	Lucite Plexiglas	duPont Rohm and Haas	Camera lenses, airplane windows, signs Molded parts, sheeting
Polyoxymethylene (POM)	Celcon Delrin	Hoechst Celanese duPont	Automotive, plumbing, appliances, electrical gears, zippers
Poly(phenylene oxide) (PPO) and PPO–HIPS blends	Noryl	General Electric	Appliances, housings, pumps, shields
Polypropylene (PP)			Carpet backing, ribbons, appliance housings
Polystyrene (PS)			Disposables: cutlery, cups, foam egg cartons
Polysulfone (PSO)	Udel	Union Carbide	Camera bodies, electrical connectors, light sockets, food appliance coatings, cookware

APPENDIX IV—*continued*

Generic name	Tradename	Manufacturer	Typical end uses
Polytetra-fluoroethylene (PTFE)	Teflon Halon	duPont Allied	Solvent resistant coatings, films and parts
	Tefzel	duPont	Pipe fittings, seals, laboratory ware, aircraft parts
Poly(vinylidene fluoride) (PVDF)	Kynar	Pennwalt	Pipe fittings, seals, laboratory ware, aircraft parts
Poly(vinyl acetate) (PVAC)			Paints, adhesives, coatings
Poly(vinyl alcohol) (PVA)	Elvanol	duPont	Coatings, adhesives, cosmetics
Poly(vinyl chloride) (PVC)			Food wrap, furniture covers, flooring, footwear
Saturated styrene–butadiene–styrene block copolymers	Kraton G	Shell Oil	Fabricated parts
Styrene–acrylonitrile (SAN)	Rovel	Uniroyal	Dentures, lenses, auto and other fabricated parts
Styrene–butadiene latex			Adhesives, coatings, binders, textile finishes
Thermotropic aromatic copolyesters	Vectra	Hoechst Celanese	Fabricated parts, interconnects, connectors
i.e. poly(benzoate–naphthoate) and poly(naphthoate–aminophenol terephthalate)	Vectran	Hoechst Celanese	Fibers in sporting goods, sailcloth, bow strings, Films for printed circuit boards

APPENDIX V. GENERAL SUPPLIERS OF MICROSCOPY ACCESSORIES

Anatech Ltd
5510 Vine Street
Alexandria, VA 22310

Bal-Tec Products, Inc.
984 Southford Road
Middlebury, CT 06762

Barry Scientific, Inc.
P.O. Box 173
Fiskdale, MA 01518

Cooke Vacuum Products, Inc.
13 Merritt Street
So. Norwalk, CT 06854

Denton Vacuum, Inc.
2 Pin Oak Ave.
Cherry Hill, NJ 08003

Diatome, USA
P.O. Box 125
Fort Washington, PA 19034

Edwards High Vacuum International
301 Ballardvale Street
Wilmington, MA 01997

Electron Microscopy Sciences
Box 251, 321 Morris Road
Fort Washington, PA 19034

EMCorp
P.O. Box 285
Chestnut Hill, MA 02167

Energy Beam Sciences, Inc.
11 Bowles Road, P.O. Box 468
Agawam, MA 01001

Ernest F. Fullam, Inc.
900 Albany Shaker Road
Latham, NY 12110-1491

ETP-USA/Electron Detectors, Inc.
1650 Holmes Street, Bldg. C
Livermore, CA 94550

FEI Company
19500 NW Gibbs Drive #100
Beaverton, OR 97006

Gatan, Inc.
6678 Owens Drive
Pleasanton, CA 94588

Kimball Physics, Inc.
Kimball Hill Road
Wilton, NH 03086

M.E. Taylor Engineering, Inc.
21604 Gentry Lane
Brookeville, MD 20833

Materials Analytical Services
2418 Blue Ridge Road
Raleigh, NC 27607

McCrone Accessories and Components
850 Pasquinelli Drive
Westmont, IL 60559

Micron, Inc.
3815 Lancaster Pike
Wilmington, DE 19805

Oxford Instruments NA
130A Baker Ave.
Concord, MA 01742

Polysciences, Inc.
400 Valley Road
Warrington, PA 18976

Raith USA, Inc.
70C Carolyn Blvd
Farmington, NY 11735

South Bay Technology, Inc.
1120 Via Callejon
San Clemente, CA 92672

SPI Supplies/Structure Probe, Inc.
P.O. Box 656, 569 East Gay Street
West Chester, PA 19381

Ted Pella, Inc.
P.O. Box 492477
Redding, CA 96049

Tousimis Research Corp.
2211 Lewis Ave.
Rockville, MD 20851

VCR Group
250 East Grand Ave. #31
South San Francisco, CA 94080

APPENDIX VI. SUPPLIERS OF OPTICAL AND ELECTRON MICROSCOPES

Amray Inc.
160 Middlesex Turnpike
Bedford, MA 01730

Cameca Instrument, Inc.
2001 West Main Street
Stamford, CT 06902

Camscan USA, Inc.
500 Thomson Park Dr. #508
Mars, PA 16046

Electroscan Corporation
66 Concord Street
Wilmington, MA 01887

Hitachi Scientific Instruments
460 E. Middlefield Road
Mountain View, CA 94043

JEOL USA, Inc.
11 Dearborn Road
Peabody, MA 01960

Leica, Inc.
111 Deer Lake Road
Deerfield, IL 60015

Nikon Inc., Instrument Group
1300 Walt Whitman Road
Melville, NY

Philips Electronic Instrument Co.
85 McKee Drive
Mahwah, NJ 07430

RJ Lee Group, Inc.
350 Hochberg Road
Monroeville, PA 15146

Topcon Technologies, Inc.
6940 Koll Center Parkway
Pleasanton, CA 94566

VG Microscopes/Fisons Instruments
32 Commerce Center
Danvers, MA 01923

Zeiss, Carl, Inc.
One Zeiss Drive
Thornwood, NY 10594

APPENDIX VII. SUPPLIERS OF X-RAY MICROANALYSIS EQUIPMENT

Advanced MicroBeam, Inc.
P.O. Box 610
4217C Kings-Graves Road
Vienna, OH 44473

Cameca Instruments, Inc.
2001 West Main Street
Stamford, CT 06902

Edax International, Inc.
91 McKee Drive
Mahwah, NJ 07430

HNU X-Ray
160 Charlemont Street
Newton Highlands, MA 02161

Kevex/Fisons Instruments
24911 Avenue Stanford
Valencia, CA 91355

Microspec. Corp.
45950 Hotchkiss Street
Freemont, CA 94539

Noran Instruments, Inc.
2551 W Beltline Highway
Middleton, WI 53562

Oxford Instr. Microanalytical
P.O. Box 2560
Oak Ridge, TN 37831

Princeton Gamma-Tech, Inc.
1200 State Road
Princeton, NJ 08540

APPENDIX VIII. SUPPLIERS OF SCANNING PROBE MICROSCOPES

Burleigh Instruments, Inc.
Burleigh Park
Fishers, NY 14453

Digital Instruments
6780 Cortona Drive
Santa Barbara, CA 93117

Park Scientific Instruments
476 Ellis Street
Mountain View, CA 94043

Technical Instrument Co.
348 Sixth Street
San Francisco, CA 94103

Topometrix
5403 Betsy Ross Drive
Santa Clara, CA 95054

Index

Page numbers in *italic* refer to illustrations